PERFECT CODES
AND RELATED STRUCTURES

PERFECT CODES
AND RELATED STRUCTURES

Tuvi Etzion
Technion, Israel

World Scientific

NEW JERSEY • LONDON • SINGAPORE • BEIJING • SHANGHAI • HONG KONG • TAIPEI • CHENNAI • TOKYO

Published by

World Scientific Publishing Co. Pte. Ltd.
5 Toh Tuck Link, Singapore 596224
USA office: 27 Warren Street, Suite 401-402, Hackensack, NJ 07601
UK office: 57 Shelton Street, Covent Garden, London WC2H 9HE

Library of Congress Control Number: 2022008894

British Library Cataloguing-in-Publication Data
A catalogue record for this book is available from the British Library.

PERFECT CODES AND RELATED STRUCTURES

ISBN 978-981-125-587-8 (hardcover)
ISBN 978-981-125-588-5 (ebook for institutions)
ISBN 978-981-125-589-2 (ebook for individuals)

For any available supplementary material, please visit
https://www.worldscientific.com/worldscibooks/10.1142/12823#t=suppl

To

Orna, Ophir, Liron, and Reut

Preface

Information theory, launched by the pioneering work of Shannon in 1948, has generated a lot of applications and ten of thousands of research papers. One can easily say, without stretching the truth, that its influence on our daily lives is pervasive. Coding theory, which is one of the important sub-areas of information theory, started with the work of Golay (1949) and Hamming (1950). This research area was motivated by engineering problems, and from 1950 until today, with the growth of digital communication, the demand for old and new techniques in coding theory has only increased. Although some basics of the theory are not very difficult, over time more and more sophisticated mathematics has been used and developed in coding theory. This has made the area of coding theory very important to electrical engineers and to computer scientists on one hand and to mathematicians on the other hand.

Two of the most important types of codes are error-correcting codes and covering codes. In the first thirty years of this research area, most work done in coding theory was related to error-correcting codes. Several textbooks and monographs were written in this research area. It was no surprise that also undergraduate and graduate courses, in this area, were developed in many universities. Forty years ago, as well, the area of covering codes started to develop, with hundreds of research papers and new applications that were found quite frequently. Perfect codes, which are the codes considered in our book, lie exactly in the intersection between error-correcting codes and covering codes. They were considered in all the books on coding theory, but were never the highlight of these books as they are in our book.

The existing books on coding theory focus mainly on codes in the Hamming space. In some books, there are one or two chapters related to other metrics, such as codes in the Lee metric or constant-weight codes in the Johnson metric. Nevertheless, the codes considered for other metrics were usually not perfect. Moreover, in these books there are one or at most two chapters devoted to perfect codes and related structures. Our book is different. First, it is devoted solely to perfect codes of various types. It is also different as we devote a few chapters to the Hamming space, but other chapters, which comprise the greater part of the book, are devoted to combinatorial designs, mixed codes, constant-weight codes, to the Lee metric, to the Grassmann scheme, to metrics related to storage devices, as well as to other metric spaces and codes related to newly important applications. Moreover, also in the Hamming space, we consider a few topics that are not covered in other books.

We tried to make the book as self-contained as possible, providing detailed proofs, many times more detailed than the ones that appear in the literature. The detailed proofs will enable to use this book also as a textbook for courses in coding theory. In many cases, no proofs for known results were provided in the cited papers and the appropriate proofs are given in our book. In some cases, we present results for the first time. Of course, we could not provide details on every important technique known in the literature and, in some cases, we offer only pointers to reference material. These references and a summary of these techniques are usually provided in the notes that form the last section of each chapter. Notably, there are many techniques to prove the nonexistence of perfect codes in the various metrics. Therefore, we concentrate on comprehensive treatment of such techniques for one metric only, the Johnson metric.

Our intension is to offer a different perspective for the area of perfect codes. For example, in many chapters there is a section devoted to diameter perfect codes. In these codes, anticodes are used instead of balls and these anticodes are related to intersecting families, an area that is part of extremal combinatorics. This is one example that shows how we direct our exposition in this book to both researchers in coding theory and mathematicians interested in combinatorics and extremal combinatorics. New perspectives for MDS codes, different from the classic ones, which lead to new directions of research on these codes are another example of how this book may appeal to both researchers in coding theory and mathematicians. Our point of view is mainly combinatorial and hence some of the algebraic approach will be omitted (also for lack of space).

The book is so that it can be used by a beginner who just wants either to learn something about prefect codes or to conduct research on perfect codes. Nevertheless, it can be also used by the more advanced reader who has some knowledge on coding theory and wants to get some information on perfect codes or to find some new lines of research in this area. Throughout the book, there are many research problems, some of which we think can be used to motivate graduate students and some which are extremely difficult. We would be very happy to see the book's reader make a breakthrough as a result of insights gleaned through reading it. This book is a monograph; as we did not introduce exercises and assignments for the reader. It was, however, written in a way that enables it to also be used as a textbook for either a basic combinatorial course in coding theory for undergraduate students in mathematics, as an advanced course for the same students, or for an advanced course in coding theory, for students in computer science or electrical engineering, which emphasizes perfect codes. Each such course should be based on different chapters of this book. This is the main reason that we have provided proofs for most of the lemmas and theorems in the book. Moreover, references are quoted only in the notes of each chapter and not along the various sections of the chapter, although some isolated proofs use results that are provided either without proofs or in the notes with or without proofs.

As a basic course for undergraduate students in mathematics, we suggest using the first seven chapters (excluding Chapter 1), which are devoted to the Hamming metric and to combinatorial designs, and Chapter 11, which focus on the Lee metric. The other chapters can be used in an advanced course for math students. As for an advanced course for students in computer science and electrical engineers, who have already completed a basic course in coding theory, we suggest taking highlights from each chapter, with the possible exception of the first chapter. In our opinion most chapters cannot be taught in only two hours. If the course compromises on thirteen weeks, we suggest splitting Chapter 3 into two weeks. We also suggest combining Chapter 4 and Chapter 5 and cover them in two classes since Chapter 4 is quite light, while Chapter 5 has lot of material.

A large part of this book is based on my own research work, which was performed over the last thirty years. My Ph.D. advisor, Abraham Lempel, guided me as I took my first steps into coding theory. My post-doc advisor, Solomon W. Golomb, introduced me to combinatorial designs and their intersection with coding theory in general and with perfect codes in particular. I was first introduced to some perfect codes problems by Gerard Cohen and Simon Litsyn. The discussions I had with them led to my first paper on perfect codes with Alexander Vardy. This was our first joint paper and

Chapter 1

Introduction

Error-correcting codes were introduced to combat errors in communication channels, storage devices, and other computerized systems. In each such system, the information may be coded in a different way and a different type of an error-correcting code should be designed for each different coding method. The information can be coded into binary words of the same length, or words of the same length over any given alphabet, or words of the same length with the same number of nonzero entries, or subspaces over some alphabet with vectors of the same length, etc.

Any coded system has its own features, but all of them can be illustrated in the same way. There is an information word of length k to be coded. There are $M(k)$ distinct such information words. Each information word x is coded into a different codeword c taken from a code $\mathcal{C}$ that has $M(k)$ codewords. This code is designed so that it can correct up to e errors. The coding of the information words into the codewords of $\mathcal{C}$ defines a bijective mapping from the set of $M(k)$ possible information words onto the set of $M(k)$ codewords of $\mathcal{C}$. The codeword c is either transmitted over a channel or stored in a storage device. The channel is noisy and the storage system is subject to errors. When received from the channel or retrieved from the storage device, the codeword c might incur some errors. Hence, the received word y might be different from the transmitted (stored) codeword c. Assume the code $\mathcal{C}$ was designed in a way that it can correct any number up to e errors. If up to e errors occurred when c was changed to y, then the decoder can detect and correct the errors and find the submitted codeword c. Since the information word x was coded to the codeword c using a bijective function, it follows that there is an inverse bijective function that can transfer c back to the information word x. This completes the coding and decoding process in the system.

Perfect codes have always drawn the attention of coding theorists and mathematician. Usually, their perfect structure yields a beautiful mathematical structure. Their optimality makes them very useful for applications in their related channel and its associated metric, and also as building blocks for many other structures. Some of these structures are good only for theory and some are used in applications. In such a metric, we are given a space $\mathcal{V}$ and a distance between any two elements of $\mathcal{V}$.

A ***metric*** d on a set $\mathcal{V}$ (called also the ***space***) is a function $d : \mathcal{V} \times \mathcal{V} \to \mathbb{Z}$, that satisfies the following three axioms:

(1) ***Identity***: For each $x, y \in \mathcal{V}$, $d(x, y) = 0$ if and only if $x = y$.
(2) ***Symmetry***: For each $x, y \in \mathcal{V}$, $d(x, y) = d(y, x)$.
(3) ***Triangle inequality***: For each $x, y, z \in \mathcal{V}$, $d(x, z) \leq d(x, y) + d(y, z)$.

The definition is trivially generalized for $d : \mathcal{V} \times \mathcal{V} \to \mathbb{R}$, but it is not presented as it will not be used in the book. The elements of the space $\mathcal{V}$ will be often called ***words*** or ***points***.

A collection $\mathcal{C}$ of points, in the discrete space $\mathcal{V}$, is a code. The code $\mathcal{C}$ is a ***perfect code*** with ***radius e***, if for any point x in the space $\mathcal{V}$, there exists a unique point in $\mathcal{C}$ whose distance from x is at most e. This definition is very strict and usually the set of parameters for which there exists a perfect code will be very limited. An equivalent definition is as follows. For a word $x \in \mathcal{V}$, the set of words that are within distance e from x is called the ***ball with radius e*** around x (or centered at x). This set is denoted by $\mathcal{B}_e(x)$, where x is called the ***center*** of the ball $\mathcal{B}_e(x)$. If the set of all balls with radius e around the codewords of a code $\mathcal{C}$ forms a partition of $\mathcal{V}$, then $\mathcal{C}$ is called a perfect code.

Assume again that we are given a code $\mathcal{C} \subseteq \mathcal{V}$, a function $d : \mathcal{V} \times \mathcal{V} \to \mathbb{Z}$ which is a metric, and each word $x \in \mathcal{V}$ is within distance e from at most one codeword $c \in \mathcal{C}$. If the size of a ball with radius e does not depend on its center x from which it is computed and the length of a word is n, then we can denote the ball with radius e by $\mathcal{B}_e(n)$ instead of $\mathcal{B}_e(x)$. Clearly, in this case where all the balls with radius e have the same size, we have that

$$|\mathcal{C}| \cdot |\mathcal{B}_e(n)| \leq |\mathcal{V}|. \tag{1.1}$$

This bound, called the ***sphere-packing bound*** (although it is a ***ball-packing bound***), is attained by a perfect code in a metric where all the balls with radius e have the same size. Moreover, when all the balls with radius e are of the same size, then (1.1) can be used as an alternative definition for a perfect code. In other words, if the size of a ball with radius e does not depend of the center of the ball, then $\mathcal{C}$ is a perfect code if

$$|\mathcal{C}| \cdot |\mathcal{B}_e(n)| = |\mathcal{V}| .$$

About fifty years after the introduction of perfect codes, it was observed that instead of looking at a ball with radius e, one can use a similar definition using a shape called anticode (which is not necessarily a ball) with diameter D to define similar codes that are called ***diameter perfect codes***. These codes will be discussed throughout the book.

There are always some trivial perfect codes, regardless of the (discrete) space $\mathcal{V}$ or the metric d. The first trivial code is the whole space that is a perfect code with radius 0. The second one is a unique point in a finite space that forms a perfect code with radius e which represents the largest distance from this unique point in the space. Depending on the space and the given radius, other trivial perfect codes might exist.

Most metrics discussed in the book can be represented by an undirected graph whose vertices represent the points of the space. Two vertices are connected by an edge if their distance in the metric is exactly one. Generally speaking, we can define a perfect code for any undirected graph Γ. A set of vertices $\mathcal{S}$ in the graph Γ forms a perfect code with radius e, in the graph, if for any vertex v in Γ there exists a unique vertex in $\mathcal{S}$ whose distance, in the graph, from v is at most e. Nevertheless, one should be careful and understand that there are metrics that cannot be represented by a graph. To be represented by a graph in this way, the metric must satisfy the following property: if for any two elements $x, y \in \mathcal{V}$, $d(x, y) = \delta$, then the shortest path between x and y in the graph has length δ. This property will be discussed in Chapter 2.

The research area of coding theory has drawn the attention of both engineers and mathematicians. Information theory started with the celebrated work of Claude Elwood Shannon in 1948, which was mainly meant for engineers. The first two papers on coding theory came in the following two years. In 1949 Marcel Jules Edouard Golay introduced codes known today as the Golay codes and in 1950 Richard Hamming introduced the family of binary codes that are known as the binary Hamming codes. All these codes fall into our family of perfect codes, which is the topic of this book. Since these early days coding theory has been developed rapidly in many directions. Even though the number of perfect codes is a relatively very small fraction of the total number of codes, interest in these codes is very strong as they have a beautiful structure. In fact, some might argue that they are the most important ones at least from a mathematical point of view.

Coding theory was developed using many mathematical tools, with many new ones developed especially for this area of research. These have also been of a great interest for mathematicians who were interested in the mathematical constructions of codes on one hand and on the mathematical structure of the constructed codes on the other hand. Their mathematical research was performed in parallel to the general research on other aspects of these codes.

This book is organized as follows. Chapter 2 through Chapter 7 are mainly concerned with codes using the Hamming metric, but the basic definitions and claims, introduced in Chapter 2, are also related to other metrics. Chapter 8 is devoted to the Johnson scheme, where the codes are binary constant-weight codes, the metric is the Johnson distance defined to be half of the Hamming distance. Chapter 9 considers nonbinary constant-weight codes, where the metric is the Hamming distance. Chapter 10 considers codes over subspaces, where other metrics are used. Chapter 11 and Chapter 12 are devoted to codes where in addition to balls, shapes which are defined by some union of unit cubes are considered. The most notable metrics for these shapes are the Lee metric and the Manhatten metric. Chapter 13 is devoted to other metrics, such as the poset metric and a metric for correction of burst errors. It also discusses the deletion channel, its distance measures and perfect codes.

The most studied metric over the years is the Hamming distance, which is defined mainly on the Hamming space. The space consists of words of length n over some finite alphabet. Usually, this alphabet is a finite field, but alphabets whose size is not a power of a prime have also been considered. There are also codes where each coordinate in the word can be over a different alphabet size. The families of codes in the Hamming space can be partitioned into two classes, linear codes and nonlinear codes. A linear code forms a linear subspace over some finite field while nonlinear codes are just sets of codewords that might not have any mathematical structure. Most books on coding theory concentrate mainly on linear codes in the Hamming space. The structure of the codes in the Hamming space, linear or nonlinear, is interesting from both combinatorial and practical points of view. Assorted codes with various properties, in the Hamming space, are discussed in the various books on coding theory. For these discussions, many definitions and basic properties that are in common for all codes should be defined and discussed.

We continue and discuss the material in each chapter of the book. Chapter 2, is devoted to these definitions and properties that are in common for all codes, but the emphasis is on codes in the Hamming space. Section 2.1 presents the basic definitions for nonlinear codes. It also considers binary codes where all the codewords have the same number of nonzero entries, called constant-weight codes. These codes are part of the Johnson scheme, which is considered in Chapter 8. The alphabet used for linear codes forms a finite field. The basic definitions of finite fields required for our exposition are defined in Section 2.2. Section 2.3 examines the definitions in the context of linear codes and also presents the specific definitions and properties of linear codes. In Section 2.4 the various definitions for perfect codes are provided and the definitions and the basic results on diameter perfect codes are presented. Most of this chapter is written to accommodate the discussions on codes in the Hamming space with the Hamming distance, but the definitions and results are written in a more general way that can be also used for other spaces and metrics. In particular, diameter perfect codes are presented in a way that can be used by most metrics.

By definition, perfect codes are combinatorial objects. As combinatorial objects they have many combinatorial properties and many combinatorial designs are embedded in them. These combinatorial structures are discussed in Chapter 3. Section 3.1 is devoted to one of the most beautiful families of such designs, the Steiner systems. Steiner systems are embedded inside perfect codes in the Hamming space and inside perfect codes in the Johnson space. They are diameter perfect codes in the Johnson space discussed in Chapter 8 and similar systems are diameter perfect codes in the space of nonbinary constant-weight words and in the Grassmann space discussed in Chapters 9 and 10, respectively. Section 3.2 is devoted to an introduction of several types of orthogonal designs, such as Latin squares, Hadamard matrices, and orthogonal arrays. Some orthogonal arrays are the nonlinear version of what are known as maximum distance separable (abbreviated MDS) codes, one of the most important families of codes in coding theory, for both theory and practice. Orthogonal arrays and MDS codes are also diameter perfect codes and, as such, they will be considered in various sections of the book. MDS codes are highly related to projective geometry, a very wide area of research in combinatorics, about which several books have been written. Projective geometries yield some important combinatorial designs and especially Steiner systems. Some of their structures are equivalent to MDS codes. Basic definitions of projective geometries that are important in our discussion will be presented in

Section 3.3. Bounds on the sizes of codes which are not prefect codes, but highly related to perfect codes and/or combinatorial structures such as combinatorial designs and codes that attain these bounds, are presented in Section 3.4. These bounds are the Plotkin bound and the Griesmer bound. The last important combinatorial structure that will be presented in this chapter are the association schemes that are the topic of Section 3.5. Most metrics discussed in this book, such as the Hamming metric and the Johnson metric, form association schemes with their associated spaces, and the formulation of the related theory is important to obtain results on perfect codes and to understand their structure.

The Hamming codes and the Golay codes are linear codes that surprisingly, form the only nontrivial linear perfect codes in the Hamming scheme. These codes are discussed in Chapter 4. Adding parity symbols for some of these codes yields extended codes that have their own importance and will be used throughout the book. The infinite family of Hamming codes and extended Hamming codes have many fascinating properties. There are a few representations for these codes, each of which can be used and applied differently. These codes and their properties are discussed in Section 4.1. The two Golay codes and their extended codes are the topic of Section 4.2. Although they represent only four codes, these codes have been heavily studied as they hold lot of interesting properties; moreover, they are also significant for practical applications. Finally, in Section 4.3 nonlinear and linear diameter perfect codes in the Hamming scheme are discussed. Other than the Hamming codes and the Golay codes, there are no known perfect codes with other parameters in the Hamming scheme. Nevertheless, the research on perfect codes in the Hamming scheme does not end with these codes.

There are many nonlinear perfect codes in the Hamming scheme. All of them have the same parameters as the Hamming codes. These codes are introduced in Chapter 5. A few constructions for such codes are given in Section 5.1. Other constructions are given in the other sections of this chapter, where properties such as the weight distribution, rank, and kernel of perfect codes are considered. In Section 5.7 a lower bound on the number of such inequivalent codes is proved. This lower bound is given for codes over any finite field $\mathbb{F}_q$, for any given q. There are no other parameters for perfect codes, over an alphabet of size q, where q is power of a prime, and the proof of this claim will be discussed in Section 5.8. The proof is based on some polynomials, called Lloyd's polynomials. The complete proof is given in many other venues and it will be omitted in our discussion.

The last section in this chapter, Section 5.9, is devoted to a variant of a celebrated combinatorial game whose solution is obtained by using perfect codes in the Hamming scheme.

Perfect codes are the most dense codes in coding theory when error-correcting codes are discussed. They are also the most sparse codes when covering codes are discussed. Error-correcting codes (covering codes, respectively), which are dense (sparse, respectively) and are "almost perfect", are the topic of Chapter 6. The concepts of density and sparse/dense codes are discussed in Section 6.1. The most important family of dense codes, that are not perfect codes, are the Preparata codes, which will be constructed in Section 6.3. These codes have many interesting properties that will be used in other sections of the book. Preparata codes are part of a family of codes called nearly-perfect codes. A code is a nearly-perfect code if it attains the Johnson bound that is proved in Section 6.2. A stronger bound is also proved in this section. They are also part of a larger family of codes, called quasi-perfect codes, which are defined in Section 6.4. Unfortunately, these codes are not as perfect as their name suggests. Therefore, we will be interested only in dense quasi-perfect codes and sparse quasi-perfect codes, respectively, depending on whether we are interested in error-correcting codes or in covering codes, respectively. Density of covering codes with radius two will be discussed in Section 6.5 and with radius three in Section 6.6.

Chapter 7 is the first one that discusses codes which are not in the Hamming scheme. The metric used is still the Hamming distance, but the space consists of words of the same length, where not all coordinates are over an alphabet of the same size. These codes are called mixed codes and the related perfect codes are called perfect mixed codes. There are many perfect mixed codes with radius one and they are the topic of Section 7.1. Constructions of codes with radius one are based on partitioning of a group into subgroups. One family of such partitions forms perfect codes where the coordinates of the words are organized in bytes. The related perfect codes are called perfect byte-correcting codes. These perfect byte-correcting codes are discussed in Section 7.2. Finally, Section 7.3 is devoted mainly to perfect mixed codes with larger radii. Most surprisingly, such nontrivial perfect codes are based on properties of the Preparata code. Structures associated with perfect mixed codes that are similar to Steiner systems will be also defined and discussed in this section. It is also shown in this section that there are diameter perfect codes for each distance if the alphabets in each coordinate is not restricted.

After the Hamming scheme, the most studied scheme is the Johnson scheme. In the Johnson space, all the words are binary having length n with a fixed number w of *ones* for all the words. Hence, these codes are usually called constant-weight codes. The Johnson distance is exactly half of the Hamming distance and this relation connects the two metrics. The nonexistence of nontrivial perfect codes in the Johnson scheme is an intriguing question in coding theory and especially in the theory of perfect codes. These codes are discussed in Chapter 8. The basic definitions for this scheme are presented in Section 8.1. To handle the codes in this scheme, we define the configuration distribution in Section 8.2. If nontrivial perfect codes exist in the Johnson scheme, then there are many Steiner systems embedded in these codes. The existence proofs for these embedded Steiner systems are provided in Section 8.3. These Steiner systems yield many necessary conditions for the existence of such codes and imply that in many graphs of this scheme, there are no nontrivial perfect codes. Tradeoff between the various parameters of perfect codes in the Johnson scheme is discussed in Section 8.4. Another approach for excluding possible perfect codes in the Johnson scheme is to consider perfect codes with a given radius e. The approach is based on the regularity of perfect codes, which is the topic of Sections 8.5, 8.6, and 8.7. The implications on the nonexistence of perfect codes with a given radius based on this approach is discussed in these sections. Diameter perfect codes in the Johnson scheme is another interesting topic. Steiner systems and their complements are one family of such codes. There may be other families, but it is conjectured that no other such family exists, as will be discussed in Section 8.8.

When nonbinary constant-weight codes are considered, there are a few families of perfect codes and diameter perfect codes. These codes are discussed in Chapter 9. Constructions of nonbinary perfect constant-weight codes are discussed in Section 9.1. Six families of nonbinary diameter perfect constant-weight codes include generalized Steiner systems, the so-called constant-weight MDS codes, and some other interesting families of codes. An introduction to these codes is given in Section 9.2. Each one of these families is discussed in one of the sections from Section 9.3 through Section 9.8. Four families of nonbinary constant-weight maximum size anticodes are defined and compared in Section 9.9.

In Chapter 10 we turn our attention to the third most studied scheme (after the Hamming and the Johnson schemes), the Grassmann scheme. In this scheme the space consists of subspaces of the same dimension from a given n-space over some finite field. The metric in this scheme, the Grassmann distance, is akin to the Johnson distance. Perfect codes do not exist in this scheme and this claim is proved in Section 10.1. Diameter perfect codes in this scheme are akin to Steiner systems in the Johnson scheme, and they are called q-Steiner systems. These codes are the topic of Section 10.2. The only known infinite family of such codes contains codes called spreads and they are the only known class of perfect byte-correcting codes mentioned in the context of perfect mixed codes and discussed in Chapter 7. Normal spreads form a family of spreads with important properties and they are discussed in Section 10.3. The set of subspaces of all dimensions from a given n-space is called the projective space. On these subspaces a metric called the subspace distance is defined. The Grassmann distance is half of the subspace distance in the same way that the Johnson distance is half of the Hamming distance. It is proved in Section 10.4 that, also in this space, there are no nontrivial perfect codes. The codes in the Grassmann space and the projective space are closely related to codes in a space whose words are matrices of the same size over a given finite field. The metric in this space is called the rank distance, which is defined as the rank of the difference between the two related matrices. This space of matrices with the rank metric is also a scheme called the bilinear forms scheme. There are also no prefect codes in this scheme, but there are diameter perfect codes. These rank-metric codes are the topic of Section 10.5. Finally, in Section 10.6 we consider another interesting type of codes, which are constant-dimension MDS codes, also called subspace-MDS codes. These codes as well as all the codes discussed in this chapter are related to network coding and this topic will also get our attention in this chapter.

A very important metric which drew attention beginning from 1970 and on is the Lee metric. This metric does not define an association scheme and it is the topic of Chapter 11. This metric is also related to another metric called the Manhattan metric, also known as the L_1 metric, the rectilinear metric, and the taxicab metric. The definitions for these metrics are given in Section 11.1. Linear codes in the Lee and the Manhattan metrics can be represented and defined with lattices. A perfect code can be represented by a tiling and a linear perfect code can be represented by a lattice tiling. The basic concepts of tilings and lattice tilings are presented in Section 11.2. Perfect codes in these metrics are constructed in Section 11.3. Diameter perfect codes are discussed and presented in Section 11.4. In Section 11.5 a lower bound on the number of different perfect codes is proved. This bound is related to nonperiodic perfect codes. Except for the parameters of the known perfect codes, it is conjectured that there are no perfect codes in these metrics. A lot of work has been done over the years to prove this conjecture. In Section 11.6 one technique to prove the nonexistence of such codes for many parameters is given and the other known results are summarized, with references provided, in Section 11.7.

A perfect code in the Lee metric is related to a tiling of the space with error balls that are formed by union of unit cubes. Such error balls (called also shapes) have a combinatorial structure and also a geometric structure. They were discussed in the context of algebraic tilings with lattices. These concepts are the topics of Chapter 12. Another technique for forming a tiling with these shapes (balls) is group splitting, which is discussed in Section 12.1. Crosses and semi-crosses are the first two shapes that will be discussed in Section 12.2. The associated codes have found applications in nonvolatile memories, which are heavily used, at the beginning of the 21st century, in computerized systems. This type of memory is discussed in Section 12.3. The type of codes associated with these memories have asymmetric types of errors. The related ball errors are called quasi-crosses and they are the topic of Section 12.4. Another shape in this context is the notched cube and its tilings are discussed in Section 12.5. This section also demonstrates some of the previous techniques and their equivalence.

In Chapter 13 other metrics are considered. Section 13.1 is devoted to the deletion channel, in which the deletions and/or insertions are defined, but perfect codes are defined when only deletions are considered. The Hamming metric is generalized to many different metrics called poset metrics, which are discussed in Section 13.2. Perfect codes with ball of radius one for these metrics are fully characterized. In computer systems, errors can come in bursts. For such errors, burst-correcting codes were designed. Section 13.3 considers perfect codes to correct one such burst.

Before we continue to the comprehensive exposition on perfect codes, let us present a few notations that will be used throughout the book (generally, they will be also defined in the appropriate place):

(1) $\mathbb{Z}$ - the set of integers.
(2) $\mathbb{R}$ - the set of real numbers.
(3) $\mathbb{Z}_m$ - the set of integers $\{0, 1, \ldots, m-1\}$. Also used for the ring of integers modulo m, which is a field if m is a prime.
(4) $\mathbb{F}_q$ - the finite field with q elements, known also as the Galois field GF(q).
(5) $\mathbb{E}_2^n$ - the set of all binary words of length n and even weight.
(6) $|A|$ - the size of the set A.

(7) $\langle A \rangle$ - the linear span of the rows of a matrix A.
(8) $\langle \mathcal{C} \rangle$ - the linear span of a set $\mathcal{C}$ of codewords.
(9) $\gcd(x, y)$ - the greatest common divisor of the nonzero integers x and y.
(10) $\mathcal{C}(G)$ - the code generated from the generator matrix G, i.e., $\mathcal{C}(G) = \langle G \rangle$.
(11) X^- - the set (or group) X without its *zero* element.
(12) A^{tr} (x^{tr}) - the transpose of the matrix A (or the vector x).
(13) For two integers i, j, where $i < j$, $[i, j]$ denotes the set of integers $\{i, i+1, \ldots, j-1, j\}$.
(14) $[n]$ - the set of integers $\{1, 2, \ldots, n\}$.
(15) $\mathbf{0}$ - an all-zero vector and also an all-zero matrix.
(16) $\mathbf{1}$ - an all-one vector.
(17) I_n - the $n \times n$ identity matrix.
(18) $\mathbf{e}_i$ - a word with exactly one nonzero entry, a *one* in the i-th coordinate.
(19) $d(x, y)$ - the distance between x and y, where the exact metric is understood from the context.
(20) $\text{wt}(x)$ - the weight of x, which is also the distance between x and the identity (*zero*) element.
(21) $p(x)$ - the parity (the sum modulo 2) of a binary vector x.
(22) For a set S, we denote by S^n, the set of all vectors of length n whose entries are taken from the set S. There are $|S|^n$ distinct vectors in the set S^n.
(23) For any string α of any length, α^n denotes a sequence obtained by a concatenation of n α's, i.e., $\alpha^n \triangleq \overbrace{\alpha\alpha\cdots\cdots\alpha}^{n \text{ times}}$.
(24) $X + Y$ - the sum of two sets (or subspaces) X and Y, i.e.,

$$X + Y \triangleq \{x + y \ : \ x \in X, \ y \in Y\} \ .$$

(25) $X \oplus Y$ - the sum of two disjoint subspaces X and Y, i.e., $X \cap Y = \{\mathbf{0}\}$, and

$$X \oplus Y \triangleq \{x + y \ : \ x \in X, \ y \in Y\} \ .$$

(26) $X \times Y$ - the cartesian product of two sets x and Y, i.e.,

$$X \times Y \triangleq \{(x, y) \ : \ x \in X, \ y \in Y\}.$$

(27) $X \otimes Y$ - the direct product of two ordered sets of subsets $\mathbb{X} = \{X_1, X_2, \ldots, X_m\}$ and $\mathbb{Y} = \{Y_1, Y_2, \ldots, Y_m\}$, is defined by

$$\mathbb{X} \otimes \mathbb{Y} \triangleq \bigcup_{i=1}^{m} (X_i \times Y_i) \ .$$

(28) k-subspace will abbreviate the term k-dimensional subspace.
(29) S_n - the symmetric group, which contains the set of $n!$ permutations of $[n]$ (or any n-set).
(30) A permutation $\pi \in S_n$ is applied on a vector $x = (x_1, x_2, \ldots, x_n)$. The permutation π can be represented as $[\pi(1), \pi(2), \ldots, \pi(n)]$, which implies that the i-th position of $\pi(x)$ will be $x_{\pi(i)}$.
(31) A permutation $\pi \in S_n$, on $x = (x_1, x_2, \ldots, x_n)$, can be also represented by its cycles decomposition $(\pi_1, \ldots, \pi_k)(\pi_{k+1}, \ldots, \pi_m)\cdots$, which implies that the elements in position π_1 of x will be moved to position π_2, the elements in position π_2 of x will be moved to position π_3, and so on, and the element in position π_k of x will be moved to position π_1. The same procedure will be applied on the cycle $(\pi_{k+1}, \ldots, \pi_m)$ and so on.

1.1 Notes

As noted in the Introduction, the area of information theory started with the seminal work of [Shannon (1948)]. The next two papers present the only known nontrivial binary linear perfect codes. The two Golay codes were introduced by [Golay (1949)]. Binary Hamming codes were presented in [Hamming (1950)].

There are many excellent books on coding theory, such as the one by [MacWilliams and Sloane (1977)] on error-correcting codes, or the one by [Cohen, Honkala, Litsyn, and Lobstein (1997)] on covering codes. Many books on error-correcting codes were written beginning in 1960 and in the following years. Of these we mention some of the important ones: [Peterson (1961); Berlkamp (1968); van Lint (1971b); Blake and Mullin (1975); Blahut (1983); Pless (1989); Lin and Costello (2004); Roth (2005)]. Last, but not the least, we point out the excellent "Handbook on Coding Theory" by [Pless and Huffman (1998)].

Chapter 2

Definitions and Preliminaries

In this chapter we present the basic concepts in coding theory used for perfect codes. Section 2.1 is devoted to nonlinear codes and also to the basic definitions of words and codes. In contrast to the research and the literature on error-correcting codes, most perfect codes are not linear, although many of them are constructed based on linear codes. Basic properties of and facts about nonlinear codes are discussed in this section. Two product constructions that will be used throughout the book will be presented in this section. Another part of this section is devoted to the family of constant-weight codes. Finally, we will define the concepts of packing radius and covering radius. These two concepts coincide only for perfect codes. A short introduction to finite fields, which are used in linear codes, is given in Section 2.2. In Section 2.3 basic definitions for linear codes are presented. These codes form the most important family of error-correcting codes. This is the family of codes that is most studied in the literature since they can be adapted for many practical applications relatively easily. They have comparably simple encoding and decoding algorithms that are useful for these applications. A few families of important perfect codes are linear and they will be discussed in the following chapters. The definitions in this chapter will be mainly used for perfect codes in the Hamming space, which is the space primarily used in the literature as well as in this book. Notwithstanding, linear codes can also be used for codes whose codewords are subspaces, for codes related to computer memories, and to codes in the Lee metric.

In Section 2.4 we finally touch on the main topic of this book, perfect codes. Their definitions will be given from a few points of view. Perfect codes attain the well-known upper bound on the size of a code, known as the sphere-packing bound. This bound is based on the sizes of the balls with a given radius in the space. In particular, the bound is very effective when all the balls, with the same radius, in the space have the same size. Analogous to the concept of a ball, which is based on a radius of a shape, there is the concept of an anticode, which is based on the diameter of the shape. In this case, an analog to perfect codes, in this context of balls and anticodes, is the concept of diameter perfect codes. These two concepts of anticodes and diameter perfect codes will be discussed in this section. Many concepts mentioned in this chapter are defined only for the Hamming metric, but most of them can be used for other metrics. Notably, most definitions of perfect codes are appropriate for all metrics and not just for the Hamming metric. The general concepts and the results when the Hamming metric is not mentioned are used for almost all the other metrics.

2.1 Nonlinear Codes

The discussion of perfect codes starts with nonlinear codes. The algebraic structure of linear codes makes it easier to handle this set of codes, which are considered in Section 2.3. Indeed, most of the literature on error-correcting codes is based on linear codes. A ***nonlinear code*** is just a collection of ***words*** (also called ***vectors***) that have the same length. These words are called ***codewords***. The codewords of the code are taken from a space $\mathcal{V}$ with words of length n, where one of the words will be identified as the *zero* (sometimes *all-zero*) word. Usually, our space will be finite, but some infinite spaces will be also discussed in Chapter 11 and in Chapter 12. The words will be over some ***alphabet*** Σ, where one of the alphabet symbols will be identified as the *zero* symbol.

For two words $x = (x_1, x_2, \dots, x_n)$, $y = (y_1, y_2, \dots, y_n)$, in the space $\mathcal{V}$ over an alphabet Σ, the ***addition*** $x + y$ is defined by

$$x + y \triangleq (x_1 + y_1, x_2 + y_2, \dots, x_n + y_n),$$

where $x_i + y_i$ is the addition defined on the alphabet Σ.

A word $x = (x_1, x_2, \dots, x_n)$ can be also over a mixed alphabet, i.e., each coordinate can by over another alphabet. If the symbol in the i-th coordinate is taken from the alphabet Σ_i we say that x is over $\Sigma_1 \times \Sigma_2 \times \cdots \times \Sigma_n$. When x and y are over $\Sigma_1 \times \Sigma_2 \times \cdots \times \Sigma_n$, in $x + y$, the addition $x_i + y_i$ is defined on the alphabet Σ_i.

Two codes $\mathcal{C}_1$ and $\mathcal{C}_2$ with codewords of length n are said to be ***isomorphic*** if there exists a permutation $\pi \in S_n$, such that $\mathcal{C}_2 = \{\pi(c) \ : \ c \in \mathcal{C}_1\}$. They are said to be ***equivalent*** if there exists a vector u of length n and a permutation $\pi \in S_n$, such that $\mathcal{C}_2 = \{u + \pi(c) \ : \ c \in \mathcal{C}_1\}$.

The ***Hamming distance*** between two words $x = (x_1, x_2, \ldots, x_n)$ and $y = (y_1, y_2, \ldots, y_n)$, over some alphabet, $d_H(x, y)$, is the number of coordinates in which x and y differ. In other words

$$d_H(x,y) \triangleq |\{i \ : \ x_i \neq y_i\}| \, .$$

The ***distance*** between two elements $x, y \in \mathcal{V}$ will be denoted by $d(x, y)$, when the metric will be understood from the context. The same will be used for the Hamming distance.

The ***minimum distance*** of a code $\mathcal{C}$, $d(\mathcal{C})$, is the smallest integer δ, such that there exist two distinct codewords $x, y \in \mathcal{C}$ for which $d(x, y) = \delta$. The minimum distance of a code cannot always be easily computed or verified. Hence, in some cases when it is said that the minimum distance of the code is at least δ, it can be larger than δ, but usually in our exposition it will be exactly δ.

For a given code $\mathcal{C} \subseteq \mathcal{V}$, the distance of a word $x \in \mathcal{V}$ from the code $\mathcal{C}$, $d(x, \mathcal{C})$, is defined as the minimum distance of x from a codeword of $\mathcal{C}$, i.e., $d(x, \mathcal{C}) = \min\{d(x, c) \ : \ c \in \mathcal{C}\}$.

The ***weight*** of a word x, $\text{wt}(x)$, is its distance from the all-zero word, i.e., $\text{wt}(x) = d(x, \mathbf{0})$. For the Hamming metric, or when there is no specified metric, $\text{wt}(x)$ is the number of nonzero coordinates in the word x. The ***support*** of a word $x = (x_1, x_2, \ldots, x_n)$, $\text{supp}(x)$, is the indices of the nonzero coordinates in x, i.e.,

$$\text{supp}(x) \triangleq \{i \ : \ x_i \neq 0\} \, .$$

Clearly, in the Hamming metric the weight of x is the size of the support of x, i.e., $\text{wt}(x) = |\text{supp}(x)|$. The ***complement*** of a binary word $x = (x_1, x_2, \ldots, x_n)$ is the word $\bar{x} = (\bar{x_1}, \bar{x_2}, \ldots, \bar{x_n})$, where $\bar{b}$ is the binary complement of $b \in \{0, 1\}$. The ***complement code*** $\bar{\mathcal{C}}$ of a binary code $\mathcal{C}$ is the code whose codewords are all the complements of the codewords of $\mathcal{C}$, i.e.,

$$\bar{\mathcal{C}} \triangleq \{\bar{c} \ : \ c \in \mathcal{C}\} \, .$$

A binary code $\mathcal{C}$ is called ***self-complement*** when $c \in \mathcal{C}$ if and only if $\bar{c} \in \mathcal{C}$, i.e., $\bar{\mathcal{C}} = \mathcal{C}$.

The set of all binary words of length n and even weight is denoted by $\mathbb{E}_2^n$ and its words are called the ***even weight words***.

A code $\mathcal{C}$ is called a ***cyclic code*** if $c \in \mathcal{C}$ implies that each cyclic shift of c is also a codeword of $\mathcal{C}$. Note, that in some places cyclic codes refer only to linear codes. Here also nonlinear codes can be called cyclic.

An $(n, M, d)_q$ ***code*** is a set of M distinct words of length n over an alphabet with q symbols whose minimum Hamming distance is at least d. The alphabet size q will be omitted whenever it is understood from the context. This omission of the alphabet size will also be done for other concepts, which follow.

Theorem 2.1. *For an $(n, M, d)_q$ code $\mathcal{C}$, $M \leq q^{n-d+1}$.*

Proof. Assume $\mathcal{C}$ is represented by an $M \times n$ matrix A, i.e., each codeword is a row in A. Consider the projection of any $n-d+1$ columns of A. If two distinct rows in this projection are equal, then the associated codewords of these two rows can differ at most in the other $d-1$ columns, and hence their distance is at most $d-1$, a contradiction. Thus, all the rows in this projection are distinct, and hence $M \leq q^{n-d+1}$. □

The bound of Theorem 2.1 is called the ***Singleton bound*** and linear codes that attain it with equality are called ***maximum distance separable codes*** (MDS codes in short) and they are probably the most important codes that combine theory and practice. These codes will also have a significant role in our discussions. Nonlinear codes that attain this bound with equality are called ***orthogonal arrays*** (see Section 3.2) and they are as important as MDS codes.

Weight distribution and distance distribution are two fundamental properties of a code from which we can find some of its other properties. For a given (n, M, d) code $\mathcal{C}$, the ***weight distribution*** of $\mathcal{C}$ is the sequence $(A_0, A_1, \ldots, A_n)$, where A_i is the number of codewords in $\mathcal{C}$ whose weight is i. The ***distance distribution*** of $\mathcal{C}$ is the sequence $(D_0, D_1, \ldots, D_n)$, where

$$D_i \triangleq \frac{|\{(c_1, c_2) \ : \ c_1, c_2 \in \mathcal{C}, \ d(c_1, c_2) = i\}|}{|\mathcal{C}|} \ .$$

In other words, the distance distribution represents the average of the number of ordered pairs of codewords for any given distance, where clearly $D_0 = 1$ by this average.

Definition 2.1. For a code $\mathcal{C}$, the ***punctured code*** $\mathcal{C}'$ with respect to the i-th coordinate is obtained from all the codewords of $\mathcal{C}$ by deleting the i-th coordinate from all the codewords.

Remark 2.1. In many of the constructions $\mathcal{C}'$ will denote a code with no connection to another code $\mathcal{C}$ used in the construction, i.e. $\mathcal{C}'$ is not a punctured code in these constructions.

The following lemma can be readily verified.

Lemma 2.1. *If $\mathcal{C}$ is an (n, M, d) code, then its punctured code $\mathcal{C}'$, with respect to the i-th coordinate, is an $(n-1, M', d-1)$ code, where $M' = M$ if $d > 1$, and $M' < M$ if and only if there exists a pair of codewords in $\mathcal{C}$ that differ only in the i-th coordinate.*

Definition 2.2. For a code $\mathcal{C}$, the ***shortened code*** with respect to the i-th coordinate, is a subset of the punctured code (with respect to the i-th coordinate), obtained from all the codewords of $\mathcal{C}$ having a *zero* in the i-th coordinate.

Lemma 2.2. *If $\mathcal{C}$ is an (n, M, d) code, then its shortened code, with respect to the i-th coordinate, is an $(n-1, M', d)$ code, where $M' \leq M$. $M' = M$ if and only if all the codewords in $\mathcal{C}$ have a* zero *in the i-th coordinate.*

Clearly, a shortened code can be defined in exactly the same way, when any other different symbol is considered instead of the *zero*.

Definition 2.3. Let $\mathcal{C}$ be a code over a space $\mathcal{V}$ and let $x \in \mathcal{V}$. The ***translate*** of $\mathcal{C}$ by the word x is the set of words

$$x + \mathcal{C} \triangleq \{x + c \ : \ c \in \mathcal{C}\},$$

where '+' is the addition (or any binary operation) defined on the space $\mathcal{V}$. This translate is a ***left translate***, and similarly a ***right translate*** is defined by

$$\mathcal{C} + x \triangleq \{c + x \ : \ c \in \mathcal{C}\}.$$

A translate whose codewords have only even weights is called an ***even translate*** and a translate whose codewords have only odd weights is called an ***odd translate***.

When $x+y = y+x$ for each two elements $x, y \in \mathcal{V}$, a right translate coincides with a left translate and it will be called a ***translate***. This will be the case in most binary operations used in the book. Translates can be defined in other spaces (and not just the Hamming space where the definition of "addition" is trivial) where instead of addition, another ***binary operation*** is defined. Such binary operations will be discussed in the related chapters.

Definition 2.4. A code $\mathcal{C}$ has the ***linear space tiling property*** if the space $\mathcal{V}$ can be partitioned into pairwise disjoint translates of $\mathcal{C}$. A code $\mathcal{C}$ has the ***space tiling property*** if the space $\mathcal{V}$ can be partitioned into pairwise disjoint codes whose size and other required parameters (such as the minimum distance) are the same as those of the code $\mathcal{C}$. Let X and Y be two subsets of $\mathcal{V}$. The pair (X,Y) is called a ***tiling*** of $\mathcal{V}$ if each element $v \in \mathcal{V}$ has a unique representation as $v = x + y$, where $x \in X$ and $y \in Y$.

Usually, most of the binary operations which will be discussed in the book are commutative operations and usually this will be our assumption. Without loss of generality (w.l.o.g. in short) the translates in Definition 2.4 will be left translates. Clearly, if $+$ is a commutative binary operation then (X,Y) is a tiling if and only if (Y,X) is tiling.

The required parameters for the space tiling property will be defined depending on the specific space and metric. It should be noted that the space $\mathcal{V}$ can be taken in different ways for the same metric. For example, a subspace $\mathcal{V}' \subset \mathcal{V}$ can be taken instead of $\mathcal{V}$ and, in particular, a code $\mathcal{V}' = \mathcal{C} \subset \mathcal{V}$ can take the role of the space, instead of $\mathcal{V}$, in the tiling. In this case, the pair (X,Y) form, the tiling of $\mathcal{C}$, where each element $c \in \mathcal{C}$ can be uniquely written as $c = x + y$, $x \in X$, and $y \in Y$. It should also be noted that the definition of tiling will be generalized later in this section. Tilings will be used in the product constructions that are described next.

Theorem 2.2. *A code $\mathcal{C} \subset \mathcal{V}$ has the linear space tiling property if and only if there exists a subset $\mathcal{A}$ such that the pair $(\mathcal{A},\mathcal{C})$ is a tiling.*

Proof. Assume first that $\mathcal{C}$ has the linear space tiling property, i.e., $\mathcal{V} = \bigcup_{i=1}^{\ell}(x_i + \mathcal{C})$, where $\{x_1, x_2, \ldots, x_\ell\} \subseteq \mathcal{V}$ and $\ell = |\mathcal{V}| \,/\, |\mathcal{C}|$. This implies that $(x_i + \mathcal{C}) \cap (x_j + \mathcal{C}) = \varnothing$, for any $1 \leq i < j \leq \ell$. Therefore, $(\bigcup_{i=1}^{\ell}\{x_i\}, \mathcal{C})$ is a tiling.

Assume now that $(\mathcal{A},\mathcal{C})$ is a tiling, i.e., for each $v \in \mathcal{V}$, there is a unique representation of v as $v = x + c$, where $x \in \mathcal{A}$ and $c \in \mathcal{C}$. This implies that for each two distinct elements $x, y \in \mathcal{A}$, we have $(x + \mathcal{C}) \cap (y + \mathcal{C}) = \varnothing$. Therefore, $\{\{x + \mathcal{C}\} \,:\, x \in \mathcal{A}\}$ forms a partition of $\mathcal{V}$ and hence $\mathcal{C}$ has the linear space tiling property. □

Codes with the space tiling property will be used as building blocks of other codes, especially with the so-called ***product constructions***. Two such constructions will be heavily used in a few chapters. To distinguish between the two constructions, they will be called by different names. The first construction is a union of direct (cartesian) products, where the ***direct (cartesian) product***, $A \times B$, for two codes A and B, is defined by

$$A \times B \triangleq \{(a,b) \;:\; a \in A,\; b \in B\}\;.$$

The Direct Product Construction

Let $\mathcal{C}$ be a code that has the space tiling property and let $\mathcal{C}_1, \mathcal{C}_2, \dots, \mathcal{C}_m$ be the partition of a subspace $\mathcal{V}$ into disjoint translates (or codes with the same parameters) of $\mathcal{C}$. Let $\mathcal{C}'$ be a code (the same or a different one) that has the space tiling property and let $\mathcal{C}'_1, \mathcal{C}'_2, \dots, \mathcal{C}'_m$ be the partition of a subspace $\mathcal{V}'$ into disjoint translates (or codes with the same parameters) of $\mathcal{C}'$. The ***direct product*** of $\mathbb{C} = \{\mathcal{C}_1, \mathcal{C}_2, \dots, \mathcal{C}_m\}$ and $\mathbb{C}' = \{\mathcal{C}'_1, \mathcal{C}'_2, \dots, \mathcal{C}'_m\}$ is the code defined by

$$\mathbb{C} \otimes \mathbb{C}' \triangleq \bigcup_{i=0}^{m} \mathcal{C}_i \times \mathcal{C}'_i = \{(x,y) \ : \ x \in \mathcal{C}_i, \ \ y \in \mathcal{C}'_i, \ \ 1 \leq i \leq m\} \ .$$

The General Product Construction

Let $\mathcal{C}$ a code of length n over $Q_1 \times Q_2 \times \cdots \times Q_n$, where $|Q_i| = q_i$. Let $\mathcal{C}^i$, $1 \leq i \leq n$, be a code of length n_i over $\mathcal{V}_i$ with the space tiling property, where $\mathcal{C}^i_1, \mathcal{C}^i_2, \dots, \mathcal{C}^i_{q_i}$ is the partition of $\mathcal{V}_i$ into disjoint translates (or codes with the same parameters) of $\mathcal{C}^i$. The ***general product*** for $\mathcal{C}$, the set of codes $\{\mathcal{C}^i\}_{i=1}^{n}$, and their translates is the code defined by

$$\{(x_1, x_2, \dots, x_n) \ : \ x_i \in \mathcal{C}^i_{c_i}, \ 1 \leq i \leq n, \ (c_1, c_2, \dots, c_n) \in \mathcal{C}\} \ .$$

In other words, the element c_i in coordinate i of the codeword $c \in \mathcal{C}$ is replaced by codewords from the c_i-th code in the partition of the i-th code in all possible combinations.

Properties of these two product constructions will be discussed when their variants will be implemented on specific metrics and codes.

We now turn our attention to constant-weight codes, which form an important family of codes in coding theory. An $(n,d,w)_q$ code is a ***constant-weight code*** over an alphabet with q symbols, whose codewords are of length n, each one has a constant weight w, and the Hamming distance between any two distinct codewords is at least d. Let $A_q(n,d,w)$ denote the maximum number of codewords in an $(n,d,w)_q$ code. When $q=2$ we can omit the alphabet size q and use (n,d,w) and $A(n,d,w)$ instead of $(n,d,w)_2$ and $A_2(n,d,w)$, respectively. The following two bounds, known as ***Johnson bounds***, are of a special interest. The next lemma is the first Johnson bound.

Lemma 2.3.

$$A(n,d,w) \leq \left\lfloor \frac{n}{n-w} A(n-1,d,w) \right\rfloor \ .$$

Proof. Let $\mathcal{C}$ be a binary (n, d, w) code with M codewords, where $M = A(n, d, w)$. Clearly, the shortened code of $\mathcal{C}$ with respect to any coordinate is an $(n-1, d, w)$ code. Each codeword $c \in \mathcal{C}$ has $n - w$ *zeroes* and hence c is considered in $n - w$ codes of these shortened codes. Therefore, the total number of codewords in all the n shortened codes is $(n - w)M$. We can also see that, obviously, one of these n codes, say $\mathcal{C}_1$, has at least an average number of codewords of the total number of codewords in all these shortened codes, i.e., at least $\frac{(n-w)M}{n}$ codewords. This code is an $(n - 1, d, w)$ code and thus,

$$\frac{n-w}{n} A(n, d, w) = \frac{n-w}{n} M \leq |\mathcal{C}_1| \leq A(n-1, d, w),$$

which implies that

$$A(n, d, w) \leq \frac{n}{n-w} A(n-1, d, w) \ ,$$

and the claim of the lemma follows. □

We can use similar arguments to the ones used in the proof of Lemma 2.3 and construct shortened codes using the *ones* instead of the *zeroes* in the codewords of the code $\mathcal{C}$. If $|\mathcal{C}| = A(n, d, w)$, then we obtain the following lemma which is the second Johnson bound.

Lemma 2.4.

$$A(n, d, w) \leq \left\lfloor \frac{n}{w} A(n-1, d, w-1) \right\rfloor \ . \tag{2.1}$$

Constant-weight codes are related to the Johnson scheme. The Johnson scheme (the concept of "scheme" will be discussed in Section 3.5) is the most studied scheme after the Hamming scheme. It can be described with binary words of length n and weight w, but it is usually defined using w-subsets. Hence, we must have a one-to-one correspondence between all the words of length n and weight w, and all the w-subsets of an n-set. One such correspondence is achieved using the following translation. The ***characteristic vector*** of a w-subset S of an n-set Q is a binary word of length n and weight w whose i-th coordinate is a *one* if and only if the i-th element of Q is contained in S. It should be noted that for a binary word x and its support $\text{supp}(x)$, the characteristic vector of $\text{supp}(x)$ is x.

Definition 2.5. If $\mathcal{C}$ is a binary $(n, M, 2e + 1)$ code, then the code $\mathcal{C}^*$ obtained by adding a parity for each codeword of $\mathcal{C}$ is called the ***extended code*** of $\mathcal{C}$.

Clearly, we have the following lemma.

Lemma 2.5. *If $\mathcal{C}$ is a binary $(n, M, 2e+1)$ code, then $\mathcal{C}^*$ is an $(n+1, M, 2e+2)$ code.*

Similarly to $A(n,d,w)$, we define $A(n,d)$ to be the maximum number of codewords in a binary code of length n and minimum distance at least d.

Theorem 2.3. *If $1 \leq 2r-1 \leq n$, then $A(n, 2r-1) = A(n+1, 2r)$.*

Proof. This is an immediate observation, by first adding a parity bit to each codeword of the binary $(n, M, 2r-1)$ code to prove that $A(n, 2r-1) \leq A(n+1, 2r)$. Thereafter, to complete the proof, puncturing of a binary $(n+1, M, 2r)$ code implies that $A(n, 2r-1) \geq A(n+1, 2r)$. □

Theorem 2.4. *If $1 \leq d \leq n$, then $A(n,d) \leq 2A(n-1,d)$.*

Proof. The proof follows immediately by considering the codewords that start with a *zero* and the codewords that start with a *one* in any one of the coordinates of a binary (n, M, d) code for which $M = A(n,d)$. These two sets of codewords form codes of length $n-1$ and minimum distance d. One of these two sets has at least $M/2$ codewords, which completes the proof. □

We now turn our attention to the size of the largest code with a given minimum distance and recall some concepts from Chapter 1. A ***ball*** with ***radius*** e (***e-ball***) of an element $v \in \mathcal{V}$ (or around an element v, or centered at v), $\mathcal{B}_e(v)$, contains the subset of elements in $\mathcal{V}$ that are within distance e from v, i.e.,

$$\mathcal{B}_e(v) \triangleq \{x \ : \ x \in \mathcal{V}, \ d(v,x) \leq e\} .$$

If all such balls in the metric are of the same size and the length of the words is n, then $\mathcal{B}_e(v)$ will also be denoted by $\mathcal{B}_e(n)$. A metric in which for each $e \geq 0$, the size of a ball with radius e does not depend on the center of the ball, will be called a ***regular metric***. Since most of the metrics that will be discussed in this book are regular, this will be our general assumption. Whenever non-regular metrics are discussed, it will be specified.

The definition of a ball leads to the most basic bound on the size of a code, the ***sphere-packing bound***. A code is called an ***e-error-correcting code*** (or an ***e-code***) if the e-balls around the codewords of $\mathcal{C}$ are nonintersecting. These definitions immediately imply the following theorem.

Theorem 2.5. *If $\mathcal{C}$ is an e-code in a finite space $\mathcal{V}$ with a metric d, then*

$$\sum_{c \in \mathcal{C}} |\mathcal{B}_e(c)| \leq |\mathcal{V}|,$$

or

$$|\mathcal{C}| \cdot |\mathcal{B}_e(n)| \leq |\mathcal{V}|$$

if the metric is regular.

The bounds in Theorem 2.5 should have been called the ***ball-packing bound***, but we will use the usual name, common to all coding theory books, which is the "sphere-packing bound". For completeness, the ***sphere*** with ***radius e*** (***e-sphere***) of a vertex $v \in \mathcal{V}$ contains the subset of vertices in $\mathcal{V}$ that are at exactly distance e to v, i.e.,

$$\{x \; : \; d(x,v) = e, \; x \in \mathcal{V}\} \; .$$

Having mentioned the term radius, we should note that for each code we are interested in two types of radii, the packing radius and the covering radius.

The packing radius of a code $\mathcal{C}$ in a space $\mathcal{V}$ is the largest integer e such that each element $x \in \mathcal{V}$ is within radius (distance) e from at most one codeword of $\mathcal{C}$, i.e., for each $x \in V$, there exists at most one codeword $c \in \mathcal{C}$ such that $d(c,x) \leq e$. In other words, the e-balls around the codewords of $\mathcal{C}$ are nonintersecting. Such a code can correct any e errors, or less, which occurred during transmission of a codeword. For each $e' \leq e$, we can also say that e' is ***a packing radius*** of $\mathcal{C}$ since the balls of radius e' around the codewords of $\mathcal{C}$ are nonintersecting. The error correction for an e-code is based on the following analysis.

Lemma 2.6. *A code $\mathcal{C}$ has a packing radius e if and only if for each word $x \in \mathcal{V}$, the ball $\mathcal{B}_e(x)$ contains at most one codeword of $\mathcal{C}$.*

Proof. Assume first that $\mathcal{C}$ has a packing radius e and assume the contrary, that $c_1, c_2 \in \mathcal{C}$ are two distinct codewords such that $c_1, c_2 \in \mathcal{B}_e(x)$ for some $x \in \mathcal{V}$. By the definition of the ball $\mathcal{B}_e(x)$, we have that $d(x, c_1) \leq e$ and $d(x, c_2) \leq e$, which contradict the fact that the packing radius of $\mathcal{C}$ is e. Hence, for any word $x \in \mathcal{V}$, the ball $\mathcal{B}_e(x)$ contains at most one codeword of $\mathcal{C}$.

Assume now that for any word $x \in \mathcal{V}$ the ball $\mathcal{B}_e(x)$ contains at most one codeword of $\mathcal{C}$. Assume the contrary, that there exists a word $x \in \mathcal{V}$ such that $x \in \mathcal{B}_e(c_1)$ and $x \in \mathcal{B}_e(c_2)$, where $c_1, c_2 \in \mathcal{C}$. Clearly, a word $y \in \mathcal{V}$

is contained in the ball $\mathcal{B}_e(x)$ if and only if $x \in \mathcal{B}_e(y)$. This implies that $c_1 \in \mathcal{B}_e(x)$ and $c_2 \in \mathcal{B}_e(x)$, a contradiction. Hence, x is within distance e from at most one codeword of $\mathcal{C}$, i.e., $\mathcal{C}$ has a packing radius e. □

Corollary 2.1. *A code $\mathcal{C}$ has packing radius e if and only if $\mathcal{C}$ is an e-code.*

Assume that a codeword c of an e-code $\mathcal{C}$ was transmitted and a word x was received, and that no more than e errors occurred in c during transmission, i.e., $d(c, x) \leq e$. Hence, since $\mathcal{C}$ is an e-code, it follows by Lemma 2.6 that the word x is in exactly one ball with radius e around a codeword of $\mathcal{C}$. This codeword is the transmitted codeword. This codeword can be found by computing the ball of radius e around x, $\mathcal{B}_e(x)$, since by Lemma 2.6, this ball cannot contain more than one codeword of $\mathcal{C}$. The codeword c will be the unique codeword in this ball.

Lemma 2.7. *If the minimum distance of a code $\mathcal{C}$ is $2e+1$, then the code $\mathcal{C}$ has packing radius e.*

Proof. Assume the contrary, that e is not a packing radius of $\mathcal{C}$, i.e., there exist two codewords $c_1, c_2 \in \mathcal{C}$ such that $\mathcal{B}_e(c_1) \cap \mathcal{B}_e(c_2) \neq \varnothing$. If $x \in \mathcal{B}_e(c_1) \cap \mathcal{B}_e(c_2)$, then $d(c_1, x) \leq e$ and $d(x, c_2) \leq e$, and by the triangle inequality

$$d(c_1, c_2) \leq d(c_1, x) + d(x, c_2) \leq 2e,$$

in contradiction to the minimum distance of $\mathcal{C}$. □

Corollary 2.2. *If the minimum distance of a code $\mathcal{C}$ is d, then the code $\mathcal{C}$ has packing radius $\left\lfloor \frac{d-1}{2} \right\rfloor$.*

Corollary 2.3. *A code $\mathcal{C}$ whose minimum distance is $2e + 1$ can correct any e errors occurring in a transmitted codeword.*

At this point it is important to note that the converse of Lemma 2.7 is not necessarily correct. The following example illustrates this scenario.

Example 2.1. Let $\mathcal{V} = \{0, 1, 2, 3\}$, $d(0, 1) = d(1, 2) = d(2, 3) = 1$, $d(0, 2) = d(1, 3) = d(0, 3) = 2$, $d(x, x) = 0$ for each $x \in \mathcal{V}$, and $d(x, y) = d(y, x)$ for each $x, y \in \mathcal{V}$. It is readily verified that d is a metric. If $\mathcal{C} = \{0, 3\}$, then $\mathcal{C}$ has packing radius $e = 1$; but the minimum distance of $\mathcal{C}$ is 2, which is less than $2e + 1 = 3$.

The metric in Example 2.1 is not regular. Nevertheless, the converse of Lemma 2.7 might also be incorrect for regular metrics. The following example illustrates it on a regular metric.

Example 2.2. Let $\mathcal{V} = \mathbb{Z}_6$, $d(i, i+1) = 1$ for each $i \in \mathbb{Z}_6$, $d(x, x) = 0$ for each $x \in \mathbb{Z}_n$, $d(i, i+2) = d(i, i+3) = 2$ for each $i \in \mathbb{Z}_6$, and $d(x, y) = d(y, x)$ for each $x, y \in \mathcal{V}$. It is readily verified that d is a metric. It is also readily verified that the metric is regular, but the code $\mathcal{C} = \{0, 3\}$ has packing radius $e = 1$, while its minimum distance is 2, which is less than $2e+1 = 3$.

The covering radius of a code $\mathcal{C}$ in a space $\mathcal{V}$ is the smallest integer R such that each element $x \in \mathcal{V}$ is within radius (distance) R from at least one codeword of $\mathcal{C}$. In other words, for each $x \in \mathcal{V}$, there exists at least one codeword $c \in \mathcal{C}$ such that $d(c, x) \leq R$. Such a code is a ***covering code*** with radius R. Similarly to the packing radius, for each $R' \geq R$ we can say that R' is ***a covering radius*** of $\mathcal{C}$ since the balls of radius R' around the codewords of $\mathcal{C}$ cover all the elements of $\mathcal{V}$. Similarly, we say that a codeword c in an e-code ***covers*** a word x in the space $\mathcal{V}$ if $d(c, x) \leq e$.

The distinction between a ***packing code*** $\mathcal{C}$ (which is an error-correcting code) and a ***covering code*** $\mathcal{C}$ is based only on whether we are interested in the packing radius of $\mathcal{C}$ or in the covering radius of $\mathcal{C}$. Generally, a ***packing*** is associated with filling of a space with copies of a shape or several shapes, where there is no intersection between any two shapes. Similarly, a ***covering*** is associated with a similar filling of the space, where each point of the space is covered by at least one shape, but distinct shapes can have a nonempty intersection. The shapes in which we will be interested are balls, anticodes, or error spheres (which will be defined later). Similarly to the sphere-packing bound of Theorem 2.5, we have the following ***ball-covering bound***, which is implied by the definitions similarly to the proof of Theorem 2.5.

Theorem 2.6. *If $\mathcal{C}$ is a code with covering radius R, in a finite space $\mathcal{V}$ with a metric d, then*

$$\sum_{c \in \mathcal{C}} |\mathcal{B}_R(c)| \geq |\mathcal{V}|$$

or

$$|\mathcal{C}| \cdot |\mathcal{B}_R(n)| \geq |\mathcal{V}|$$

if the metric is regular.

Corollary 2.4. *A code $\mathcal{C}$ in a finite space $\mathcal{V}$ attains the bound of Theorem 2.5 with equality if and only if $\mathcal{C}$ attains the bound of Theorem 2.6 with equality, where e of Theorem 2.5 is equal to R of Theorem 2.6.*

In other words, Corollary 2.4 is implied by the fact that if each word $x \in \mathcal{V}$ is contained in exactly one ball with radius e centered in a codeword c of a code $\mathcal{C}$, then $\mathcal{C}$ meets the bounds of Theorems 2.5 and 2.6.

2.2 Finite Fields

Finite fields play a major role in coding theory, especially for linear codes, but also in codes based on subspaces and in some constructions of nonlinear codes. Two concepts which are related to finite fields and their definition lead to the definition of a finite field are a group and a ring.

Definition 2.6. A pair $(\mathcal{G}, \circ)$ is called a ***group*** if $\mathcal{G}$ is a nonempty set, $\circ$ is a binary operation defined on $\mathcal{G}$, and the following three properties are satisfied:

(1) $(a \circ b) \circ c = a \circ (b \circ c)$ for all $a, b, c \in \mathcal{G}$.
(2) There is an ***identity*** element $e \in \mathcal{G}$ such that $a \circ e = e \circ a = a$ for all $a \in \mathcal{G}$.
(3) For each $a \in \mathcal{G}$ there exists an ***inverse*** element $a^{-1} \in \mathcal{G}$ such that $a \circ a^{-1} = a^{-1} \circ a = e$.

The group $(\mathcal{G}, \circ)$ is called an ***abelian group*** (or a ***commutative group***) if $a \circ b = b \circ a$ for all $a, b \in \mathcal{G}$.

The group $(\mathcal{G}, \circ)$ is a ***cyclic group*** if there exists an element $a \in \mathcal{G}$, such that each $b \in \mathcal{G}$ is equal to $a^i \triangleq \overbrace{a \circ a \circ \cdots \circ a}^{i \text{ times}}$ for some integer i. The element a is called a ***generator*** of the group.

A special interest is in the group $\mathbb{Z}_m$, $m \geq 2$, that contains the set $\{0, 1, \ldots, m-1\}$ of integers, where the binary operation is addition modulo m. The elements of $\mathbb{Z}_m$ can also be considered as the m distinct residues modulo m.

Definition 2.7. A triple $(\mathcal{R}, +, \cdot)$ is called a ***ring*** if $\mathcal{R}$ is a nonempty set, $+$ and $\cdot$ are two binary operations defined on $\mathcal{R}$, and the following four properties are satisfied:

(1) $(\mathcal{R}, +)$ is an abelian group.

(2) $(a \cdot b) \cdot c = a \cdot (b \cdot c)$ for all $a, b, c \in \mathcal{R}$.
(3) There is a unique element $1 \in \mathcal{R}$ such that $a \cdot 1 = 1 \cdot a = a$ for all $a \in \mathcal{R}$.
(4) $a \cdot (b + c) = a \cdot b + a \cdot c$ and $(a + b) \cdot c = a \cdot c + b \cdot c$ for all $a, b, c \in \mathcal{R}$.

The identity element of the group $(\mathcal{R}, +)$ is denoted by 0.

The ring $(\mathcal{R}, +, \cdot)$ is called a ***commutative ring*** if $a \cdot b = b \cdot a$ for all $a, b \in \mathcal{R}$.

Note that $(\mathcal{R} \setminus \{0\}, \cdot)$ is not necessarily a group since it might not have an inverse for each element of $\mathcal{R} \setminus \{0\}$.

Definition 2.8. A ring $(\mathbb{F}, +, \cdot)$ is called a ***field*** if the pair $(\mathbb{F} \setminus \{0\}, \cdot)$ is an abelian group. The element 0 is the identity element of the abelian group $(\mathbb{F}, +)$ and 1 is the identity element of the abelian group $(\mathbb{F} \setminus \{0\}, \cdot)$.

We denote the set $\mathcal{G} \setminus \{0\}$, where $\mathcal{G}$ is a group (also for a ring or a field) by $\mathcal{G}^-$. The group $(\mathbb{F}, +)$ is called the ***additive group*** of the field and the group $(\mathbb{F}^-, \cdot)$ is called the ***multiplicative group*** of the field.

Our main interest is in ***finite fields***, i.e., fields with a finite number of elements. All such fields with the same number of elements are isomorphic and they are called ***Galois fields***. The number of elements in such a field is q, where q is a power of a prime and it is denoted by $\text{GF}(q)$ or $\mathbb{F}_q$. The abelian group $(\mathbb{F}_q^-, \cdot)$ is a cyclic group.

The ring of integers modulo m will be denoted by $\mathbb{Z}_m$ (as the group $\mathbb{Z}_m$). Addition and multiplication in the ring is performed modulo m. This ring is a field if p is a prime integer. It contains the set of integers $\{0, 1, \ldots, p-1\}$ (or equivalently the set of p distinct residues modulo p) where addition and multiplication are performed modulo p.

The finite field $\mathbb{F}_{q^k}$, where q is a power of a prime, has q^k elements. The multiplicative group of $\mathbb{F}_{q^k}^-$ is a cyclic group with a generator α. The generator α is a root of some irreducible polynomial

$$c(x) = x^k - \sum_{i=1}^{k} c_i x^{k-i}, \quad c_i \in \mathbb{F}_q$$

called a ***primitive polynomial*** and each one of its root α is called a ***primitive element***. The elements of $\text{GF}(q^k)$ can be represented as the q^k vectors of length k over $\mathbb{F}_q$. For two elements α^i, α^j, represented by the vectors $x = (x_1, x_2, \ldots, x_k) \in \mathbb{F}_q^k$ and $y = (y_1, y_2, \ldots, y_k) \in \mathbb{F}_q^k$, respectively, we have that $\alpha^i \cdot \alpha^j = \alpha^{i+j}$, where superscripts are taken modulo $q^k - 1$, and

$$\alpha^i + \alpha^j = x + y = (x_1 + y_1, x_2 + y_2, \ldots, x_k + y_k) = \alpha^\ell \ ,$$

where α^ℓ is represented by the vector $(x_1 + y_1, x_2 + y_2, \ldots, x_k + y_k) \in \mathbb{F}_q^k$.

Since α is a root of $c(x)$, it follows that $0 = c(\alpha) = \alpha^k - \sum_{i=1}^k c_i\alpha^{k-i}$ and $\alpha^k = \sum_{i=1}^k c_i\alpha^{k-i}$. The element $\alpha^0 = 1$ is represented by the vector $(00\cdots001)$, the element α by the vector $(00\cdots010)$, and so on, where α^{k-1} is represented by the vector $(10\cdots000)$. The element α^k is represented by the vector $(c_1, c_2, \ldots, c_k)$. Similarly, if $\alpha^i = (a_1, a_2, \ldots, a_k)$, then $\alpha^{i+1} = (a_2, \ldots, a_k, 0)$ when $a_1 = 0$ and if $a_1 \neq 0$, then $\alpha^{i+1} = (a_2, \ldots, a_k, 0) + a_1\alpha^k = (a_2, \ldots, a_k, 0) + (a_1c_1, a_1c_2, \ldots, a_1c_k)$. The irreducible polynomial $c(x)$ is a primitive polynomial if each of its roots (primitive elements) generates the field, i.e., the $q^k - 1$ powers of any root α, of $c(x)$, are distinct elements as q-ary vectors in this computation.

The representation of the elements of GF(q^k) by the q-ary vectors of length k, over $\mathbb{F}_q$, induces a bijection between $\mathbb{F}_{q^k}$ and $\mathbb{F}_q^k$. This bijection is used to simplify many results and to simplify some representations of codes in general and perfect codes in particular.

Finally, in many cases we are required to take the elements of a group $\mathcal{G}$ (a ring $\mathcal{R}$, a field $\mathbb{F}$, or a subspace X, respectively) without its additive identity. The structure without the identity (the *zero* element) will be denoted, as was defined before, by $\mathcal{G}^-$ ($\mathcal{R}^-$, $\mathbb{F}^-$, X^-, respectively). In the literature it is frequently denoted by G^* ($\mathcal{R}^*$, $\mathbb{F}^*$, X^*, respectively), but the different notation, which is used in the book, serves to distinguish it from the extended code $\mathcal{C}^*$ of a code $\mathcal{C}$.

2.3 Linear Codes

An $[n, k]_q$ ***(linear) code*** is a linear subspace of dimension k over $\mathbb{F}_q^n$, i.e., a linear subspace, whose dimension is k, from the set of all words (vectors) of ***length*** n over $\mathbb{F}_q$. Later on, the abbreviation a ***k-subspace*** will be used frequently instead of a subspace of dimension k or a k-dimensional subspace.

An $[n, k]_q$ code $\mathcal{C}$ can be represented by two matrices. The first one is a ***generator matrix*** G, which is a $k \times n$ matrix over $\mathbb{F}_q$, whose rows form a basis for the code, i.e., the linear span of the rows of G is $\mathcal{C}$. We also denote by $\mathcal{C}(G)$ the code generated from the generator matrix G, i.e., $\mathcal{C}(G) = \langle G \rangle$, where $\langle A \rangle$ is the ***linear span*** of the rows from the matrix A. The second matrix is a ***parity-check matrix*** H, which is an $(n - k) \times n$ matrix over $\mathbb{F}_q$, whose rows form a basis for the ***dual subspace*** $\mathcal{C}^\perp$ of the code $\mathcal{C}$. The dimension $r = n - k$ of this dual subspace is called the

redundancy of the code.

A generator matrix of an $[n, k]_q$ code is in ***standard form*** if its first k columns form an identity matrix of order k, i.e.,

$$G = [\ I_k \mid A\ ],$$

where I_k is the $k \times k$ identity matrix. The related parity-check matrix is given by

$$H = \left[\ -A^{\text{tr}} \mid I_{n-k}\ \right].$$

It is readily verified that with this representation we have that

$$G \cdot H^{\text{tr}} = \mathbf{0}$$

and

$$H \cdot G^{\text{tr}} = \mathbf{0},$$

where $\mathbf{0}$ is an all-zero matrix of the appropriate size and A^{tr} is the ***transpose*** of the matrix A.

One can use a generator matrix with more than k rows for an $[n, k]_q$ code or a parity-check matrix with more than $r = n-k$ rows, by using some redundant rows. Usually, these redundant rows will not be necessary, but in some cases they will be required. The following proposition is a simple observation.

Proposition 2.1. *The parity-check matrix H of an $[n, k]_q$ code $\mathcal{C}$ is a generator matrix of an $[n, n-k]_q$ code.*

If G is the generator matrix of an $[n, k]_q$ code $\mathcal{C}$, then the $[n, n-k]_q$ code whose generator matrix is the parity-check matrix H of $\mathcal{C}$ is called the ***dual code*** of $\mathcal{C}$. The dual code of $\mathcal{C}$ is denoted by $\mathcal{C}^{\perp}$. A code $\mathcal{C}$ is called ***self-dual*** if $\mathcal{C} = \mathcal{C}^{\perp}$.

There is another representation of the parity-check matrix. Let α be a primitive element in $\mathbb{F}_{q^r}$ and let $H = [h_1, h_2, \ldots, h_n]$ be an $r \times n$ parity check-matrix for the code $\mathcal{C}$. Assume that h_j is the q-ary representation of the element α^{i_j}, $1 \leq j \leq n$, in $\mathbb{F}_{q^r}$. The parity-check matrix can be written as $H = [\alpha^{i_1}, \alpha^{i_2}, \ldots, \alpha^{i_n}]$. Finally, note that the word $x = (x_1, x_2, \ldots, x_n) \in \mathbb{F}_q^n$ is a codeword in $\mathcal{C}$ if and only if $H \cdot x^{\text{tr}} = 0$.

A set S of k coordinates in a code $\mathcal{C}$ (not necessarily linear), over a q-set Q, are called ***systematic*** if in the projection on these k coordinates of $\mathcal{C}$, each of the q^k vectors of length k over Q appears exactly once. Clearly, in an $[n, k]_q$ code whose generator matrix is in standard form, the first k

coordinates are systematic. By definition one can easily verify the following lemma.

Lemma 2.8. *A set of k coordinates in the generator matrix G of an $[n,k]_q$ code $\mathcal{C}$ are systematic coordinates if and only if the related k vector columns of G are linearly independent.*

A code $\mathcal{C}$ is a ***systematic code*** if it has k systematic coordinates. Clearly, all linear codes are systematic.

For an $[n,k]_q$ code $\mathcal{C}$ and a word $x \in \mathbb{F}_q^n$, the translate

$$x + \mathcal{C} \triangleq \{x + c \; : \; c \in \mathcal{C}\},$$

is called a ***coset*** of $\mathcal{C}$.

Let y and z be a two words in the coset $x + \mathcal{C}$. Clearly, $y = x + c_1$ and $z = x + c_2$, where $c_1, c_2 \in \mathcal{C}$, and $z - y = c_2 - c_1$. Since the code $\mathcal{C}$ is linear, it follows that $c_2 - c_1 \in \mathcal{C}$. This implies the following lemma.

Lemma 2.9. *The words y and z are in the same coset of an $[n,k]_q$ linear code $\mathcal{C}$ if and only if $y - z \in \mathcal{C}$.*

Corollary 2.5. *If $\mathcal{C}$ is a linear code, then z is a word in the coset $x + \mathcal{C}$ if and only if $-z \in x + \mathcal{C}$.*

Corollary 2.6. *The words y and z are in the same coset of an $[n,k]_q$ linear code $\mathcal{C}$ if and only if $y + z \in \mathcal{C}$.*

Proposition 2.2. *The cosets of an $[n,k]_q$ code $\mathcal{C}$ form a partition of $\mathbb{F}_q^n$, where each coset has q^k distinct words of length n.*

Proof. Clearly, for any word $x \in \mathbb{F}_q^n$ and two distinct codewords $c_1, c_2 \in \mathcal{C}$, we have that $x + c_1 \neq x + c_2$. This implies that each coset has q^k distinct words of length n.

If $y \in x_1 + \mathcal{C}$ and $y \in x_2 + \mathcal{C}$, then $y = x_1 + c_1 = x_2 + c_2$, where $c_1, c_2 \in \mathcal{C}$. If $z \in x_1 + \mathcal{C}$, then $z = x_1 + c_3$, where $c_3 \in \mathcal{C}$. Hence, $z = y - c_1 + c_3 = x_2 + c_2 - c_1 + c_3 = x_2 + c_4$, where $c_4 \in \mathcal{C}$. This implies that $z \in x_2 + \mathcal{C}$ and, therefore, any two cosets are either disjoint or coincide. Thus, the disjoint cosets of $\mathcal{C}$ form a partition of $\mathbb{F}_q^n$. □

Corollary 2.7. *Any two cosets of an $[n,k]_q$ code are either equal or disjoint.*

Corollary 2.8. *A linear code $\mathcal{C}$ has the linear space tiling property.*

A set S of representatives from the cosets of a linear codes $\mathcal{C}$ is a set which contains exactly one word from each coset of $\mathcal{C}$.

Corollary 2.9. *If $\mathcal{A}$ is a set of q^{n-k} representatives of the q^{n-k} distinct cosets of an $[n,k]_q$ code $\mathcal{C}$, then $(\mathcal{A},\mathcal{C})$ is a tiling.*

The ***coset leader*** of a coset $x+\mathcal{C}$ is a word of minimum weight in the coset. If there are a few words with minimum weight, then one of them is chosen randomly to be the coset leader. The following claim is implied from Corollary 2.9.

Corollary 2.10. *If $\mathcal{A}$ is a set of q^{n-k} coset leaders of an $[n,k]_q$ code $\mathcal{C}$, then $(\mathcal{A},\mathcal{C})$ is a tiling.*

Definition 2.9. Let $\mathcal{C}$ be a linear code, over $\mathbb{F}_q$, with an $r \times n$ parity-check matrix H. For any word $x = (x_1, x_2, \ldots, x_n) \in \mathbb{F}_q^n$, the ***syndrome*** of x, $\mathcal{S}(x)$, is defined by

$$\mathcal{S}(x) = H \cdot x^{\text{tr}} \ .$$

Clearly, the syndromes are column vectors of length r, the redundancy of the code. Hence, there are q^r possible distinct syndromes. The first important property related to the syndromes is about the syndromes of the codewords. The value of these syndromes can be verified from the definition of the parity-check matrix of a code $\mathcal{C}$.

Proposition 2.3. *The syndrome of a codeword in a linear code is equal to the all-zero vector.*

The syndromes have some properties that are very useful in correcting errors that occur during the transmission of the information words using a linear code. A certain set of syndromes are also very important in answering the question whether a linear code is a perfect code.

Definition 2.10. An $[n,k,d]_q$ ***code*** is an $[n,k]_q$ code whose minimum Hamming distance is at least d.

Corollary 2.11. *The minimum distance d on an $[n,k,d]_q$ code $\mathcal{C}$ is the minimum number of linearly dependent columns of its parity-check matrix H.*

Proof. The claim follows immediately from the fact that $c \in \mathcal{C}$ if and only if $H \cdot c^{\text{tr}} = \mathbf{0}$ and hence the minimum number of linearly dependent columns of H is the minimum weight of a nonzero codeword in $\mathcal{C}$. $\square$

The $[n,k,d]_q$ code $\mathcal{C}$ has a generator matrix G and a parity-check matrix H. The code $\mathcal{C}$ is used to transmit information words of length k over $\mathbb{F}_q$, via a channel that accepts words of length n. An information word $z = (z_1, z_2, \ldots, z_k)$ is transformed into a codeword $c = (c_1, c_2, \ldots, c_n)$ of length n, where $c = z \cdot G$. Since c is generated as a linear combination of rows from G and the rows of H span a subspace orthogonal to the linear span of the rows of G, it follows, as also implied by Proposition 2.3, that $\mathcal{S}(c) = H \cdot c^{\text{tr}} = \mathbf{0}$. Assume that in the channel, an error ε has occurred in the codeword c and instead of the codeword c, the word $c+\varepsilon$ was received. The syndrome of $c+\varepsilon$ is

$$\mathcal{S}(c+\varepsilon) = H \cdot (c+\varepsilon)^{\text{tr}} = H \cdot c^{\text{tr}} + H \cdot \varepsilon^{\text{tr}} = H \cdot \varepsilon^{\text{tr}} .$$

This implies that if it is assumed that only a set $\mathcal{E}$ of errors can occur and each of the elements in the set $\mathcal{E}$ has a different syndrome, then using the value of the syndrome of the received word we have the syndrome of the error. This syndrome should be unique to this error and hence we can find the exact error and recover the codeword that was transmitted over the channel. This implies the following observation.

Corollary 2.12. *In a linear e-code all the syndromes of the distinct words with weight at most e are distinct.*

Finally, to conclude this section we will prove that the two concepts of weight distribution and distance distribution coincide for linear codes.

Theorem 2.7. *If* $(A_0, A_1, \ldots, A_n)$ *and* $(D_0, D_1, \ldots, D_n)$ *are the weight distribution and the distance distribution, respectively, of an* $[n,k]_q$ *code* $\mathcal{C}$, *then* $A_i = D_i$ *for each* $0 \leq i \leq n$.

Proof. Assume that for a given i, $0 \leq i \leq n$, $A_i = t$, i.e., $\mathcal{C}$ has t codewords $c_1, c_2, \ldots, c_t$ of weight i. Hence, if $c \in \mathcal{C}$, then $c+c_1, c+c_2, \ldots, c+c_t$ are distinct codewords for which $d(c, c+c_j) = i$ for each $1 \leq j \leq t$ and there is no other codeword c' such that $d(c,c') = i$. In other words, there are exactly t pairs in the set $\{(c,c') \,:\, d(c,c') = i,\ c' \in \mathcal{C}\}$. Therefore, we have that

$$|\{(x,y) \,:\, x,y \in \mathcal{C},\ d(x,y) = i\}| = |\{(c,c_j) \,:\, c \in \mathcal{C},\ 1 \leq j \leq t\}| = t\,|\mathcal{C}|$$

and hence $D_i = t = A_i$. $\square$

2.4 Definitions of Perfect Codes

We are now in a position to present the definitions of perfect codes and their generalizations. The most simple definition relates to metrics that can be described in terms of a graph. Given an undirected graph $\Gamma = (\mathcal{V}, E)$, where $\mathcal{V}$ is a set of vertices (the space) and E is the edge set of Γ, we define the following simple metric Γ. First we note that Γ denotes the graph and we refer to it also as the metric defined by that graph. But, the distance between two vertices x, y of the graph will be denoted by $d_\Gamma(x, y)$ and in the metric their distance will be denoted by $d(x, y)$. The distance between two vertices $u, v \in \mathcal{V}$, $d_\Gamma(u, v)$ is the length of the shortest path between two vertices u and v in Γ (the number of edges in this path). The length δ of this shortest path, between $u, v \in \mathcal{V}$, is also the distance for the metric Γ defined by the graph, i.e., $d_\Gamma(u, v) = \delta$. It is easy to verify that this definition of distance based on the length of the shortest path in the graph Γ is a metric. A code (set of vertices) $\mathcal{C}$ in the graph is an ***e*-perfect code** if for each $v \in \mathcal{V}$ there exists exactly one codeword $c \in \mathcal{C}$ such that $d_\Gamma(v, c) \leq e$. In other words, a code $\mathcal{C}$ is an e-perfect code if the balls with radius e around the codewords (vertices) of $\mathcal{C}$ form a partition of the vertices in $\mathcal{V}$. Given a space $\mathcal{V}$ and a metric d on $\mathcal{V}$, can d be represented by a graph $\Gamma = (\mathcal{V}, E)$, where $E = \{\{x, y\} \ : \ x, y \in \mathcal{V}, \ d(x, y) = 1\}$? The answer is very simple. This representation is possible if for any two elements x and y of the space $\mathcal{V}$ such that $d(x, y) = \delta$, the shortest path between x and y in the graph Γ has length δ. This immediately implies that Γ must be a connected graph to define a metric and in particular to define a perfect code via the graph Γ. The following lemma asserts that if $d(x, y) = \delta$, then this distance between x and y, in the graph, cannot be smaller than δ.

Lemma 2.10. *Let $d : \mathcal{V} \times \mathcal{V} \to \mathbb{Z}$ be a metric and let $\Gamma = (\mathcal{V}, E)$ be the graph defined for this metric, i.e., $d(x, y) = 1$ if and only if $\{x, y\} \in E$, and assume that Γ is a connected graph. Let x and y be two distinct vertices in $\mathcal{V}$. If $d(x, y) = \delta$, then the length of the shortest path between x and y, in Γ, is at least δ.*

Proof. Assume the contrary, that $d(x, y) = \delta$ but $d_\Gamma(x, y) = \delta - \epsilon < d(x, y)$, i.e., $\epsilon > 0$. Assume that $\delta - \epsilon$ is the length of the shortest path in Γ with this required property, i.e., if for $u, v \in \mathcal{V}$, $d_\Gamma(u, v) = \delta'$ and $d(u, v) > \delta'$, then $\delta - \epsilon \leq \delta'$. Since, by definition, for $u, v \in \mathcal{V}$, $d_\Gamma(u, v) = 1$ if and only if $d(u, v) = 1$, it follows that $\delta - \epsilon > 1$. Consider now the path of length $\delta - \epsilon$ between x and y in Γ. This path contains another $\delta - \epsilon - 1$ vertices. Let z be the first of these vertices in this path, i.e., $d_\Gamma(x, z) = 1$ and $d_\Gamma(z, y) = \delta - \epsilon - 1$. Clearly, by definition $d(x, z) = 1$ and by the triangle inequality, we have that $\delta = d(x, y) \leq d(x, z) + d(z, y)$ and hence $d(z, y) \geq \delta - 1$. This implies that $d_\Gamma(z, y) = \delta - \epsilon - 1 < \delta - 1 \leq d(z, y)$, which contradicts the fact that $\delta - \epsilon$ is the shortest path in Γ with the required property, and the claim of the lemma is proved. □

Lemma 2.10 implies that if Γ is a connected graph defined by the metric d, then the distance in Γ between any two vertices is at least their distance in the metric. Is the converse also correct? The answer is no and as an example we can consider the metric defined in Example 2.1 or the metric defined in Example 2.2. This implies that the distance in the graph Γ is not necessarily equal to the distance in the metric, from which Γ was defined, and the metric cannot be represented by a graph for these two examples. Another scenario in which this property is not satisfied is when Γ is not a connected graph. Such an example will be discussed in Chapter 9. In the case when the graph is not a connected graph, we will also have to be careful in the definition for an e-perfect code.

A metric d on a finite space $\mathcal{V}$ is called a ***graphic metric*** if it can be represented by a graph $\Gamma = (\mathcal{V}, E)$, where $\{x, y\} \in E$ if and only if $d(x, y) = 1$, and for each two vertices $x, y \in \mathcal{V}$, $d_\Gamma(x, y) = \delta$ if and only if $d(x, y) = \delta$.

Fortunately, most metrics in our context are graphic. The formal definitions that follow, however, can serve for all metrics and they are equivalent to the definitions given for a connected graph.

Lemma 2.11. *If $\mathcal{C}$ an e-code in a graphic metric, then* $d(\mathcal{C}) \geq 2e + 1$.

Proof. Since $\mathcal{C}$ is an e-code, it follows by Corollary 2.1 that $\mathcal{C}$ has packing radius e. Let $c_1, c_2 \in \mathcal{C}$ and assume that $d(c_1, c_2) < 2e + 1$. Since d is a graphic metric, it follows that there exist a word $x \in \mathcal{V}$ such that $d(x, c_1) \leq e$ and $d(x, c_2) \leq e$, contradicting Lemma 2.6. Thus, $d(\mathcal{C}) \geq 2e + 1$. □

We continue with the metric considered by its formal definition (not by the related graph). There are two basic definitions for a perfect code, which will now be presented. Although in most of our book the space $\mathcal{V}$ is finite, these two definitions can also serve in the case where the space $\mathcal{V}$ is infinite.

Definition 2.11. A code $\mathcal{C} \subseteq \mathcal{V}$ is an e-***perfect code*** with respect to a metric $d : \mathcal{V} \times \mathcal{V} \to \mathbb{Z}$ if, for each element $x \in \mathcal{V}$, there exists a unique codeword $c \in \mathcal{C}$ such that $d(c, x) \leq e$.

Definition 2.12. A code $\mathcal{C} \subseteq \mathcal{V}$ is an e-***perfect code*** with respect to a metric $d : \mathcal{V} \times \mathcal{V} \to \mathbb{Z}$ if the set of balls $\{\mathcal{B}_e(c) \ : \ c \in \mathcal{C}\}$ form a partition of $\mathcal{V}$.

It can easily be verified that a code that satisfies Definitions 2.11 and 2.12 also meets the sphere-packing bound and the ball-covering bound when $\mathcal{V}$ is finite. These two bounds can also be used to supply alternative definitions of an e-perfect code, but only when $\mathcal{V}$ is a finite space.

Definition 2.13. A code $\mathcal{C} \subseteq \mathcal{V}$ is an e-***perfect code*** with respect to a metric $d : \mathcal{V} \times \mathcal{V} \to \mathbb{Z}$, where $\mathcal{V}$ is a finite space, if it meets the bound of Theorem 2.5.

Definition 2.14. A code $\mathcal{C} \subseteq \mathcal{V}$ is an R-***perfect code*** with respect to a metric $d : \mathcal{V} \times \mathcal{V} \to \mathbb{Z}$, where $\mathcal{V}$ is a finite space, if it meets the bound of Theorem 2.6.

The equivalence of all these four definitions for a perfect code is summarized as follows.

Theorem 2.8. *A code $\mathcal{C} \subseteq \mathcal{V}$ satisfies Definition 2.11 if and only if $\mathcal{C}$ satisfies Definition 2.12. Furthermore, if $\mathcal{V}$ is a finite space, then Definitions 2.11, 2.12, 2.13, and 2.14 are equivalent.*

If the code is linear, then a perfect code can be defined using the parity-check matrix H. An $[n, k]_q$ code is e-perfect if the syndromes related to all possible errors of weight at most e are distinct and contains all the q^{n-k} column vectors of length $n - k$.

Lemma 2.12. *If $\mathcal{C}$ is a linear e-perfect code in a space $\mathcal{V}$ and $\mathcal{A}$ is a set which contains all the words in $\mathcal{V}$ of weight e or less, then $(\mathcal{A}, \mathcal{C})$ is a tiling.*

Proof. If $\mathcal{C}$ is a linear e-perfect code, then all the coset leaders have weight at most e. If two x and y words of $\mathcal{V}$ are in the same coset of $\mathcal{C}$, then by Corollary 2.6 their sum $x + y$ is a codeword in $\mathcal{C}$. If x and y have weight at most e, then their sum $x + y$ has weight at most $2e$ and hence by $d(\mathcal{C}) \leq 2e$, contradicting Lemma 2.11. Hence, x and y are in a different cosets of $\mathcal{C}$, which by Corollary 2.10 implies that the pair $(\mathcal{A}, \mathcal{C})$ is a tiling. □

Corollary 2.13. *If $\mathcal{C}$ is a linear e-perfect code, then each coset contains exactly one word from each ball $\mathcal{B}_e(c)$, for each $c \in \mathcal{C}$.*

There are scenarios where the criterion (for different syndromes) will be the most convenient one for proving that a code is perfect. When the set of errors in an $[n,k]_q$ code over a space $\mathcal{V}$ is confined to a set S and the syndromes associated with the words in S are disjoint and contain all the q^{n-k} vectors of length $n-k$, then the code $\mathcal{C}$ is a perfect code. This type of code will be discussed in some sections later on in the book.

If the metric is regular, i.e., for each $e \geq 1$, all the balls of radius e are of the same size, and the sphere-packing bound is attained with equality by a code $\mathcal{C}$, i.e., $\mathcal{C}$ is an e-perfect code for some e, then there are three properties that $\mathcal{C}$ satisfies:

(1) The size of the code that is given by the sphere-packing bound,
(2) its packing radius is e, and
(3) its covering radius is also e.

Nevertheless, to verify that a given code $\mathcal{C}$, in a regular metric, is an e-perfect code, we need only verify that any two of these three properties are satisfied. In other words, if we prove that the packing radius of the code is e and its covering radius is also e, then the code is e-perfect and we do not need to compute its size. If we compute the size of $\mathcal{C}$ and it attains the sphere-packing bound for radius e and we also prove that either the packing radius is e or the covering radius is e, then it implies that the other radius is also e and hence the code $\mathcal{C}$ is e-perfect.

If the metric is regular, and we can partition the space $\mathcal{V}$ into copies of these related balls, then the code formed by the centers of all these balls of the partition is e-perfect. This claim is readily verified from Definition 2.12. It also leads to the following simple observation.

Theorem 2.9. *Assume $\mathcal{C}$ is an e-perfect code in a space $\mathcal{V}$ with a regular metric d, and $\circ$ is a binary operation between the elements of $\mathcal{V}$. If for each $x \in \mathcal{V}$ we have $\mathcal{B}_e(x) = x \circ \mathcal{B}_e(0)$, then $(\mathcal{C}, \mathcal{B}_e(0))$ is a tiling of $\mathcal{V}$, where 0 is the identity element of $\mathcal{V}$. In other words, $\mathcal{C}$ has the linear space tiling property and $\mathcal{B}_e(n)$ also has the linear space tiling property, where n is the length of the elements in $\mathcal{V}$.*

There are definitions of codes that are "almost" perfect. We will mention two types of such definitions. The first one is for ***quasi-perfect codes***, where both the packing radius and the covering radius are considered. Such a code has packing radius e and covering radius $e+1$. Quasi-perfect codes will be considered in this book only for the Hamming space (although they are defined for other metrics too). This family of codes will be discussed in Chapter 6. The reason that we consider quasi-perfect codes only in the Hamming space is that although called quasi-perfect, these codes are not "almost" perfect, as it looks based on the two radii whose difference is one. For most of them, not "almost" all words are within distance e from exactly one codeword, i.e., too many words do not have such a codeword. This is related to the fact that these codes might not be dense, a property that will be defined in Chapter 6. Fortunately, there are families of quasi-perfect codes that are dense, e.g., nearly-perfect codes (see Section 6.2). All these concepts will be also discussed in Chapter 6. We should note that if we consider quasi-perfect codes from a covering point of view, they should be sparse. A code can be dense from a packing point of view and sparse from a covering point of view if and only if it is a perfect code.

The second definition of codes that are "almost" perfect, relates to ***diameter perfect codes***, where an anticode takes the role of a ball. This family of codes, which forms a natural generalization for the family of perfect codes, will be discussed throughout this book and we now define it.

The most basic concept in an error-correcting code is its minimum distance. The minimum distance d dictates the number of errors that can be corrected when the code is used in practice. This distance is related to the packing radius $\lfloor \frac{d-1}{2} \rfloor$ of the code (see Corollary 2.2). The balls with this packing radius around the codewords of the code are disjoint.

Any given ball (in any space) has a radius; a related parameter of a ball is its diameter. For example, in Euclidian space, the diameter of a ball with radius e is $2e$. Nevertheless, the diameter is a concept that is independent of the radius and exists also in shapes that are not balls. The ***diameter*** of a shape (set) $\mathcal{S} \subseteq \mathcal{V}$, defined with a metric $d : \mathcal{V} \times \mathcal{V} \to \mathbb{Z}$, is the maximum distance between any two elements of $\mathcal{S}$. In this definition of the diameter, the maximum distance between any two elements is considered, in contrast to the minimum distance that is considered for error-correcting codes. This leads to the next definition.

Definition 2.15. An ***anticode*** $\mathcal{A}$ with ***diameter*** D is a set of elements taken from a space $\mathcal{V}$ with a distance $d : \mathcal{V} \times \mathcal{V} \to \mathbb{Z}$, where the maximum distance between the codewords of $\mathcal{A}$ is at most D. Elements of an anticode will be called ***anticodewords*** to distinguish them from codewords.

Theorem 2.10. *In any metric, an e-ball is an anticode whose diameter is at most* $2e$.

Proof. Let $y, z \in \mathcal{B}_e(x)$ be two distinct words in a the ball centered at x, i.e., $d(y,x) \leq e$ and $d(x,z) \leq e$. Hence, by the triangle inequality,

$$d(y,z) \leq d(y,x) + d(x,z) \leq 2e,$$

which implies that an e-ball is an anticode whose diameter is at most $2e$. □

Any ball has a center, while anticodes that are not balls usually do not have a center. Sometimes, however, anticodes must have a point that can fulfil the role of a center in a ball. This point can be chosen arbitrarily and it is called the ***balanced point*** of the anticode. Once chosen, it should be the same for all the translates of the anticode, i.e., if b is the balanced point of the anticode $\mathcal{A}$, then $x+b$ is the balanced point of its translate $x+\mathcal{A}$. The importance of the balanced point is, for example, when we consider a tiling of a space $\mathcal{V}$ with an anticode $\mathcal{A}$ and the set of points $\mathcal{T}$. In this $(\mathcal{T}, \mathcal{A})$ tiling, the balanced points, in the translates of $\mathcal{A}$ in the tiling, assume the role of the centers of the balls as they coincides with the points of $\mathcal{T}$.

We are interested in an as large as possible anticode with diameter D. As noted, codes and anticodes are defined as sets of elements from the space $\mathcal{V}$. The distinction between the two concepts is that in codes, the important parameter is the minimum distance, while in anticodes the important parameter is the maximum distance. By Theorem 2.10, a ball with radius e is an anticode with diameter at most $2e$. A ball, however, is not necessarily the largest possible anticode with diameter $2e$. Moreover, a ball might not be a maximal anticode. Finding the largest anticode for any given metric is an interesting problem. It will be discussed later for some of the regular metrics in which we are interested. At this point we will concentrate on a bound that will be used as a generalization for the sphere-packing bound. This bound is called the ***code-anticode bound***. In contrast to the sphere-packing bound, the proof of this bound might require separate proofs for different metrics. In the present section it will be proved for most of the metrics discussed in this book. For other metrics, separate proofs will be given in Sections 8.8, 9.2, and 10.2.

A metric $d : \mathcal{V} \times \mathcal{V} \to \mathbb{Z}$ on a space $\mathcal{V}$ (not necessarily a group space) with a binary operation $\circ$, i.e., for each two elements $x, y \in \mathcal{V}$, $x \circ y \in \mathcal{V}$, is ***right distance invariant*** if, for each three elements $x, y, z \in \mathcal{V}$, $d(x \circ z, y \circ z) = d(x, y)$. Similarly, d is ***left distance invariant*** if, for each three elements $x, y, z \in \mathcal{V}$, $d(z \circ x, z \circ y) = d(x, y)$. The metric d

is ***distance invariant*** if it is both right distance invariant and left distance invariant. Throughout this book all the metrics that we consider are distance invariant if there exists a binary operation $\circ$ (an exception will be mentioned in the notes of Chapter 13). It should be noted that there are spaces and metric with no binary operation between the elements, but there are other transformations applied on the elements which also make the metric d distance invariant in the same way which is required in this section. The existence of a binary operation $\circ$, associated with the space $\mathcal{V}$ and the metric $d : \mathcal{V} \times \mathcal{V} \to \mathbb{Z}$, has some applications. For example, to compute $d(x, y)$ for $x, y \in \mathcal{V}$, when the metric is right distance invariant, we have to note that

$$d(x, y) = d(x \circ x^{-1}, y \circ x^{-1}) = d(0, y \circ x^{-1}) = \text{wt}(y \circ x^{-1}) .$$

It is important to note that for this equation, we have used the inverse of the element x. This is essential and such an inverse exists for some of our spaces, but not for all of them.

Lemma 2.13. *If d is a right (or left) distance invariant metric in a finite space $\mathcal{V}$ with a binary operation $\circ$, then for each $x, y \in \mathcal{V}$, there exists a unique $\alpha \in \mathcal{V}$ such that $y = \alpha \circ x$. Similarly, for each $x, y \in \mathcal{V}$, there exists a unique $\alpha \in \mathcal{V}$ such that $y = x \circ \alpha$.*

Proof. Let $d : \mathcal{V} \times \mathcal{V} \to \mathbb{Z}$ be a right distance invariant metric on a finite space $\mathcal{V}$. Let α_1, α_2 be two distinct elements of $\mathcal{V}$. Since the metric d with the operation $\circ$ is right distance invariant and $d(\alpha_1, \alpha_2) > 0$, it follows that for each $x \in \mathcal{V}$, $d(\alpha_1 \circ x, \alpha_2 \circ x) > 0$, i.e., $\alpha_1 \circ x \neq \alpha_2 \circ x$. Thus, since $\mathcal{V}$ is a finite space, it follows that for each $x, y \in \mathcal{V}$, there exists a unique $\alpha \in \mathcal{V}$ such that $y = \alpha \circ x$. This also implies that x has a left inverse x^{-1}.

Now assume the contrary, that for some $x \in \mathcal{V}$ there exist two distinct elements $\alpha_1, \alpha_2 \in \mathcal{V}$ such that $x \circ \alpha_1 = x \circ \alpha_2$. Since x has a left inverse, it follows that $\alpha_1 = \alpha_2$, a contradiction. Therefore, for each $x, y \in \mathcal{V}$, there exists a unique $\alpha \in \mathcal{V}$ such that $y = x \circ \alpha$.

Similarly, if d is a left distance invariant metric, then for each $x, y \in \mathcal{V}$, there exists a unique $\alpha \in \mathcal{V}$ such that $y = x \circ \alpha$, and for each $x, y \in \mathcal{V}$, there exists a unique $\alpha \in \mathcal{V}$ such that $y = \alpha \circ x$. □

Corollary 2.14. *Let $d : \mathcal{V} \times \mathcal{V} \to \mathbb{Z}$ be a metric on a finite space $\mathcal{V}$ with a binary operation $\circ$. If d is a right or left distance invariant metric, then each element $x \in \mathcal{V}$ has a right and a left inverse.*

The consequence of Lemma 2.13 can be obtained also if for each element of $\mathcal{V}$ there exists a right inverse and a left inverse, where there is no requirement for a distance invariant metric. The result was obtained when $\mathcal{V}$ together with its binary operation might not define a group. Lemma 2.13 was written in a way that it is readily verified that only some of the conditions are required, while in reality the metrics used in this book satisfy more properties than the ones required by the lemma. The following lemma, called the ***local inequality lemma*** is the key result required for our next definition of diameter perfect codes. It has several proofs, depending on the space and the metric being considered. Nevertheless, there are spaces and metrics for which this lemma is not satisfied and hence it cannot be used.

The Local Inequality Lemma

Let $\mathcal{C}_\mathcal{D}$ be a code in a finite space $\mathcal{V}$ with a metric $d : \mathcal{V} \times \mathcal{V} \to \mathbb{Z}$, where the distances between the codewords in $\mathcal{C}_\mathcal{D}$ are taken from a subset $\mathcal{D}$. Let $\mathcal{A}$ be a nonempty subset of $\mathcal{V}$ and let $\mathcal{C}'_\mathcal{D} \subseteq \mathcal{A}$ be the largest code in $\mathcal{A}$, where the distances between the codewords of $\mathcal{C}'_\mathcal{D}$ are taken from $\mathcal{D}$. Then

$$\frac{|\mathcal{C}_\mathcal{D}|}{|\mathcal{V}|} \leq \frac{|\mathcal{C}'_\mathcal{D}|}{|\mathcal{A}|} \ . \tag{2.2}$$

Equation (2.2) in the local inequality lemma will be referred to as the ***local inequality bound***. The local inequality lemma implies the code-anticode bound, given in the following statement.

Corollary 2.15. *Assume that for a metric d on a space $\mathcal{V}$, the set of possible distances in $\mathcal{V}$ is Δ. Assume further that the conditions of the local inequality lemma on $\mathcal{D}$, $\mathcal{C}_\mathcal{D}$, $\mathcal{A}$, and $\mathcal{C}'_\mathcal{D}$ are satisfied, and also (2.2) is satisfied. If $\mathcal{A}$ is an anticode with maximum distance D and $\mathcal{C}$ is a code with minimum distance $D+1$, then*

$$|\mathcal{C}| \cdot |\mathcal{A}| \leq |\mathcal{V}| \ . \tag{2.3}$$

Proof. In the local inequality lemma, let $\mathcal{D} \triangleq \Delta \setminus \{1, 2, \ldots, D\}$ and let $\mathcal{A}$ be a maximum size anticode with distances taken from $\{0, 1, 2, \ldots, D\}$. Since the minimum distance of $\mathcal{C}'_\mathcal{D}$ is at least $D+1$ and $\mathcal{C}'_\mathcal{D} \subset \mathcal{A}$, it follows that $\mathcal{C}'_\mathcal{D}$ contains exactly one codeword. As a consequence, it immediately implied from (2.2) in the local inequality lemma that

$$|\mathcal{C}_\mathcal{D}| \cdot |\mathcal{A}| \leq |\mathcal{V}| \ ,$$

and the claim of the corollary follows. □

A code $\mathcal{C}$ that attains (2.3) with equality is called a ***D-diameter perfect code***. The following theorem implies a tight connection between perfect codes and diameter perfect codes.

Theorem 2.11. *An e-perfect code in a graphic metric is a* $(2e)$*-diameter perfect code.*

Proof. Let $\mathcal{C}$ be an e-perfect code. By Theorem 2.10, the ball with radius e is an anticode with diameter at most $2e$. Assume the contrary, that $c_1, c_2 \in \mathcal{C}$ are two distinct codewords for which $d(c_1, c_2) \leq 2e$. Since the metric is graphic, it follows that there exists a path whose length is at most $2e$ between c_1 and c_2 and hence there exists an element x such that $d(c_1, x) \leq e$ and $d(x, c_2) \leq e$, which implies that $\mathcal{C}$ is not e-perfect, a contradiction. This implies that the minimum distance of $\mathcal{C}$ is $2e+1$ and also that the diameter of the ball with radius e is $2e$.

Clearly, $\mathcal{C}$ attains the sphere-packing bound with equality and hence it also attains (2.3) with equality. Thus, an e-perfect code in a graphic metric is a $(2e)$-diameter perfect code. □

Corollary 2.16. *An e-perfect code in a graphic metric has minimum distance* $2e+1$.

The following lemma is an instant of the local inequality lemma for a certain family of metrics that covers a large number of the spaces with their defined metrics and binary operations.

Lemma 2.14. *Let d be a right (left) distance invariant metric in a finite space $\mathcal{V}$ with a binary operation $\circ$. Let $\mathcal{C}_\mathcal{D}$ be a code in $\mathcal{V}$, where the distances between codewords in $\mathcal{C}_\mathcal{D}$ are taken from a subset $\mathcal{D}$. Let $\mathcal{A}$ be a subset of $\mathcal{V}$ and let $\mathcal{C}'_\mathcal{D} \subseteq \mathcal{A}$ be the largest code in $\mathcal{A}$ with distances taken from $\mathcal{D}$. Then*

$$\frac{|\mathcal{C}_\mathcal{D}|}{|\mathcal{V}|} \leq \frac{|\mathcal{C}'_\mathcal{D}|}{|\mathcal{A}|} . \tag{2.4}$$

Proof. Assume that $d : \mathcal{V} \times \mathcal{V} \rightarrow \mathbb{Z}$ is a right distance invariant metric in a finite space $\mathcal{V}$. Let $S \triangleq \{(c, v) \;:\; c \in \mathcal{C}_\mathcal{D},\; v \in \mathcal{V},\; c \circ v \in \mathcal{A}\}$. Since by Corollary 2.14, each element in $\mathcal{V}$ has a left inverse, it follows by Lemma 2.13 that for a given codeword $c \in \mathcal{C}_\mathcal{D}$ and an element $\alpha \in \mathcal{A}$, there exists exactly one element $v \in \mathcal{V}$ such that $\alpha = c \circ v$. Therefore, $|S| = |\mathcal{C}_\mathcal{D}| \cdot |\mathcal{A}|$.

Since the metric is right distance invariant, it follows that for each $u \in \mathcal{V}$, the set $\mathcal{C}_u \triangleq \{c \circ u \;:\; c \in \mathcal{C}_\mathcal{D}\}$ has the same distances between its codewords as in $\mathcal{C}_\mathcal{D}$, i.e., the distances between codewords of $\mathcal{C}_u$ are taken from the set $\mathcal{D}$. Together with the fact that $\mathcal{C}'_\mathcal{D}$ is the largest code in $\mathcal{A}$, where the distances between codewords in $\mathcal{C}'_\mathcal{D}$ are taken from the set $\mathcal{D}$, it follows that for any given word $u \in \mathcal{V}$, the set $S_u \triangleq \{(c, u) \ : \ c \in \mathcal{C}_\mathcal{D}, \ c \circ u \in \mathcal{A}\}$ has at most $|\mathcal{C}'_\mathcal{D}|$ codewords, i.e., $|S_u| \leq |\mathcal{C}'_\mathcal{D}|$. Clearly, by the definition of S_u, for each two distinct elements $u_1, u_2 \in \mathcal{V}$, we have that $S_{u_1} \cap S_{u_2} = \varnothing$, and also $S = \bigcup_{v \in \mathcal{V}} S_v$, and hence $|S| \leq |\mathcal{C}'_\mathcal{D}| \cdot |\mathcal{V}|$.

Thus, since $|S| = |\mathcal{C}_\mathcal{D}| \cdot |\mathcal{A}|$, it follows that $|\mathcal{C}_\mathcal{D}| \cdot |\mathcal{A}| \leq |\mathcal{C}'_\mathcal{D}| \cdot |\mathcal{V}|$ and the claim of the lemma is proved. □

The local inequality lemma can be applied to many metrics that are discussed in this book. The given proof of Lemma 2.14 can be applied on a large number of such metrics, but unfortunately, it cannot be applied to all these metrics since not all the metrics on the given spaces have the necessary binary operation $\circ$ (for example, an inverse to the obvious binary operation does not always exist). In some cases, such a binary operation does not exist. As an example, the requirements of Lemma 2.14 does not hold for the Johnson scheme and the Grassmann scheme. A specific proof for each of these metrics, for the local inequality lemma, will be given in the relevant sections. Moreover, for each metric it will be required to show either that the conditions of Lemma 2.14 are satisfied or to provide another proof for the local inequality lemma.

Are there diameter perfect codes that are not perfect codes? Many such codes will be presented in this book. The bound in (2.3) of Corollary 2.15 is called the ***code-anticode bound***. By Theorem 2.11 it is a generalization and an improvement on the sphere-packing bound.

The code-anticode bound is also a generalization and improvement of the Johnson bounds (Lemmas 2.3 and 2.4) as will be proved in Section 8.8. We can also generalize the definition of diameter perfect codes for infinite spaces as it will be discussed in Section 11.4.

For the proof of the next result, recall that the distances between words in a translate $\mathcal{A}'$ of a set $\mathcal{A}$ are exactly the same distances as the ones in $\mathcal{A}$.

Corollary 2.17. *Let $\mathcal{C}$ be a code in $\mathcal{V}$ whose minimum distance is $D+1$ and let $\mathcal{A}$ be a related anticode with maximum distance D. The code $\mathcal{C}$ is a D-diameter perfect code and $\mathcal{A}$ is a maximum size anticode if and only if each translate of $\mathcal{A}$ in $\mathcal{V}$ contains exactly one codeword of $\mathcal{C}$.*

Proof. Assume first that $\mathcal{A}$ is a maximum size anticode with diameter D

and $\mathcal{C}$ is a D-diameter perfect code for which $|\mathcal{C}| \cdot |\mathcal{A}| = |\mathcal{V}|$. Define

$$S \triangleq \{(c, x \circ \mathcal{A}) \;:\; c \in \mathcal{C},\; x \in \mathcal{V},\; c \in x \circ \mathcal{A}\} \,. \tag{2.5}$$

Given a point $v \in \mathcal{V}$, v can be associated with each point of $\mathcal{A}$ and hence it is contained in $|\mathcal{A}|$ distinct translates of $\mathcal{A}$. Since each point in $\mathcal{V}$ is contained in $|\mathcal{A}|$ distinct translates of $\mathcal{A}$, it follows that the same is true for each codeword $c \in \mathcal{C}$ and hence $|S| = |\mathcal{C}| \cdot |\mathcal{A}|$, which immediately implies that $|S| = |\mathcal{V}|$.

Since $\mathcal{A}$ has diameter D and $d(\mathcal{C}) = D+1$, it follows that each translate of $\mathcal{A}$ in $\mathcal{V}$ contains at most one codeword from $\mathcal{C}$. The balanced point of $\mathcal{A}$ can coincide with each point of $\mathcal{V}$ and hence the number of distinct translates of $\mathcal{A}$ in $\mathcal{V}$ is $|\mathcal{V}|$. Since each such translate of $\mathcal{A}$ can contain at most one codeword of $\mathcal{C}$, it follows that each translate of $\mathcal{A}$ is contained in at most one pair of S. Furthermore, $|S| = |\mathcal{V}|$ implies that S contains exactly $|\mathcal{V}|$ pairs and hence it follows that each translate of $\mathcal{A}$ in $\mathcal{V}$ contains exactly one codeword of $\mathcal{C}$.

Assume now that each translate of the anticode $\mathcal{A}$ in $\mathcal{V}$ contains exactly one codeword from $\mathcal{C}$. Define again S as in (2.5). Since each translate of $\mathcal{A}$ in $\mathcal{V}$ contains exactly one codeword and there are exactly $|\mathcal{V}|$ translates of $\mathcal{A}$ in $\mathcal{V}$, it follows that $|S| = |\mathcal{V}|$. Since each codeword of $\mathcal{C}$ is contained in exactly $|\mathcal{A}|$ translates of $\mathcal{A}$ in $\mathcal{V}$, it follows that $|S| = |\mathcal{C}| \cdot |\mathcal{A}|$. Thus, $|\mathcal{C}| \cdot |\mathcal{A}| = |\mathcal{V}|$, i.e., $\mathcal{C}$ is a D-diameter perfect code and $\mathcal{A}$ is a maximum size anticode with diameter $D+1$. □

The concept of an e-perfect code is associated with the computation of a ball with radius e. Finding the size of such a ball is not always simple and it depends on the space $\mathcal{V}$ and the metric d, but fortunately it is rather simple in the most metrics discussed in this book, i.e., the Hamming metric, the Johnson metric, and the Lee metric. It is slightly more difficult in metrics such as the Grassmann metric. The concept of a D-diameter perfect code is related to the computation of the largest anticode with diameter D. This computation is much more difficult than the computation of the size of the related ball. This computation of the maximum size anticode is related to computation of a maximum t-intersecting family, which is a set S of elements from a space $\mathcal{V}$, where each two elements S have an intersection whose size is at least t. The size of the largest anticode will be discussed throughout the book for the various metrics.

The result of Corollary 2.15 can be strengthened with the following consequence from the local inequality lemma.

Corollary 2.18. *Assume that for a metric d on a space $\mathcal{V}$, the set of possible distances in $\mathcal{V}$ is Δ. Assume further that the conditions of the local inequality lemma on $\mathcal{D}$, $C_{\mathcal{D}}$, $\mathcal{A}$, and $C'_{\mathcal{D}}$ are satisfied, and also (2.2) is satisfied. If $\mathcal{A}$ is a set with distances taken from a set $\mathcal{D}$ and $\mathcal{C}$ is a set with distances taken from the set $\Delta \setminus \mathcal{D}$, then*

$$|\mathcal{C}| \cdot |\mathcal{A}| \leq |\mathcal{V}| \ .$$

The result of Corollary 2.18 implies another generalization for the concept of a perfect code. This new concept is also a generalization of a diameter perfect code. A pair $[\mathcal{C}, \mathcal{A}]$ in a space $\mathcal{V}$ with a metric d is called a ***perfect set*** if $|\mathcal{C}| \cdot |\mathcal{A}| = |\mathcal{V}|$ and the set of distances between distinct elements in $\mathcal{A}$ is disjoint from the set of distances between distinct elements in $\mathcal{C}$. Clearly, a diameter perfect code is a perfect set, but a perfect set does not have to be a diameter perfect code.

Problem 2.1. Develop a theory for perfect sets with various metrics. Distinguish between perfect sets and diameter perfect codes.

In the discussion on perfect codes we have to consider elements that cover other elements. Recall that a codeword c in an e-perfect code $\mathcal{C}$ ***covers*** the word $v \in \mathcal{V}$ if $d(c, x) \leq e$. When no code is specified, we say that an element x ***covers*** (contains) an element y if $y \subset x$, when the elements can be represented as subsets. Clearly, such a cover when we consider subsets is also a cover with respect to the codewords represented by these subsets, but not the converse (this means that a codeword c in an e-perfect code $\mathcal{C}$ can cover x, i.e., $d(c, x) \leq e$, but c does not cover (contain) x when c and x are considered as subsets). When the word "cover" is used, the related meaning should be understood from the context.

Finally, a perfect code is related to the concept of tiling mentioned before. This concept can be generalized as follows. A ***tiling*** (of a finite space $\mathcal{V}$) is a set

$$\mathcal{T} \triangleq \{S_1, S_2, \ldots, S_m\}$$

such that $S_i \cap S_j = \varnothing$ for $1 \leq i < j \leq m$, and $\mathcal{V} = \bigcup_{i=1}^{m} S_i$. Such a tiling can also be used in the direct product construction and the general product construction. This definition will be further generalized in Chapter 11 for the infinite space $\mathcal{V} = \mathbb{Z}^n$. If $\mathcal{C}$ is an e-perfect code, then the set

$$\{\mathcal{B}_e(c) \ : \ c \in \mathcal{C}\},$$

i.e., the balls with radius e around the codewords of $\mathcal{C}$ form a partition of $\mathcal{V}$, i.e., form a tiling of $\mathcal{V}$. Do we have similar tilings for diameter perfect codes? This question will be discussed during our exposition on anticodes and diameter perfect codes in various metrics.

2.5 Notes

Definitions for finite fields, linear codes, and nonlinear codes can be found in all the books on coding theory mentioned in Section 1.1.

Section 2.1. Theorem 2.1 is credited to [Singleton (1964)] who called those codes that meet the bound MDS codes. The bound, however, was published at the same time in [Golomb and Posner (1964)]. The latter obtained their results a couple of years earlier and presented them in technical reports. MDS code will be used throughout the book, as designs in Chapter 3, as diameter perfect codes in Chapter 4, and as building blocks for nonbinary diameter constant-weight codes in Chapter 9. The Singleton bound will be adapted to nonbinary constant-weight codes in Chapter 9, to rank-metric codes and to subspace codes in Chapter 10, and to burst-correcting codes in Chapter 13. Related names will be given to codes which meet the corresponding bounds.

The concept of tiling is strongly related to perfect codes. A perfect code defines a tiling and some tilings with specified properties can be constructed from perfect codes with related properties. These tilings in binary spaces were considered in [Cohen, Litsyn, Vardy, and Zémor (1996)] and further investigated in [Etzion and Vardy (1998); Östergård and Vardy (2004)].

The two product constructions have appeared with many variants in many papers throughout the years. They will be used in many chapters throughout the book. Their use in coding theory and especially for perfect codes can be attributed for example to [Phelps (1983, 1984b)] who also gave a general construction for codes that are not necessarily perfect [Phelps (1984a)]. Direct product constructions were given for error-correcting codes and for covering codes. We will present a combined construction for error-correcting codes and for covering codes in Chapter 6. Finally, the two important Johnson bounds were proved for the first time by [Johnson (1972)].

Section 2.2. For finite field, the reader can consult the excellent book by [Lidl and Niederreiter (1997)].

Section 2.4. There are a few surveys on perfect codes and we mention the ones of [van Lint (1975)] and [Heden (2008)]. The concept of perfect codes in graphs was introduced in [Biggs (1973)]. His work was mainly presented for distance-transitive graphs. Many other related results on perfect codes in graphs were subsequently published: [Biggs (1974); Heden (1974); Hammond and Smith (1975); Hammond (1976); Cameron, Thas, and Payne (1976); Thas (1977); Smith (1980); Kratochvil (1985, 1986); Etienne (1987); Kratochvil (1988); Mollard (2011)].

It should be noted that perfect codes in graphs are also connected to algebraic graph theory for which there are two excellent books [Biggs (1993)] and [Godsil and Royle (2001)].

The concept of diameter perfect codes was introduced in [Ahlswede, Aydinian, and Khachatrian (2001)], where they first proved the local inequality lemma for the Johnson scheme. Their proof will be presented in Section 8.8. They claimed that the proof can be generalized for each metric whose graph admits a transitive group of automorphisms, but no proof was given in the paper. The proof that we give for this lemma (Lemma 2.14), which can also be applied to other different metrics, is different and does not depend on the structure of the graph (e.g., a distance-regular graph). It depends only on the basic properties of the metric (right invariant or left invariant, a binary operation $\circ$ with an inverse for each element of $\mathcal{V}$, etc.). Our proof does not hold for all metrics (even not for all metrics based on distance-regular graphs) and hence different alternative proofs should be generated and will be given for some metrics, discussed in this book. The code-anticode bound (Corollary 2.15) was first proved in the seminal work of [Delsarte (1973)] for metrics defined by distance-regular graphs that are related to association schemes (see Section 3.5). The proofs given in [Ahlswede, Aydinian, and Khachatrian (2001)] and in this chapter are simpler and more general than the one presented in [Delsarte (1973)]. Another interesting variant of the lemma was introduced in [Krotov, Östergård, and Pottonen (2016)]. The variant of the proof given for Lemma 2.14 is similar to the ones presented in [Etzion (2011); Buzaglo and Etzion (2015)]. The metric discussed in [Buzaglo and Etzion (2015)] is the Kendall τ-metric on the set of all permutation on n elements. Another proof for the code-anticode bound when the space is the set of permutations and the metric is the L_∞ was given by [Tamo and Schwartz (2010)]. For these two metrics, the binary operation is a multiplication of permutations. This operation in these metrics form examples of right distance invariant metrics which are not left distance invariant. These are examples of metrics which have some important applications from a practical point of view and some interesting and not standard properties from a theoretical point of view. It will not be discussed in our book, but the exposition in [Tamo and Schwartz (2010);

Schwartz and Tamo (2011); Buzaglo and Etzion (2015)] suggests that there are many interesting questions related to these metrics and perfect codes.

The search for maximum size anticodes is an interesting topic in itself. As was pointed out in [Ahlswede, Aydinian, and Khachatrian (2001)], it is related to finding the maximum size of intersecting families. Some relevant references for each metric will be cited in the related chapters. Anticodes have some other applications in coding theory, e.g., to construct some codes that attain some bounds in coding theory (see Section 3.4). They also have an important role in associative memories [Yaakobi and Bruck (2019)]. They are considered also in some important space metrics which are not discussed in our book. For example, when the space is the set of permutations on n elements S_n there are many metrics defined on this set. Anticodes for the L_∞ metric were considered in [Schwartz and Tamo (2011)] and anticodes for the Kendall τ-metric were discussed in [Buzaglo and Etzion (2015)]. Other coding problems on anticodes in the Euclidian space were considered, for example, in [Blackburn, Etzion, Martin, and Paterson (2010)].

Chapter 3

Combinatorial Designs and Bounds

This chapter is devoted to combinatorial designs and bounds on the sizes of codes. Combinatorial designs are highly related to perfect codes and will be considered throughout the book. The bounds which will be considered are for parameters where perfect codes cannot exist. The bounds which will be presented will be attained by combinatorial designs and anticodes and hence the topics of this chapter are tied together. This chapter is a relatively short introduction for these topics.

Combinatorial designs form a branch in combinatoric with an extensive theory, many applications in variety of areas, such as cryptography, designs of experiments, software testing, group testing, biostatistics, and, of course, coding theory and, especially, perfect codes in various metrics. Some of these designs resemble perfect codes and indeed some are diameter perfect codes. Other are embedded in perfect codes or used as building blocks for perfect codes. The first part of this chapter is devoted to some basic definitions and results on combinatorial designs that will be used in the chapters that follow. The bounds that are attained by the different types of perfect codes (the sphere-packing bound or the code-anticode bound) cannot be attained for most parameters for different reasons. There are bounds that improve on these two bounds and they are the topic of the second part of this chapter.

Section 3.1 will be devoted to Steiner systems that are used in many chapters and will be proved to be diameter perfect codes in Chapter 8. Orthogonal designs presented in Section 3.2. They form diameter perfect codes as will be discussed in Section 4.3. Section 3.3 will be devoted to projective geometries that are very useful in construction of block designs and codes and also in techniques to obtain upper bounds on the sizes of codes. Section 3.4 is devoted to some upper bounds on codes size, which are important in our context. These bounds include codes for which the distance is very large compared to their length, and a generalization of the Singleton bound in the binary case. Structures such as Hadamard matrices and anticodes can be used to meet these bounds and they will also be discussed. Section 3.5 offers a brief introduction to association schemes, which can be used in coding theory as foundations to find bounds on the largest possible size of codes in some metrics.

3.1 Steiner Systems and Generalized Steiner Systems

Block designs form the main structures in combinatorial designs. They appear in this book in various connections to perfect codes. Some of them will form diameter perfect codes, some of them will be embedded in perfect codes, and some will be used to form codes that meet certain bounds. Nonexistence results based on block designs will also be presented. The most celebrated family of block designs with respect to perfect codes are Steiner systems.

Definition 3.1. A ***Steiner system*** $\mathrm{S}(t,k,n)$ is a pair (Q,B), where Q is an n-set, whose elements are called ***points***, and B is a collection of k-subsets of Q, called ***blocks***, such that each t-subset of Q is contained in exactly one block of B.

There are a few trivial necessary conditions for the existence of Steiner systems.

Lemma 3.1. *The number of blocks in a Steiner system* $\mathrm{S}(t,k,n)$ *is* $\binom{n}{t}/\binom{k}{t}$.

Proof. The number of distinct t-subsets in a block (a k-subset) is $\binom{k}{t}$. The total number of distinct t-subsets in an n-set is $\binom{n}{t}$. Since each t-subset is contained in exactly one block, it follows that the number of blocks in a Steiner system $\mathrm{S}(t,k,n)$ is $\binom{n}{t}/\binom{k}{t}$. □

Using characteristic vectors, sets are transferred into vectors and sets of blocks into codes. For a Steiner system, this implies the following result.

Proposition 3.1. *The characteristic vectors of the blocks in a Steiner system* $\mathrm{S}(t,k,n)$ *form a binary* $(n,2(k-t+1),k)$ *constant-weight code with* $\binom{n}{t}/\binom{k}{t}$ *codewords. This code meets a trivial packing bound and also meets a trivial covering bound for constant-weight codes (packing and covering of* t*-subsets by* k*-subsets of an* n*-set).*

Steiner systems by their definition behave like perfect codes as we will see later on in the book. When we consider a transmission of the characteristic vectors of a Steiner system $\mathrm{S}(t,k,n)$ and exactly $k-t$ errors, of *ones* that were changed to *zeroes*, occurred, then we receive a word of length n and weight t. Any such word of length n and weight t can be uniquely decoded into the original characteristic vector of length n and weight k. Therefore, these (Steiner systems) are constant-weight codes that correct asymmetric errors. They form perfect codes in the sense that any word of weight t can be received, when exactly $k-t$ errors occurred. Any such word of weight t can be received after transmission and it is uniquely decoded into a codeword of weight k.

The next result can be compared with the Johnson bound of Lemma 2.4.

Lemma 3.2. *If there exists a Steiner system* $\mathrm{S}(t,k,n)$, $t>1$, *then there exists a Steiner system* $\mathrm{S}(t-1,k-1,n-1)$.

Proof. Let $\mathcal{S} = (Q,B)$ be a Steiner system $\mathrm{S}(t,k,n)$, where $t>1$ and $Q=\{1,2,\ldots,n\}$. Define the system $\mathcal{S}' = (Q',B')$, where $Q'=\{1,2,\ldots,n-1\}$ and

$$B' \triangleq \{X \cap Q' \ : \ X \in B, \ \ n \in X\} \ .$$

One can easily verify that $\mathcal{S}'$ is a Steiner system $\mathrm{S}(t-1,k-1,n-1)$. □

Corollary 3.1. *A necessary condition that a Steiner system* $\mathrm{S}(t,k,n)$ *exists is that all the numbers* $\frac{\binom{n-i}{t-i}}{\binom{k-i}{t-i}}$, $0 \le i \le t-1$, *are integers.*

Given a Steiner system $\mathrm{S}(t,k,n)$, the Steiner system $\mathrm{S}(t-1,k-1,n-1)$ is called the ***derived system***.

We are interested in parameters for which Steiner systems exist. As in perfect codes, there are trivial Steiner systems $\mathrm{S}(t,k,k)$ and $\mathrm{S}(k,k,n)$. The next question is whether the necessary conditions of Corollary 3.1 are also sufficient? There are parameters for which these necessary conditions are also sufficient. For example, if k divides n, then it is easy to construct a Steiner system $\mathrm{S}(1,k,n)$, and Lemma 3.1 and Corollary 3.1 imply that such a system exists only if k divides n. Let us now consider the existence of a Steiner system $\mathrm{S}(t,k,n)$ when $k=t+1$. When $t=2$ and $k=3$, Corollary 3.1 implies that 3 divides $\binom{n}{2}$ and 2 divides $n-1$. This immediately implies that n is odd and 6 divides $(n-1)n$. Clearly, it follows that $n \equiv 1$ or 3 (mod 6). Such systems are relatively easily constructed. If $t=3$, $k=4$, then a Steiner system $\mathrm{S}(3,4,n)$ exists if and only if $n \equiv 2$ or $4 \pmod 6$. The necessary conditions are derived easily from Corollary 3.1, while constructions for all parameters are slightly more difficult. Similarly, a Steiner system $\mathrm{S}(2,4,n)$ exists if and only if $n \equiv 1$ or $4 \pmod{12}$. There are many constructions for infinite families of Steiner systems $\mathrm{S}(2,k,n)$, $k > 4$ and except for a finite number of exceptions, the necessary conditions are also sufficient.

There are a few families of Steiner systems that are based on finite geometries. These geometries are discussed in Section 3.3. Such systems are the Steiner systems $\mathrm{S}(2,q,q^n)$ (affine geometries), $\mathrm{S}(3,q+1,q^n+1)$ (spherical geometries), $\mathrm{S}(2,q+1,(q^n-1)/(q-1))$ (projective geometries), $\mathrm{S}(2,q+1,q^3+1)$ (unitals), for each prime power q and $n \geq 2$.

For $t \geq 5$, only 12 nontrivial Steiner systems were constructed until 2021. These systems are the Steiner systems $\mathrm{S}(5,6,12)$, $\mathrm{S}(5,6,24)$, $\mathrm{S}(5,8,24)$, $\mathrm{S}(5,7,28)$, $\mathrm{S}(5,6,36)$, $\mathrm{S}(5,6,48)$, $\mathrm{S}(5,6,72)$, $\mathrm{S}(5,6,84)$, $\mathrm{S}(5,6,108)$, $\mathrm{S}(5,6,132)$, $\mathrm{S}(5,6,168)$, $\mathrm{S}(5,6,244)$. For $t = 4$, there are no new systems, except for the related derived systems.

Problem 3.1. Present new constructions for Steiner systems $\mathrm{S}(t,k,n)$, where $t > 3$.

Problem 3.2. Present a construction for an infinite family of Steiner systems $\mathrm{S}(t,k,n)$, where $t > 3$.

Of special interest is the Steiner system $\mathrm{S}(5,6,12)$ for which there are many different constructions. The reader is encouraged to find some constructions for this system. The following method is a construction for this system, but as it has a random step there is no proof that this system is always generated. This interesting construction that seems to work is the following one. Let $\Gamma = (\mathcal{V}, E)$ be the graph whose set of vertices $\mathcal{V}$ consists of the $\binom{12}{6} = 924$ vertices represented by the 6-subsets of a 12-set. Let $E \triangleq \{\{x,y\} \ : \ x,y \in \mathcal{V}, \ |x \cap y| = 5\}$. Apply the following simple algorithm. At each step we have a set of vertices U and a set of 6-subsets S, where initially $U \triangleq \mathcal{V}$ and $S \triangleq \varnothing$. At the general step of the algorithm search for a vertex $u \in U$ whose degree in the subgraph of Γ induced by U is minimal (if a few such vertices exist, then choose one of them in random). Set $S := S \cup \{u\}$ and $U := U \setminus (\{u\} \cup \{x \ : \ \{x,u\} \in E\})$. The algorithm terminates when U is an empty set.

Problem 3.3. Prove that when the algorithm terminates, the set S is a Steiner system $\mathrm{S}(5,6,12)$ or show an instance of the algorithm where S is not a Steiner system.

It is interesting to note that when a similar algorithm is applied on the 5-subsets of an 11-set, the outcome is not necessarily the Steiner system S(4, 5, 11).

Based on probabilistic arguments, it was proved that for any given pair (t, k), for large enough n, the necessary conditions of Corollary 3.1 are also sufficient. Unfortunately, this n is an astronomic integer, beyond our imagination, and cannot be of any help for any practical use. Nevertheless, this probabilistic proof has closed a previous debate on whether the necessary conditions of Corollary 3.1 are sufficient and infinitely many Steiner systems $\mathrm{S}(t, k, n)$ exist for any given pair (t, k), where $t < k$.

Steiner systems form a small subfamily of a larger family of designs called block designs. A ***block design*** $\mathrm{S}_\lambda(t, k, n)$ is a pair (Q, B), where Q is an n-set whose elements are called points and B is a collection of k-subsets, called blocks, of Q, such that each t-subset of Q is contained in exactly λ blocks of B. The questions about the block designs, with $\lambda > 1$, which are embedded in codes have always been interesting for all codes, but we will not discuss them.

It should be noted that any block design $\mathrm{S}_\lambda(t, k, n)$ can be described by a matrix called the ***incidence matrix*** of the design. This incidence matrix A has n rows indexed by the points Q and $\lambda\binom{n}{t}/\binom{k}{t}$ columns indexed by the blocks of B. The matrix A is a $(0, 1)$ matrix, where $A_{i,j} = 1$ if and only if the i-th point of Q is contained in the j-th block of B. This incidence matrix and its transpose have many interesting properties that also depend on the design, but we will not go further in this direction, except for one more definition.

Let $\mathcal{S} = (Q, B)$ be a block design $\mathrm{S}_\lambda(t, k, n)$ with an incidence matrix A. The ***dual design*** is the design defined by the incidence matrix A^{tr}.

Steiner systems have a role in perfect codes since they are embedded in many such codes as we will see in the following chapters. General block designs that are not Steiner systems are of less importance in our context.

If a Steiner system behaves like a perfect code (and it is a diameter perfect code as will be proved in Section 8.8), it is natural and important to ask whether such systems have the space tiling property. For this we have the following definition. A ***large set*** of Steiner systems $\mathrm{S}(t, k, n)$ is a set of pairwise disjoint Steiner systems $\mathrm{S}(t, k, n)$ such that their union is the set of all $\binom{n}{k}$ k-subsets of the n-set Q. Do such large sets exist? The answer is yes, for at least some parameters. A simple example is a large set of Steiner systems $\mathrm{S}(1, 2, n)$, n even, known also as a ***one-factorization*** of the complete undirected graph K_n on n vertices. For even n, let $\mathbb{Z}_n$ be the vertices of K_n and let $\{\{x, y\} \,:\, 0 \leq x < y < n\}$ be the set of edges. Define

$$\mathcal{F} \triangleq \{\mathcal{F}_0, \mathcal{F}_1, \ldots, \mathcal{F}_{n-2}\},$$

where for $0 \leq i \leq n-2$,

$$\mathcal{F}_i \triangleq \{\{x, y\} \,:\, 0 \leq x < y < n-1,\ x + y \equiv i \pmod{n-1}\}$$

$$\cup\{\{i/2 \pmod{n-1}, n-1\}\} .$$

It is readily verified that $\mathcal{F}$ is a large set of Steiner systems $\mathrm{S}(1, 2, n)$.

Steiner systems are associated with binary constant-weight codes as implied by Proposition 3.1. There is a similar definition for nonbinary constant-weight codes.

Definition 3.2. A ***generalized Steiner system*** $\mathrm{GS}(t, k, n, q)$ is a constant-weight code $\mathcal{C}$, over an alphabet Q of size q, whose length is n and weight k for each codeword, such that:

(1) The minimum Hamming distance of $\mathcal{C}$ is $2(k-t)+1$.
(2) Each word v of length n and weight t over Q is covered by exactly one codeword $c \in \mathcal{C}$, i.e., $d(v, c) = k - t$.

The two required properties of Definition 3.2 form a generalization for the Steiner systems in terms of exact covering and minimum Hamming distance as implied by Proposition 3.1. Note that the minimum distance $2(k-t)+1$, which is required by the first property, is the largest possible distance if $q > 2$. The reason is that if we consider two words of length n and weight t that differ in exactly one coordinate, which is possible since $q > 2$, then two different codewords c_1 and c_2 should cover them. The Hamming distance between two such words c_1 and c_2, of weight k, is at most $1 + 2(k-t)$, since they share at least t coordinates with nonzero symbols from which exactly $t-1$ have the same symbols.

What are the necessary conditions for the existence of such systems? For which parameters do they exist? We start by generalizing the necessary conditions of Corollary 3.1 for the existence of a Steiner system. The proofs are similar to the ones for a Steiner system and are left as an exercise for the reader.

Lemma 3.3. *The number of blocks in a generalized Steiner system* $\mathrm{GS}(t, k, n, q)$ *is* $\frac{\binom{n}{t}}{\binom{k}{t}}(q-1)^t$.

Lemma 3.4. *If there exists a generalized Steiner system* $\mathrm{GS}(t,k,n,q)$, $t>1$, *then there exists a generalized Steiner system* $\mathrm{GS}(t-1,k-1,n-1,q)$.

Corollary 3.2. *A necessary condition for a generalized Steiner system* $\mathrm{GS}(t,k,n,q)$ *to exist is that all the numbers* $\frac{\binom{n-i}{t-i}}{\binom{k-i}{t-i}}(q-1)^{t-i}$, $0 \le i \le t-1$, *are integers.*

To consider a more specific case, let $t = 1$ which is a trivial case for Steiner systems, but it is not as trivial for generalized Steiner system. In this case we will consider a simple lower bound and a simple upper bound on n, given w and q, for a generalized Steiner system $\mathrm{GS}(1,w,n,q)$.

Theorem 3.1. *If there exists a generalized Steiner system* $\mathrm{GS}(1,w,n,q)$, *then* $n \le \frac{wA(n,2w-2,w)}{q-1}$ *and* $n \ge 1+(w-1)(q-1)$.

Proof. Since the minimum Hamming distance of a generalized Steiner system $\mathrm{GS}(1,w,n,q)$ is $2w-1$, it follows that two codewords can share at most one coordinate with a (distinct) nonzero symbol and hence the supports of the codewords form a binary constant-weight code of weight w and minimum distance $2w-2$. Hence, the size of a generalized Steiner system $\mathrm{GS}(1,w,n,q)$ is at most $A(n,2w-2,w)$, and the number of nonzero entries in all the codewords is at most $wA(n,2w-2,w)$. Since each coordinate must have exactly one codeword which each of the $q-1$ nonzero alphabet letters we must have $n \le \frac{wA(n,2w-2,w)}{q-1}$.

Now, there are $q-1$ codewords which have a nonzero symbol in the first coordinate, each one having a different symbol in the first coordinate, and other than the first coordinate they do not share any of the other $w-1$ coordinates with nonzero symbols. Hence, we have that $n \ge 1+(w-1)(q-1)$. □

If $t = 2$ and $k = 3$. If $q = 2$, then a generalized Steiner system $\mathrm{GS}(2,3,n,2)$ is simply a Steiner system $\mathrm{S}(2,3,n)$ and it exists if and only if $n \equiv 1$ or $3 \pmod 6$. When $q > 2$, the divisibility conditions are slightly different and also the minimum distance of the code should be taken into account. The following theorem does not take into account the requirement of the minimum distance.

Theorem 3.2. *A necessary condition for the existence of a generalized Steiner system* $\mathrm{GS}(2,3,n,q)$ *is that* $n \ge q+1$ *and for* $q \not\equiv 1 \pmod 6$ *we have the following requirements.*

(c.1) *If* $q \equiv 4 \pmod 6$, *then* $n \equiv 1 \pmod 2$.
(c.2) *If* $q \equiv 3$ *or* $5 \pmod 6$, *then* $n \equiv 0$ *or* $1 \pmod 3$.
(c.3) *If* $q \equiv 0$ *or* $2 \pmod 6$, *then* $n \equiv 1$ *or* $3 \pmod 6$.

If $q \equiv 1 \pmod 6$, *then* $n \geq q+1$ *is a necessary condition for the existence of* $\mathrm{GS}(2,3,n,q)$.

Proof. Let $\mathcal{C}$ be a generalized Steiner system $\mathrm{GS}(2,3,n,q)$ and consider first the set of $q-1$ codewords

$$\mathcal{T} \triangleq \{(1,i,x_3,\ldots,x_n) \ : \ i \in \mathbb{Z}_q^-\}.$$

Since the minimum Hamming distance of $\mathcal{C}$ is 3, it follows that the third nonzero symbol in each codeword of $\mathcal{T}$ must be in a different coordinate. This implies that $n-2 \geq q-1$, i.e., $n \geq q+1$.

Now note that with the assignment $i=1$ in Corollary 3.2, we have that 2 divides $(n-1)(q-1)$, i.e., either n or q should be an odd integer. The assignment $i=0$ in Corollary 3.2 implies that 6 divides $(n-1)n(q-1)^2$.

If $q \equiv 1 \pmod 6$, then 6 divides $q-1$ and hence no more conditions are required.

If $q \equiv 4 \pmod 6$, then 3 divides $q-1$ and hence we should only require that n be odd and, therefore, (c.1) is proved.

If $q \equiv 3$ or $5 \pmod 6$, then 3 does not divide $q-1$ and hence it should divide $(n-1)n$, which implies that $n \equiv 0$ or $1 \pmod 3$ and, therefore, (c.2) is proved.

If $q \equiv 0$ or $2 \pmod 6$, then n must be odd, since 2 does not divide $q-1$, and 6 must divide $(n-1)n$, which implies that $n \equiv 1$ or $3 \pmod 6$, and, therefore, (c.3) is proved. □

There are many constructions for generalized Steiner systems $\mathrm{GS}(t,k,n,q)$, and especially for generalized Steiner systems $\mathrm{GS}(2,3,n,q)$, but in contrast to the binary case, in the nonbinary case, there is no complete solution to generalized Steiner systems $\mathrm{GS}(2,3,n,q)$ for all $q>2$. Moreover, for a given pair (t,k), where $t<k$, there are many associated open problems. These systems play an important role in nonbinary diameter perfect constant-weight codes as will be discussed in Chapter 9. On the other hand, the knowledge on these codes is not as much as known on Steiner system and research in this direction seems to be appealing. The following problems represent a small sample of such open problems (more problems appear in Chapter 9).

Problem 3.4. Prove that the necessary conditions for the existence of generalized Steiner systems $\mathrm{GS}(2,3,n,q)$ are also sufficient for any given q, with a possible small number of exceptions. Can these exceptions be specified?

The necessary conditions by Corollary 3.2 for the existence of a generalized Steiner system $\mathrm{GS}(3,4,n,3)$ implies that $n \equiv 1$ or $2 \pmod 3$. There are known constructions for generalized Steiner systems $\mathrm{GS}(3,4,n,3)$ only when $n \equiv 2$ or $4 \pmod 6$. Moreover, $\mathrm{GS}(3,4,n,q)$, where $q > 3$, has not been considered yet as of 2021.

Problem 3.5. Are the necessary conditions for the existence of a generalized Steiner system $\mathrm{GS}(3,4,n,3)$, where $n \equiv 1$ or $2 \pmod 3$, also sufficient? Present constructions for generalized Steiner systems $\mathrm{GS}(3,4,n,3)$, to cover the case when $n \equiv 1$ or $5 \pmod 6$.

Problem 3.6. Consider the existence question of generalized Steiner systems $\mathrm{GS}(3,4,n,q)$, where $q > 3$.

When $t > 3$ we do not have any construction for generalized Steiner systems $\mathrm{GS}(t,k,n,q)$. This leads to the following problems.

Problem 3.7. Analyze the necessary conditions of Corollary 3.2 for the existence of generalized Steiner systems $\mathrm{GS}(t,k,n,q)$ and present them in a simpler way as was done in Theorem 3.2.

Problem 3.8. Construct generalized Steiner systems $\mathrm{GS}(t,k,n,q)$, for $4 \leq t < k < n$ and $q > 2$.

Another problem which was not touched as it looks to be very difficult is whether there exist large sets of generalized Steiner systems. This problem might be not so difficult for some parameters and hence we have the following research problem.

Problem 3.9. Construct large sets of generalized Steiner systems and especially large sets of generalized Steiner systems $\mathrm{GS}(1,k,n,q)$, for some k, n, and q, and large sets of generalized Steiner systems $\mathrm{GS}(2,3,n,q)$, for some n and q.

Finally, Steiner systems have found applications in many problems related to coding theory and cryptography. We would like to see similar applications for generalized Steiner system.

Problem 3.10. Find problems in coding theory where generalized Steiner systems can be applied. Find nonbinary codes used in coding theory where generalized Steiner systems are embedded. Find applications of generalized Steiner systems in cryptography.

3.2 Orthogonal Designs

Another family of combinatorial designs are the orthogonal designs. There are a few types of orthogonal designs, some of which are of interest in the context of perfect codes and other related codes.

An ***orthogonal array*** $\text{OA}_\lambda(t,n,q)$ is a $(\lambda q^t)\times n$ matrix A with elements taken from a q-set Q such that each projection of t columns from A contains each t-tuple with elements from Q exactly λ times. When $\lambda = 1$, the orthogonal array is denoted by $\text{OA}(t,n,q)$ and is called an ***orthogonal array of index unity***. It is easy to verify that the rows of such an orthogonal array form a $(n, q^t, n-t+1)_q$ code. These orthogonal arrays will be our main interest among all orthogonal arrays. The parameters of these codes meet the Singleton bound of Theorem 2.1. This is summarized in the following theorem.

Theorem 3.3. *An orthogonal array* $\text{OA}(t,n,q)$ *is equivalent to an* $(n, q^t, n-t+1)_q$ *code that meets the Singleton bound with equality.*

Since by Theorem 3.3 an orthogonal array $\text{OA}(t,n,q)$ is equivalent to an $(n, q^t, n-t+1)_q$ code, it follows that shortening can be applied on orthogonal arrays. Therefore, by Lemma 2.2 we have the following two consequences.

Corollary 3.3. *If there exists an* $\text{OA}(t,n,q)$*, then there exists an* $\text{OA}(t-1,n-1,q)$.

Corollary 3.4. *If there is no* $\text{OA}(t,n,q)$*, then there is no* $\text{OA}(t+\delta,n+\delta,q)$ *for each* $\delta \geq 0$.

A ***Latin square*** of order n is an $n \times n$ matrix with entries from an n-set Q, such that each row and each column of the matrix is a permutation of Q. A pair of $n \times n$ squares A and B with entries from Q are said to be ***orthogonal*** if the set of ordered pairs $\{(A(i,j), B(i,j)) \; : \; 1 \leq i,j \leq n\}$ contains all the n^2 possible ordered pairs of Q.

Lemma 3.5. *The number of pairwise orthogonal Latin squares of order* n *is at most* $n-1$.

Proof. Assume that there exists a set of k pairwise orthogonal Latin squares of order n. W.l.o.g. assume that the first row of each square is $(1, 2, \ldots, n)$. Note that in the second row, the first entry cannot be a *one*. Moreover, since the pair (i,i), $1 \leq i \leq n$, appears in the pairs associated with the first row of each two squares, it follows that two distinct squares cannot have the same element in the first entry of the second row. Therefore, there are $n-1$ possible assignments for this entry, i.e., $k \leq n-1$. □

Theorem 3.4. *A set of k pairwise orthogonal $n \times n$ Latin squares exists if and only if there exists an orthogonal array* $\mathrm{OA}(2, k+2, n)$.

Proof. Given k pairwise orthogonal Latin squares of order n, let A be the $n^2 \times (k+2)$ matrix constructed as follows. The rows of A are indexed by (i, j), $1 \leq i, j \leq n$. The last two entries of the (i, j)-th row are (i, j). The entry in the (i, j)-th row of the r-th column of A, $1 \leq r \leq k$, will be the (i, j)-th entry of the r-th Latin square. It is easy to verify that the array A is an orthogonal array $\mathrm{OA}(2, k+2, n)$.

Let A be an orthogonal array $\mathrm{OA}(2, k+2, n)$. A set of k pairwise disjoint Latin squares is constructed in reverse order. The last two columns of A have all the pairs (i, j), where $1 \leq i, j \leq n$ and hence we can order the rows of A such that in the (i, j)-th row of A the pair (i, j) will be in the last two columns. The (i, j)-th entry in the r-th Latin square, $1 \leq r \leq k$, is defined by the entry of the (i, j)-th row of the r-th column in A. □

Corollary 3.5. *If there exists an* $\mathrm{OA}(2, n, q)$, *then* $n-1 \leq q$.

Applying also Corollary 3.3 we have the following consequence.

Corollary 3.6. *If there exists an* $\mathrm{OA}(3, n, 2)$, *then* $n \leq 4$.

Example 3.1. The following 8×4 array is an $\mathrm{OA}(3, 4, 2)$.

$$\begin{matrix} 0\,0\,0\,0 \\ 0\,0\,1\,1 \\ 0\,1\,0\,1 \\ 0\,1\,1\,0 \\ 1\,0\,0\,1 \\ 1\,0\,1\,0 \\ 1\,1\,0\,0 \\ 1\,1\,1\,1 \end{matrix}.$$

Corollary 3.6 is generalized with the following lemma.

Lemma 3.6. *Assume there exists an* $\mathrm{OA}(3, n, q)$, *where* $q \geq 2$.

(1) *If q is even, then $n \leq q+2$.*
(2) *If q is odd, then $n \leq q+1$.*

Proof. Note first that if there exists an $\mathrm{OA}(3, q+\delta, q)$, where $\delta \geq 3$, then there exists an $\mathrm{OA}(3, q+3, q)$. Assume the contrary, that there exists an $\mathrm{OA}(3, n, q)$, where $n = q+3$. This implies by Corollary 3.3 that there exists an $\mathrm{OA}(2, q+2, q)$. Hence, by Theorem 3.4, there exists a set of q orthogonal Latin squares of order q, contradicting Lemma 3.5 and therefore $n \leq q+2$.

Assume now that there exists an $\mathrm{OA}(3, q+2, q)$. Let A be such an orthogonal array. Clearly, each symbol appears exactly q^2 times in each column of A, each pair of symbols is contained exactly q times in any projection of an ordered pair of columns of A, and each ordered triple of symbols is contained exactly once in the projection of three columns of A.

Clearly, a permutation on the symbols in a column of A yield an equivalent $\mathrm{OA}(3, q+2, q)$. Hence, w.l.o.g. the first row of A consists only of *zeroes*. This implies that each other row of A contains at most two *zeroes*. The number of pairs of columns is $\binom{q+2}{2}$ and each pair of *zeroes* must appear in another $q-1$ rows (in addition to the first row) in each projection of a pair of columns. Hence, the total number of other rows that contain a pair of *zeroes* is $(q-1)\binom{q+2}{2}$. Each of the $q+2$ columns contains q^2 *zeroes* and hence all the rows, except for the first row of *zeroes*, contain exactly $(q+2)(q^2-1)$ *zeroes*. Since $2(q-1)\binom{q+2}{2} = (q+2)(q^2-1)$, it follows that each row of A, other than the first row, either contains two *zeroes* or contains no *zeroes*. Hence, there are $q^3 - 1 - (q-1)\binom{q+2}{2}$ rows with no *zeroes*. Since $q^3 - 1 - (q-1)\binom{q+2}{2} > 0$, it follows that we can assume w.l.o.g. that the second row of A has no *zeroes* and that all the symbols in this row are *ones* (since the symbols in each column can be permuted independently of the other columns). Two *ones* in the first two columns should appear with a *zero*, in any given column, exactly in one row. Since each such row contains exactly two *zeroes*, it follows that the number of columns, $q+2$, is divisible by 2, i.e., q is even. Thus, the claim of the lemma follows. □

Lemma 3.6 is generalized with the following theorem.

Theorem 3.5. *Assume there exists an* $\mathrm{OA}(t, n, q)$, *where* $3 \leq t \leq q$.

(1) *If* q *is even, then* $n \leq q+t-1$;
(2) *If* q *is odd, then* $n \leq q+t-2$.

Proof. By applying Corollary 3.3 on $\mathrm{OA}(t, n, q)$ $t-3$ times yields an $\mathrm{OA}(3, n-t+3, q)$. By Lemma 3.6, this implies that if q is even, then $n-t+3 \leq q+2$, i.e., $n \leq q+t-1$; and if q is odd, then $n-t+3 \leq q+1$, i.e., $n \leq q+t-2$. □

For $t = 2$, the result of Theorem 3.5 does not hold since the proof is obtained by shortening the orthogonal array $t-3$ times. The required bound for $t = 2$ is given in Corollary 3.5. In Theorem 3.5 the cutoff point is for $t \le q$. In the next two theorems the cutoff point is $q \le t$.

Theorem 3.6. *If there exists an* $\mathrm{OA}(t, n, q)$, *where* $q \le t$, *then* $n \le t + 1$.

Proof. Assume the contrary, that $q \le t$, A is an $\mathrm{OA}(t, t+2, q)$, and distinguish between three cases, depending on whether $q = 2$, $q = 3$, or $q > 3$. To prove the theorem it suffices to find a contradiction in each case.

Case 1. $q = 2$.

By applying Corollary 3.3 $t-2$ times on $\mathrm{OA}(t, t+2, q)$ we obtain an $\mathrm{OA}(2, 4, 2)$, contradicting Corollary 3.5.

Case 2. $q = 3$.

W.l.o.g. assume that the first 15 rows of the $3^t \times (t+2)$ array A are as follows:

$$
\begin{array}{l}
0 \cdots 0\,0\,0\,0\,0\,0 \\
0 \cdots 0\,0\,0\,1\,1\,1 \\
0 \cdots 0\,0\,0\,2\,2\,2 \\
0 \cdots 0\,0\,1\,0\,\bullet\,\bullet \\
0 \cdots 0\,0\,1\,\bullet\,0\,\bullet \\
0 \cdots 0\,0\,1\,\bullet\,\bullet\,0 \\
0 \cdots 0\,0\,2\,0\,\bullet\,\bullet \\
0 \cdots 0\,0\,2\,\bullet\,0\,\bullet \\
0 \cdots 0\,0\,2\,\bullet\,\bullet\,0 \\
0 \cdots 0\,1\,0\,0\,\bullet\,\bullet \\
0 \cdots 0\,1\,0\,\bullet\,0\,\bullet \\
0 \cdots 0\,1\,0\,\bullet\,\bullet\,0 \\
0 \cdots 0\,2\,0\,0\,\bullet\,\bullet \\
0 \cdots 0\,2\,0\,\bullet\,0\,\bullet \\
0 \cdots 0\,2\,0\,\bullet\,\bullet\,0
\end{array},
$$

where the $\bullet$ stands for either 1 or 2. For a given pair (a, b), where $a, b \in \{0, 1, 2\}$, and a or b is a *zero*, there are five words of the form $(\overbrace{0 \cdots\cdots 0}^{t-1 \text{ times}}, a, b)$. Consider now the four rows with two $\bullet$'s and a *zero* in the last entry. These four rows share $t-2$ identical coordinates (the first $t-3$ coordinates and the last one with *zeroes*) and hence their bullets should be assigned the four distinct options of *ones* and *twos*. Two of these four assignments (11 and 22), however, will create in two of these four rows the same projection of t columns with the second or the third row, a contradiction.

The arguments hold for $t \geq 5$ and to complete the proof we have to show that there is no OA$(3,5,3)$ and no OA$(4,6,3)$. The nonexistence of OA$(3,5,3)$ is implied by Lemma 3.6 and hence by Corollary 3.4 it implies that there is no OA$(4,6,3)$. This also implies that there is no OA$(t,t+2,q)$, but the longer proof will serve us as introduction to the more complicated case when $q > 3$.

Case 3. $q > 3$.

As in Case 2, we analyze the rows of the $q^t \times (t+2)$ array A. Consider all the rows of A with at least $t-2$ *zeroes* in the first $t-1$ columns. W.l.o.g. assume that the first q rows of this form are in the set

$$A_1 \triangleq \{(\overbrace{0\cdots\cdots 0}^{t-1 \text{ times}}, i, i, i) \ : \ 0 \leq i \leq q-1\} \ .$$

For the other rows of A, there are $t-1$ different possible positions for the only nonzero symbol in the first $t-1$ columns. This nonzero symbol can be any one of the $q-1$ nonzero symbols. Therefore, there are $(q-1)(t-1)$ distinct possible vectors formed from the first $t-1$ positions in the rows whose projection on their first $t-1$ columns has weight $t-2$. For each such vector v, of length $t-1$, each of the alphabet symbols can appear as the last symbol in the row associated with v (whose prefix of length $t-1$ is v). Therefore, there are $q(q-1)(t-1)$ rows in A whose projection on the first $t-1$ columns has weight $t-2$. This set of $q(q-1)(t-1)$ rows in A will be denoted by A_2. Thus, there are $q+q(q-1)(t-1)$ rows in A with at least $t-2$ *zeroes* in the first $t-1$ columns and these rows form the set $A_1 \cup A_2$.

Consider now the set A_3 of $3(q-1)(t-1)$ rows of A_2 with a *zero* in one of the last three columns (note that each such row cannot have more than one *zero* symbol in these three columns since this would form a projection with t *zeroes*, which already appear in the all-zero row of A_1).

Consider now the set A'_3 of the other $(q-3)(q-1)(t-1)$ rows of A_2, i.e., $A'_3 \triangleq A_2 \setminus A_3$. Let α be the nonzero symbol that appears most frequently in the last column of A'_3. Since each of the $q-1$ nonzero symbols occurs exactly once in each of the last three columns in each set of q rows having the same $(t-1)$-tuple in the first $t-1$ columns, it follows that α (which occurs most frequently) occurs at least $(q-3)(t-1)$ times in these rows of the set A'_3 (note that there is no symbol β that can occur twice in the last three entries of a row from A_2, since A_1 contains any projection of $t-2$ *zeroes* in the first $t-1$ columns and the same symbol β in two of the last three columns.).

There are $(q-2)^2$ pairs of symbols that do not contain a *zero* and an α. To prevent a repeat of two identical t-tuples in A'_3, distinct pair of symbols ((β, β) should be avoided) should be assigned before α in the rows of A'_3 whose last entry is α. Note also that there is no *zero* in the last three entries of these rows. Hence, the $(q-2)^2$ pairs are assigned to $q-2$ rows of A_1 (the pairs (β, β) were assigned to $q-2$ rows of A_1) and the $(q-3)(t-1)$ rows of A'_3 whose last entry is α. This implies that $(q-2)+(q-3)(t-1) \leq (q-2)^2$, which is equivalent to $0 \leq (q-3)(q-t-1)$. Since $q > 3$, this implies that $0 \leq q-t-1$, i.e., $t+1 \leq q$, a contradiction. □

Theorem 3.7. *There exists a set of* $n-1$ *pairwise orthogonal Latin squares of order* n *if and only if there exists a Steiner system* $\mathrm{S}(2, n+1, n^2+n+1)$.

Proof. Assume first that there exists a set of $n-1$ pairwise orthogonal Latin squares of order n over the alphabet $[n]$. Add two more $n \times n$ squares that are not Latin squares. The first one has the i-th symbol in all the entries of the i-th row and the second one has the i-th symbol in all the entries of the i-th column, where $1 \leq i \leq n$. Clearly, the $n+1$ squares are pairwise orthogonal. We form a system $\mathcal{S} = (Q, B)$, where $Q = \{(i,j) \ : \ 1 \leq i,j \leq n\} \cup [n+1]$. Enumerating the squares by $A_1, A_2, \ldots, A_{n+1}$, we form a set of n^2+n blocks, n blocks per square, as follows. The ℓ-th block, $1 \leq \ell \leq n$, for the i-th square A_i is

$$\{i\} \cup \{(j,m) \ : \ A_i(j,m) = \ell\}.$$

To these n^2+n blocks we add the block $[n+1]$. It is readily verified now that these n^2+n+1 blocks form a Steiner system $\mathrm{S}(2, n+1, n^2+n+1)$. For example, if in contrast, there exists two blocks that contain the pairs $\{(j_1, m_1), (j_2, m_2)\}$, then there exist two squares A_{i_1}, A_{i_2}, where $1 \leq i_1 < i_2 \leq n+1$, for which $A_{i_1}(j_1, m_1) = A_{i_1}(j_2, m_2) = \ell_1$ and $A_{i_2}(j_1, m_1) = A_{i_2}(j_2, m_2) = \ell_2$, a contradiction to the orthogonality of A_{i_1} and A_{i_2}, which completes the first direction of the proof.

Assume now that there exists a Steiner system $\mathrm{S}(2, n+1, n^2+n+1)$, $\mathcal{S} = (Q, B)$, where $Q = \{(i,j) \ : \ 1 \leq i,j \leq n\} \cup [n+1]$ and w.l.o.g. we assume that $[n+1]$ is one of the blocks. This implies that each other block contains at most one point from $[n+1]$. Each point of $[n+1]$ must be paired with each one of the points of Q exactly in one block and hence each point of $[n+1]$ is contained in exactly n blocks in addition to the block $[n+1]$. Consider the n such block which contain the point $n+1$. The other n^2 points in these blocks are the n^2 distinct pairs of $Q \times Q$. Therefore, we can permute the points in Q such that B will contain the n blocks

$$\{n+1,(i,1),(i,2),\ldots,(i,n)\},\ 1 \le i \le n. \tag{3.1}$$

Now, consider the n blocks, different from $[n+1]$, which contain the point n. For a given j, $1 \le j \le n$, there is no block which contains the pair of points $\{(j,m_1),(j,m_2)\}$, where $1 \le m_1 < m_2 \le n$, since this pair of points is contained in the blocks defined in (3.1). Therefore, we can permute again the points of Q such that B will contain the same blocks as in (3.1) and also n blocks

$$\{n,(1,j),(2,j),\ldots,(n,j)\},\ 1 \le j \le n. \tag{3.2}$$

Define now $n \times n$ squares from the blocks of $\mathcal{S}$ in reverse order to the blocks defined from $n+1$ pairwise orthogonal squares, from which $n-1$ are Latin squares, which form a Steiner system $\mathrm{S}(2,n+1,n^2+n+1)$ in the first part of the proof. The $2n$ blocks defined in (3.1) and in (3.2) will be associated with two squares, one in which all the entries of the i-th row contain the i-th symbol and a second in which all the entries of the i-th column contain the i-th symbol for each $1 \le i \le n$. The remaining $n^2 - n$ blocks of B, excluding $[n+1]$, define $n-1$ pairs of orthogonal Latin squares. The $2n$ blocks defined in (3.1) and in (3.2) implies that these squares are Latin squares and the fact that each pair of points is contained in exactly one block of $\mathcal{S}$ implies that any two such squares are orthogonal. Note also that the symbol $j \in [n+1]$, which is not in a pair, defines the index of the square. Thus, $n-1$ pairwise orthogonal Latin square of order n were formed and the proof was completed. □

Corollary 3.7. *There exists an orthogonal array* $\mathrm{OA}(2,n+1,n)$ *if and only if there exists a Steiner system* $\mathrm{S}(2,n+1,n^2+n+1)$.

Let $\mathcal{S}=(Q,B)$ be a Steiner system $\mathrm{S}(2,n+1,n^2+n+1)$ and A be its incidence matrix. The transpose of A, A^{tr}, is also an incidence matrix for a design. By Lemma 3.1, The number of blocks in $\mathcal{S}$ is $\frac{\binom{n^2+n+1}{2}}{\binom{n+1}{2}} = n^2+n+1$ and each point of Q is contained in $\frac{n^2+n}{n} = n+1$ blocks of B. Moreover, by the definition of a Steiner system $\mathrm{S}(2,n+1,n^2+n+1)$, two distinct blocks cannot contain the same pair of points. Hence, two distinct points cannot be in the intersection of the same pair of blocks. This implies the following important result.

Theorem 3.8. *The dual design of a Steiner system* $\mathrm{S}(2,n+1,n^2+n+1)$ *is also a Steiner system* $\mathrm{S}(2,n+1,n^2+n+1)$.

Does there exist a set of $n-1$ pairwise orthogonal Latin squares of order n? Such a set exists whenever n is a power of a prime. This set is not always unique and one such set can be constructed as follows. Let Q be the set of one-subspaces of $\mathbb{F}_q^3$, where q is a power of a prime, i.e., each one-subspace is a point. Let B be the set of blocks, where each block contains the one-subspaces contained in a distinct two-subspace of $\mathbb{F}_q^3$, i.e., each two-subspace form a block. We claim that $\mathcal{S} = (Q, B)$ is a Steiner system $\mathrm{S}(2, q+1, q^2+q+1)$. A one-subspace of $\mathbb{F}_q^3$ contains $q-1$ nonzero elements of $\mathbb{F}_q^3$ and hence there are $\frac{q^3-1}{q-1} = q^2+q+1$ one-subspaces in $\mathbb{F}_q^3$, i.e., $|Q| = q^2+q+1$. A two-subspace contains q^2-1 nonzero elements of $\mathbb{F}_q^3$ and hence it contains $\frac{q^2-1}{q-1} = q+1$ one-subspaces. Therefore, each block contains exactly $q+1$ points of Q. Each two distinct one-subspaces define a unique two-subspace and hence each two points of Q are contained in a unique block of B, which implies that $\mathcal{S}$ is a Steiner system $\mathrm{S}(2, q+1, q^2+q+1)$. This implies the following theorem.

Theorem 3.9. *If q is a power of a prime, then there exists a Steiner system $\mathrm{S}(2, q+1, q^2+q+1)$.*

By Theorem 3.7, a Steiner system $\mathrm{S}(2, q+1, q^2+q+1)$ exists if and only if there exists a set with $q-1$ pairwise orthogonal Latin squares of order q. Moreover, the proof of Theorem 3.7 describes how to form this set of Latin squares from a Steiner system $\mathrm{S}(2, q+1, q^2+q+1)$. This construction and Theorems 3.3 and 3.4, tie together the relations between Latin squares, Steiner systems, orthogonal arrays, and some codes that meet the Singleton bound.

A related concept is the strength of a code of length n over an alphabet with q letters. The ***strength*** of a code over an alphabet of size q, is the largest t such that each projection of t columns contains each of the q^t possible t-tuples in the same number of rows. Clearly, such a code with M codewords is an orthogonal array $\mathrm{OA}_\lambda(t, n, q)$, where $\lambda = \frac{M}{q^t}$.

Recall that by Theorem 2.1 an MDS code is an $[n, k, d]_q$ code $\mathcal{C}$ for which $k = n - d + 1$, i.e., the code $\mathcal{C}$ attains the Singleton bound of Theorem 2.1 with equality. In such a code $\mathcal{C}$, any set of k columns in the generator matrix are linearly independent and hence they form a set of systematic coordinates. This implies that in each projection of k columns, each k-tuple over $\mathbb{F}_q$ is contained exactly once. Thus, such an $[n, k, d]_q$ code is an $\mathrm{OA}(k, n, q)$.

Theorem 3.10. *If $\mathcal{C}$ is an $[n, k, d]$ code, then the following properties are equivalent:*

(1) *the code $\mathcal{C}$ is an $[n, k, n-k+1]$ MDS code;*
(2) *every k columns of a generator matrix G of $\mathcal{C}$ are linearly independent;*
(3) *every $n-k$ columns of a parity-check matrix H of $\mathcal{C}$ are linearly independent;*
(4) *the dual code $\mathcal{C}^\perp$ is an $[n, n-k, k+1]$ MDS code.*

Proof. We will prove that (1) $\Rightarrow$ (2), (2) $\Rightarrow$ (3), (3) $\Rightarrow$ (4), and (4) $\Rightarrow$ (1).

Assume that $\mathcal{C}$ is an $[n, k, n-k+1]$ code and let G be a generator matrix of $\mathcal{C}$. Let S be any set of k columns of $\mathcal{C}$. If these k columns are linearly dependent, then the $k \times k$ matrix of G defined by the projection of the columns of S on G is a singular matrix. Hence, there exists a nontrivial linear combination of the rows of G in which the k entries of S are *zeroes*. Therefore, there exists a nonzero codeword $c \in \mathcal{C}$, with *zeroes* in these entries, i.e., the weight of c is at most $n-k$, which contradicts the minimum distance of $\mathcal{C}$. Thus, every k columns of a generator matrix G of $\mathcal{C}$ are linearly independent. Hence, (1) $\Rightarrow$ (2).

Assume now that every k columns of a generator matrix G are linearly independent. Recall that the structure of the generator matrix G and the parity-check matrix H of the code can be taken as

$$G = [\ I_k \mid A\ ],$$

and

$$H = [\ -A^{\text{tr}} \mid I_{n-k}\ ].$$

In other words, if S is a set of k columns indices, in G, which are linearly independent, then the columns whose indices are $[n] \setminus S$, in H are also linearly independent since we can always write an equivalent generator matrix whose projection on the columns of S is an identity matrix. Thus, since every k columns of G are linearly independent, it follows that every $n-k$ columns of H are linearly independent. Therefore, (2) $\Rightarrow$ (3).

Assume now that every $n-k$ columns of a parity-check matrix H of $\mathcal{C}$ are linearly independent. Let $\mathcal{C}^\perp$ be that dual code of $\mathcal{C}$. Similarly to the previous part of the proof, it follows that in G, which is the parity-check matrix of $\mathcal{C}^\perp$, each k columns are linearly independent. Therefore, the minimum number of linearly dependent columns in G is $k+1$. Thus, by Corollary 2.11, $\mathcal{C}^\perp$ is an $[n, n-k, k+1]$ MDS code. This implies that (3) $\Rightarrow$ (4).

Now assume that $\mathcal{C}^\perp$ is an $[n, n-k, k+1]$ MDS code. By the previous parts of the proof, this implies that every $n-k$ columns of the generator matrix of $\mathcal{C}^\perp$ are linearly independent. Therefore, every $n-k$ columns of the parity-check matrix of $\mathcal{C}$ are linearly independent, i.e., the minimum number of dependent columns in the parity-check matrix of $\mathcal{C}$ is $n-k+1$. Thus, $\mathcal{C}$ is an $[n, k, n-k+1]$ MDS code. As a consequence (4) $\Rightarrow$ (1).

Thus, all four properties laid out in the theorem are equivalent. □

Note that the only distinction between the concepts of MDS codes and orthogonal arrays is that orthogonal arrays do not have to be linear subspace. The following conjecture is known as the ***MDS conjecture***.

Conjecture 3.1. *If $d \geq 3$, then for a prime power q there exists an $[n, k, d]_q$ MDS code if and only if $n \leq q+1$ and $2 \leq k \leq q-1$, unless q is even and $k \in \{3, q-1\}$, in which case $n \leq q+2$.*

The last family of orthogonal designs that we define is Hadamard matrices, which are $n \times n$ matrices over the set of real numbers having the largest possible determinants, $n^{n/2}$, among all $n \times n$ matrices with real entries between -1 and 1. The equivalent definition for these matrices is the one required in our exposition.

Definition 3.3. A ***Hadamard matrix*** of order n is an $n \times n$ binary matrix H over $\{-1, +1\}$ such that $H^{\text{tr}} H = nI_n$.

A ***normalized Hadamard matrix*** is a Hadamard matrix in which the first row and the first column have only $+1$'s. One can readily verify that each Hadamard matrix can be made a normalized Hadamard matrix by multiplying each row and column starting with a -1 by -1.

Theorem 3.11. *If a Hadamard matrix of order n exists, then n is 1, 2, or a multiple of 4.*

Proof. Clearly, Hadamard matrices of orders 1 and 2 exist. It is also easy to verify that a Hadamard matrix of order 3 does not exist. If $n > 3$ and a Hadamerd matrix H of order n exists, w.l.o.g. assume that H is a normalized Hadamard matrix and consider the first three rows of A. The triples, generated by the columns, in these three rows can be $(+1, +1, +1)^{\text{tr}}$, $(+1, +1, -1)^{\text{tr}}$, $(+1, -1, +1)^{\text{tr}}$, or $(+1, -1, -1)^{\text{tr}}$. Assume that there are i_1 triples of the form $(+1, +1, +1)^{\text{tr}}$, i_2 triples of the form $(+1, +1, -1)^{\text{tr}}$, i_3 triples of the form $(+1, -1, +1)^{\text{tr}}$, and i_4 triples of the form $(+1,-1,-1)^{\text{tr}}$. Since $H^{\text{tr}}H = nI_n$ implies that each two distinct rows are orthogonal, i.e., their inner product is equal to *zero*, it follows that

$$i_1 + i_2 - i_3 - i_4 = 0, \quad \text{from the first row and second row,}$$

$$i_1 - i_2 + i_3 - i_4 = 0, \quad \text{from the first row and third row,}$$

$$i_1 - i_2 - i_3 + i_4 = 0, \quad \text{from the second row and third row.}$$

The solution for this set of three equations implies that $i_1 = i_2 = i_3 = i_4$, i.e., n is divisible by 4. □

There are many constructions for Hadamard matrices and it is conjectured that for each n divisible by 4, there exists such a matrix. There are many applications for Hadamard matrices in coding theory as well as in other areas.

The most celebrated construction of Hadamard matrices is Sylvester's construction, which is a simple doubling construction. Let H be any $n \times n$ Hadamard matrix. The following matrix

$$\begin{bmatrix} H & H \\ H & -H \end{bmatrix}$$

is a $(2n) \times (2n)$ Hadamard matrix.

This construction immediately yields a Hadamard matrix for each power of 2 order, but it is also applied to other orders for which Hadamard matrices are known to exist.

The Sylvester's construction can be generalized as follows. Let A by an $n \times n$ Hadamard matrix and B be an $m \times m$ Hadamerd matrix. The Kronecker product of A and B defined by

$$\begin{bmatrix} a_{11}B & a_{12}B & \cdots & a_{1n}B \\ a_{21}B & a_{22}B & \cdots & a_{2n}B \\ \vdots & \vdots & \ddots & \vdots \\ a_{n1}B & a_{n2}B & \cdots & a_{nn}B \end{bmatrix}$$

is an $(nm) \times (nm)$ Hadamard matrix. It generalizes the Sylvester's construction by taking the 2×2 Hadamard matrix

$$A = \begin{bmatrix} 1 & 1 \\ 1 & -1 \end{bmatrix}$$

in the Kronecker product.

Let H_n be a normalized Hadamard matrix of order n and let A_n be the code obtained from the matrix H_n by replacing each $+1$ by a *zero* and each -1 by a *one*, where the rows of the matrix are the codewords of A_n. The matrix A_n is called a ***binary Hadamard matrix*** and its first row and first column contain only *zeroes*. Since any two distinct rows of H_n are orthogonal, it follows that they agree in $\frac{n}{2}$ coordinates and do not agree in $\frac{n}{2}$ coordinates. Hence, we have that A_n forms a binary $(n, n, \frac{n}{2})$ code. The code A_n is just one of a few codes that can be derived from a Hadamard matrix. The most obvious codes are the following four codes, called ***Hadamard codes***, and which are obtained from the normalized Hadamard matrix H_n and its associated binary Hadamard matrix A_n.

(1) The code Λ_n consists of the rows of A_n and their complements.
(2) The code Υ_n is the punctured code, obtained from the rows of A_n, with respect to the first coordinate.
(3) The code Ψ_n is the punctured code of Λ_n, with respect to the first coordinate, i.e., it is constructed from the codewords of Υ_n and their complements.
(4) The code Φ_n is the shortened code of Υ_n, with respect to any coordinate.

One can easily verify the parameters of the Hadamard codes given in the following theorem.

Theorem 3.12. *Assume H_n is an $n \times n$ normalized Hadamard matrix.*

(1) *Λ_n is a binary $(n, 2n, \frac{n}{2})$ code.*
(2) *Υ_n is a binary $(n-1, n, \frac{n}{2})$ code.*
(3) *Ψ_n is a binary $(n-1, 2n, \frac{n}{2}-1)$ code.*
(4) *Φ_n is a binary $(n-2, \frac{n}{2}, \frac{n}{2})$ code.*

Hadamard codes will play a pivotal role in constructing codes which meet the Plotkin bound presented in Section 3.4.

3.3 Projective Geometries

Projective geometries were already mentioned in the connection of some infinite families of Steiner systems. These geometries are also very important in connection to other families of codes. One example are codes that meet the bound of Theorem 2.1 and also meet the bound implied by

Conjecture 3.1 for all parameters. In other words, the work on the MDS conjecture can be explained in terms of projective geometry.

Definition 3.4. A ***finite projective geometry*** consists of a set of ***points*** and a set of ***lines*** with the following four axioms:

- Every line contains at least three points.
- Every two distinct points are contained in exactly one line.
- If p, q, r, are distinct points on a line L_1 and s, t, and r are distinct points on another line L_2, then the line L_3, which contains the points p and s, and the line L_4, which contains the points q and t, contain a common point.
- Given a line L, there are two points not on L, and, for each point p, there are two lines that do not contain p.

A ***subspace*** in the projective geometry $\mathbb{P}$ is a set S of points, where all the points on a line L that contains two distinct points p and q of S, are contained in S. In other words, if a line L contains two points of S, then S contains all the points of L. A ***hyperplane*** H is a subspace, where the only subspace that contains H and does not equal to H is the set of all points in $\mathbb{P}$. A set of points S is called ***independent*** if each point $x \in S$ is not contained in the smallest subspace that contains $S \setminus \{x\}$. The ***dimension*** of a subspace S is m, where $m+1$ is the size of the largest set of independent points in S.

The most important example of a finite projective geometry is the ***projective geometry*** $\text{PG}(m,q)$, where q is a power of a prime and $m \geq 2$. The ***points*** of $\text{PG}(m,q)$ are $(m+1)$-tuples $(a_0, a_1, \ldots, a_m)$, $a_i \in \mathbb{F}_q$, where $(a_0, a_1, \ldots, a_m)$ is considered to be the same point as $(\lambda a_0, \lambda a_1, \ldots, \lambda a_m)$, for any $\lambda \in \mathbb{F}_q^-$. A ***line*** through the two distinct points $(a_0, a_1, \ldots, a_m)$ and $(b_0, b_1, \ldots, b_m)$ consists of the points $(\lambda a_0 + \mu b_0, \lambda a_1 + \mu b_1, \ldots, \lambda a_m + \mu b_m)$, where $\lambda, \mu \in \mathbb{F}_q$ and $\lambda \neq 0$ or $\mu \neq 0$. A hyperplane in $\text{PG}(m,q)$ is an $(m-1)$-subspace that consists of all the points $(a_0, a_1, \ldots, a_m)$ that satisfy a homogeneous linear equation

$$\lambda_0 a_0 + \lambda_1 a_1 + \cdots + \lambda_m a_m = 0, \quad \lambda_i \in \mathbb{F}_q \ ,$$

for some choice of $\lambda_0, \lambda_1, \ldots, \lambda_m$, where at least one of λ_i's is not *zero*. This hyperplane is isomorphic to a $\text{PG}(m-1,q)$ and is denoted by $[\lambda_0, \lambda_1, \ldots, \lambda_m]$. Clearly, the hyperplane $[\lambda_0, \lambda_1, \ldots, \lambda_m]$ is the same hyperplane as the hyperplane $[\mu\lambda_0, \mu\lambda_1, \ldots, \mu\lambda_m]$, for each $\mu \in \mathbb{F}_q^-$.

We will make now a connection between MDS codes and the projective geometries $\mathrm{PG}(m,q)$. The following concept is defined in projective geometry.

Definition 3.5. An ***n*-arc** in $\mathrm{PG}(k-1,q)$ is a set of n points, where every k of them are (linearly) independent, i.e., no k points lie in a hyperplane $\mathrm{PG}(k-2,q)$, where $3 \leq k \leq n$. An n-arc in $\mathrm{PG}(k-1,q)$ is called ***complete*** if and only if it is not contained in an $(n+1)$-arc of $\mathrm{PG}(k-1,q)$. If $k=3$, then an n-arc is called an ***n*-cap**.

The following two results, which are proved directly from Definition 3.5 and Theorem 3.10.

Theorem 3.13. *The set* $K = \{g_1, g_2, \ldots, g_n\}$ *is an* n*-arc in* $\mathrm{PG}(k-1,q)$*, where* g_i *is a column vector of length* k*, if and only if the* $k \times n$ *matrix* $G = [g_1\ g_2\ \cdots\ g_n]$ *is a generator matrix of an* $[n,k,n-k+1]_q$ *MDS code.*

Theorem 3.14. *If* $K = \{g_1, g_2, \ldots, g_n\}$ *is an* n*-arc in* $\mathrm{PG}(k-1,q)$*, defining the* $[n,k,n-k+1]_q$ *MDS code* $\mathcal{C}$ *with generator matrix* $G = [g_1\ g_2\ \cdots\ g_n]$*, then there exists an* n*-arc* $\tilde{K} = \{h_1, h_2, \ldots, h_n\}$ *in* $\mathrm{PG}(n-k-1,q)$ *such that* $\tilde{K}$ *defines the dual* $[n,n-k,k+1]_q$ *MDS code* $\mathcal{C}^{\perp}$ *via the parity-check matrix* $H = [h_1\ h_2\ \cdots\ h_n]$ *of* $\mathcal{C}$*, i.e., the generator matrix of* $\mathcal{C}^{\perp}$*.*

Theorem 3.5 considers bounds on the parameters of $\mathrm{OA}(t,n,q)$, which imply the following bounds on the parameters of projective geometries.

Theorem 3.15. *If* $\mathcal{C}$ *is a nontrivial* $[n,k,n-k+1]_q$ *MDS code, where* $k \geq 3$ *and* q *is odd, then* $n \leq q+k-2$*. In other words, for any* n*-arc in* $\mathrm{PG}(k-1,q)$*,* q *odd, we have that* $n \leq q+k-2$*.*

If $\mathcal{C}$ *is a nontrivial* $[n,k,n-k+1]_q$ *MDS code, where* q *is even, then* $n \leq q+k-1$*. In other words, for any* n*-arc in* $\mathrm{PG}(k-1,q)$*,* q *even, we have that* $n \leq q+k-1$*.*

We continue with the concept of a projective plane that is a finite projective geometry of dimension 2, but can be defined independently.

Definition 3.6. A ***projective plane*** consists of a set of points and a set of lines with the following four axioms.

- There is a line with at least three points.
- Every two distinct points are contained in exactly one line.
- Every two distinct lines intersect at exactly one point.

- There are four points for which no three are collinear (i.e., no three points are on the same line).

Example 3.2. On the left side of Fig. 3.2 we have a structure that satisfies the first three axioms of a projective plane, but does not satisfy the last axiom. On the right side of Fig. 3.2 we have the well-known ***Fano plane***, which is a PG(2, 2). The Fano plane has seven points and seven lines (the circle is also a line).

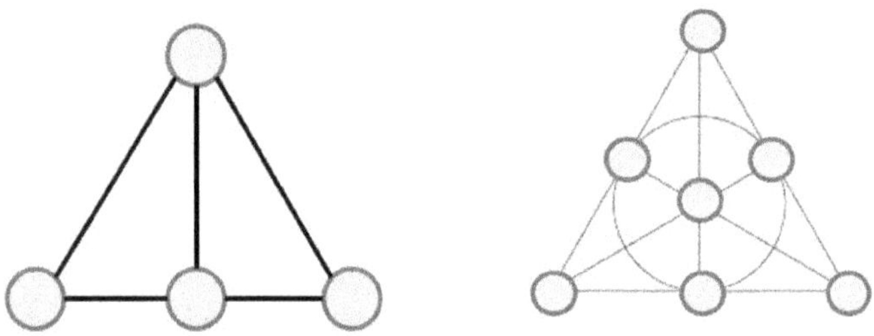

Fig. 3.1 A non-projective plane and the Fano plane PG(2, 2).

Lemma 3.7. *The number of points in a projective plane is $n^2 + n + 1$ and the number of lines in a projective plane is also $n^2 + n + 1$, where each line contains $n + 1$ points and each point is contained in $n + 1$ lines.*

Proof. Consider a point p, and a line L that does not contain p. Any two points are contained in exactly one line and, hence, each point on L and the point p are contained in a unique line. Moreover, every two lines intersect at exactly one point. This implies that the number of lines that contain p equals the number of points on L. Hence, if $n + 1$ is the number of lines that contain p, then L contains $n + 1$ points.

Since each point q, $q \neq p$, and p are contained in a line, it follows that this set of $n + 1$ lines containing p contains all the points of the projective plane. Moreover, since the number of lines that contain p is $n + 1$ and the same arguments hold for any line like L, which does not contain p, it follows that each line which does not contain p, contains $n + 1$ points.

Assume q is another point not on L (it exists since each line, which contains p, contains at least three points). Since p is contained in $n + 1$ lines and each such line contains at least three points, it follows by simple enumeration that there exists another point not on L and not on the line containing p and q and, hence, by the same arguments, the line containing p and q also contains of $n+1$ points. Thus, each line contains $n+1$ points.

To summarize, there are $n+1$ lines that contains p, each of which contains n distinct points (excluding p). Thus, the total number of points in the projective plane is $(n+1)n+1 = n^2+n+1$. Since each pair of points defines a distinct line, it follows that the number of lines is $\frac{\binom{n^2+n+1}{2}}{\binom{n+1}{2}} = n^2+n+1$. This also implies that each line contains $n+1$ points (as was proved) and each point is contained in $n+1$ lines. $\square$

A projective plane with n^2+n+1 lines and n^2+n+1 points, where each line contains $n+1$ points and each point is contained in $n+1$ lines, is a ***projective plane of order*** n.

Theorem 3.16. *A projective plane of order* $n \geq 2$ *exists if and only if there exists a Steiner system* $\mathrm{S}(2, n+1, n^2+n+1)$.

Proof. Given a projective plane $\mathbb{P}$ of order n with point set Q, we construct the following system $S = (Q, B)$, where the set of points is Q and the points of each line in $\mathbb{P}$ define a block in B. By Lemma 3.7 each such block contains $n+1$ points. Clearly, by the axiom of Definition 3.6, that in a projective plane, every two distinct points are contained in exactly one line, such a system S is a Steiner system $\mathrm{S}(2, n+1, n^2+n+1)$.

Assume now that $S = (Q, B)$ is a Steiner system $\mathrm{S}(2, n+1, n^2+n+1)$, $n \geq 2$, and define a system $\mathbb{P}$ with points and lines. Let Q be the set points, in $\mathbb{P}$, and the points of each block in B defines a line in $\mathbb{P}$. To prove that $\mathbb{P}$ is a projective plane, it suffices to show that the four axioms of a projective plane, presented in Definition 3.6, are satisfied.

Clearly, each line contains at least three points, since $n \geq 2$, and therefore the first axiom is satisfied.

By the definition of a Steiner system, every two points Q are contained in exactly one block of B and hence each two points of $\mathbb{P}$ are contained in a unique line of $\mathbb{P}$ and hence the second axiom is satisfied.

Two lines cannot intersect at more than one point since this implies that a pair of points are contained in two blocks of S. Since each block has size $n+1$ and each given point p is paired with any other point q in exactly one block, it follows that the number of blocks (lines) that contain a given point p is $\frac{n^2+n}{n} = n+1$. Each of the other points is contained in exactly one of these blocks (lines). A block (line) L that does not contain the point p has an intersection size at most one with each of these blocks (lines). Since L has exactly $n+1$ points, it follows that it must intersect each of them in one point. Therefore, L intersects each of the lines of $\mathbb{P}$. Since L can be taken arbitrarily, it follows that every two lines intersect at exactly one point. This implies that the third axiom of Definition 3.6 is satisfied.

Let L_1 and L_2 be two lines that contain the point p. Let a_1, a_2 be two other distinct points on L_1 and a_3, a_4 be two other distinct points on L_2. Clearly, a_1, a_2, a_3, a_4 are four distinct points for which no three are collinear. Hence, the last axiom of Definition 3.6 is satisfied.

Thus, a projective plane of order $n \geq 2$ exists if and only if there exists a Steiner system $\mathrm{S}(2, n+1, n^2+n+1)$. □

The third axiom in the proof of Theorem 3.16 can be proved in many different ways. For example, one can use the dual design of the Steiner system $\mathrm{S}(2, n+1, n^2+n+1)$. By Theorem 3.8, the dual design is also a Steiner system $\mathrm{S}(2, n+1, n^2+n+1)$, and hence the arguments used for the points are also true for the lines and vice versa.

A related concept to projective geometry is the affine geometry. An ***affine geometry*** of order n is obtained from a projective geometry $\mathbb{P}$ of order n by deleting the points of any fixed hyperplane H, which will be called ***hyperplane of infinity***, from all the subspaces of $\mathbb{P}$. The obtained sets from the subspaces of the projective geometry are the subspaces of the affine geometry. The affine geometry obtained from $\mathrm{PG}(m, q)$ is denoted by $\mathrm{EG}(m, q)$. It is most convenient to delete from $\mathrm{PG}(m, q)$ the hyperplane H whose points starts with $a_0 = 0$. It implies that $\mathrm{EG}(m, q)$ will consist of the set of q^m points $\{(1, a_1, a_2, \ldots, a_m) \; : \; a_i \in \mathbb{F}_q, \; 1 \leq i \leq m\}$.

An ***affine plane*** is obtained in this way from a projective plane $\mathbb{P}$, by deleting any line L of $\mathbb{P}$, and all the points of L from all the lines of $\mathbb{P}$. Since a projective plane of order n is equivalent to a Steiner system $\mathrm{S}(2, n+1, n^2+n+1)$, if follows that an affine plane of order n is equivalent to a Steiner system $\mathrm{S}(2, n, n^2)$ as will be proved in the next theorem.

Theorem 3.17. *A Steiner system* $\mathrm{S}(2, n+1, n^2+n+1)$ *exists if and only if a Steiner system* $\mathrm{S}(2, n, n^2)$ *exists.*

Proof. Let $\mathcal{S}$ be a Steiner system $\mathrm{S}(2, n+1, n^2+n+1)$. The number of blocks in this system, $\mathcal{S}$, is

$$\frac{\binom{n^2+n+1}{2}}{\binom{n+1}{2}} = n^2 + n + 1 \, .$$

Let $B = \{x_1, x_2, \ldots, x_{n+1}\}$ be a block in $\mathcal{S}$. The point x_i, $1 \leq i \leq n+1$, is paired with each one of the other n^2+n points in exactly one block. Since two blocks intersect in at most one point (as otherwise there will be a pair of points contained in two distinct blocks), it follows that x_i is contained in exactly $\frac{n^2+n}{n} = n+1$ blocks. Let $\mathcal{B}_\ell$, $1 \leq \ell \leq n+1$, be the set of $n+1$ blocks which contain x_ℓ. The pair $\{x_r, x_j\}$, $1 \leq r < j \leq n+1$ is contained in exactly one block which is B (since no pair is contained in more than one block). Hence, $\mathcal{B}_r \cap \mathcal{B}_j = \{B\}$ and points from $x_1, x_2, \ldots, x_{n+1}$ are contained in exactly $(n+1)n = n^2+n$ blocks in addition to B (since the additional n blocks of each one are different and the $\mathcal{B}_\ell$'s intersect only in B). Since the number of blocks in the system is n^2+n+1, it follows that these n^2+n blocks, which contain points from $\{x_1, x_2, \ldots, x_{n+1}\}$ and intersect B in one point, are all the blocks in $\mathcal{S}$ in addition to B. Since all these n^2+n blocks intersect B in one point and B was taken arbitrarily, it follows that any two blocks in $\mathcal{S}$ have a nonempty intersection. Now, let $\mathcal{S}'$ be the system constructed from $\mathcal{S}$ by removing the block B and the points $x_1, x_2, \ldots, x_{n+1}$ from all the blocks in $\mathcal{S}$. Since, all the blocks of $\mathcal{S}$, except for B, contain exactly one of the x_r's, $1 \leq r \leq n+1$, it follows that in $\mathcal{S}'$ the size of each block now is n. Since, $\mathcal{S}$ has n^2+n+1 points and $n+1$ points were removed from $\mathcal{S}$ to form $\mathcal{S}'$, it follows that in $\mathcal{S}'$ there are exactly n^2 points. Since each pair of elements appears in exactly one block of $\mathcal{S}$ and only the x_i's, and the block B which contains only these points, were removed from the system, it follows that each pair of the n^2 points appears in exactly one block of $\mathcal{S}'$ as in $\mathcal{S}$.

Thus, $\mathcal{S}'$ is a Steiner system $\text{S}(2, n, n^2)$.

Now, let $\mathcal{S}$ be a Steiner system $\text{S}(2, n, n^2)$. The number of blocks in this system, $\mathcal{S}$, is

$$\frac{\binom{n^2}{2}}{\binom{n}{2}} = n^2 + n \ .$$

Let $B = \{x_1, x_2, \ldots, x_n\}$ be a block in $\mathcal{S}$. The point x_i, $1 \leq i \leq n$, is also paired with any one of the other n^2-n points, not in B, in a block of $\mathcal{S}$. Since two blocks intersect in at most one point, it follows that x_i is contained in exactly $\frac{n^2-1}{n-1} = n+1$ blocks. Let $\mathcal{B}_\ell$, $1 \leq \ell \leq n$, be the set of blocks which contain x_ℓ. The pair $\{x_r, x_j\}$, $1 \leq r < j \leq n$, is contained in exactly one block which is B (since no pair appears in more than one block). Hence, $\mathcal{B}_\ell \cap \mathcal{B}_j = \{B\}$ and $x_1, x_2, \ldots, x_n$ are contained in exactly $n \cdot n = n^2$ blocks in addition to B (since any two such blocks which contain two different x_is are different). Therefore, there are $n^2+n-(n^2+1)=n-1$ blocks in $\mathcal{S}$ which do not intersect B.

Let Q be the set of points of $\mathcal{S}$. Since each point of $Q \setminus \{x_1, x_2, \ldots, x_n\}$ is contained with each point of $\{x_1, x_2, \ldots, x_n\}$ in exactly one block of $\mathcal{S}$, it follows that each point of $Q \setminus \{x_1, x_2, \ldots, x_n\}$ is contained the same number of times in the set $\{B' \, : \, B' \in \mathcal{B}_\ell, \; 1 \leq \ell \leq n\}$. In $\mathcal{S}$, each point of Q appears in the same number of blocks and hence the $n-1$ blocks which do not intersect B, and contain the n^2-n distinct points of $Q \setminus \{x_1, x_2, \ldots, x_n\}$, contain each point of $Q \setminus \{x_1, x_2, \ldots, x_n\}$ exactly once. This implies that these $n-1$ blocks are pairwise disjoint.

Let $\mathcal{P}_1$ be the set of blocks which consists of the block B and the $n-1$ pairwise disjoint blocks which do not intersect it. Since B was taken arbitrarily, it follows that $\mathcal{S}$ can be partitioned into $n+1$ pairwise nonintersecting sets $\mathcal{P}_i$, $1 \leq i \leq n+1$, such that each set contains n pairwise disjoint blocks of $\mathcal{S}$. Note, that each block of $\mathcal{P}_i$ cannot participate in another such set since this set is unique for each block as for B.

Define the system $\mathcal{S}'$ with the points $Q \cup \{1, 2, \ldots, n+1\}$,

$$\{\{i\} \cup B' \; : \; 1 \leq i \leq n+1, \; B' \in \mathcal{P}_i\} \cup \{1, 2, \ldots, n+1\} \; .$$

Since each $\mathcal{P}_i$, $1 \leq i \leq n+1$, contains each of the n^2 points of Q in exactly one block, it follows that each pair of points containing exactly one point from $\{1, 2, \ldots, n+1\}$ appears exactly once in $\mathcal{S}'$. The pairs from $[n+1]$ appear in the block $\{1, 2, \ldots, n+1\}$. The other pairs are exactly the ones which were in $\mathcal{S}$. Hence, $\mathcal{S}'$ is a Steiner system $\mathrm{S}(2, n+1, n^2+n+1)$. □

Theorem 3.3, Corollary 3.7, Theorems 3.7, 3.16, and 3.17 imply the following consequence.

Corollary 3.8. *The following structures are equivalent:*

(1) *A projective plane of order* n.
(2) *An affine plane of order* n.
(3) *A set of* $n-1$ *pairwise orthogonal Latin squares of order* n.
(4) *A Steiner system* $\mathrm{S}(2, n+1, n^2+n+1)$.
(5) *A Steiner system* $\mathrm{S}(2, n, n^2)$.
(6) *An orthogonal array* $\mathrm{OA}(2, n+1, n)$.
(7) *An* $(n+1, n^2, n)_n$ *code.*

The most simple projective plane of order q is $\mathrm{PG}(2, q)$, whose construction is implied by Theorem 3.9 and Theorem 3.16. For each prime p only one projective plane of order p is known, although there might be other non-isomorphic ones. There are orders, for higher powers of primes, for which there exist other nonisomorphic projective planes. This is in contrast to the finite projective geometry of dimension $m \geq 3$, where PG(m, q) is the unique finite projective geometry of dimension m. Moreover, there are no known projective planes of orders that are not powers of primes, and, for infinitely many values of such orders, their nonexistence was proved. Since n divides $n^2 + n$ and $\binom{n+1}{2}$ divides $\binom{n^2+n+1}{2}$, it follows that the necessary conditions of Corollary 3.1 are satisfied for the existence of a Steiner system S$(2, n+1, n^2+n+1)$. Thus, for such an order n where the existence of a projective plane was ruled out, there is no Steiner system S$(2, n+1, n^2+n+1)$. This implies that the necessary conditions of Corollary 3.1 for the existence of Steiner systems S$(2, k', n')$ are not sufficient in infinitely many cases. To conclude we present the two main open problems concerning projective planes.

Problem 3.11. Are there projective planes of order n, where n is not a power of a prime?

Problem 3.12. Are there nonisomorphic projective planes of order p, where p is a prime?

3.4 The Plotkin Bound and the Griesmer Bound

The sphere-packing bound and the code-anticode bound are two bounds on the size of codes that are attained by perfect codes and diameter perfect codes, respectively. The code-anticode bound is a generalization of the sphere-packing bound. Moreover, it will be shown in Section 4.3 that the code-anticode bound is also important in the context of the Singleton bound (Theorem 2.1). There are other important bounds on the size of codes that involve some of the already mentioned structures (e.g., anticodes and Hadamard matrices). Two such bounds, the Plotkin bound and the Griesmer bound, are discussed below.

The Plotkin Bound

By its nature, the sphere-packing bound is attained with equality for nontrivial perfect codes whose minimum distance is smaller than half of the possible maximum distance in the space with the given metric. For the Hamming metric, if the distance is larger than half of the length, then another bound is designed and it is attained by using codes from a construction whose main ingredients are Hadamard matrices.

Theorem 3.18. *If there exists a binary* (n, M, d) *code* C *for which* $n < 2d$, *then*

$$M \leq 2\left\lfloor \frac{d}{2d-n} \right\rfloor .$$

Proof. Calculate the sum

$$\sum_{x\in C}\sum_{y\in C} d(x,y) \tag{3.3}$$

in two ways. First, since $d(x,y) \geq d$ if $x \neq y$, the sum is equal to at least $M(M-1)d$. On the other hand, let A be the $M \times n$ matrix whose rows are the codewords. Suppose the i-th column of A contains η_i *zeroes* and $M - \eta_i$ *ones*. This column contributes $2\eta_i(M - \eta_i)$ to the sum in (3.3), so that the sum equals

$$\sum_{i=1}^{n} 2\eta_i(M-\eta_i) . \tag{3.4}$$

If M is even, this expression is maximized if for each i, $\eta_i = M/2$, and with this assignment, the sum in (3.4) equals $\frac{nM^2}{2}$. Thus, the sum in (3.3) equals at most $\frac{nM^2}{2}$ and we have

$$M(M-1)d \leq \frac{nM^2}{2},$$

which is equivalent to

$$M \leq \frac{2d}{2d-n} . \tag{3.5}$$

But since M is even, it follows that

$$M \leq 2\left\lfloor \frac{d}{2d-n} \right\rfloor .$$

On the other hand, if M is odd, the expression in (3.4) is maximized if $\eta_i = (M-1)/2$ or $\eta_i = (M+1)/2$, and the sum in (3.3) equals at most $\frac{n(M^2-1)}{2}$ and, instead of (3.5), we have that

$$M \leq \frac{n}{2d-n} = \frac{2d}{2d-n} - 1 .$$

Since $\lfloor 2x \rfloor \leq 2\lfloor x \rfloor + 1$, this implies that

$$M \leq \left\lfloor \frac{2d}{2d-n} \right\rfloor - 1 \leq 2\left\lfloor \frac{d}{2d-n} \right\rfloor .$$

□

The consequence of Theorem 3.18 is the Plotkin bound for all possible sets of parameters.

Corollary 3.9. *If d is even and $2d > n$, then*

$$A(n,d) \le 2\left\lfloor \frac{d}{2d-n} \right\rfloor , \tag{3.6}$$

and

$$A(2d,d) \le 4d . \tag{3.7}$$

If d is odd and $2d+1 > n$, then

$$A(n,d) \le 2\left\lfloor \frac{d+1}{2d+1-n} \right\rfloor , \tag{3.8}$$

and

$$A(2d+1,d) \le 4d+4 . \tag{3.9}$$

Proof. Equation (3.6) is proved in Theorem 3.18.

By Theorems 2.4 and 3.18, we have that

$$A(2d,d) \le 2A(2d-1,d) \le 2\left\lfloor \frac{d}{2d-(2d-1)} \right\rfloor = 4d$$

and (3.7) is proved.

If d is odd, then by Theorems 2.3 and 3.18, we have that

$$A(n,d) = A(n+1,d+1) \le 2\left\lfloor \frac{d+1}{2(d+1)-(n+1)} \right\rfloor = \left\lfloor \frac{d+1}{2d+1-n} \right\rfloor$$

and (3.8) is proved.

Finally, Theorem 2.3 and (3.7) imply that

$$A(4\delta+3, 2\delta+1) = A(4\delta+4, 2\delta+2) \le 8\delta+8$$

and (3.9) is proved. □

Let $\mathcal{C}_1$ be a binary (n_1, M_1, d_1) code and let $\mathcal{C}_2$ be a binary (n_2, M_2, d_2) code. Assume further that the M_1 codewords of $\mathcal{C}_1$ are ordered by $\alpha_1, \alpha_2, \ldots, \alpha_{M_1}$ and the M_2 codewords of $\mathcal{C}_2$ are ordered by $\beta_1, \beta_2, \ldots, \beta_{M_2}$.

The (s,t)-***concatenated code*** $\mathcal{C}$, of $\mathcal{C}_1$ and $\mathcal{C}_2$, denoted by $s\mathcal{C}_1 \odot t\mathcal{C}_2$, is defined as follows

$$s\mathcal{C}_1 \odot t\mathcal{C}_2 \triangleq \{(\overbrace{\alpha_i \cdots \alpha_i}^{s \text{ times}} \overbrace{\beta_i \cdots \beta_i}^{t \text{ times}}) \,:\, \alpha_i \in \mathcal{C}_1,\ \beta_i \in \mathcal{C}_2,\ 1 \le i \le \min\{M_1, M_2\}\}.$$

With the same notation the s-***concatenated code*** of $\mathcal{C}_1$, denoted by $s\mathcal{C}_1$, is defined by

$$s\mathcal{C}_1 \triangleq \{(\overbrace{\alpha_i\alpha_i\cdots\alpha_i}^{s\text{ times}}) \ : \ \alpha_i \in \mathcal{C}_1, \ 1 \le i \le M_1\}.$$

Similarly, we can define a concatenation of more than two codes, where each one is concatenated a few times.

It is easy to verify the parameters of the (s,t)-concatenated code $s\mathcal{C}_1 \odot t\mathcal{C}_2$, presented in the following result.

Lemma 3.8. *The (s,t)-concatenated code $s\mathcal{C}_1 \odot t\mathcal{C}_2$ is a code of length $s \cdot n_1 + t \cdot n_2$, minimum Hamming distance $s \cdot d_1 + t \cdot d_2$, with $\min\{M_1, M_2\}$ codewords, i.e., it is an $(s \cdot n_1 + t \cdot n_2, \min\{M_1, M_2\}, s \cdot d_1 + t \cdot d_2)$ code.*

Theorem 3.19. *Provided that enough Hadamard matrices exist, there are codes that attain the Plotkin bound with equality, i.e., meet the bounds in Corollary 3.9.*

Proof. If d is even and $d \le n < 2d$, define $\kappa = \left\lfloor \frac{d}{2d-n} \right\rfloor$, and

$$s = d(2\kappa + 1) - n(\kappa + 1), \quad t = \kappa n - d(2\kappa - 1). \tag{3.10}$$

Clearly, s and t are nonnegative integers and

$$n = (2\kappa - 1)s + (2\kappa + 1)t, \quad d = \kappa s + (\kappa + 1)t \ .$$

If n is even, then by (3.10), s and t are also even. If n is odd and κ is even, then t is even. If n is odd and κ is odd, then s is even. Let

$$\mathcal{C} \triangleq \frac{s}{2}\Phi_{4\kappa} \odot \frac{t}{2}\Phi_{4\kappa+4}, \ \text{ if } n \text{ is even}$$

$$\mathcal{C} \triangleq s\Upsilon_{2\kappa} \odot \frac{t}{2}\Phi_{4\kappa+4}, \ \text{ if } n \text{ is odd, and } \kappa \text{ is even}$$

$$\mathcal{C} \triangleq \frac{s}{2}\Phi_{4\kappa} \odot t\Upsilon_{2\kappa+2}, \ \text{ if } n \text{ is odd, and } \kappa \text{ is odd.}$$

By Lemma 3.8, we have that $\mathcal{C}$ has length n, minimum distance d, and $\mathcal{C}$ contains $2\kappa = 2\left\lfloor \frac{d}{2d-n} \right\rfloor$ codewords, and, hence, $\mathcal{C}$ attains (3.6) with equality.

The Hadamard code Λ_{2n} is a $(2n, 4n, n)$ code that attains (3.7) with equality.

Finally, codes that attain (3.8) and (3.9), respectively, for odd d, are derived by any puncturing of the codes, with distance $d + 1$, which attain (3.6) and (3.7), respectively, with equality. For this part of the proof Theorem 2.3 has to be used. □

Note that to have codes that attain the Plotkin bound with equality for all parameters, certain Hadamard matrices are required. For (3.7), a Hadamard matrix of order $2d$ is required. To attain (3.6), Hadamard matrices of orders 4κ and $4\kappa + 4$, as defined in the proof of Theorem 3.19, are required. Moreover, to attain (3.6), a Hadamard matrix of order 2κ is also required if κ is even, and a Hadamard matrix of order $2\kappa + 2$ is also required if κ is odd. These Hadamard matrices are taken from the sets of such matrices which were constructed during the years.

The Griesmer Bound

Lemma 3.6 and Theorems 3.5 and 3.6 imply that an $[n, k, n-k+1]_q$ MDS code, which obviously meets the Singleton bound, can exist only if n is relatively very small compared to q. If Conjecture 3.1 is true, then n is at most $q+2$ in the best case. Hence, it is desirable to find better bounds on n, especially for a small alphabet size. The next bound resolves this problem for binary linear codes. Let $N(k, d)$ be the length of the shortest binary linear code of dimension k and minimum Hamming distance d.

Theorem 3.20.

$$N(k, d) \geq d + N\left(k-1, \left\lceil \frac{d}{2} \right\rceil\right) .$$

Proof. Let $\mathcal{C}$ be an $[N(k, d), k, d]$ code, with a $k \times N(k, d)$ generator matrix G. W.l.o.g.

$$G = \left[\begin{array}{c|c} 0 \cdots 0 & 1 \cdots 1 \\ \hline & \\ G_1 & \text{X} \\ & \end{array}\right],$$

where the first row of G has weight d. Consider the $(k-1) \times (N(k, d) - d)$ matrix G_1. We claim that the rank of G_1 is $k-1$. Assume the contrary, that the rank of G_1 is less than $k-1$. This implies that by using a linear combination of the last $k-1$ rows of G, we can form another generator matrix G' for $\mathcal{C}$ with the same first row as in G and whose second row starts with $N(k, d) - d$ *zeroes*. Adding the first two rows of G' implies a codeword of weight less than d, a contradiction. Therefore, G_1 is a generator matrix of an $[N(k, d) - d, k-1, d_1]$ code. Let $(x, y) \in \mathcal{C}$, where $\text{wt}(x) = d_1$ and $y \in \mathbb{F}_2^d$. Clearly, by adding (x, y) to the first row of G we have that $(x, \bar{y}) \in \mathcal{C}$, and hence

$$d_1 + \text{wt}(y) \geq d,$$

$$d_1 + d - \mathrm{wt}(y) \geq d.$$

Adding these two equations implies that $2d_1 \geq d$, i.e., $d_1 \geq \lceil d/2 \rceil$. Therefore,

$$N(k-1, \lceil d/2 \rceil) \leq N(k,d) - d \ ,$$

which completes the proof. □

Theorem 3.21.

$$N(k,d) \geq \sum_{i=0}^{k-1} \left\lceil \frac{d}{2^i} \right\rceil \ .$$

Proof. By applying Theorem 3.20 iteratively, we have after the second iteration that

$$N(k,d) \geq d + N\left(k-1, \left\lceil \frac{d}{2} \right\rceil\right) \geq d + \left\lceil \frac{d}{2} \right\rceil + N\left(k-2, \left\lceil \frac{d}{4} \right\rceil\right) \ .$$

After k iterations, we have that

$$N(k,d) \geq \sum_{i=0}^{k-2} \left\lceil \frac{d}{2^i} \right\rceil + N\left(1, \left\lceil \frac{d}{2^{k-1}} \right\rceil\right) = \sum_{i=0}^{k-1} \left\lceil \frac{d}{2^i} \right\rceil \ .$$

□

Let S_k be the code whose $k \times (2^k - 1)$ generator matrix $G(S_k)$ consists of all nonzero binary column vectors of length k.

Theorem 3.22. *For each $k \geq 2$, the code S_k is a $[2^k - 1, k, 2^{k-1}]$ code.*

Proof. The columns of $G(S_k)$ consist of all nonzero binary k-tuples and hence each row has weight 2^{k-1}. Any set of k nontrivial linear combinations of rows from $G(S_k)$, which are linearly independent, also contains each nonzero binary k-tuple as a column of a related $k \times (2^k - 1)$ generator matrix. Hence, the weight of each such row is also 2^{k-1}. Thus, the weight of a nonzero codeword of S_k is 2^{k-1} and hence S_k is a $[2^k - 1, k, 2^{k-1}]$ code. □

The code S_k is called the ***simplex code*** of order k. By appending a *zero* to each codeword of S_k, the resulting codewords form a binary Hadamard matrix of order 2^k, which is isomorphic to the one obtained by applying iteratively the Sylvester construction from the Hadamard matrix of order 1. Let $G(S_k, r)$ be the generator matrix of the code obtained by concatenating horizontally r copies of $G(S_k)$, i.e., generator matrix of the r-concatenated code rS_k. Using the same arguments as in Lemma 3.8, the following parameters of the code, obtained from $G(S_k, r)$, are derived.

Theorem 3.23. *The code $\mathcal{C}$, whose generator matrix is $G(S_k, r)$, is an $[r(2^k - 1), k, r \cdot 2^{k-1}]$ code.*

Theorem 3.24. *For each $r \geq 1$, the $[r(2^k - 1), k, r \cdot 2^{k-1}]$ code obtained from the generator matrix $G(S_k, r)$ meets the Griesmer bound.*

Proof. By Theorem 3.21 we have that

$$N(k, r \cdot 2^{k-1}) \geq \sum_{i=0}^{k-1} \frac{r \cdot 2^{k-1}}{2^i} = \sum_{i=0}^{k-1} (r \cdot 2^{k-i-1}) = r \sum_{i=0}^{k-1} 2^i = r(2^k - 1) ,$$

which completes the proof. □

Let $G(\mathcal{A})$ be a binary $k \times m$ generator matrix of an anticode $\mathcal{A}$. Assume further that $G(\mathcal{A})$ can have linearly dependent rows, i.e., the dimension of the anticode can be less than k. All the 2^k linear combinations of the k rows of $G(\mathcal{A})$ form a linear anticode of length m whose diameter δ is the maximum weight of an anticodeword in $\mathcal{A}$. If the dimension of the code is less than k some words are contained more than once as anticodewords in $\mathcal{A}$. If rank $G(\mathcal{A}) = \rho \leq k$, then each anticodeword of $\mathcal{A}$ is contained $2^{k-\rho}$ times in $\mathcal{A}$. If rank $G(\mathcal{A}) = k$, then clearly all the 2^k anticodewords of $\mathcal{A}$ are distinct.

Lemma 3.9. *Let $G(S_k)$ be the generator matrix of the simplex code S_k and $G(\mathcal{A})$ be the $k \times m$ generator matrix, with m distinct columns, of an anticode $\mathcal{A}$ whose diameter is δ. By removing the columns of the matrix $G(\mathcal{A})$ from the columns of the matrix $G(S_k)$, we obtain a $k \times (2^k - 1 - m)$ generator matrix of a $[2^k - 1 - m, \kappa, 2^{k-1} - \delta]$ code $\mathcal{C}$, where $\kappa \leq k$.*

Proof. The length and dimension of $\mathcal{C}$ are readily verified from its definition. As for the minimum distance of the code, note that as mentioned before, each nonzero codeword of S_k has weight 2^{k-1}. From at least one of these codewords, say c, δ *ones* are contained in the columns projected by $\mathcal{A}$. The codeword c is replaced by c' in $\mathcal{C}$ after the coordinates projected by $\mathcal{A}$ are removed and hence c' has weight $2^{k-1} - \delta$. If $\mathcal{C}$ has a codeword whose weight is smaller than $2^{k-1} - \delta$, then it implies that $\mathcal{A}$ must have an anticodeword whose weight is larger than δ, a contradiction to the diameter of $\mathcal{A}$. This completes the proof. □

The construction of the code $\mathcal{C}$ in Lemma 3.9 can be generalized for the $[r(2^k-1), k, r \cdot 2^{k-1}]$ code of Theorem 3.23. The parameters of the constructed code are given in the following lemma whose proof is a combination of the proofs in Lemma 3.9 and Theorem 3.23.

Lemma 3.10. *Let $G(\mathcal{A})$ be the $k \times m$ generator matrix of an anticode $\mathcal{A}$, where each k-tuple appears at most r times as a column in $G(\mathcal{A})$. Assume further that the diameter of $\mathcal{A}$ is δ. By removing the columns of $G(\mathcal{A})$ from the columns of $G(\mathcal{S}_k, r)$ (a column vector v which appears in $G(\mathcal{A})$ ρ times, where $\rho < r$, is removed from ρ arbitrary entries of $G(\mathcal{S}_k, r)$ that contain the column vector v), we obtain a $k \times (2^k - 1 - m)$ generator matrix of an $[r(2^k-1) - m, \kappa, r \cdot 2^{k-1} - \delta]$ code $\mathcal{C}$, where $\kappa \leq k$.*

We continue with a specific family of anticodes to be used in Lemma 3.10. Let $\mathcal{A}(k,m)$ be a $[\sum_{j=1}^{m}(2^{\ell_j}-1), k, \sum_{j=1}^{m} 2^{\ell_j - 1}]$ code (used as an anticode), where $1 \leq \ell_j < \ell_{j+1} < k$, $1 \leq j \leq m-1$, whose generator matrix is $G(k,m)$. Such a generator matrix $G(k,m)$ of $\mathcal{A}(k,m)$ is formed by a concatenation of m generator matrices of simplex codes of m distinct orders, $\ell_1, \ell_2, \ldots, \ell_m$. Since each one of these m simplex codes has a dimension less than k, it follows that there are repeated anticodewords in $\mathcal{A}(k,m)$, where redundant rows are added to have k rows in each generator matrix. Note that $G(k,m)$ is not unique for a given choice of $\ell_1, \ell_2, \ldots, \ell_m$, since the order of the rows in each one of the m distinct generator matrices of the distinct simplex codes might be different and also the redundant rows can be taken in different ways. This concatenation is represented by a $k \times \sum_{j=1}^{m}(2^{\ell_j}-1)$ matrix, where no vector column, of length k, is contained more than s times in $G(k,m)$. It is obvious now that the concatenation of the m generator matrices can yield different anticodes. Note further that if $\ell_j < k$, then each row is contained exactly $2^{k-\ell_j}$ times in the generator matrix of the related simplex code $\mathcal{S}_{\ell_j}$. Nevertheless, this does not imply the same property for $\mathcal{A}(k,m)$.

Theorem 3.25. *Let $\mathcal{C}$ be the code whose generator matrix is $G(\mathcal{S}_k, r)$ from which a generator matrix $G(k,m)$ of $\mathcal{A}(k,m)$ was removed. If $r \geq 2$, then $\mathcal{C}$ is an $[r(2^k-1) - \sum_{j=1}^{m}(2^{\ell_j}-1), k, r \cdot 2^{k-1} - \sum_{j=1}^{m} 2^{\ell_j-1}]$ code, which meets the Griesmer bound.*

Proof. The length of $\mathcal{C}$ and its minimum distance are trivial observations from the definition of $\mathcal{C}$ as in Lemma 3.10. Now, note that since $1 \leq \ell_j < \ell_{j+1}$ for $1 \leq j \leq m-1$, it follows that $m < k$. Since all the ℓ_j's are distinct and less than $k < m$, it follows that $\sum_{j=1}^{m}(2^{\ell_j}-1) \leq 2^k - 1$.

Moreover, it also implies that from one copy of $G(S_k)$ at least 2^{k-1} distinct columns remain in the generator matrix of $\mathcal{C}$, and hence the dimension of $\mathcal{C}$ is equal to the dimension of S_k, i.e., to k. To complete the proof, it is sufficient to show that $\mathcal{C}$ meets the Griesmer bound. By Theorem 3.21, we have that

$$N(k, r \cdot 2^{k-1} - \sum_{j=1}^{m} 2^{\ell_j - 1}) \geq \sum_{i=0}^{k-1} \left\lceil \frac{r \cdot 2^{k-1} - \sum_{j=1}^{m} 2^{\ell_j - 1}}{2^i} \right\rceil$$

$$= \sum_{i=0}^{k-1} (r \cdot 2^{k-i-1}) - \sum_{j=1}^{m} \sum_{i=0}^{\ell_j - 1} 2^{\ell_j - i - 1} = r(2^k - 1) - \sum_{j=1}^{m} (2^{\ell_j} - 1) ,$$

where one should note in the computation that if $\ell_j - 1 < i$ for $1 \leq j \leq \eta$ and all the ℓ_j's are distinct, then

$$\left\lceil \frac{\beta \cdot 2^{\nu} - \sum_{j=1}^{\eta} 2^{\ell_j - 1}}{2^i} \right\rceil = \beta \cdot 2^{\nu - i} ,$$

where β is a positive integer and $\nu \geq i$. □

Example 3.3. Let $k = 4$, $r = 4$, $m = 3$, $\ell_1 = 1$, $\ell_2 = 2$, and $\ell_3 = 3$, be the parameters in Theorem 3.25.

The anticode with these parameters is not unique. Let two such anticodes be $\mathcal{A}_1(4, 3)$ and $\mathcal{A}_2(4, 3)$ generated by the generator matrices $G(\mathcal{A}_1(4, 3))$ and $G(\mathcal{A}_2(4, 3))$, respectively.

$$G(\mathcal{A}_1(4,3)) = \begin{bmatrix} 1\,0\,1\,1\,0\,0\,0\,1\,1\,1\,1 \\ 0\,1\,0\,1\,0\,1\,1\,0\,0\,1\,1 \\ 0\,1\,1\,0\,1\,0\,1\,0\,1\,0\,1 \\ 1\,0\,0\,0\,1\,0\,1\,0\,1\,0\,1 \end{bmatrix}$$

and

$$G(\mathcal{A}_2(4,3)) = \begin{bmatrix} 1\,0\,1\,1\,0\,0\,0\,1\,1\,1\,1 \\ 0\,1\,0\,1\,0\,1\,1\,0\,0\,1\,1 \\ 0\,0\,1\,1\,1\,0\,1\,0\,1\,0\,1 \\ 0\,1\,0\,1\,1\,1\,0\,1\,0\,0\,1 \end{bmatrix} .$$

When we remove the columns of $G(\mathcal{A}_1(4, 3))$ from the columns of the 4×60 generator matrix $G(S_4, 4)$, we obtain the following 4×49 generator matrix of a $[49, 4, 25]$ code:

$$\begin{bmatrix} 00011000000011111110000000111111100000001111111 \\ 0011100011110000111000111100001111000111100001111 \\ 0100101100110011011011001100110011011001100110011 \\ 1011010101010101101101010101010101101010101010101 \end{bmatrix}.$$

When we delete the columns of $G(\mathcal{A}_2(4,3))$ from the columns of the 4×60 generator matrix $G(S_4, 4)$, we obtain the following 4×49 generator matrix of a $[49, 4, 25]$ code:

$$\begin{bmatrix} 0000111000000011111110000000111111100000001111111 \\ 0011011000111000111000111100001111000111100001111 \\ 0101101011011001001011001100110011011001100110011 \\ 1001110101001011010101010101010101101010101010101 \end{bmatrix}.$$

It should be noted that the construction implied by Theorem 3.25 can be further generalized in many different ways to obtain many more families of codes that meet the Griesmer bound.

3.5 Association Schemes

A $(\mathcal{V}, \mathcal{R})$ ***association scheme*** with n ***classes*** consists of a finite set $\mathcal{V}$ with t points, and a set $\mathcal{R}$ with $n+1$ relations, $\mathcal{R} = \{\mathcal{R}_0, \mathcal{R}_1, \dots, \mathcal{R}_n\}$, defined on $\mathcal{V}$. These relations satisfy the following properties:

- Each $\mathcal{R}_i$ is symmetric, i.e., $(x, y) \in \mathcal{R}_i$ implies $(y, x) \in \mathcal{R}_i$.
- For every $x, y \in \mathcal{V}$, $(x, y) \in \mathcal{R}_i$ for exactly one i.
- $\mathcal{R}_0 = \{(x, x) \ : \ x \in \mathcal{V}\}$ is the identity relation.
- If $(x, y) \in \mathcal{R}_k$, then the number of elements $z \in \mathcal{V}$ such that $(x, z) \in \mathcal{R}_i$ and $(y, z) \in \mathcal{R}_j$ is a constant $p_{i,j}^k$ (called ***the intersection number***) that depends on i, j, k but not on the particular choice of x and y.

Let Γ be a connected graph with v vertices, with no loops or multiple edges, and let $\mathcal{V}$ be the set of vertices in Γ. The maximum distance, say n, between any two vertices in Γ is called the ***diameter*** of the graph. The graph Γ is called ***distance-regular*** (or ***metrically-regular*** or ***perfectly-regular***) if, for any $x, y \in \mathcal{V}$ with $d(x, y) = k$, the number of vertices $z \in \mathcal{V}$ such that $d(x, z) = i$ and $d(y, z) = j$ is a constant $p_{i,j}^k$ independent of the choice of x and y. Clearly, from such a distance-regular graph whose diameter is n, we obtain an association scheme with n classes. This scheme is called a ***metric (association) scheme***.

We also denote $p^0_{i,i} = t_i$ and $|\mathcal{V}| = t$; t_i is called the ***valency*** of $\mathcal{R}_i$ and it is the number of vertices (points) in $\mathcal{V}$ at distance i from any point $x \in \mathcal{V}$. The degree of a vertex in the graph Γ is t_1. It is easy to verify, that for any association scheme, the following conditions hold:

$$p^i_{i,0} = 1, \; p^j_{i,0} = 0 \;\text{ and }\; p^0_{i,j} = 0 \text{ for } i \neq j.$$

$$\sum_{j=0}^{n} p^k_{i,j} = t_i \;\text{ and }\; p^k_{i,j} t_k = p^i_{k,j} t_i.$$

Let $(\mathcal{V}, \mathcal{R})$ be a metric association scheme with a distance function d defined on a set $\mathcal{V}$, and a set $\mathcal{R}$ with $n+1$ relations, i.e., $\mathcal{R} = \{\mathcal{R}_0, \mathcal{R}_1, \ldots, \mathcal{R}_n\}$, where $\mathcal{R}_i = \{(x,y) \,:\, x, y \in \mathcal{V},\; d(x,y) = i\}$. By the triangle inequality, we have that

$$p^k_{i,j} = 0, \quad \text{if } i+j<k \text{ or } i+k<j \text{ or } j+k<i.$$

The most important scheme is the Hamming scheme. Other important schemes are the Johnson scheme, the Grassmann scheme, and the bilinear forms scheme. These four schemes will be considered in our exposition. In all schemes, the relations are defined by the distance d, i.e.,

$$\mathcal{R}_i \triangleq \{(x,y) \,:\, x, y \in \mathcal{V},\; d(x,y) = i\} \,.$$

In the following chapters we will discuss perfect codes in the Hamming scheme $\mathcal{H}_q(n)$, which consists of vectors of length n over the finite field $\mathbb{F}_q$. The scheme can be further generalized to $\mathcal{H}_m(n)$, for any integer $m \geq 2$, which consists of vectors of length n over an alphabet with m symbols. The metric used in this scheme is the Hamming distance. One can easily verify that $\mathcal{H}_m(n)$ with the Hamming distance is indeed an association scheme. When $j \geq i$ and $m = 2$, if $i+j-k$ is odd, then we have that $p^k_{i,j} = 0$; when $j \geq i$ and $m = 2$, if $i+j-k$ is even, then

$$p^k_{i,j} = \binom{k}{(i+j-k)/2} \binom{n-k}{(j-i+k)/2} \,.$$

When $m > 2$, the computation of the intersection numbers are left as an exercise. These numbers are important and sometimes are used to prove some properties of perfect codes and to prove bounds on the sizes of codes, but this will be beyond our discussions in this book. As for other metrics, the question whether they form an association scheme will be discussed in the related chapters.

3.6 Notes

Block designs and Steiner systems are covered in many books. We mention only two such books. The first is [Beth, Jungkickel, and Lenz (1999)] and the second important good reference, is the Handbook of Combinatorial Designs [Colbourn and Dinitz (2007)]. The latter covers all types of combinatorial designs as well as orthogonal designs.

Section 3.1. It was proved by [Keevash (2014)] that for fixed (t, k), $0 < t < k$, if n is large enough, the necessary conditions of Corollary 3.1 for the existence of a Steiner system $S(t, k, n)$ are also sufficient. Unfortunately, this n is extremely large and it is beyond our imagination. The proof is based on probabilistic arguments and it is not easy to follow. Another (simpler) probabilistic proof, but still quite complicated, was provided by [Glock, Kühn, Lo, and Osthus (2016)].

Problem 3.13. Give a simpler proof for the existence of $S(t, k, n)$ for any given pair (t, k), $0 < t < k$, if n is large enough and the necessary conditions of Corollary 3.1 are satisfied. Give a proof for the existence of $S(t, k, n)$ for any given pair (t, k), $0 < t < k$, if n is large enough and the necessary conditions of Corollary 3.1 are satisfied, but n is smaller than in the known proofs.

The proof for the existence of Steiner quadruple systems $S(3, 4, n)$ if and only if $n \equiv 2$ or $4 \pmod 6$ was provided in [Hanani (1960)]. The necessary conditions for $S(2, k, n)$ and their existence with a finite number of exceptions for each k were proved in [Wilson (1972a,b, 1975)].

For large sets, a reference book on one-factorizations is [Wallis (1997)]. Large sets for $S(1, k, n)$, whenever k divides n are equivalent to one-factorizations of the complete k-uniform hypergraph. The existence of such one-factorizations was proved in [Baranyai (1975)]. Most existence proofs for large sets of Steiner triple systems, $S(2, 3, n)$ were worked out by [Lu (1983, 1984)], with the last six cases proved in [Teirlinck (1991)]. A shorter proof for this existence problem was provided by [Ji (2005)]. A partition of the 4-subsets of a 13-set into disjoint Steiner system $S(2, 4, 13)$ was done in [Chouinard (1983)]. It was proved in [Keevash (2018)] that if n is large enough, then the necessary conditions of Corollary 3.1 for the existence of a Steiner system $S(t, k, n)$ are also sufficient for the existence of a large set of $S(t, k, n)$, for any given pair (t, k), $0 < t < k$. As expected, this large n is much larger than the one required for the existence of the related

Steiner system, and hence the proof is useless for any practical purpose. Finally, [Etzion and Hartman (1991)] provide a construction of $2^m - 5$ pairwise disjoint Steiner quadruple systems $S(3, 4, 2^m)$, where $2^m - 4$ such Steiner systems imply the existence of a large set. A set of distinct Steiner systems $S(3, 4, 2^m)$, $m \geq 3$, in which each quadruple is contained in exactly two of the systems was constructed in [Etzion (1996b)]. Generalization for this result was presented in [Etzion and Zhou (2021)]. An improvement for these results implies a solution to the following open problem.

Problem 3.14. Construct a large set of Steiner systems $S(3, 4, n)$ for some values of n.

Generalized Steiner systems were introduced in [Etzion (1997)] who gave the first construction of such designs. The necessary conditions for the existence of generalized Steiner systems $GS(2, 3, n, 3)$ are identical to the necessary conditions for the existence of group divisible designs [Hanani (1975)]. The distinction between the two structures is that in group divisible designs there is no requirement for a minimum distance. Therefore, these two structures differ in their constructions. More precisely, any construction for generalized Steiner systems $GS(2, 3, n, 3)$ is also a construction for a group divisible design, but not the reverse. For example it was proved in [Etzion (1997)] that the generalized Steiner system $GS(2, 3, n, 3)$ exists if and only if $n \equiv 0$ or $1 \pmod 3$, where $n \geq 4$. This work has motivated lot of further research, e.g., [Phelps and Yin (1997); Svanström (1999a); Chen, Ge, and Zhu (1999); Wilson and Phelps (1999); Chen, Ge, and Zhu (2000); Ge (2000); Wu, Ge and Zhu (2001); Wu and Zhu (2001); Ge (2002); Ji, Wu and Zhu (2005); Cao, Ji, and Zhu (2007); Chee and Ling (2007); Chee, Dau, Ling, and Ling (2008); Wu and Fan (2009); Zhang, Zhang and Ge (2012); Zhang and Ge (2013); Chee, Ge, Zhang and Zhang (2015)]. As an example, in [Chee, Dau, Ling, and Ling (2010)] the following theorem was proved.

Theorem 3.26. *For all sufficiently large n satisfying that w divides $n(q-1)$, there exists a generalized Steiner system* $GS(1, k, n, q)$.

Steiner system have an important role in analyzing binary constant-weight codes which are discussed in Chapter 8. Generalized Steiner system have similar role in nonbinary constant-weight codes which are discussed in Chapter 9.

Section 3.2. A good book for all orthogonal designs including Hadamard matrices is [Geramita and Seberry (1979)]. Orthogonal arrays and their applications are covered in another book [Hedayat, Sloane, and Stufken (1999)]. Latin squares and their applications can be found in [Keedwell and Dénes (2015)] and another important book on orthogonal designs is [Raghavarao (1971)].

It is conjectured that except for a set of $n-1$ pairwise orthogonal Latin squares for a prime power n, whose construction was presented in this section of the chapter, there are no other values for which such a set exists. [Euler (1782)] conjectured that when $n \equiv 2 \pmod 4$, there are no such pairs of orthogonal Latin squares of order n. The conjecture has been proven to be true for $n = 6$ by [Tarry (1900, 1901)] while a shorter proof was given by [Stinson (1984)], i.e.,

Theorem 3.27. *There is no pair of orthogonal Latin squares of order 6.*

Nevertheless, the conjecture was found to be false for any other order different from 6 [Bose, Shrikhande, and Parker (1960)]. The most remarkable result in this direction is the Bruck-Ryser Theorem [Bruck and Ryser (1949)].

Theorem 3.28. *If $n \equiv 1$ or $2 \pmod 4$, then a necessary condition for the existence of a set with $n-1$ pairwise orthogonal Latin squares of order n is that there exist two integers x and y such that $n = x^2 + y^2$.*

Theorem 3.28 excludes many values, e.g., there is no set with $n-1$ pairwise orthogonal Latin squares of order n, for $n = 14, 21, 22, 30, 33, 38, 42, 46,$ and so on.

Lemma 3.6, Theorem 3.5, and Theorem 3.6 were proved in [Bush (1952)]. The MDS conjecture (Conjecture 3.1) was stated in [Segre (1955)] has been considered in many papers. Some remarkable results in this direction that were obtained in [Ball (2012)] and in [Ball and de Beule (2012)]. For example [Ball (2012)] proved that the conjecture holds for q which is a prime number. Linear codes (MDS codes) that meet the bounds given in the conjecture were constructed for all parameters. They can be found in most books on coding theory mentioned above. In other words, we have the following theorem.

Theorem 3.29. *If $d \geq 3$, then there exists an $[n,k,d]_q$ MDS code when $n \leq q+1$ for all prime power q and $2 \leq k \leq q-1$. If q is a power of 2 and $k \in \{3, q-1\}$, then there exists an $[n,k,d]_q$ MDS code when $n \leq q+2$.*

Parameters of known orthogonal arrays for an alphabet that is not a power of a prime can be found in the books on orthogonal arrays [Geramita and Seberry (1979); Hedayat, Sloane, and Stufken (1999); Raghavarao (1971)]. The nonexistence results by [Ball (2012); Ball and de Beule (2012)] were obtained using projective geometry, and are applied only for linear codes. There are no related results for nonlinear orthogonal arrays of index unity.

Problem 3.15. Find techniques to prove the nonexistence of orthogonal arrays of index unity where the alphabet size is not a power of a prime. In particular, find techniques to prove the nonexistence of orthogonal arrays of index unity for the same parameters where the nonexistence proof of MDS codes was presented by techniques from projective geometry.

Hadamard matrices were defined in [Hadamard (1893)], where it was proved that any real $n \times n$ matrix A, with real entries between -1 and $+1$, satisfies $|\det A| \leq n^{n/2}$. Hadamard matrices meet this bound. These matrices have many applications in various areas of communication, information theory, and computer science. They are part of a family of matrices called weighing matrices. Coding with Hadamard matrices and related weighing matrices were done, for example, by [Etzion, Vardy, and Yaakobi (2013)]. The current order, which is divisible by 4 (as of 2021), for which no Hadamard matrix is known, is 668 after the previous order of 428 was constructed in [Kharaghani and Tayfeh-Rezaie (2004)].

Problem 3.16. Prove that for every positive integer$n$ divisible by 4 there exists a Hadamard matrix of order n.

Hadamard matrices can be connected to other types of designs. For example, we can present the family of difference sets.

An (n, k, λ) ***difference set***, $D = \{d_1, d_2, \ldots, d_k\}$, is a collection of k residues modulo the positive integer n such that for each residue $\alpha \not\equiv 0 \pmod{n}$, the equation

$$d_j - d_i \equiv \alpha \pmod{n}$$

has exactly λ solutions. If $\lambda = 1$, one can use the cyclic shifts of D to form a cyclic Steiner system $\mathrm{S}(2, k, n)$. The following theorem is proved by using observations from the definition and some algebraic manipulations.

Theorem 3.30. *If D is an (n, k, λ) difference set, then we have that $4m - 1 \leq n \leq m^2 + m + 1$, where $m = k - \lambda$.*

The two extremes in the bounds of Theorem 3.30 are related to some important difference sets. If $n = m^2 + m + 1$, then the difference set is equivalent to a cyclic Steiner system $\mathrm{S}(2, m+1, m^2+m+1)$, i.e., a projective plane of order m. If $n = 4m - 1$, then the difference set is sometimes called a ***Hadamard difference set*** (there are other difference sets called Hadamard differences sets (see [Jungnickel (1992)]). This family of difference sets has parameters $(4m-1, 2m-1, m-1)$ and there are three types of such difference sets.

(1) $n = 2^t - 1$ and this type is related to M-sequences.
(2) $n = 4t - 1$ is a prime and this type is related to quadratic residues.
(3) $n = p(p+2)$, where p and $p+2$ are primes and this type is related to twin primes.

Given such a difference set, it is quite straightforward to generate a related binary Hadamard matrix by taking the cyclic shifts of all the characteristic vectors of the difference set, and adding a parity-check symbol to each such characteristic vector. To these rows in the matrix we add the all-zero vector. The obtained matrix is a binary Hadamard matrix over the alphabet $\{0, 1\}$ from which a Hadamard matrix over $\{-1, +1\}$ is easily obtained. M-sequences of length $2^t - 1$ are very important in communication and coding as well as in other areas. The collection of the cyclic shifts of one such sequence with the all-zero word form an isomorphic structure to the finite field $\mathbb{F}_{2^t}$. Moreover, these $2^t - 1$ cyclic shifts together with the all-zero sequence form the simplex code S_t. An excellent book on these sequences and related shift-register sequences is [Golomb (2017)]. Shift-register sequences and digital sequences are highly important in many modern communications applications. In particular, autocorrelation and cross-correlation properties of sequences are related to difference sets. The Hadamard matrices constructed via difference sets can be used in the Sylvester's construction and in the Kronecker product construction to form Hadamard matrices for many infinite families of parameters. For more information on difference sets we direct the reader to the survey in [Jungnickel (1992)] and the books of [Baumert (1971)] and [Ding (2014)].

The concept of difference sets was generalized to difference families, where there are several sets $D_1, D_2, \ldots$, each of size k, where each residue modulo n occurs as a difference in exactly λ of these sets. If $\lambda = 1$, then such a difference family forms a cyclic Steiner system $\mathrm{S}(2, k, n)$. Constructions and bounds for such family of designs can be found, for example, in [Wilson (1972c); Buratti (1993); Bitan and Etzion (1995)]. This family of designs is also related to a family of codes called optical orthogonal codes or constant-weight cyclically permutable codes [Chung, Salehi, and Wei (1989); A, Gyorfi, and Massey (1992); Chung and Kumar (1990); Bitan and Etzion (1995); Moreno, Zhang, Kumar, and Zionviev (1995)].

Section 3.3. There are many books on projective geometries, most notable are [Albert and Sandler (1968); Dembowski (1968); Hughes and Piper (1973); Hirschfeld (1998)]. Projective geometries are highly related to coding theory and in particular to MDS codes and to the area of network coding which was developed at the start of the 21st century. A survey on these geometries and their connection with coding theory and network coding can be found in [Etzion and Storme (2016)]. The projective geometry PG(m, q), where $m > 2$ is unique, but for $m = 2$ there might be many nonisomorphic projective planes. The projective planes of order $n = 2, 3, 4, 5, 7, 8$ are unique, while there are four nonisomorphic projective planes of order 9, at least 22 of order 16, at least 193 of order 25, and more than 500,000 of order 49. For more information on nonisomorphic projective planes and other related properties of projective planes we refer the readers to [Bruck and Ryser (1949); Albert and Sandler (1968); Hughes and Piper (1973); Hiramine (1989); Lam, Thiel, and Swiercz (1989); Assmus and Key (1990); Lam, Kolesova, and Thiel (1991); Czerwinski and Oakden (1992); Dempwolff (1994)].

Section 3.4. The Plotkin bound (Theorem 3.18 and Corollary 3.9) was proved in [Plotkin (1960)]. The construction with Hadamard matrices that meet the bound is contained in the work of [Levenshtein (1961)].

The Griesmer bound was introduced in [Griesmer (1960)]. Codes meeting the bound have been developed for years. The ideas presented in this section and summarized in Theorem 3.25 are based on the results of [Solomon and Stiffler (1965); Belov, Logachev, and Sandimirov (1974); Alltop (1976)]. More codes meeting the Griesmer bound can be found, for example, in [Helleseth (1981); Helleseth and van Tilborg (1981); Helleseth (1983)]. The Griesmer bound is easily generalized for $[n, k, d]_q$ codes, where it takes the form

$$n \geq \sum_{i=0}^{k-1} \left\lceil \frac{d}{q^i} \right\rceil .$$

Codes attaining this bound were considered, for example, in [Hamada, Helleseth, and Ytrehus (1993)].

Section 3.5. The area of association schemes started with the work of [Bose and Nair (1939)], but the name of the concept was only coined later in [Bose and Shimamoto (1952)]. Their work was based on block designs. It was extended in [Bose and Mesner (1959)] into its algebraic structure and introduced as the Bose-Mesner algebra. The most important contribution to the theory of association schemes, in the context of coding theory, however, was presented by Delsarte in his seminal work [Delsarte (1973)]. Delsarte made the connection between the theory of association schemes, design theory, and coding theory. The representation in this section was taken from [Mounits, Etzion and Litsyn (2007)] who used it to obtain upper bounds on the sizes of codes. An excellent presentation of association schemes for coding theory is given in [MacWilliams and Sloane (1977)].

Finally, distance-regular graphs are the topic of the book by [Brouwer, Cohen, and Neumaier (1989)]. A later survey on the connection between association schemes and coding theory was done in [Delsarte and Levenshtein (1998)]. The work on graphs that represent metrics in the connection of association schemes is highly connected to algebraic graph theory and the books of [Biggs (1993); Godsil and Royle (2001)] can be used for the related results.

Chapter 4

Linear Perfect Codes

In this chapter we discuss the known linear perfect codes in the Hamming scheme. There are two families of nontrivial linear perfect codes. The first family of codes, the Hamming codes, is an infinite family over any power of a prime q and the second family consists of two codes that are the binary and ternary Golay codes. In Section 4.1 we concentrate on the Hamming codes and the extended Hamming codes. We start with the binary codes and discuss some of their properties. Many of these properties are generalized to $\mathbb{F}_q$, $q > 2$, but the proofs of these properties are slightly more complicated for $q > 2$. In Section 4.2 the two Golay codes and their extended codes are presented. Since these codes are two isolated codes, their presentation will be short and we direct the readers to appropriate literature on these interesting codes. Finally, Section 4.3 is devoted to diameter perfect codes in the Hamming scheme, which are MDS codes in the linear case and orthogonal arrays in the nonlinear case. Before starting the exposition it should be mentioned that in addition to the usual trivial perfect codes, the binary code of length $n = 2e + 1$ which contains the all-zero codeword and the all-one codeword, is also a trivial e-perfect code.

4.1 Hamming Codes

The most celebrated perfect codes are the Hamming codes and, especially, the binary Hamming codes. The binary $[2^r - 1, 2^r - r - 1, 3]$ ***Hamming code***, $\mathcal{H}(r)$, has a very simple parity-check matrix $\begin{bmatrix} h_0 \; h_1 \; h_2 \cdots \cdots \cdots h_{2^r-2} \end{bmatrix}$ consisting of the $2^r - 1$ distinct nonzero column vectors of length r. This implies that all the $2^r - 1$ syndromes related to the $2^r - 1$ possible errors are distinct. Clearly, there is no other way to choose $2^r - 1$ distinct column vectors, to obtain a 1-perfect code. Hence, the linear 1-perfect code of length $2^r - 1$ is unique. The parity-check matrix of $\mathcal{H}(r)$ is the generator matrix of the simplex code S_r and hence the Hamming code of length $2^r - 1$ is the dual code of the simplex code of length $2^r - 1$. The first property of these codes is derived from the following general property of all possible perfect codes (linear or nonlinear) in the binary Hamming scheme.

Theorem 4.1. *The nonzero codewords of minimum weight in a binary perfect* $(n, M, 2e+1)$ *code, which contains the all-zero codeword, form a Steiner system* $\mathrm{S}(e+1, 2e+1, n)$.

Proof. If the all-zero word is a codeword in a binary e-perfect $(n, M, 2e+1)$ code $\mathcal{C}$, then it covers all the words of weight at most e and hence the minimum weight codewords have weight $2e+1$. These codewords must cover all the words of weight $e+1$ and each of these words of weight $e+1$ must be covered exactly once. Hence, these codewords of weight $2e+1$ in $\mathcal{C}$ form a Steiner system $\mathrm{S}(e+1, 2e+1, n)$. □

Corollary 4.1. *The codewords of weight three in* $\mathcal{H}(r)$ *form a Steiner system* $\mathrm{S}(2, 3, 2^r - 1)$.

A triple $\{i, j, k\}$ in the Steiner system $\mathrm{S}(2, 3, 2^r - 1)$ of $\mathcal{H}(r)$ is associated with a codeword of weight three in $\mathcal{H}(r)$, and with three columns of the parity-check matrix h_i, h_j, h_k, such that $h_i + h_j + h_k = \mathbf{0}$. Hence, we have that the set $\{\mathbf{0}, h_i, h_j, h_k\}$ is a two-dimensional subspace of $\mathbb{F}_2^r$. Therefore, for the linear code, this Steiner system is defined by the nonzero vectors of the two-dimensional subspaces of $\mathbb{F}_2^r$, which are the one-dimensional subspaces of $\mathrm{PG}(r-1, 2)$.

The following definition generalizes Definition 2.5 which was given for binary codes.

Definition 4.1. Let $\mathcal{C}$ be an $(n, M, d)_q$ code. The code $\mathcal{C}^*$ defined from $\mathcal{C}$ by adding another symbol to the end of each codeword of $\mathcal{C}$ is called the ***extended code*** of $\mathcal{C}$ if $\mathcal{C}^*$ is an $(n+1, M, d+1)_q$ code.

For any binary perfect code $\mathcal{C}$ with minimum distance $2e+1$, ***the extended code*** $\mathcal{C}^*$ is a code whose minimum distance is $2e+2$ and this code is defined by adding a parity bit to all the codewords of $\mathcal{C}$. Therefore, all codewords in an extended binary perfect code have even weight.

The binary $[2^r, 2^r - r - 1, 4]$ ***extended Hamming code***, $\mathcal{H}^*(r)$, is obtained from $\mathcal{H}(r)$ by adding a parity bit at the end of each codeword.

The parity-check matrix for $\mathcal{H}^*(r)$ is

$$\begin{bmatrix} h_0 & h_1 & h_2 & \cdots & \cdots & \cdots & h_{2^r-2} & \mathbf{0} \\ 1 & 1 & 1 & \cdots & \cdots & \cdots & 1 & 1 \end{bmatrix},$$

where $\begin{bmatrix} h_0 & h_1 & h_2 & \cdots & \cdots & \cdots & h_{2^r-2} \end{bmatrix}$ is a parity-check matrix of $\mathcal{H}(r)$.

Are there linear codes with the same parameters, which are not isomorphic to $\mathcal{H}^*(r)$? The answer is clearly no. By Lemma 2.1, puncturing $\mathcal{H}^*(r)$ on any coordinate yields a code with the parameters of $\mathcal{H}(r)$. It can be extended in a unique way (adding a parity bit). This implies that since $\mathcal{H}(r)$ is unique, also $\mathcal{H}^*(r)$ is unique.

The 2^{r+1} cosets of $\mathcal{H}^*(r)$ are formed from coset leaders. These coset leaders have weight 0 (the code itself), weight one, and weight two. There are 2^r cosets with a coset leader of weight one and hence $2^r - 1$ cosets with a coset leader of weight two. The words in the cosets with a coset leader of weight one have odd weights and hence they are called ***odd cosets***. The code itself and the cosets with a coset leader of weight two have only words with even weight and hence they are called ***even cosets***. Clearly, there are 2^r even cosets that contain all the 2^{2^r-1} inary words of length 2^r and even weight, and 2^r odd cosets that contain all the 2^{2^r-1} binary words of length 2^r and odd weight. The same arguments will hold also for the translates of any extended binary nonlinear 1-perfect code.

Theorem 4.2. *The nonzero codewords of minimum weight in an* $(n, M, 2e+2)$ *extended binary perfect code containing the all-zero codeword form a Steiner system* $\mathrm{S}(e+2, 2e+2, n)$.

Proof. Let $\mathcal{C}^*$ be an $(n, M, 2e+2)$ extended binary perfect code and $\mathcal{C}$ be its punctured code. Clearly, $\mathcal{C}$ is an e-perfect code and hence, by Theorem 4.1, its codewords with weight $2e+1$ form a Steiner system $\mathrm{S}(e+1, 2e+1, n-1)$ whose size is $\binom{n-1}{e+1}/\binom{2e+1}{e+1}$. Therefore, there are $\binom{n-1}{e+1}/\binom{2e+1}{e+1}$ codewords in $\mathcal{C}$ with weight $2e+1$. The related codewords of weight $2e+2$ in $\mathcal{C}^*$ cover (contain) all the words of weight $e+2$ with a *one* in the punctured coordinate. Each codeword of $\mathcal{C}$ with weight $2e+1$ also covers $\binom{2e+1}{e+2}$ words of length $n-1$ with weight $e+2$. Hence, it remains to cover by $\mathcal{C}$

$$\binom{n-1}{e+2} - \frac{\binom{n-1}{e+1}}{\binom{2e+1}{e+1}} \binom{2e+1}{e+2}$$

words of length $n-1$ and weight $e+2$. Since $\mathcal{C}$ is an e-perfect code, it follows that each of these words is covered exactly once by the codewords of weight $2e+2$ in $\mathcal{C}$ (which are also codewords of $\mathcal{C}^*$ of weight $2e+2$ with a *zero* in the punctured coordinate of $\mathcal{C}^*$). Hence, each word of length n and weight $e+2$ is covered exactly once by a codeword of weight $2e+2$ in $\mathcal{C}^*$. Thus, the codewords of $\mathcal{C}^*$ with weight $2e+2$ form a Steiner system $\mathrm{S}(e+2, 2e+2, n)$. □

Corollary 4.2. *The codewords of weight four in $\mathcal{H}^*(r)$ form a Steiner system* $\mathrm{S}(3,4,2^r)$.

Another important way to represent $\mathcal{H}(r)$ is to use a parity-check matrix that yields a cyclic code. Let α be a primitive element of $\mathbb{F}_{2^r}$ and let H be the following parity-check matrix

$$H = \begin{bmatrix} \alpha^0 & \alpha^1 & \alpha^2 & \cdots & \cdots & \cdots & \alpha^{2^r-2} \end{bmatrix}$$

where α^i is represented by a column vector of length r based on the binary representation of α^i in $\mathbb{F}_{2^r}$. Clearly, H contains each nonzero column vector of length r exactly once and hence it is a parity-check matrix of $\mathcal{H}(r)$. Moreover, this representation yields a cyclic code because if x is a codeword, i.e., $H \cdot x^{\mathrm{tr}} = \mathbf{0}$ and y is a cyclic shift (by one position to the right) of x, then $H \cdot y^{\mathrm{tr}} = \alpha(H \cdot x^{\mathrm{tr}}) = \mathbf{0}$.

By Corollary 4.2, the codewords of weight four in $\mathcal{H}^*(r)$ form a Steiner system $\mathrm{S}(3,4,2^r)$. This Steiner system is a Steiner quadruple system called a ***boolean Steiner quadruple system*** and will be denoted now by B_0. It can also be defined as follows.

$$B_0 \triangleq \{\{x,y,z,w\} \,:\, x,y,z,w \in \mathbb{F}_2^r,\; x+y+z+w=\mathbf{0},\;\; |\{x,y,z,w\}|=4\}\,.$$

There are other types of boolean Steiner quadruple systems and the systems of these types are related to the even cosets of $\mathcal{H}^*(r)$. There are 2^r-1 such cosets (excluding the code itself) and each is associated with one such system. Given a nonzero vector v of length r, its system B_v is the union of the following two sets

$$\{\{x,y,z,w\} \,:\, x+y+z+w=v,\;\; |\{x,y,z,w\}|=4,\;\; x,y,z,w \in \mathbb{F}_2^r\}$$

$$\{\{x,y,z,w\} \,:\, x+y=z+w=v,\;\; |\{x,y,z,w\}|=4,\;\; x,y,z,w \in \mathbb{F}_2^r\}.$$

The following proposition can be easily verified.

Proposition 4.1. *Each quadruple $\{x,y,z,w\}$ of $\mathbb{F}_2^r$ is contained in at least one of the B_v's. If $x+y+z+w=v$, where $x,y,z,w \in \mathbb{F}_2^r$ and $|\{x,y,z,w\}|=4$, then the quadruple is contained only in B_v. If $x+y+z+w=\mathbf{0}$, where $x,y,z,w \in \mathbb{F}_2^r$ and $|\{x,y,z,w\}|=4$, then the quadruple is contained in exactly four of the B_v's, B_0, B_i, B_j, and B_k, where $i=x+y$, $j=x+z$, and $k=x+w$.*

These 2^r boolean Steiner quadruple systems are very important in constructions of sets with pairwise disjoint Steiner quadruple systems. They are also highly related to the following two observations.

Lemma 4.1. *In any coset of $\mathcal{H}^*(r)$, whose coset leader has weight two, there are exactly 2^{r-1} words of weight two.*

Proof. Assume $\{i,j\}$ is a coset leader of and even coset of $\mathcal{H}^*(r)$. Since by Corollary 4.2 the codewords of weight four in $\mathcal{H}^*(r)$ form a Steiner system $S(3,4,2^r)$, it follows that $\mathcal{H}^*(r)$ contains exactly $2^{r-1}-1$ codewords of the form $\{i,j,\alpha,\beta\}$ (this is also implied since each $\alpha \in \mathbb{F}_2^r \setminus \{i,j\}$ is contained in exactly one such codeword). For each such codeword $\{i,j,\alpha,\beta\}$, $\{\alpha,\beta\}$ is clearly a word in the coset $\{i,j\}+\mathcal{H}^*(r)$. Thus, $\{i,j\}$ and the $2^{r-1}-1$ words implied from codewords of the form $\{i,j,\alpha,\beta\}$ yield 2^{r-1} words of weight two in the coset. There are no more words of weight two in the coset since any other word of weight two intersects two of these 2^{r-1} words in exactly one coordinate and their addition implies a word of weight two in $\mathcal{H}^*(r)$, which contradicts the minimum distance of $\mathcal{H}^*(r)$. □

Lemma 4.2. *Each word $x \in \mathbb{F}_2^{2^r}$ satisfies one of the following three conditions:*

- *x is a codeword in $\mathcal{H}^*(r)$;*
- *there exists exactly one codeword in $\mathcal{H}^*(r)$ with distance one from x;*
- *there exist exactly 2^{r-1} codewords in $\mathcal{H}^*(r)$ with distance two from x.*

Proof. We distinguish between a word x of odd weight and a word x of even weight.

If x has odd weight, then it is contained in an odd coset $x+\mathcal{H}^*(r)$ and hence its distance is one from exactly one codeword c of $\mathcal{H}^*(r)$, such that $\mathbf{e}_i = c+x$, for some $1 \le i \le 2^r$, is the coset leader of weight one in the coset $x+\mathcal{H}^*(r)$.

If x has even weight, then either x is a codeword of $\mathcal{H}^*(r)$ or x is not a codeword of $\mathcal{H}^*(r)$. If x is not a codeword of $\mathcal{H}^*(r)$, then it is contained in the coset $x+\mathcal{H}^*(r)$. By Lemma 4.1 this coset contains 2^{r-1} words of weight two $x_1, x_2, \ldots, x_{2^{r-1}}$, and by Lemma 2.9 for each i, $x+x_i$ is a codeword of $\mathcal{H}^*(r)$ with distance two from x. There are no other codewords in $\mathcal{H}^*(r)$ at distance two from x since such a codeword would imply more words of weight two in the coset $x+\mathcal{H}^*(r)$, contradicting Lemma 4.1. □

Corollary 4.3. *Each word $x \in \mathbb{F}_2^{2^r-1}$ satisfies one of the following two conditions:*

- *x is a codeword in $\mathcal{H}(r)$;*
- *there exists a codeword in $\mathcal{H}(r)$ with distance one from x, and exactly $2^{r-1}-1$ codewords in $\mathcal{H}(r)$ with distance two from x.*

The q-ary Hamming code is a $[(q^r-1)/(q-1), (q^r-1)/(q-1)-r, 3]$ code over $\mathbb{F}_q$. It generalizes the binary Hamming code. Some of the properties mentioned for the binary Hamming codes can also be generalized.

The parity-check matrix H of the code has $(q^r-1)/(q-1)$ columns, which are the $(q^r-1)/(q-1)$ points of the projective geometry $\text{PG}(r-1,q)$. This is also a straightforward generalization from the binary case. In other words, H contains one representative from each $q-1$ nonzero column vectors of length r, from which each two are linearly dependent. There is a wide range of such possible representatives, each of which may serve to prove different properties of the code and could also be used for different constructions. This leads to the following theorem.

Theorem 4.3. *For each $r > 1$, there exists a 1-perfect Hamming code over $\mathbb{F}_q$. This code has length $(q^r-1)/(q-1)$, redundancy r, and minimum Hamming distance 3.*

Similarly to the binary case, we can represent the parity-check matrix of the code in a cyclic way, where the columns represent the elements $\alpha^0, \alpha^1, \alpha^2, \ldots, \alpha^{\ell-1}$, $\ell = \frac{q^r-1}{q-1}$ and α is a primitive element in $\mathbb{F}_{q^r}$. The vector representation of these $\frac{q^r-1}{q-1}$ elements coincides with the points of $\text{PG}(r-1,q)$.

Unfortunately, some properties of the binary case do not generalize to the q-ary case. The most important such property is that for most parameters, there is no extended q-ary Hamming code. The only exception is when $q = 2^\ell$, $\ell > 1$, where the only extended q-ary Hamming code is a $[q+2, q-1, 4]$ code.

Theorem 4.4. *If α is a primitive element in $\mathbb{F}_q$, $q = 2^\ell$, then the following matrix*

$$H = \begin{bmatrix} 1 & 1 & 1 & \cdots & 1 & 1\,0\,0 \\ \alpha^0 & \alpha^1 & \alpha^2 & \cdots & \alpha^{q-2} & 0\,1\,0 \\ \alpha^0 & \alpha^2 & \alpha^4 & \cdots & \alpha^{2q-4} & 0\,0\,1 \end{bmatrix} \tag{4.1}$$

is a parity-check matrix of a $[q+2, q-1, 4]$ code.

Proof. The length of the code and its dimension are immediate results from the definition of the parity-check matrix H. To complete the proof of the claim, it suffices to show that each two columns of H or each three columns of H are linearly independent. This will imply, by Corollary 2.11, that the minimum distance of the code is at least four. This can be done in the most naive way as follows. It is readily verified that each two columns of H are linearly independent. Clearly, the last three columns are linearly independent. Moreover, it is readily verified that taking any two of the last three columns with one of the first $q-1$ columns results in a set of three linearly independent column vectors.

Consider now three columns from the first $q-1$ columns. Assume that there exist three nonzero elements $\beta_1, \beta_2, \beta_3 \in \mathbb{F}_q^-$ and integers $0 \leq i < j < k \leq q-2$ such that

$$\beta_1 \begin{pmatrix} 1 \\ \alpha^i \\ \alpha^{2i} \end{pmatrix} + \beta_2 \begin{pmatrix} 1 \\ \alpha^j \\ \alpha^{2j} \end{pmatrix} + \beta_3 \begin{pmatrix} 1 \\ \alpha^k \\ \alpha^{2k} \end{pmatrix} = \begin{pmatrix} 0 \\ 0 \\ 0 \end{pmatrix}. \tag{4.2}$$

This implies that

$$\beta_1 + \beta_2 + \beta_3 = 0,$$

$$\beta_1 \alpha^i + \beta_2 \alpha^j + \beta_3 \alpha^k = 0,$$

$$\beta_1 \alpha^{2i} + \beta_2 \alpha^{2j} + \beta_3 \alpha^{2k} = 0.$$

The first of these three equations implies that $\beta_3 = -(\beta_1 + \beta_2)$ and plugging this into the other two equations yields

$$\beta_1(\alpha^i - \alpha^k) = \beta_2(\alpha^k - \alpha^j),$$

$$\beta_1(\alpha^{2i} - \alpha^{2k}) = \beta_2(\alpha^{2k} - \alpha^{2j}).$$

Plugging the solution for β_1 in the first equation into the second equation implies that

$$\beta_2 \frac{\alpha^{2i} - \alpha^{2k}}{\alpha^i - \alpha^k} = \beta_2 \frac{\alpha^{2k} - \alpha^{2j}}{\alpha^k - \alpha^j}$$

and since $i \neq k$ and $j \neq k$, this is equivalent to

$$\alpha^i + \alpha^k = \alpha^k + \alpha^j,$$

i.e., $i = j$ and since there is symmetry for i, j, and k, in (4.2), it follows that $i = j = k$, a contradiction.

It remains to consider only the case of one column from the last three columns and two columns from the first $q-1$ columns. We distinguish between three cases depending on the unique column taken from the last three columns.

Case 1. This column is $\begin{pmatrix} 1 \\ 0 \\ 0 \end{pmatrix}$.

In this case it suffices to show that there is no nontrivial linear combination

$$\beta_1 \begin{pmatrix} \alpha^i \\ \alpha^{2i} \end{pmatrix} + \beta_2 \begin{pmatrix} \alpha^j \\ \alpha^{2j} \end{pmatrix} = \begin{pmatrix} 0 \\ 0 \end{pmatrix},$$

where $0 \leq i < j \leq q-2$. The existence of such a nontrivial linear combination implies that $\beta_2 = -\beta_1 \alpha^{i-j}$ and $\beta_2 = -\beta_1 \alpha^{2i-2j}$ and, as a consequence, $i = j$, a contradiction.

Case 2. This column is $\begin{pmatrix} 0 \\ 0 \\ 1 \end{pmatrix}$.

In this case we have to show that there is no nontrivial linear combination

$$\beta_1 \begin{pmatrix} 1 \\ \alpha^i \end{pmatrix} + \beta_2 \begin{pmatrix} 1 \\ \alpha^j \end{pmatrix} = \begin{pmatrix} 0 \\ 0 \end{pmatrix},$$

where $0 \leq i < j \leq q-2$. The existence of such a nontrivial linear combination implies that $\beta_2 = -\beta_1$ and $\beta_2 = -\beta_1 \alpha^{i-j}$ and, as a consequence, $i = j$, a contradiction.

Case 3. This column is $\begin{pmatrix} 0 \\ 1 \\ 0 \end{pmatrix}$.

In this case we have to show that there is no nontrivial linear combination

$$\beta_1 \begin{pmatrix} 1 \\ \alpha^{2i} \end{pmatrix} + \beta_2 \begin{pmatrix} 1 \\ \alpha^{2j} \end{pmatrix} = \begin{pmatrix} 0 \\ 0 \end{pmatrix},$$

where $0 \leq i < j \leq q-2$. The existence of such a nontrivial linear combination implies that, $\beta_2 = -\beta_1$ and $\beta_2 = -\beta_1 \alpha^{2i-2j}$ and, as a consequence, $\alpha^{2i} = \alpha^{2j}$. Since q is even, it follows that this is possible only if $j = i$, a contradiction.

Thus, the minimum number of linearly dependent column vectors of H is four and hence the minimum Hamming distance of the code, whose parity-check matrix is H, is 4. □

If q is not a power of two, i.e., q is odd, then the following three column vectors (which are column vectors of H) in (4.1)

$$\begin{pmatrix} 1 \\ \alpha^0 \\ \alpha^0 \end{pmatrix}, \begin{pmatrix} 1 \\ \alpha^{(q-1)/2} \\ \alpha^{q-1} \end{pmatrix} = \begin{pmatrix} 1 \\ \alpha^{(q-1)/2} \\ \alpha^0 \end{pmatrix}, \begin{pmatrix} 0 \\ 1 \\ 0 \end{pmatrix}$$

are clearly linearly dependent and the matrix in (4.1) is not a parity-check matrix of a code whose minimum distance is 4.

Does there exist a system like a Steiner system formed by the words of minimum weight from a nonbinary perfect codes? The answer is yes, if we use generalized Steiner systems instead of Steiner systems. The following theorem is a straightforward generalization of Theorem 4.1.

Theorem 4.5. *The nonzero codewords of minimum weight in an e-perfect* $(n, M, 2e+1)_q$ *code which containing the all-zero codeword form a generalized Steiner system* $\mathrm{GS}(e+1, 2e+1, n, q)$.

4.2 Golay Codes

Except for the Hamming code, there are only two other nontrivial linear perfect codes in the Hamming scheme $\mathcal{H}_q(n)$, q a prime power, called ***the Golay codes***. The binary Golay code will be denoted by $\mathcal{G}_{23}$ and its extended code by $\mathcal{G}_{24}$. The ternary Golay code will be denoted by $\mathcal{G}_{11}$ and its extended code is $\mathcal{G}_{12}$. The nonexistence of other perfect codes will be discussed in Section 5.8.

The binary extended Golay code denoted by $\mathcal{G}_{24}$ has the following 12×24 generator matrix

$$G_{24} = \begin{bmatrix}
1\,1\,0\,0\,0\,0\,0\,0\,0\,0\,0\,0\,0\,1\,1\,0\,1\,1\,1\,0\,0\,0\,1\,0 \\
1\,0\,1\,0\,0\,0\,0\,0\,0\,0\,0\,0\,0\,0\,1\,1\,0\,1\,1\,1\,0\,0\,0\,1 \\
1\,0\,0\,1\,0\,0\,0\,0\,0\,0\,0\,0\,0\,1\,0\,1\,1\,0\,1\,1\,1\,0\,0\,0 \\
1\,0\,0\,0\,1\,0\,0\,0\,0\,0\,0\,0\,0\,0\,1\,0\,1\,1\,0\,1\,1\,1\,0\,0 \\
1\,0\,0\,0\,0\,1\,0\,0\,0\,0\,0\,0\,0\,0\,0\,1\,0\,1\,1\,0\,1\,1\,1\,0 \\
1\,0\,0\,0\,0\,0\,1\,0\,0\,0\,0\,0\,0\,0\,0\,0\,1\,0\,1\,1\,0\,1\,1\,1 \\
1\,0\,0\,0\,0\,0\,0\,1\,0\,0\,0\,0\,0\,1\,0\,0\,0\,1\,0\,1\,1\,0\,1\,1 \\
1\,0\,0\,0\,0\,0\,0\,0\,1\,0\,0\,0\,0\,1\,1\,0\,0\,0\,1\,0\,1\,1\,0\,1 \\
1\,0\,0\,0\,0\,0\,0\,0\,0\,1\,0\,0\,0\,1\,1\,1\,0\,0\,0\,1\,0\,1\,1\,0 \\
1\,0\,0\,0\,0\,0\,0\,0\,0\,0\,1\,0\,0\,0\,1\,1\,1\,0\,0\,0\,1\,0\,1\,1 \\
1\,0\,0\,0\,0\,0\,0\,0\,0\,0\,0\,1\,0\,1\,0\,1\,1\,1\,0\,0\,0\,1\,0\,1 \\
0\,0\,0\,0\,0\,0\,0\,0\,0\,0\,0\,0\,1\,1\,1\,1\,1\,1\,1\,1\,1\,1\,1\,1
\end{bmatrix}.$$

The binary extended Golay code has 4096 codewords and hence all its properties can be verified easily by a computer search. Nevertheless, below we present proofs for some properties of the codes.

By replacing each *zero* by a $+1$ and each *one* by a -1, the last twelve columns of G_{24} form a 12×12 Hadamard matrix. This immediately implies, by the structure of G_{24}, that the inner product of any two rows of G_{24} is equal to *zero*, i.e., any two rows are orthogonal. Hence, the parity-check matrix H_{24} of $\mathcal{G}_{24}$ is G_{24}, i.e., $H_{24} = G_{24}$ and, therefore, $\mathcal{G}_{24}$ is a self-dual code. It is also a self-complement code since the sum of all the rows of G_{24} is the all-ones word. Given two binary codewords whose weight is divisible by 4, to be orthogonal they should share *ones* in an even number of coordinates. Hence, the addition of these two codewords is also a codeword whose weight is divisible by 4. Therefore, since all the rows of G_{24} have weights divisible by 4 and the code is self-dual, it follows that all codewords of $\mathcal{G}_{24}$ have weights divisible by 4.

Lemma 4.3. *The extended Golay code $\mathcal{G}_{24}$ is invariant under the following permutation π of the 24 coordinates*

$$\pi \triangleq (1,13)(2,14)(3,24)(4,23)(5,22)\cdots(10,17)(11,16)(12,15)$$

Proof. The permutation π sends the first row of G_{24} into the word

$$010100011101110000000000,$$

which is the sum of rows 1, 3, 7, 8, 9, 11, 12 of G_{24}. Due to the cyclic structure of the two halves of G_{24}, excluding the 1-st and 13-th columns, the same arguments, on the sum of rows and the permutation π, hold for the first eleven rows. The permutation π applied on the last row of G_{24} is the complement of the last row of G_{24}, which completes the proof. □

Corollary 4.4. *The word (x, y), where $x = (x_1, x_2, \ldots, x_{12})$, $y = (y_1, y_2, \ldots, y_{12})$ is a codeword in $\mathcal{G}_{24}$ if and only if $(x', y') \in \mathcal{G}_{24}$, where $x' = (y_1, y_2, y_{12}, y_{11}, \ldots, y_3)$ and $y' = (x_1, x_2, x_{12}, x_{11}, \ldots, x_3)$.*

Lemma 4.4. *The extended Golay code $\mathcal{G}_{24}$ does not contain a codeword of weight 4.*

Proof. By Corollary 4.4 and the structure of G_{24}, if there exists a codeword of weight 4 in $\mathcal{G}_{24}$, then we can consider only a codeword of the form (x, y), $x, y \in \mathbb{F}_2^{12}$, and either $\text{wt}(x) = 0$, $\text{wt}(y) = 4$ or $\text{wt}(x) = \text{wt}(y) = 2$. Clearly, by the structure of G_{24}, we cannot have $\text{wt}(x) = 0$, $\text{wt}(y) = 4$ (the last row of G_{24} implies that if $\text{wt}(x) = 0$, then either $\text{wt}(y) = 0$ or $\text{wt}(y) = 12$).

Any codeword (x, y) for which $\mathrm{wt}(x) = 2$ is obtained by adding one or two of the first eleven rows of G_{24} and possibly also the last row. Hence, the weight of y can only be six (again as a consequence of the structure of these rows in G_{24}). □

Corollary 4.5. *The codewords of $\mathcal{G}_{24}$ have weights 0, 8, 12, 16, and 24.*

The Golay code $\mathcal{G}_{23}$ is a $[23, 12]$ code obtained from $\mathcal{G}_{24}$ by puncturing any column of $\mathcal{G}_{24}$.

Corollary 4.6. *The minimum Hamming distance of the $[24, 12]$ binary extended Golay code, $\mathcal{G}_{24}$, is 8 and the minimum Hamming distance of the $[23, 12]$ Golay code, $\mathcal{G}_{23}$, is 7.*

The size of a ball with radius 3 for a binary word of length 23 is

$$1 + \binom{23}{1} + \binom{23}{2} + \binom{23}{3} = 2^{11}$$

and since by Corollary 4.6, the Golay code $\mathcal{G}_{23}$ is a $[23, 12, 7]$ code, it follows by the sphere-packing bound that $\mathcal{G}_{23}$ is a 3-perfect code.

By Theorem 4.1, the codewords of weight seven in $\mathcal{G}_{23}$ form a Steiner system $\mathrm{S}(4, 7, 23)$ and, by Theorem 4.2, the codewords of weight eight in $\mathcal{G}_{24}$ form a Steiner system $\mathrm{S}(5, 8, 24)$. Let A_i be the number of codewords of weight i in $\mathcal{G}_{23}$. By considering either that $\mathcal{G}_{23}$ is a perfect code or the Steiner systems $\mathrm{S}(4, 7, 23)$ and $\mathrm{S}(5, 8, 24)$, respectively, embedded in $\mathcal{G}_{23}$ and in $\mathcal{G}_{24}$, respectively, and the fact that the codes are self-complements, we have that for $\mathcal{G}_{23}$

$$A_0 = A_{23} = 1,\ A_7 = A_{16} = 253,\ A_8 = A_{15} = 506,\ A_{11} = A_{12} = 1288,$$

and, for each other i, we have that $A_i = 0$. The weight distribution of $\mathcal{G}_{24}$ is easily derived from the weight distribution of $\mathcal{G}_{23}$.

Consider now any codeword of weight eight in $\mathcal{G}_{24}$ and permute the columns of G_{24} such that all the *ones* of this codeword are in the first eight coordinates. Since $H_{24} = G_{24}$, it follows that the first seven columns of the related new parity-check matrix (and also the generator matrix) are linearly independent since otherwise, by Corollary 2.11, the code will have a codeword of weight at most seven. Therefore, the 8-th column of $\mathcal{G}_{24}$ (also of each codeword in $\mathcal{G}_{24}$ and of each row in G_{24}) is the sum of the first seven columns. Partitioning the codewords by the value of the first seven coordinates yields $2^7 = 128$ sets, each with $2^5 = 32$ codewords (by simple enumeration since no such sub-code can have more than 32 codewords by the Plotkin bound for a code of length 16 and minimum distance 8, as implied by (3.7)). Consider now the $7 \cdot 32 = 224$ codewords of weight two with a *one* in the first coordinate and the 32 codewords that start with eight *zeroes*. By puncturing the first eight coordinates in these 256 codewords, we obtain a code of length 16, whose minimum Hamming distance is 6, with 256 codewords. This code is called the Nordstorm–Robinson code or the Preparata code of length 16. This code will be discussed in Chapter 6.

We continue with the next code, the ternary Golay code $\mathcal{G}_{11}$. The generator matrix of the ternary [12, 6, 6] extended Golay code $\mathcal{G}_{12}$ is

$$\begin{bmatrix} 1\,0\,0\,0\,0\,0\,0\,1\,1\,1\,1\,1 \\ 0\,1\,0\,0\,0\,0\,1\,0\,1\,2\,2\,1 \\ 0\,0\,1\,0\,0\,0\,1\,1\,0\,1\,2\,2 \\ 0\,0\,0\,1\,0\,0\,1\,2\,1\,0\,1\,2 \\ 0\,0\,0\,0\,1\,0\,1\,2\,2\,1\,0\,1 \\ 0\,0\,0\,0\,0\,1\,1\,1\,2\,2\,1\,0 \end{bmatrix}.$$

Similarly to the binary Golay code, one can analyze this generator matrix and obtain the properties of this code, the ternary extended Golay code $\mathcal{G}_{12}$ and its punctured code $\mathcal{G}_{11}$. One property is that the supports of the codewords of weight six in $\mathcal{G}_{12}$ form a Steiner system S(5, 6, 12), while the supports of the codewords of weight five in $\mathcal{G}_{11}$ form a Steiner system S(4, 5, 11). By Theorem 4.5, these codewords also form a generalized Steiner system GS(3, 5, 11, 3).

4.3 Diameter Perfect Codes

We continue and consider diameter perfect codes (linear and nonlinear) in the Hamming scheme. Let $\mathcal{A}_q(n, D)$ be the largest anticode with diameter D in the Hamming scheme $\mathcal{H}_q(n)$. Note that q can be any integer greater than one. The following theorem, which is given without a proof, determines the size of the largest possible anticode.

Theorem 4.6. *For $q \geq 2$ and $D < n$ we have that*

$$|\mathcal{A}_q(n, D)| = |\mathcal{B}_e(n - D + 2e)| \cdot q^{D-2e}$$

where

$$e = \begin{cases} \left\lfloor \frac{D}{2} \right\rfloor & \text{if } (D+1)q \leq 2n \\ \left\lfloor \frac{n-D+1}{q-2} \right\rfloor & \text{if } (D+1)q > 2n \end{cases}. \tag{4.3}$$

Theorem 4.7. *An OA(t, k, q) is a $(k - t)$-diameter perfect code.*

Proof. By Theorem 3.3, an OA(t, k, q) is a $(k, q^t, k-t+1)_q$ code $\mathcal{C}$. Define

$$\mathcal{A} \triangleq \{(a_1, a_2, \ldots, a_k) \ : \ a_i = 0, \ 1 \le i \le t, \ a_i \in \mathbb{Z}_q, \ t+1 \le i \le k\}.$$

Clearly, $|\mathcal{A}| = q^{k-t}$, and the maximum distance in $\mathcal{A}$ is $k-t$, i.e., $\mathcal{A}$ is an anticode of length k and diameter $k-t$ over $\mathbb{Z}_q$. This implies that

$$|\mathcal{C}| \cdot |\mathcal{A}| = q^t q^{k-t} = q^k \ .$$

Therefore, an OA(t, k, q) and the anticode $\mathcal{A}$ satisfy the code-anticode bound ((2.3) in Corollary 2.15). Thus, an OA(t, k, q) is a $(k-t)$-diameter perfect code. □

Corollary 4.7. *An $[n, k, n-k+1]$ MDS code is a linear $(n-k)$-diameter perfect code.*

Theorem 4.8. *If $\mathcal{C}$ is a D-diameter perfect code in $\mathcal{H}_q(n)$, then its extended code $\mathcal{C}^*$ is a $(D+1)$-diameter perfect code in $\mathcal{H}_q(n+1)$.*

Proof. Let $\mathcal{C}$ be a D-diameter perfect code, in $\mathcal{H}_q(n)$, and let $\mathcal{A}$ be an associated maximum size anticode with diameter D, over $\mathbb{Z}_q^n$. Define

$$\mathcal{A}^* \triangleq \{(a, \alpha) \ : \ a \in \mathcal{A}, \ \alpha \in \mathbb{Z}_q\}.$$

Since the diameter of $\mathcal{A}$ is D, it follows that the diameter of $\mathcal{A}^*$ is $D+1$ and, clearly, $|\mathcal{A}^*| = q|\mathcal{A}|$. Since $\mathcal{C}$ is a D-diameter perfect code and $\mathcal{A}$ is its related maximum size anticode, it follows by the code-anticode bound that $|\mathcal{C}| \cdot |\mathcal{A}| = q^n$. Moreover, clearly $|\mathcal{C}^*| = |\mathcal{C}|$ and hence $|\mathcal{C}^*| \cdot |\mathcal{A}^*| = q^{n+1}$. Since $\mathcal{C}^*$ is the extended code of $\mathcal{C}$, it follows by Definition 4.1 that $d(\mathcal{C}^*) = d(\mathcal{C}) + 1 = D + 2$. Thus, by the code-anticode bound, $\mathcal{C}^*$ is a $(D+1)$-diameter perfect code. □

Our discussion in Section 4.1 and Section 4.2 (and also regarding Theorem 4.10 and Theorem 5.13 presented in Chapter 5) implies that parameters of extended perfect codes in $\mathcal{H}_q(n)$, q being a prime power, are as in the following linear codes.

(1) The binary $[2^r, 2^r - r - 1, 4]$ extended Hamming code, where $r \ge 2$.
(2) The binary $[24, 12, 8]$ extended Golay code.
(3) The ternary $[12, 6, 6]$ extended Golay code.
(4) The $[q+2, q-1, 4]$ extended Hamming code over $\mathbb{F}_q$, $q = 2^m$, $m \ge 2$.

Theorem 4.9. *In $\mathcal{H}_q(n)$, q a prime power, there are no diameter perfect codes except for the codes with the parameters of the Hamming codes, the extended Hamming codes, the Golay codes, the extended Golay codes, and the MDS codes.*

Proof. Let $\mathcal{C}$ be a D-diameter perfect $(n, M, D+1)$ code and let e be the parameter (as in (4.3)) of the anticode $\mathcal{A}$ with diameter D denoted by $\mathcal{A}_q(n, D)$. Assume that $\mathcal{C}$ is not an MDS code and $\mathcal{C}$ is not a perfect code, i.e., $e > 0$ and $D > 2e$. By the code-anticode bound of Corollary 2.15 and since $\mathcal{C}$ is a D-diameter perfect code, it follows that

$$M = \frac{q^n}{|\mathcal{A}_q(n, D)|},$$

and, in view of Theorem 4.6, we have that

$$|\mathcal{A}_q(n, D)| = |\mathcal{B}_e(n - D + 2e)| \cdot q^{D-2e}.$$

Puncturing $\mathcal{C}$ in any coordinate, yields by Lemma 2.1 an $(n-1, M, D)$ code $\mathcal{C}'$. The code $\mathcal{C}'$ is a $(D-1)$-diameter perfect code since, by Theorem 4.6, there exists an anticode $\mathcal{A}$ in $\mathcal{H}_q(n-1)$ of diameter $D-1$ and size $|\mathcal{B}_e(n - D + 2e)| \cdot q^{D-2e-1}$. This anticode and the code $\mathcal{C}'$ meet the code-anticode bound of Corollary 2.15 since

$$|\mathcal{C}'| \cdot |\mathcal{A}| = M \cdot |\mathcal{B}_e(n - D + 2e)| \cdot q^{D-2e-1} = q^{n-1} .$$

Continuing this argument iteratively yields a $(2e)$-diameter perfect code whose length is $\eta = n - D + 2e$ and whose minimum distance is $\delta = 2e + 1$. This is an e-perfect code with the parameters of the Hamming code or the Golay codes, where for length $\eta + 1$ we have the associated extended codes. Clearly, there are no doubly extended perfect codes (extended codes for the extended perfect codes), which completes the proof. □

Note, that while an extended code cannot be extended, it is possible to extend many diameter perfect codes (MDS codes and orthogonal arrays) a few times. Another interesting fact is that $[q+1, q-1, 3]_q$ code is an MDS code and also a 1-perfect code over $\mathbb{F}_q$. Finally, Theorem 4.9 and Theorem 5.13 (presented in the Chapter 5 which follows) imply that we know all the parameters of perfect codes in $\mathcal{H}_q(n)$, where q is a prime power. We probably know all the parameters of MDS codes too (see Conjecture 3.1). There is still a lot about nonlinear codes, i.e., orthogonal arrays, that we do not know. Moreover, over an alphabet that is not a prime power, there are some parameters where it has yet to be proven that perfect codes cannot exist.

4.4 Notes

Section 4.1. Binary Hamming codes were introduced by Hamming [Hamming (1950)]. The codes over nonbinary alphabets were introduced in [Shapiro and Slotnick (1959)].

As we easily proved, both the $[2^r - 1, 2^r - r - 1, 3]$ Hamming code and its $[2^r, 2^r - r - 1, 4]$ extended code are unique. But, what about their shortened codes? It is not difficult to prove that these codes are also unique. Moreover, the question can be generalized further. It is easy to verify that the largest dimension k of a $[2^r - 1 - \ell, k, 3]$ code, where $0 \leq \ell \leq 2^{r-1} - 1$, is $k = 2^r - 1 - r - \ell$, and to obtain a code with these parameters one has to shorten the $[2^r - 1, 2^r - r - 1, 3]$ Hamming code ℓ times. All these codes can also be obtained by removing any ℓ columns from the parity-check matrix of the Hamming code. Depending on the ℓ deleted columns, different codes, which are not necessarily isomorphic, are generated. The same question can be asked about the $[2^r, 2^r - r - 1, 4]$ extended Hamming code, where it is also important in the context of codes for semiconductor memories [Davydov and Tombak (1991)]. It is easily verified that the largest dimension k of a $[2^r - \ell, k, 4]$ code, where $0 \leq \ell \leq 2^{r-1} - 1$, is $k = 2^r - 1 - r - \ell$, and to obtain a code with these parameters one has to shorten the $[2^r, 2^r - r - 1, 4]$ extended Hamming code ℓ times. Nevertheless, there are also other codes with some of these parameters that cannot be generated by deleting columns from the parity-check matrix of the $[2^r, 2^r - r - 1, 4]$ binary extended Hamming code. Note that this is a distinction between the Hamming code and the extended Hamming code. It is also easy to verify that the covering radius of the extended Hamming code is 2 since the covering radius of the Hamming code is 1. It was proved in [Davydov and Tombak (1989a)] that there are exactly three families of codes with these parameters that also have covering radius 2. These families are the $[2^r, 2^r - r - 1, 4]$ extended Hamming code; a $[5 \cdot 2^{r-4}, 5 \cdot 2^{r-4} - r - 1, 4]$ code, where $r \geq 7$; and a $[9 \cdot 2^{r-5}, 9 \cdot 2^{r-5} - r - 1, 4]$ code, where $r \geq 9$.

Each binary Hamming code has an extended code. For nonbinary Hamming codes, there is only one small family of extended codes. This is stated in the following theorem.

Theorem 4.10. *An extended Hamming code over $\mathbb{F}_q$, $q > 2$, exists only when q is a power of two and the extended code is a $[q + 2, q - 1, 4]$ code.*

Theorem 4.10 was proved in [Hill (1978)] by using techniques from projective geometry. A $[q + 2, q - 1, 4]$ code when q is a power of 2 was presented in Theorem 4.4. It should be noted that a $[q + 2, q - 1, 4]$ code is also an MDS code.

The boolean Steiner quadruple systems were defined and used in [Etzion and Hartman (1991)] to form a set with a very large number of disjoint Steiner quadruple systems.

Section 4.2. The Golay codes were introduced by [Golay (1949)]. It should be mentioned that the code was found earlier in the context of football pools. In football pools there are n games of football with three possible results for each game, the first team wins, the second team wins, or a draw. A Finnish football pools specialist, Juhani Virtakallio, published in issue 27/1947 of the Finnish magazine Veikkaaja (Veikkaus-Lotto), on August 1, 1947, a system for the football pools which comprising the 729 codewords of the ternary Golay codes. These codewords form the smallest number of guesses required to guess at least nine outcomes of eleven football games correctly.

Many properties of the binary Golay code including its weight distribution, designs embedded in the code, and the Nordstorm–Robinson code embedded in the code, were found by [Goethals (1971)]. More properties were observed in [Goldberg (1986)]. There are also other ways to construct the code, e.g., the ones in [Pasquier (1980); Peng and Farrell (2006)]. The uniqueness of the linear Golay codes were proved in a sequence of papers by [Pless (1968, 1992)]. The proof that there are no nonlinear codes with the parameters of the Golay code was done in [Delsarte and Goethals (1975)]. Another simpler approach to prove the uniqueness of the ternary Golay code was given in [Drápal (2002)]. The Golay codes are, without doubts, the most researched codes, taking into account all specific codes. Decoding of the Golay codes was considered by many authors, e.g., [Pless (1986); Conway and Sloane (1986); Snyders and Be'ery (1989); Vardy and Be'ery (1991); Amrani, Be'ery, Vardy, Sun, and van Tilborg (1994)]. The proof that $\mathcal{G}_{24}$ is unique is based on the fact that the Steiner system $S(5, 8, 24)$ embedded in the code is unique [Pless (1968)]. The supports of the codewords of weight 6 in $\mathcal{G}_{12}$ form a Steiner system $S(5, 6, 12)$. The uniqueness of this system implies the uniqueness of $\mathcal{G}_{12}$ [Pless (1968)]. Some sub-codes of $\mathcal{G}_{24}$ are also unique as was proved in [Dodunekov and Encheva (1993)]. Finally, the Nordstrom–Robinson was found by [Nordstrom and Robinson (1967)] and independently later by [Semakov and Zinoviev (1969)].

Section 4.3. The analysis on diameter perfect codes which was done in [Ahlswede, Aydinian, and Khachatrian (2001)]. The size of a maximum anticode was proved in [Ahlswede, and Khachatrian (1998)].

Chapter 5

Nonlinear Perfect Codes

The only infinite family of perfect codes in the Hamming scheme is the family of 1-perfect codes. Therefore, during the years the main research in this direction was on constructions of other nonlinear 1-perfect codes with the same parameters having some desired properties. This is the topic of Chapter 5. In Section 5.1 three constructions for binary 1-perfect codes and one construction for a nonbinary alphabet, which is also useful for other metrics, are presented. In Section 5.7 we present a construction that yields a large number of nonequivalent 1-perfect codes. The construction is for the nonbinary case, but it can be also implemented for binary codes. Moreover, the ideas of this construction for binary codes are presented in all the other sections. The number of binary codes known in other constructions is at most slightly larger than the number of codes generated by this construction presented in Section 5.7. Since the difference in the number is very small, the simpler construction was chosen for Section 5.7. The presented construction is based of a switching method that is also useful when examining the intersection numbers, the ranks, and the kernels of 1-perfect codes, which are discussed in Sections 5.3, 5.5, and 5.6, respectively. Section 5.4 is devoted to the intersection numbers of linear codes.

In Section 5.2 the weight distribution and the distance distribution of 1-perfect codes are considered. It is shown that all 1-perfect codes have the same distance distribution and weight distribution (depending on whether the all-zero word is in the code). In Section 5.8 we discuss the nonexistence of perfect codes with other parameters.

In Section 5.9 we present an application of 1-perfect codes to the well-known mathematical game of hat guessing. We emphasize again that most of the results presented in this chapter are for binary codes, but they can be generalized for any finite field $\mathbb{F}_q$. In particular, the topics of Section 5.2

through Section 5.6 are presented only for the binary alphabet.

5.1 Constructions of Nonlinear Perfect Codes

The first known construction of nonlinear 1-perfect codes is defined as follows. Let $\mathcal{C}_n$ be a 1-perfect code of length $n = 2^r - 1$. Let $f : \mathcal{C}_n \to \{0,1\}$ be an arbitrary mapping such that $f(\mathbf{0}) = 0$ and $f(c_1) + f(c_2) \neq f(c_1 + c_2)$ for some $c_1, c_2, c_1 + c_2 \in \mathcal{C}_n$. This last condition is given to obtain a nonlinear code if $\mathcal{C}_n$ is a linear code.

Theorem 5.1. *The code $\mathcal{C}_{2n+1}$, defined by*

$$\mathcal{C}_{2n+1} \triangleq \{(v, v + c, p(v) + f(c)) \ : \ v \in \mathbb{F}_2^n, \ c \in \mathcal{C}_n\},$$

is a binary 1-perfect code.

Proof. Clearly, $\mathcal{C}_{2n+1}$ is a code of length $2n + 1 = 2^{r+1} - 1$ and $|\mathcal{C}_{2n+1}| = 2^n |\mathcal{C}_n| = 2^{2^r - 1} 2^{2^r - 1 - r} = 2^{2^{r+1} - 2 - r}$, which is the required number of codewords. Hence, to complete the proof, it suffices to show that the minimum Hamming distance of the code is 3. Assume that $c_1' = (v_1, v_1 + c_1, p(v_1) + f(c_1))$ and $c_2' = (v_2, v_2 + c_2, p(v_2) + f(c_2))$ are two distinct codewords of $\mathcal{C}_{2n+1}$. We distinguish between four cases, depending on the value of $d(v_1, v_2)$.

Case 1. $d(v_1, v_2) = 0$.

Clearly, $v_1 = v_2$ and hence $c_1 \neq c_2$, i.e., $d(c_1, c_2) \geq 3$, which implies that $d(v_1 + c_1, v_2 + c_2) = d(c_1, c_2) \geq 3$ and, therefore, $d(c_1', c_2') \geq 3$.

Case 2. $d(v_1, v_2) = 1$.

This implies that v_1 and v_2 have different parity, i.e., $p(v_1) \neq p(v_2)$. If $c_1 = c_2$, then $d(v_1 + c_1, v_2 + c_2) = d(v_1, v_2) = 1$, $d(p(v_1) + f(c_1), p(v_2) + f(c_2)) = d(p(v_1), p(v_2)) = 1$, and, therefore, $d(c_1', c_2') = 3$. If $c_1 \neq c_2$, then $d(c_1, c_2) \geq 3$, which implies that $d(v_1 + c_1, v_2 + c_2) \geq 2$ and hence $d(c_1', c_2') \geq 3$.

Case 3. $d(v_1, v_2) = 2$.

Assume first that $c_1 = c_2$, which implies that $d(v_1 + c_1, v_2 + c_2) = d(v_1, v_2) = 2$ and hence $d(c_1', c_2') \geq 4$. If $c_1 \neq c_2$, then $d(c_1, c_2) \geq 3$, which implies that $d(v_1 + c_1, v_2 + c_2) \geq 1$ and, therefore, $d(c_1', c_2') \geq 3$.

Case 4. $d(v_1, v_2) \geq 3$

This immediately implies that $d(c_1', c_2') \geq 3$.

□

Let $\mathcal{C}^*(r)$ be an extended 1-perfect code of length 2^r, and let

$\mathcal{C}_1^*, \mathcal{C}_2^*, \ldots, \mathcal{C}_{2^r}^*$ be the even translates of $\mathcal{C}^*(r)$. The next construction is based on the direct product construction.

Theorem 5.2. *If $(i_1, i_2, \ldots, i_{2^r})$ is a permutation of $[2^r]$, then the code defined by*

$$\hat{\mathcal{C}} \triangleq \{(x,y) \;:\; x \in \mathcal{C}_j^*, \;\; y \in \mathcal{C}_{i_j}^*, \;\; 1 \leq j \leq 2^r\}.$$

is an extended 1-perfect code of length 2^{r+1}.

Proof. Clearly, $\hat{\mathcal{C}}$ is a code of length 2^{r+1} and $\left|\hat{\mathcal{C}}\right| = 2^r\left|\mathcal{C}^*(r)\right|^2 = 2^r 2^{2^r-1-r} 2^{2^r-1-r} = 2^{2^{r+1}-2-r}$, which is the required number of codewords. Hence, to complete the proof, it suffices to show that the minimum Hamming distance of the code is 4. Assume that $c_1 = (x_1, y_1)$ and $c_2 = (x_2, y_2)$ are two distinct codewords of $\hat{\mathcal{C}}$ and distinguish between two cases, depending on whether x_1 and x_2 are codewords of the same translate $\mathcal{C}_j^*$ or not.

Case 1. x_1 and x_2 are codewords of the same translate $\mathcal{C}_j^*$.

If $x_1 \neq x_2$, then since x_1 and x_2 are words of the same tanslate, it follows that $d(x_1, x_2) \geq 4$, which implies that $d(c_1, c_2) \geq 4$. If $x_1 = x_2$, then y_1 and y_2 are two distinct words in the tanslate $\mathcal{C}_{i_j}^*$ and hence $d(y_1, y_2) \geq 4$, which implies that $d(c_1, c_2) \geq 4$.

Case 2. x_1 and x_2 are words of two distinct tanslates $\mathcal{C}_j^*$ and $\mathcal{C}_\ell^*$, where $j \neq \ell$.

This implies that also y_1 and y_2 are words in two distinct tanslates of $\mathcal{C}^*(r)$. Therefore, $d(x_1, x_2) \geq 2$ and $d(y_1, y_2) \geq 2$, and hence $d(c_1, c_2) \geq 4$. □

The next two constructions are variants of the general product construction. For the next construction, let $\mathcal{C}^*(r)$ be an extended 1-perfect code of length $n = 2^r$ for some $r \geq 2$. For each $b \in \mathcal{C}^*(r)$, let Q_b be a code of length $n = 2^r$, minimum distance 2, over an alphabet with $m + 1 = 2^\ell$ symbols. An example of such a code is

$$Q_b \triangleq \left\{(x_1, \ldots, x_n) \;:\; x_i \in \mathbb{Z}_{m+1}, \; 1 \leq i \leq n, \; x_n \equiv \sum_{i=1}^{n-1} x_i \pmod{m+1}\right\},$$

where clearly $|Q_b| = (m+1)^{n-1}$. Let $\mathcal{C}_1^0, \mathcal{C}_2^0, \ldots, \mathcal{C}_{m+1}^0$ be the even tanslates of an extended 1-perfect code $\mathcal{C}^*$ of length $m + 1 = 2^\ell$, where $\mathcal{C}_1^0 = \mathcal{C}^*$. Let $\mathcal{C}_1^1, \mathcal{C}_2^1, \ldots, \mathcal{C}_{m+1}^1$ be the odd tanslates of an extended 1-perfect code of length $m + 1$.

Theorem 5.3. *The code defined by*

$$\hat{\mathcal{C}} \triangleq \{(c_1, \ldots, c_n) \;:\; b = (b_1, \ldots, b_n) \in \mathcal{C}^*(r), \;\; (j_1, \ldots, j_n) \in Q_b, \;\; c_i \in \mathcal{C}_{j_i}^{b_i}\}$$

is an extended 1-perfect code of length $n(m+1) = 2^{r+\ell}$.

Proof. Since $n = 2^r$ is the length of $\mathcal{C}^*(r)$ and $m+1 = 2^\ell$ is the length of the extended 1-perfect code $\mathcal{C}^*$, i.e., it follows that the length of the code $\hat{\mathcal{C}}$ is $n(m+1) = 2^{r+\ell}$. The number of codewords in $\hat{\mathcal{C}}$ is

$$|\mathcal{C}^*(r)| \cdot |Q_b| \cdot |\mathcal{C}^*|^n = 2^{2^r-1-r}(2^\ell)^{n-1}(2^{2^\ell-1-\ell})^n = 2^{2^{r+\ell}-1-(r+\ell)},$$

which is the required number of codewords. Hence, to complete the proof, it suffices to show that the minimum Hamming distance of the code is 4.

Let $c = (c_1, c_2, \ldots, c_n)$ and $c' = (c'_1, c'_2, \ldots, c'_n)$ be two distinct codewords of $\hat{\mathcal{C}}$. These two codewords were constructed based on the two codewords $b = (b_1, b_2, \ldots, b_n)$ and $b' = (b'_1, b'_2, \ldots, b'_n)$ of $\mathcal{C}^*(r)$, respectively, and the two codewords $j = (j_1, j_2, \ldots, j_n)$ and $j' = (j'_1, j'_2, \ldots, j'_n)$, respectively, of Q_b and $Q_{b'}$, respectively.

We distinguish now between two cases, depending on whether $b = b'$ or $b \neq b'$.

Case 1. $b = b'$.

If $j = j'$, then for some i, $1 \leq i \leq n$, we have that $c_i \neq c'_i$, where $c_i, c'_i \in \mathcal{C}^{b_i}_{j_i}$, and since $d(\mathcal{C}^{b_i}_{j_i}) = 4$, it follows that $d(c_i, c'_i) \geq 4$, which implies that $d(c, c') \geq 4$.

If $j \neq j'$, then since $d(Q_b) = 2$, it follows that j and j' differ in at least two coordinates i and s, where $c_i \in \mathcal{C}^{b_i}_{j_i}$, $c'_i \in \mathcal{C}^{b_i}_{j'_i}$, $c_s \in \mathcal{C}^{b_s}_{j_s}$, $c'_s \in \mathcal{C}^{b_s}_{j'_s}$, and hence $d(c_i, c'_i) \geq 2$, $d(c_s, c'_s) \geq 2$, which implies that $d(c, c') \geq 4$.

Case 2. $b \neq b'$.

Since $b, b' \in \mathcal{C}^*(r)$, it follows that $d(b, b') \geq 4$, i.e., b and b' differ in at least four coordinates i, k, s, and t. The corresponding pairs of sub-codewords (c_i, c'_i), (c_k, c'_k), (c_s, c'_s), and (c_t, c'_t), where each of these eight sub-codewords is of length $m+1$, are associated with different odd and even tanslates of $\mathcal{C}^*$. Each pair differs in at least one coordinate and hence $d(c, c') \geq 4$.

Thus, $d(\hat{\mathcal{C}}) \geq 4$ and the proof is completed. □

The last construction that will be presented is for any alphabet and not just a binary one. This simple construction, for 1-perfect codes in the Hamming scheme, is also very effective, for example, in constructions of perfect codes in the Lee metric (see Section 11.3). For this construction, two 1-perfect codes in the Hamming scheme will be used. The first code $\mathcal{C}^1$ is a 1-perfect code of length $n = \frac{q^r-1}{q-1}$ over an alphabet with q symbols, which has a total of q^r translates, including $\mathcal{C}^1$ itself. The second code $\mathcal{C}^2$ is a 1-perfect code of length $\ell = \frac{q^{rs}-1}{q^r-1}$ over an alphabet with q^r symbols.

Let $\mathcal{C}_i^1$, $1 \leq i \leq q^r$, be the i-th translate of $\mathcal{C}^1$, where $\mathcal{C}_1^1 = \mathcal{C}^1$. Define the following code $\hat{\mathcal{C}}$:

$$\hat{\mathcal{C}} \triangleq \{(x_{i_1}, x_{i_2}, \ldots, x_{i_\ell}) \;:\; (i_1, i_2, \ldots, i_\ell) \in \mathcal{C}^2, \; x_{i_t} \in \mathcal{C}_{i_t}^1\} \,.$$

Theorem 5.4. *The code $\hat{\mathcal{C}}$ is a q-ary 1-perfect code of length $\frac{q^{rs}-1}{q-1}$.*

Proof. Clearly, the length of the codewords from $\hat{\mathcal{C}}$ is $n\ell = \frac{q^r-1}{q-1}\frac{q^{rs}-1}{q^r-1} = \frac{q^{rs}-1}{q-1}$. By the sphere-packing bound, the size of $\mathcal{C}^1$ is $\frac{q^n}{1+(q-1)n} = q^{n-r}$ and the size of $\mathcal{C}^2$ is $\frac{q^{r\ell}}{1+(q^r-1)\ell} = q^{r\ell-rs}$. Hence,

$$\left|\hat{\mathcal{C}}\right| = \left|\mathcal{C}^2\right| \cdot \left|\mathcal{C}^1\right|^\ell = q^{r\ell-rs} q^{(n-r)\ell} = q^{n\ell-rs} = \frac{q^{n\ell}}{1+(q-1)n\ell} \,,$$

which is the required number of codewords. Therefore, to complete the proof, it suffices to show that the minimum Hamming distance of the code $\hat{\mathcal{C}}$ is three.

Let $(x_{i_1}, x_{i_2}, \ldots, x_{i_\ell})$ and $(y_{j_1}, y_{j_2}, \ldots, y_{j_\ell})$ be two distinct codewords of $\hat{\mathcal{C}}$. We distinguish now between two cases depending on whether $(i_1, \ldots, i_\ell) \neq (j_1, \ldots, j_\ell)$ or $(i_1, \ldots, i_\ell) = (j_1, \ldots, j_\ell)$.

Case 1. $(i_1, \ldots, i_\ell) \neq (j_1, \ldots, j_\ell)$.

Since $(i_1, \ldots, i_\ell), (j_1, \ldots, j_\ell) \in \mathcal{C}^2$ and $d(\mathcal{C}^2) = 3$, it follows that $d((i_1, \ldots, i_\ell), (j_1, \ldots, j_\ell)) \geq 3$. W.l.o.g., we can assume that $i_1 \neq j_1$, $i_2 \neq j_2$, and $i_3 \neq j_3$, and hence $x_{i_1} \neq y_{j_1}$, $x_{i_2} \neq y_{j_2}$, and $x_{i_3} \neq y_{j_3}$, which imply that $d((x_{i_1}, \ldots, x_{i_\ell}), (y_{j_1}, \ldots, y_{j_\ell})) \geq 3$.

Case 2. $(i_1, \ldots, i_\ell) = (j_1, \ldots, j_\ell)$

This implies that there exists a t, $1 \leq t \leq \ell$, such that $x_{i_t} \neq y_{j_t}$ and since $i_t = j_t$ (i.e., $\mathcal{C}_{i_t}^1 = \mathcal{C}_{j_t}^1$), $x_{i_t}, y_{j_t} \in \mathcal{C}_{i_t}^1$ and $d(\mathcal{C}_{i_t}^1) = 3$, it follows that $d(x_{i_t}, y_{j_t}) \geq 3$ and, therefore, $d((x_{i_1}, \ldots, x_{i_\ell}), (y_{j_1}, \ldots, y_{j_\ell})) \geq 3$.

Case 1 and Case 2 imply that $d(\hat{\mathcal{C}}) \geq 3$ and thus we have that $\hat{\mathcal{C}}$ is a q-ary 1-perfect code of length $\frac{q^{rs}-1}{q-1}$. □

There are many other constructions for 1-perfect codes, besides the four above-presented ones. In particular, in Section 5.7, a construction for many nonequivalent 1-perfect codes will be presented. The construction is based on one of the most simple and effective ways to construct 1-perfect codes, known as the switching method. The construction will work on any finite field $\mathbb{F}_q$ and it will be demonstrated in its most general way. The idea behind this construction in the binary case will be also demonstrated in most of the following sections of this chapter.

5.2 Weight and Distance Distribution

It should be noted first that all the q-ary 1-perfect codes have the same weight distribution. This is an obvious observation from the following definitions and observations presented for binary codes. Let A_i be the number of codewords of weight i in a binary 1-perfect code $\mathcal{C}$ of length $n = 2\nu + 1$. Counting the covered words of weight i in $\mathbb{F}_2^n$, by codewords of $\mathcal{C}$, yields

$$(n - i + 1)A_{i-1} + A_i + (i + 1)A_{i+1} = \binom{n}{i} . \tag{5.1}$$

Obviously, if $\mathbf{0} \in \mathcal{C}$, then $A_0 = 1$ and $A_1 = 0$, whereas if $\mathbf{0} \notin \mathcal{C}$, then $A_0 = 0$ and $A_1 = 1$. This implies that all binary 1-perfect codes have one of two possible weight distributions. Assume $\mathbf{0} \in \mathcal{C}$ and let B_i be the number words of weight i in a tanslate $x + \mathcal{C}$, where $x \notin \mathcal{C}$. Then

$$A_i + nB_i = \binom{n}{i} . \tag{5.2}$$

If $\Delta_i = A_i - B_i$, then applying (5.1) for both $\mathcal{C}$ and $x + \mathcal{C}$, we have that for $0 \leq i \leq 2\nu$

$$i\Delta_i = -\Delta_{i-1} - (n - i + 2)\Delta_{i-2} , \tag{5.3}$$

where $\Delta_0 = 1$ and $\Delta_1 = -1$. There is a unique solution for the recurrence in (5.3) with these initial conditions:

$$\Delta_i = \begin{cases} \binom{\nu}{\lfloor i/2 \rfloor} & i \equiv 0, 3 \pmod 4 \\ -\binom{\nu}{\lfloor i/2 \rfloor} & i \equiv 1, 2 \pmod 4 \end{cases} . \tag{5.4}$$

Equations (5.2) and (5.4) imply and explicit expression for A_i and B_i.

Proposition 5.1. *The weight distribution of a 1-perfect code $\mathcal{C}$ of length n is given by*

$$A_i = \frac{\binom{n}{i} + n\Delta_i}{n + 1}.$$

The weight distribution of a translate $x + \mathcal{C}$, $x \notin \mathcal{C}$ is given by

$$B_i = \frac{\binom{n}{i} - \Delta_i}{n + 1},$$

where $0 \leq i \leq n$.

Corollary 5.1. *If the all-zero word is a codeword of the 1-perfect code $\mathcal{C}$, then the all-one word is also a codeword in $\mathcal{C}$.*

The analysis of binary 1-perfect codes can be generalized for all q-ary 1-perfect codes. This analysis is a good exercise for the reader. We end this section with the distance distribution of a 1-perfect code.

Theorem 5.5. *The distance distribution and the weight distribution of a 1-perfect code (over any alphabet) coincide. In other words, for each* i, $0 \leq i \leq n$, $D_i = A_i$, *where* A_i *is the distance distribution of a 1-perfect code containing the all-ones word.*

Proof. Let $\mathcal{C}$ be a 1-perfect code for which $\mathbf{0} \in \mathcal{C}$, let $c \in \mathcal{C}$ be any codeword, and let $D_i(c)$ be the number of codewords at distance i from c. Clearly, $\mathbf{0} \in c+\mathcal{C}$ and hence by Corollary 5.1, $\mathbf{1} \in c+\mathcal{C}$ which implies that the translate $\mathbf{1} + c + \mathcal{C}$ is a 1-perfect code for which $\mathbf{0} \in \mathbf{1} + \mathbf{c} + \mathcal{C}$, and therefore its weight distribution is the same as the weight distribution of $\mathcal{C}$. Moreover, $D_i(c)$ is equal to A_i in $c+\mathcal{C}$. Since this analysis is the same for each codeword c in $\mathcal{C}$, it follows that the distance distribution of $\mathcal{C}$ coincides with the weight distribution of $\mathcal{C}$. □

The same arguments as in Theorem 5.5 hold for all perfect codes in the Hamming scheme. Unfortunately, as already mentioned, there are no such codes except for codes with the parameters of the Hamming codes and the Golay codes.

5.3 Intersection Numbers

Let $\mathcal{C}_1$ and $\mathcal{C}_2$ be two distinct binary 1-perfect codes of length $n = 2^r - 1$. What is the maximum possible cardinality of their intersection $\mathcal{C}_1 \cap \mathcal{C}_2$?

For a positive integer $n = 2\nu+1$, let $\mathcal{U}$ be a nonempty subset of $\mathbb{F}_2^n$ such that there exist two disjoint perfect coverings of $\mathcal{U}$ with balls of radius one. Namely, let $\mathcal{A}$ and $\mathcal{B}$ be two distinct sub-codes of $\mathcal{U}$, such that any vector of $\mathcal{U}$ is within distance one from a unique codeword of $\mathcal{A}$ and a unique codeword of $\mathcal{B}$ and all the words within radius one from $\mathcal{A}$ and from $\mathcal{B}$ are contained in $\mathcal{U}$. Each such sub-code is said to ***perfectly cover*** $\mathcal{U}$.

Lemma 5.1.

$$|\mathcal{U}| \geq (\nu+1)2^{\nu+1} \ .$$

Proof. W.l.o.g. assume that $\mathbf{0} \in \mathcal{A}$. Assume also that $\mathcal{A} \cap \mathcal{B} = \varnothing$, since otherwise $\mathcal{U}$ is not minimal. Hence, $\mathbf{0} \notin \mathcal{B}$ and $\mathcal{B}$ contains a unique codeword of weight one. Let A_i and B_i denote the number of codewords of weight i

in $\mathcal{A}$ and $\mathcal{B}$, respectively. Counting the number of words of weight i in $\mathcal{U}$, we have for $1 \leq i \leq 2\nu$,

$$(n-i+1)A_{i-1}+A_i+(i+1)A_{i+1} = (n-i+1)B_{i-1}+B_i+(i+1)B_{i+1} \ . \quad (5.5)$$

If $\Delta_i = A_i - B_i$, then (5.5) implies the following recurrence

$$i\Delta_i = -\Delta_{i-1} - (n-i+2)\Delta_{i-2} \ . \quad (5.6)$$

We have already established that $A_0 = 1$, $A_1 = 0$ and $B_0 = 0$, $B_1 = 1$, and hence, $\Delta_0 = 1$ and $\Delta_1 = -1$. The unique solution of the recurrence (5.6) with these initial conditions is provided in (5.4). Obviously,

$$|\mathcal{U}| = (1+n)\,|\mathcal{A}| = (1+n)\sum_{i=0}^{n} A_i \ .$$

Hence

$$\frac{|\mathcal{U}|}{1+n} = \sum_{\substack{i=0 \\ i\equiv 0,3 \pmod 4}}^{n} A_i + \sum_{\substack{i=0 \\ i\equiv 1,2 \pmod 4}}^{n} A_i \geq \sum_{\substack{i=0 \\ i\equiv 0,3 \pmod 4}}^{n} A_i$$

$$= \sum_{\substack{i=0 \\ i\equiv 0,3 \pmod 4}}^{n} (\Delta_i + B_i) \geq \sum_{\substack{i=0 \\ i\equiv 0,3 \pmod 4}}^{n} \Delta_i = \sum_{\substack{i=0 \\ i\equiv 0,3 \pmod 4}}^{n} \binom{\nu}{\lfloor i/2 \rfloor} .$$

Substituting $j = \lfloor i/2 \rfloor$ in this inequality yields

$$|\mathcal{U}| \geq (1+n)\sum_{j=0}^{\nu} \binom{\nu}{j} = (\nu+1)2^{\nu+1} \ .$$

□

Corollary 5.2. *Let $\mathcal{C}_1$ and $\mathcal{C}_2$ be two distinct 1-perfect codes of length $n = 2^r - 1$. Then*

$$|\mathcal{C}_1 \cap \mathcal{C}_2| \leq 2^{n-r} - 2^{\nu} \ ,$$

where $\nu = (n-1)/2$.

Proof. For distinct 1-perfect codes $\mathcal{C}_1$ and $\mathcal{C}_2$ we may always take $\mathcal{A} = \mathcal{C}_1 \setminus \mathcal{C}_2$ and $\mathcal{B} = \mathcal{C}_2 \setminus \mathcal{C}_1$, to be codes that perfectly cover the same subset $\mathcal{U}$ of $\mathbb{F}_2^n$. Therefore, the upper bound follows immediately by Lemma 5.1. □

We now construct two codes $\mathcal{C}_1$ and $\mathcal{C}_2$, such that the cardinality of $\mathcal{C}_1 \cap \mathcal{C}_2$ attains the upper bound of Corollary 5.2. Let $\mathcal{H}(r)$ be the Hamming code of length $n = 2\nu + 1 = 2^r - 1$, and let H be its parity-check matrix. Further, assume that the columns of H, i.e., $h_1, h_2, \ldots, h_n$, are arranged such that for some column vector $z = h_n$,

$$h_i + h_{i+\nu} = z \quad \text{for all } i = 1, 2, \ldots, \nu\ . \tag{5.7}$$

Let $\mathcal{C}_1$ be a coset of $\mathcal{H}(r)$ such that the syndrome $\mathcal{S}(c) = Hc^{\text{tr}} = z$ for all $c \in \mathcal{C}_1$. Define

$$\mathcal{A} \triangleq \{(x, x, p(x)) \ :\ x \in \mathbb{F}_2^\nu\}, \qquad \mathcal{B} \triangleq \{(x, x, p(x) + 1) \ :\ x \in \mathbb{F}_2^\nu\}\ . \tag{5.8}$$

Let $\mathcal{C}_2 = (\mathcal{C}_1 \setminus \mathcal{B}) \cup \mathcal{A}$; this removal of $\mathcal{B}$ and its replacement by $\mathcal{A}$ is the basic operation in the switching method.

Proposition 5.2. *The code $\mathcal{C}_2$ is a 1-perfect code.*

Proof. Obviously $\mathcal{A} \subset \mathcal{H}(r)$, and $\mathcal{B} \subset \mathcal{C}_1$, where $\mathcal{B}$ is a coset of $\mathcal{A}$, which is disjoint to $\mathcal{H}(r)$. Hence, $|\mathcal{C}_2| = |\mathcal{C}_1| - |\mathcal{B}| + |\mathcal{A}| = 2^{n-r}$. Indeed, $d(\mathcal{A}) = \min_{a_1,a_2 \in \mathcal{A}} d(a_1, a_2) = 3$ and $d(\mathcal{C}_1 \setminus \mathcal{B}) = 3$. Now, let $v = a + c$, where $a \in \mathcal{A}$ and $c \in \mathcal{C}_1$. Clearly, since $\mathcal{A} \subset \mathcal{H}(r)$, it follows that for the syndrome of a, $\mathcal{S}(a)$, we have $\mathcal{S}(a) = 0$, and for the syndrome of v, $\mathcal{S}(v)$, we have $\mathcal{S}(v) = \mathcal{S}(a) + \mathcal{S}(c) = \mathcal{S}(c) = z$. Hence, if $\text{wt}(v) \leq 2$, then either $v = (\mathbf{0}, \mathbf{0}, 1)$ or $v = (u, u, 0)$ where $\mathbf{0}, u \in \mathbb{F}_2^\nu$ and u is a vector of weight one; but, then $c = a + v$ is either $(x, x, p(x) + 1)$ or $(x + u, x + u, p(x))$, for some $x \in \mathbb{F}_2^\nu$. Since $p(x) = p(x + u) + 1$, it follows that, in both cases, $c \in \mathcal{B}$. Thus, $d(\mathcal{A}, \mathcal{C}_1 \setminus \mathcal{B}) \triangleq \min_{a \in \mathcal{A}, c \in \mathcal{C}_1 \setminus \mathcal{B}} d(a, c) \geq 3$, and, therefore, $\mathcal{C}_2$ is a 1-perfect code. $\square$

By the construction, $|\mathcal{C}_1 \cap \mathcal{C}_2| = |\mathcal{C}_1| - |\mathcal{B}| = 2^{n-r} - 2^\nu$. Thus, Proposition 5.2 shows that the upper bound of Corollary 5.2 is attainable for all n. Another consequence of Proposition 5.2 is very important in all our discussions of 1-perfect codes.

Corollary 5.3. *The sets $\mathcal{A}$ and $\mathcal{B}$ defined in (5.8) perfectly cover the same subset of $\mathbb{F}_2^n$.*

Proof. This follows immediately from the fact that both $(\mathcal{C}_1 \setminus \mathcal{B}) \cup \mathcal{A}$ and $(\mathcal{C}_1 \setminus \mathcal{B}) \cup \mathcal{B}$ are perfect. $\square$

The method implied by Proposition 5.2 is called the ***switching method***. Since the set $\mathcal{A}$ defined in (5.8) is a linear sub-code of $\mathcal{H}(r)$, it follows that $\mathcal{H}(r)$ can be partitioned into cosets of $\mathcal{A}$, i.e., $\mathcal{H}(r)$ is a union of disjoint cosets of $\mathcal{A}$. Each such coset $x + \mathcal{A}$ can be replaced by $x + \mathcal{B}$ to obtain another 1-perfect code. This idea will be generalized and used in Section 5.7 to construct many inequivalent 1-perfect codes over $\mathbb{F}_q$. There are many other such isomorphic sub-codes of the Hamming code that are isomorphic to $\mathcal{A}$. Furthermore, we will show that there are many sub-codes that are disjoint. This will enable us to construct many 1-perfect codes with desired properties.

Having settled the largest possible intersection question, we now consider its natural counterpart: What is the minimum possible cardinality of the (nonempty) intersection of two 1-perfect codes? Clearly, $|\mathcal{C}_2 \cap \mathcal{H}(r)| = |\mathcal{A}| = 2^\nu$; however, the intersection of these two codes is not the smallest possible. We presently construct a 1-perfect code $\mathcal{C}_2'$, such that the cardinality of $\mathcal{C}_2 \cap \mathcal{C}_2'$ is less than 2^ν.

Let W be a subspace of $\mathbb{F}_2^r$ of dimension $r-1$, such that $z \notin W$. Then one way to arrange the columns of H so that (5.7) is satisfied is to take

$$\{h_1, h_2, \ldots, h_\nu\} = W \setminus \{\mathbf{0}\},$$

$$\{h_{\nu+1}, h_{\nu+2}, \ldots, h_n\} = z + W. \tag{5.9}$$

If (5.9) is employed, then the order of all columns in H is determined by the order in which the nonzero elements of W are listed in $\{h_1, h_2, \ldots, h_n\}$. The latter is, however, completely arbitrary. We may further restrict the order of columns in H as follows. Set $z' = h_\nu$ and arrange the columns $h_1, h_2, \ldots, h_{\nu-1}$ such that

$$h_i + h_{i+\nu'} = z' \quad \text{for all } i = 1, 2, \ldots, \nu', \tag{5.10}$$

where $\nu' = (\nu - 1)/2$. It follows from (5.7) and (5.10) that we also have $h_{i+\nu} + h_{i+\nu+\nu'} = (z + h_i) + (z + h_{i+\nu'}) = z'$, for all $i = 1, 2, \ldots, \nu'$. Furthermore, $h_{n-1} + h_n = z'$, in view of (5.7). Hence define

$$\mathcal{A}' \triangleq \{(x, x, p(x+y) + \alpha, y, y, \alpha, \alpha) \ : \ x, y \in \mathbb{F}_2^{\nu'}, \ \alpha \in \mathbb{F}_2\}.$$

$$\mathcal{B}' \triangleq \{(x, x, p(x+y) + \alpha + 1, y, y, \alpha, \alpha) \ : \ x, y \in \mathbb{F}_2^{\nu'}, \ \alpha \in \mathbb{F}_2\}. \tag{5.11}$$

By the construction, $\mathcal{A}' \subset \mathcal{H}(r)$. Let $\mathcal{C}_1'$ be a coset of $\mathcal{H}(r)$, such that the syndrome of all the vectors in $\mathcal{C}_1'$ is z'. Then, obviously, $\mathcal{B}' \subset \mathcal{C}_1'$. Denote $\mathcal{C}_2' = (\mathcal{C}_1' \setminus \mathcal{B}') \cup \mathcal{A}'$. To see that $\mathcal{C}_2'$ is a 1-perfect code, note that $\mathcal{A}'$ and $\mathcal{B}'$ are isomorphic to $\mathcal{A}$ and $\mathcal{B}$, and hence perfectly cover the same subset of $\mathbb{F}_2^n$ by Corollary 5.3.

Proposition 5.3. $|\mathcal{C}_2 \cap \mathcal{C}_2'| = 2^{\nu'+1}$.

Proof. Clearly, $\mathcal{H}(r)$, $\mathcal{C}_1$, and $\mathcal{C}_1'$ are disjoint. Thus, $\mathcal{C}_2 \cap \mathcal{C}_2' = \mathcal{A} \cap \mathcal{A}'$. Note that both $\mathcal{A}$ and $\mathcal{A}'$ are linear, and hence so is $\mathcal{A} \cap \mathcal{A}'$. The word $a_0 \in \mathbb{F}_2^n$ of weight 3 with nonzero elements at positions ν, $n-1$, n is in $\mathcal{A} \cap \mathcal{A}'$. For $i = 1, 2, \ldots, \nu'$, let $a_i \in \mathbb{F}_2^n$ be a word of weight 4 with nonzeroes at positions $1, i+\nu', i+\nu, i+\nu'+\nu$. This implies again that $a_i \in \mathcal{A} \cap \mathcal{A}'$. Since the words $a_0, a_1, \ldots, a_{\nu'}$ have disjoint supports, it follows that they are linearly independent. Furthermore, it may be readily verified that these vectors generate $\mathcal{A} \cap \mathcal{A}'$. □

Proposition 5.3 establishes the intersection of cardinality $2^{\nu'+1}$. This is not the minimum possible intersection for most parameters. For $n = 7$ it is easy to find two isomorphic 1-perfect codes for which the intersection is $\{\mathbf{0}, \mathbf{1}\}$. For larger n, we find a smaller intersection.

It is obvious that the intersection problem, in general, has the same answer for 1-perfect codes and for extended 1-perfect codes. We will use this simple fact later. This is stated formally in the following lemma.

Lemma 5.2. *Two 1-perfect codes of length $2^r - 1$ with intersection number η exist if and only if there exist two extended 1-perfect codes of length 2^r with intersection number η.*

We now use a combination of the directed product construction in Theorem 5.2 and the switching method to construct two extended 1-perfect codes with intersection number 2. Let $\mathcal{H}^*(r)$ be an extended Hamming code of length 2^r, and let $\hat{\mathcal{H}}_0^*(r), \hat{\mathcal{H}}_1^*(r), \ldots, \hat{\mathcal{H}}_{2^r-1}^*(r)$ be the even cosets of $\mathcal{H}^*(r)$ in $\mathbb{E}_2^{2^r}$. Thus, $\hat{\mathcal{H}}_0^*(r), \hat{\mathcal{H}}_1^*(r), \ldots, \hat{\mathcal{H}}_{2^r-1}^*(r)$ is a partition of $\mathbb{E}_2^{2^r}$ into extended 1-perfect codes. Hence, the code

$$\mathcal{C}^* \triangleq \{(x, y) \ : \ x, y \in \hat{\mathcal{H}}_i^*(r) \ \text{ for some } i = 0, 1, \ldots, 2^r - 1\} \tag{5.12}$$

is an extended 1-perfect code of length 2^{r+1} obtained by the direct product construction presented in Theorem 5.2, with π being the identity permutation. Furthermore, it can easily be verified that $\mathcal{C}^*$ is a linear code, and hence it must be $\mathcal{H}^*(r+1)$. W.l.o.g., assume that the parity-check matrix of $\mathcal{C}^*$ is given by

$$H_{r+1} = \begin{bmatrix} 0\,0\,0\,0 \cdots 1\,1 \\ \vdots\;\vdots\;\vdots\;\vdots\;\ddots\;\vdots\;\vdots \\ 0\,0\,1\,1 \cdots 1\,1 \\ 0\,1\,0\,1 \cdots 0\,1 \\ 1\,1\,1\,1 \cdots 1\,1 \end{bmatrix}.$$

That is, the columns of H_{r+1} are all the $(r+1)$-tuples that end with a *one*, ordered lexicographically. Indeed, it is easy to see that

$$H_{r+1} = \left[\begin{array}{c|c} 0 \cdots 0 & 1 \cdots 1 \\ \hline H_r & H_r \end{array}\right],$$

where H_r is the parity-check matrix for $\mathcal{H}^*(r) = \hat{\mathcal{H}}_0^*(r)$. Thus, the code defined by the parity-check matrix H_{r+1} is consistent with the parity-check matrix for the code C^* defined in (5.12). Notice that all the words in a given coset of $\mathcal{H}^*(r)$ have the same syndrome with respect to H_r. That is, for all $i = 0, 1, \ldots, 2^r - 1$, we have

$$s_i \triangleq \mathcal{S}(x) = H_r x^{\text{tr}} \quad \text{for all} \quad x \in \hat{\mathcal{H}}_i^*(r), \tag{5.13}$$

and we say that s_i is the syndrome of $\hat{\mathcal{H}}_i^*(r)$.

We now modify the extended Hamming code $\mathcal{C}$ in (5.12) in an appropriate manner by using the switching method. Let

$$\mathcal{A}^* \triangleq \{(x, x) \ : \ x \in \hat{\mathcal{H}}_i^*(r) \ \text{ for some } i = 0, 1, \ldots, 2^r - 1\} \,. \tag{5.14}$$

Comparing (5.12) and (5.14), we see that $\mathcal{A}^*$ is a sub-code of $\mathcal{C}^*$. Furthermore, since the codes $\hat{\mathcal{H}}_0^*(r), \hat{\mathcal{H}}_1^*(r), \hat{\mathcal{H}}_2^*(r), \ldots, \hat{\mathcal{H}}_{2^r-1}^*(r)$ form a partition of $\mathbb{E}_2^{2^r}$, we can write

$$\mathcal{A}^* = \{(x, x) \ : \ x \in \mathbb{E}_2^{2^r}\},$$

which implies that $\mathcal{A}^*$ is just an extended code of $\mathcal{A}$ in (5.8). Pick a fixed integer j in the range $1 \le j \le 2^r$, and let $\mathcal{B}^* = (e_j, e_j) + \mathcal{A}^*$. Then $\mathcal{B}^*$ is the extended code of $\mathcal{B}$ in (5.8). By the switching method, this implies that the code

$$\mathcal{C}_1^* = (\mathcal{C}^* \setminus \mathcal{A}^*) \cup \mathcal{B}^*$$

is also an extended 1-perfect code. Note that $\mathcal{C}_1^*$ does not contain the all-zero word; however, the translate $\mathcal{C}_2^* = (e_j, e_j) + \mathcal{C}_1^*$ does. This translate is an extended 1-perfect code, which can be written as $\mathcal{C}_2^* = \mathcal{A}^* \cup \mathcal{D}^*$, where

$$\mathcal{D}^* = \{(x+e_j, y+e_j) \ : \ x, y \in \hat{\mathcal{H}}_i^*(r), \text{ and } x \neq y, \text{ for some } i = 0, 1, \ldots, 2^r-1\}.$$

Now, let π be the permutation that fixes the last 2^r coordinates of $\mathcal{C}_2^*$ and affects the cyclic shift by one position on the first 2^r coordinates. If $\mathcal{C}_3^* \triangleq \pi(\mathcal{C}_2^*)$, then, obviously, $\mathcal{C}_3^*$ is also an extended 1-perfect code and we have the following.

Theorem 5.6. *The intersection number of* $\mathcal{C}_2^*$ *and* $\mathcal{C}_3^*$ *is* $\eta(\mathcal{C}_2^*, \mathcal{C}_3^*) = 2$.

Proof. Consider the structure of a word $(x, y) \in \mathcal{C}_2^*$, for some $x, y \in \mathbb{F}_2^{2^r}$. Clearly, $\mathrm{wt}(x) \equiv \mathrm{wt}(y) \equiv 0 \pmod 2$ if and only if $x = y$ and $(x, y) \in \mathcal{A}^*$, while $\mathrm{wt}(x) \equiv \mathrm{wt}(y) \equiv 1 \pmod 2$ if and only if $(x, y) \in \mathcal{D}^*$. Since the permutation π preserves the weight of x and y, we have

$$\mathcal{C}_2^* \cap \mathcal{C}_3^* = (\mathcal{A}^* \cap \pi(\mathcal{A}^*)) \cup (\mathcal{D}^* \cap \pi(\mathcal{D}^*)). \tag{5.15}$$

A vector $x \in \mathbb{F}_2^{2^r}$ is equal to its own cyclic shift by one position if and only if $x \in \{\mathbf{0}, \mathbf{1}\}$. Hence, $\mathcal{A}^* \cap \pi(\mathcal{A}^*) = \{\mathbf{0}, \mathbf{1}\}$. We now show that $\mathcal{D}^* \cap \pi(\mathcal{D}^*) = \varnothing$. First, notice that for each $(x, y) \in \mathcal{D}^*$, we have

$$H_r x^{\mathrm{tr}} = H_r y^{\mathrm{tr}} = s_i + H_r(e_j)^{\mathrm{tr}} \tag{5.16}$$

for some $i = 0, 1, \ldots, 2^r - 1$, where s_i is taken from (5.13). On the other hand, it can be shown that if $(x, y) \in \pi(\mathcal{D}^*)$, then $H_r x^{\mathrm{tr}} \neq H_r y^{\mathrm{tr}}$. Indeed, let $(x', y') \in \mathcal{D}^*$ be the pre-image of (x, y) under π. That is, $y = y'$ and x is the cyclic shift of x' by one position. Then $H_r y^{\mathrm{tr}} = H_r(y')^{\mathrm{tr}} = H_r(x')^{\mathrm{tr}}$ by (5.16). Now, both x' and its cyclic shift x have odd weight, and, therefore,

$$(0101 \;\cdots\; 01)(x')^{\mathrm{tr}} \neq (0101 \;\cdots\; 01)x^{\mathrm{tr}} \,.$$

Since $(0101 \cdots 01)$ is a row of H_r, it follows that $H_r x^{\mathrm{tr}} \neq H_r(x')^{\mathrm{tr}} = H_r y^{\mathrm{tr}}$. Comparing this with (5.16), we conclude that $\mathcal{D}^* \cap \pi(\mathcal{D}^*) = \varnothing$. In conjunction with (5.15), this implies that $\mathcal{C}_2^* \cap \mathcal{C}_3^* = \mathcal{A}^* \cap \pi(\mathcal{A}^*) = \{\mathbf{0}, \mathbf{1}\}$, and, therefore, $\eta(\mathcal{C}_2^*, \mathcal{C}_3^*) = 2$. □

It follows from Theorem 5.6 and Corollary 5.2 that the intersection number of any two distinct 1-perfect codes $\mathcal{C}_2$, $\mathcal{C}_3$ of length $n = 2^r - 1$ is in the range

$$2 \leq \eta(\mathcal{C}_2, \mathcal{C}_3) \leq 2^{2^r - r - 1} - 2^{2^{r-1} - 1} \tag{5.17}$$

and both the lower biund and the upper bound are achievable for all $r \geq 3$. Since binary 1-perfect codes are self-complement, their intersection numbers must be even. Thus, a natural question is which even integers in the range of (5.17) are intersection numbers of 1-perfect codes of length $2^r - 1$? Since the code $\mathcal{A}$ in (5.8) is a linear sub-code of the Hamming code, using the switching method technique iteratively and using a similar proof to the one in the proof of Proposition 5.2, we obtain intersection numbers of the form

$$\kappa 2^{2^{r-1} - 1} \quad \text{for all} \quad \kappa = 1, 2, \ldots, 2^{2^{r-1} - r} - 1 \,.$$

The same technique is also used iteratively in Section 5.7 to obtain nonequivalent 1-perfect codes over $\mathbb{F}_q$. Furthermore, using Theorem 5.2 and the switching method, we can obtain many more intersection numbers.

5.4 Intersection Numbers of Linear Codes

A variant of the intersection problem asks for all the intersection numbers of *linear* 1-perfect codes, namely, the Hamming codes of length $2^r - 1$. In what follows, we provide a complete solution to this problem.

Let $\mathcal{H}_1(r)$, $\mathcal{H}_2(r)$ be two Hamming codes of length $n = 2^r - 1$. Since Hamming codes are unique, $\mathcal{H}_1(r)$ and $\mathcal{H}_2(r)$ are necessarily isomorphic. Since both codes are linear, their intersection number is necessarily a power of 2. For $r = 3$ and $n = 7$, it is easy to find specific permutations such that $\eta(\mathcal{H}_1(r), \mathcal{H}_2(r)) = 2, 4$ or 8. For example, let $\mathcal{H}_1(r)$ be the code defined by the parity-check matrix whose columns are ordered lexicographically, and let $\mathcal{H}_2(r)$ be a code defined by the parity-check matrix

$$\begin{bmatrix} 0\,0\,1\,1\,0\,1\,1 \\ 0\,1\,1\,0\,1\,1\,0 \\ 1\,0\,0\,0\,1\,1\,1 \end{bmatrix} \text{ or } \begin{bmatrix} 0\,0\,1\,1\,0\,1\,1 \\ 0\,1\,1\,0\,1\,0\,1 \\ 1\,0\,0\,0\,1\,1\,1 \end{bmatrix} \text{ or } \begin{bmatrix} 0\,0\,0\,1\,1\,1\,1 \\ 0\,1\,1\,0\,0\,1\,1 \\ 1\,1\,0\,0\,1\,0\,1 \end{bmatrix}, \tag{5.18}$$

respectively. We will show that a similar situation occurs for all $r \geq 3$; namely, all the powers of 2 in the range $2^{n-2r}, 2^{n-2r+1}, \ldots, 2^{n-r-1}$ are attainable as intersection numbers of distinct Hamming codes of length $n = 2^r - 1$. Note that a smaller intersection between distinct Hamming codes is not possible.

Let H_1, H_2 be parity-check matrices of the Hamming codes of length $n = 2^r - 1$, $\mathcal{H}_1(r)$ and $\mathcal{H}_2(r)$, respectively. Then $\mathcal{C} = \mathcal{H}_1(r) \cap \mathcal{H}_2(r)$ is a linear code, whose parity-check matrix is given by

$$H = \left[\frac{H_1}{H_2}\right]. \tag{5.19}$$

For the sake of brevity, we henceforth write $H = H_1||H_2$ to denote the structure of (5.19). It is obvious that rank $H \leq 2r$, since H_1 and H_2 each have r rows and, therefore,

$$\eta(\mathcal{H}_1(r), \mathcal{H}_2(r)) = |\mathcal{C}| = 2^{n-\text{rank } H} \geq 2^{n-2r}.$$

It is also obvious that $\eta(\mathcal{H}_1(r), \mathcal{H}_2(r)) \leq 2^{n-r-1}$ if the codes $\mathcal{H}_1(r)$ and $\mathcal{H}_2(r)$ are distinct.

Lemma 5.3. *For each $r \geq 3$, there exist two Hamming codes $\mathcal{H}_1(r)$, $\mathcal{H}_2(r)$ of length $n = 2^r - 1$ such that $\eta(\mathcal{H}_1(r), \mathcal{H}_2(r)) = 2^{n-2r}$.*

Proof. As $\eta(\mathcal{H}_1(r), \mathcal{H}_2(r)) = 2^{n-\text{rank } H}$, we need to construct parity-check matrices H_1 and H_2 for the codes $\mathcal{H}_1(r)$ and $\mathcal{H}_2(r)$ such that rank $(H_1||H_2) = 2r$. We first show that there exists a $(2r) \times (2r)$ binary matrix $A_r = A_1||A_2$, where A_1, A_2 are two $r \times (2r)$ binary matrices whose columns are distinct and nonzero, such that rank $A_r = 2r$. For $r = 3$, such a matrix is given by

$$A_3 = \begin{bmatrix} 1\,0\,0\,0\,1\,1 \\ 0\,1\,0\,1\,0\,1 \\ 0\,0\,1\,1\,1\,0 \\ \hline 1\,0\,0\,1\,0\,1 \\ 0\,1\,0\,1\,1\,1 \\ 0\,0\,1\,0\,1\,1 \end{bmatrix} .$$

For $r \geq 4$, we can construct A_r recursively as follows. Suppose that $A_{r-1} = A_1'||A_2'$, and take

$$A_r = \left[\frac{A_1}{A_2}\right] = \left[\begin{array}{c|c|c} 1 & 0 \cdots 0 & 0 \\ \hline \mathbf{0} & A_1' & x \\ \hline \mathbf{1} & A_2' & \mathbf{0} \\ \hline 1 & 0 \cdots 0 & 1 \end{array}\right], \tag{5.20}$$

where x is any nonzero $(r-1)$-tuple that does not appear as a column of A_1'. It is easy to see from (5.20) that if A_{r-1} is a nonsingular matrix of rank $2(r-1)$, then A_r is a nonsingular matrix of rank $2r$. Now, since the columns of A_1 and A_2 are nonzero and distinct, it follows that these matrices can be extended, in an arbitrary manner, to parity-check matrices H_1 and H_2, respectively, of two Hamming codes of length $2^r - 1$. By this construction, we have that rank $(H_1||H_2) = \text{rank } (A_1||A_2) = 2r$. □

Theorem 5.7. *For each $r \geq 3$, there exist two Hamming codes $\mathcal{H}_1(r)$, $\mathcal{H}_2(r)$ of length $n = 2^r - 1$, such that*

$$\eta(\mathcal{H}_1(r), \mathcal{H}_2(r)) = 2^{n-t} \quad \textit{for} \quad t = r+1, r+2, \ldots, 2r \ .$$

Proof. The proof is by induction of r. The induction basis for $r = 3$ is established in (5.18). Assume that, for each $t = r, r+1, \ldots, 2(r-1)$, there exist parity-check matrices H_1' and H_2' of two Hamming codes of length $2^{r-1} - 1$, such that rank $(H_1'||H_2') = t$. Take

$$H_1 = \left[\begin{array}{c|c|c} 0 \cdots 0 & 1 & 1 \cdots 1 \\ \hline H_1' & \mathbf{0} & H_1' \end{array}\right], \qquad H_2 = \left[\begin{array}{c|c|c} 0 \cdots 0 & 1 & 1 \cdots 1 \\ \hline H_2' & \mathbf{0} & H_2' \end{array}\right] .$$

It is easy to see that H_1, H_2 are parity-check matrices of isomorphic Hamming codes of length $2^r - 1$, and that

$$\text{rank } (H_1||H_2) = \text{rank } (H_1'||H_2') + 1 = t + 1.$$

Thus, all ranks in the range $t + 1 = r+1, r+2, \ldots, 2r-1$ are attainable. Finally, the rank of $2r$ is also attainable by Lemma 5.3, which completes the induction step. □

5.5 Full-Rank Perfect Codes

A code $\mathcal{C} \in \mathbb{F}_2^n$ is of ***full rank*** if rank $\mathcal{C} = n$, or equivalently $\langle \mathcal{C} \rangle = \mathbb{F}_2^n$, where $\langle \mathcal{C} \rangle$ is the linear span of the words in $\mathcal{C}$. Does there exist a full-rank 1-perfect code? In this section such codes will be constructed. The constructed method will imply that there exist 1-perfect codes of length $n = 2^r - 1$ and any rank in the range between $2^r - r - 1$ and $2^r - 1$.

Let $h_1, h_2, \ldots, h_n$ be the columns of H, the parity-check matrix of the Hamming code $\mathcal{H}(r)$ of length $n = 2^r - 1$, $n = 2\nu + 1$, arranged in some fixed order. Let z be any nonzero vector in $\mathbb{F}_2^r$. Obviously, there is a unique $\ell \in [n]$ such that $z = h_\ell$. We denote this index ℓ by $\varphi(z)$. Further, the vector z induces a partition of the columns $h_1, h_2, \ldots, h_{\varphi(z)-1}, h_{\varphi(z)+1}, \ldots, h_n$ into ν pairs (h_i, h_j), such that $h_i + h_j = z$ (let $j = \phi_z(i)$ and $i = \phi_z(j)$). Requiring, in addition to the above, that $i < j$ makes the partition unique. More precisely, there is a unique set $I_z \subset [n] \setminus \{\varphi(z)\}$ of cardinality ν, such that $h_i + h_{\phi_z(i)} = z$ and $i < \phi_z(i)$ for all $i \in I_z$. With this notation we may define for each $z \in \mathbb{F}_2^r \setminus \{\mathbf{0}\}$, the sets $\mathcal{A}(z)$ and $\mathcal{B}(z)$ as follows,

$$\mathcal{A}(z) \triangleq \left\{ (x_1, x_2, \ldots, x_n) \in \mathbb{F}_2^n \ : \ \forall i \in I_z, \ x_i = x_{\phi_z(i)} \text{ and } x_{\varphi(z)} = \sum_{i \in I} x_i \right\},$$

$$\mathcal{B}(z) \triangleq \left\{ (x_1, \ldots, x_n) \in \mathbb{F}_2^n \ : \ \forall i \in I_z, \ x_i = x_{\phi_z(i)} \text{ and } x_{\varphi(z)} = 1 + \sum_{i \in I} x_i \right\},$$

where the summation determining $x_{\varphi(z)}$ is performed modulo 2. By the construction, $\mathcal{A}(z) \subset \mathcal{H}(r)$ for all z. One can easily verify that $\mathcal{A}(z_1)$ and $\mathcal{B}(z_1)$ are isomorphic to $\mathcal{A}(z_2)$ and $\mathcal{B}(z_2)$, respectively, for all $z_1, z_2 \in \mathbb{F}_2^r$. They are also isomorphic to the sets $\mathcal{A}$ and $\mathcal{B}$, respectively, defined in (5.8). By Corollary 5.3, for all z, $\mathcal{A}(z)$ and $\mathcal{B}(z)$ perfectly cover the same subset of $\mathbb{F}_2^n$.

We are now in a position to describe the method for the construction of full-rank 1-perfect codes. We shall construct these codes from the Hamming code $\mathcal{H}(r)$ by the a "cut and paste" method. That is, some r disjoint subsets of $\mathcal{H}(r)$, isomorphic to the code $\mathcal{A}(z)$, are removed from $\mathcal{H}(r)$. Then cosets of these subsets are pasted in their place so that the resulting code is perfect and has full rank.

Let $r \geq 4$, and for a positive integer $k \leq r$, let $z_1, z_2, \ldots, z_k$ be some k linearly independent vectors in $\mathbb{F}_2^r$. Since a word of $\mathcal{B}(z)$ differs from some codeword of $\mathcal{H}(r)$ only in coordinate $\varphi(z)$ and $d(\mathcal{H}(r)) = 3$, it follows that

$$\mathcal{B}(z_1) \cap \mathcal{B}(z_2) = \varnothing \tag{5.21}$$

for any $z_1 \neq z_2$. Nevertheless, it follows from the proof of Proposition 5.3 that $|\mathcal{A}(z_1) \cap \mathcal{A}(z_2)| = 2^{0.5(\nu+1)}$ for all $z_1 \neq z_2$. Thus, the first task is to find k codewords $c_1, c_2, \ldots, c_k$ in $\mathcal{H}(r)$, such that the sets $c_1 + \mathcal{A}(z_1)$, $c_2 + \mathcal{A}(z_2)$,...,$c_k + \mathcal{A}(z_k)$ are disjoint.

Lemma 5.4. *There exist $c_1, c_2, \ldots, c_k \in \mathcal{H}(r)$, such that*

$$(c_i + \mathcal{A}(z_i)) \cap (c_j + \mathcal{A}(z_j)) = \varnothing,$$

for any distinct $i, j \in [k]$.

Proof. Define a mapping ξ from the nonzero vectors of $\mathbb{F}_2^r$ onto the vectors of weight one in $\mathbb{F}_2^n$ as follows:

$$\forall z \in \mathbb{F}_2^r \setminus \{\mathbf{0}\},\ \xi(z) = (x_1, x_2, \ldots, x_n), \quad \text{where } x_i = \begin{cases} 1 & i = \varphi(z) \\ 0 & i \neq \varphi(z) \end{cases}.$$

Using this notation, set

$$c_1 = \xi(z_1) + \xi(z_1 + z_2 + z_3) + \xi(z_1 + z_2 + z_4) + \xi(z_1 + z_3 + z_4),$$

$$c_2 = \xi(z_1) + \xi(z_2) + \xi(z_1 + z_3 + z_4) + \xi(z_2 + z_3 + z_4), \tag{5.22}$$

$$c_4 = \xi(z_1) + \xi(z_2) + \xi(z_3) + \xi(z_4) + \xi(z_1 + z_2 + z_3)$$

$$+\xi(z_1 + z_2 + z_4) + \xi(z_1 + z_3 + z_4) + \xi(z_2 + z_3 + z_4).$$

If $k < 4$, then to obtain the vectors c_1 and c_2, we can complete $z_1, z_2, \ldots, z_k$ with some $z_{k+1}, z_{k+2}, \ldots, z_4$ such that z_1, z_2, z_3, z_4 are linearly independent. Henceforth, let $j \in [k] \setminus \{1, 2, 4\}$. If j is odd, define

$$c_j \triangleq \sum_{i=1}^{j} \xi(z_i) + \xi(z_1 + z_2 + \cdots + z_j). \tag{5.23}$$

Otherwise, for even j set

$$c_j \triangleq \sum_{i=1}^{j} \xi(z_i) + \xi(z_1 + z_2 + \cdots + z_{j/2}) + \xi(z_{j/2+1} + z_{j/2+2} + \cdots + z_j). \tag{5.24}$$

Note that $\mathcal{S}(\xi(z)) = H \cdot \xi(z)^{\text{tr}} = z$. Thus, one can readily verify that $\mathcal{S}(c_j) = \mathbf{0}$ for all j, and the vectors $c_1, c_2, \ldots, c_k$ are indeed in $\mathcal{H}(r)$. As $z_1, z_2, \ldots, z_k$ are linearly independent, the weight of c_j is just the number of summands in (5.22), (5.23), and (5.24). Counting these shows that $c_1, c_2, \ldots, c_k$ are all of even weight.

Now assume, for contradiction, that $(c_i + \mathcal{A}(z_i)) \cap (c_j + \mathcal{A}(z_j)) \neq \varnothing$. Then $c_i + x = c_j + y$ for some $x = (x_1, x_2, \ldots, x_n) \in \mathcal{A}(z_i)$ and $y = (y_1, y_2, \ldots, y_n) \in \mathcal{A}(z_j)$. The parity of $x + y$ is given by

$$p(x+y) = \sum_{i=1}^{n} x_i + \sum_{i=1}^{n} y_i = x_{\varphi(z_i)} + y_{\varphi(z_j)}, \tag{5.25}$$

where all summations are performed modulo 2, and the second equality follows from the definitions of $\mathcal{A}(z)$. W.l.o.g. assume that $i < j$ and let $c_i = (a_1, a_2, \ldots, a_n)$, $c_j = (a'_1, a'_2, \ldots, a'_n)$. It follows from (5.22), (5.23), and (5.24) that

$$a_{\varphi(z_i)} = a'_{\varphi(z_i)} = 1, \quad a_{\varphi(z_j)} = 0, \quad a'_{\varphi(z_j)} = 1, \quad a_{\varphi(z_i+z_j)} = a'_{\varphi(z_i+z_j)} = 0.$$

Hence, we have that

$$x_{\varphi(z_i)} = y_{\varphi(z_i)}, \quad x_{\varphi(z_j)} = 1 + y_{\varphi(z_j)}, \quad x_{\varphi(z_i+z_j)} = y_{\varphi(z_i+z_j)}. \tag{5.26}$$

Substituting (5.26) into (5.25) yields $p(x+y) = x_{\varphi(z_i)} + x_{\varphi(z_j)} + 1$. Since $\phi_{z_i}(\varphi(z_j)) = \varphi(z_i + z_j)$, it folllows that $x_{\varphi(z_j)} = x_{\varphi(z_i+z_j)} = y_{\varphi(z_i+z_j)}$, where the first equality follows from the definition of $\mathcal{A}(z)$ and the second from (5.26). Thus, $p(x+y) = x_{\varphi(z_i)} + y_{\varphi(z_j)} = x_{\varphi(z_i)} + x_{\varphi(z_j)} + 1 = x_{\varphi(z_i)} + y_{\varphi(z_i+z_j)} + 1$. Similarly, since $\phi_{z_j}(\varphi(z_i)) = \varphi(z_i + z_j)$, it follows that $y_{\varphi(z_i+z_j)} = y_{\varphi(z_i)} = x_{\varphi(z_i)}$ and, therefore, $p(x+y) = 1$. Since, however, both c_i and c_j are of even weight, it follows that $p(x+y) = p(c_i + c_j) = 0$, which is a contradiction. Thus, $(c_i + \mathcal{A}(z_i)) \cap (c_j + \mathcal{A}(z_j)) = \varnothing$. □

Define

$$\tilde{\mathcal{A}} \triangleq \bigcup_{j=1}^{k} (c_j + \mathcal{A}(z_j)),$$

$$\tilde{\mathcal{B}} \triangleq \bigcup_{j=1}^{k} (c_j + \mathcal{B}(z_j)),$$

and

$$\mathcal{C} \triangleq (\mathcal{H}(r) \setminus \tilde{\mathcal{A}}) \cup \tilde{\mathcal{B}}.$$

Let $\mathcal{V}_{\tilde{\mathcal{A}}}$, respectively $\mathcal{V}_{\tilde{\mathcal{B}}}$, be the set of all words in $\mathbb{F}_2^n$ within distance one from some codeword of $\tilde{\mathcal{A}}$, respectively $\tilde{\mathcal{B}}$. By Corollary 5.3 we have $\mathcal{V}_{\tilde{\mathcal{A}}} = \mathcal{V}_{\tilde{\mathcal{B}}}$. Since $\mathcal{H}(r)$ is linear, $c_1, c_2, \ldots, c_k \in \mathcal{H}(r)$, and $\mathcal{A}(z_i) \subset \mathcal{H}(r)$ for each $1 \leq i \leq k$, it follows that $\tilde{\mathcal{A}} \subset \mathcal{H}(r)$ and, therefore, $\mathbb{F}_2^n \setminus \mathcal{V}_{\tilde{\mathcal{A}}}$ is perfectly covered by $\mathcal{H}(r) \setminus \tilde{\mathcal{A}}$. Thus, any $x \in \mathbb{F}_2^n$ is within distance one from some codeword of $\mathcal{C}$. It follows from (5.21) that $|\tilde{\mathcal{B}}| = k2^\nu$. Hence, if $c_1, c_2, \ldots, c_k$ are such that $c_1+\mathcal{A}(z_1)$, $c_2+\mathcal{A}(z_2), \ldots, c_k+\mathcal{A}(z_k)$ are disjoint, then $|\tilde{\mathcal{A}}| = |\tilde{\mathcal{B}}|$. Thus, $|\mathcal{C}| = |\mathcal{H}(r)| - |\tilde{\mathcal{A}}| + |\tilde{\mathcal{B}}| = 2^{n-r}$ and $\mathcal{C}$ is perfect.

Proposition 5.4. *rank* $\mathcal{C} = n - r + k$.

Proof. Since $\left|\mathcal{H}(r) \setminus \tilde{\mathcal{A}}\right| = 2^{n-r} - k2^\nu > 2^{n-r-1}$, it follows that more than half the codewords of $\mathcal{H}(r)$ are contained in $\mathcal{C}$. Hence, $\mathcal{H}(r) \subset \langle \mathcal{C} \rangle$. Let v_1, v_2, v_{n-r} be a basis for $\mathcal{H}(r)$. Since $\mathcal{B}(z) = \xi(z) + \mathcal{A}(z)$, it follows that using $\langle v_1, v_2, \ldots, v_{n-r} \rangle$ and $\langle c_j + \mathcal{B}(z_j) \rangle$ we can generate $\xi(z_j)$. Thus, the vectors $v_1, v_2, \ldots, v_{n-r}, \xi(z_1), \xi(z_2), \ldots, \xi(z_k)$ are in $\langle \mathcal{C} \rangle$. To see that these vectors are linearly independent, assume to the contrary that

$$x = \sum_{i=1}^{n-r} \alpha_i v_i + \sum_{i=1}^{k} \beta_i \xi(z_i) = \mathbf{0}$$

for some $\alpha_1, \alpha_2, \ldots, \alpha_{n-r}, \beta_1, \beta_2, \ldots, \beta_k \in \mathbb{F}_2$. The syndrome, however, is then

$$\mathcal{S}(x) = \sum_{i=1}^{n-r} \alpha_i s(v_i) + \sum_{i=1}^{k} \beta_i s((\xi(z_i)) = \sum_{i=1}^{k} \beta_i z_i = \mathbf{0},$$

which contradicts the fact that $z_1, z_2, \ldots, z_k$ are linearly independent. To see that $\langle \mathcal{C} \rangle$ is generated by $v_1, v_2, \ldots, v_{n-r}, \xi(z_1), \xi(z_2), \ldots, \xi(z_k)$, note that $v_1, v_2, \ldots, v_{n-r}$ and $\xi(z_j)$ generate all the vectors in $c_j + \mathcal{B}(z_j)$. □

Setting $k = r$ in the foregoing construction produces a binary 1-perfect code of full rank. Hence, we have the following result.

Corollary 5.4. *For any* $r \geq 4$, *there exists a full-rank 1-perfect code of length* $2^r - 1$.

5.6 Kernels of Perfect Codes

The ***kernel*** of a code $\mathcal{C} \subseteq \mathbb{F}_2^n$, denoted by ker $\mathcal{C}$, is the set $\mathcal{K} \in \mathbb{F}_2^n$ such that any vector in $\mathcal{K}$ leaves $\mathcal{C}$ invariant under translation. In other words, x is an element of the kernel of $\mathcal{C}$ if and only if $x+\mathcal{C} = \mathcal{C}$. In this section we examine the possible dimensions of kernels of binary 1-perfect codes. The techniques used in this section are similar to the ones used in Section 5.5, but some concepts will be introduced in a slightly different, but equivalent, way.

Lemma 5.5. *If* $\mathcal{C}$ *is a code of size* 2^k, $\mathcal{K} = \ker\ \mathcal{C}$, *and* $\mathbf{0} \in \mathcal{C}$, *then*

(1) *The kernel* $\mathcal{K}$ *is a linear sub-code of the code* $\mathcal{C}$.
(2) *The code* $\mathcal{C}$ *is the union of cosets of the kernel* $\mathcal{K}$.
(3) *If* $\mathcal{C}$ *is linear, then* $\mathcal{K} = \mathcal{C}$; *otherwise,* $\dim \mathcal{K} \leq k - 2$.

Proof.

(1) Since $\mathbf{0} + \mathcal{C} = \mathcal{C}$, it follows that $0 \in \mathcal{K}$. Assume now that $x, y \in \mathcal{K}$ and $c \in \mathcal{C}$. Since $y \in \mathcal{K}$ and $c \in \mathcal{C}$, it follows that $y + c \in \mathcal{C}$. Since $x \in \mathcal{K}$ and $y+c \in \mathcal{C}$, it follows that $x+y+c \in \mathcal{C}$. This implies that $x+y \in \mathcal{K}$ and hence $\mathcal{K}$ is a linear code.
(2) If $c \in \mathcal{C}$ and $c \notin \mathcal{K}$, then $c + \mathcal{K} \subset \mathcal{C}$ and since $\mathcal{K}$ is a linear code, it follows that $c+\mathcal{K}$ is a coset of $\mathcal{K}$ and hence $(c+\mathcal{K}) \cap \mathcal{K} = \varnothing$. Moreover, since $c + \mathcal{K}$ is a coset of $\mathcal{K}$ in $\mathcal{C}$, it follows that for $c_1, c_2 \in \mathcal{C}$, either $c_1 + \mathcal{K} = c_2 + \mathcal{K}$ or $(c_1 + \mathcal{K}) \cap (c_2 + \mathcal{K}) = \varnothing$. This implies that $\mathcal{C}$ is the union of cosets of $\mathcal{K}$.
(3) If $\mathcal{C}$ is linear, then for each two codewords $c_1, c_2 \in \mathcal{C}$, we have that $c_1+c_2 \in \mathcal{C}$, which implies that $\mathcal{K} = \mathcal{C}$. If $\mathcal{C}$ is not linear, then $\dim \mathcal{K} < k$. If we assume, on the contrary, that $\dim \mathcal{K} = k - 1$, then since $\mathcal{C}$ is a union of cosets of $\mathcal{K}$, it follows that $\mathcal{C} = \mathcal{K} \cup (x+\mathcal{K})$ for some $x \in \mathcal{C} \setminus \mathcal{K}$. This implies that $\mathcal{C}$ is a linear code and hence $\mathcal{K} = \mathcal{C}$, a contradiction. Thus, $\dim \mathcal{K} < k - 1$.

□

Lemma 5.6. *If* $\mathcal{C}$ *is a nonlinear 1-perfect code of length* $2^r - 1$, *and* $\mathcal{K} = \ker\ \mathcal{C}$, *then* $\dim \mathcal{K} \in \{1, 2, \ldots, 2^r - r - 3\}$.

Proof. If $\mathcal{C}$ is a linear 1-perfect code, then $\dim \mathcal{K} = \dim \mathcal{C} = 2^r - r - 1$. Hence, if $\mathcal{C}$ is a nonlinear 1-perfect code, then by Lemma 5.5 we have that $\dim \mathcal{K} \leq 2^r - r - 3$. By Corollary 5.1, we have that $\mathbf{1} \in \mathcal{C}$, and hence $\dim \mathcal{K} \geq 1$. □

Recall that for a subset $\mathcal{C} \subseteq \mathbb{F}_2^n$, with odd minimum distance $d(\mathcal{C})$, $\mathcal{C}^*$ denote the set of $\mathbb{F}_2^{n+1}$ obtained by adding a parity symbol for all the words of $\mathcal{C}$.

Lemma 5.7. *If* $\mathcal{K}$ *is the kernel of a code* $\mathcal{C}$, *then* $\mathcal{K}^*$ *is the kernel of* $\mathcal{C}^*$.

Proof. If $x \in \mathcal{K}$, then $x + c \in \mathcal{C}$ for all $c \in \mathcal{C}$. Clearly,

$$(x, p(x)) + (c, p(c)) = (x + c, p(x) + p(c)) = (x + c, p(x + c)) \in \mathcal{C}^*$$

and hence $(x, p(x)) \in \mathcal{K}^*$. Obviously, if $x \notin \mathcal{K}$, then $(x, b) \notin \mathcal{K}^*$, for $b \in \mathbb{F}_2$. Thus, the kernel of $\mathcal{C}^*$ is $\mathcal{K}^*$. □

Lemma 5.8. *Let $\mathcal{C}_1$, $\mathcal{C}_2$ be two 1-perfect codes of length n with kernels $\mathcal{K}_1$ and $\mathcal{K}_2$, respectively, where $k_1 = \dim \mathcal{K}_1 < \frac{n-1}{2}$ and $k_2 = \dim \mathcal{K}_2$. If*

$$\mathcal{C} \triangleq (\mathcal{C}_1 \times \mathcal{C}_2^*) \cup \bigcup_{i=1}^{n} ((\mathbf{e}_i + \mathcal{C}_1) \times (\mathbf{e}_i + \mathcal{C}_2)^*) ,$$

then the kernel of $\mathcal{C}$ is $\mathcal{K}_1 \times \mathcal{K}_2^*$ and its dimension is $k_1 + k_2$.

Proof. Let $\mathcal{K}$ be the kernel of $\mathcal{C}$. By the definition of the kernel, it can be immediately verified that $\mathcal{K}_1 \times \mathcal{K}_2^*$ is contained in $\mathcal{K}$. It is also clear that the dimension of $\mathcal{K}_1 \times \mathcal{K}_2^*$ is $k_1 + k_2$. Therefore, to complete the proof it suffices to show that there is no other word in the kernel of $\mathcal{C}$.

Let $(x, y^*) \in \mathcal{K}$, where $x \in \mathbb{F}_2^n$, and assume, on the contrary that $x \notin \mathcal{K}_1$. Therefore, $x + \mathcal{C}_1 = \mathbf{e}_i + \mathcal{C}_1$ for some i, i.e., $\mathbf{e}_i + x + \mathcal{C}_1 = \mathcal{C}_1$. This implies that $e_i + x \in \mathcal{K}_1$. In general, for each j, $\mathbf{e}_j + x + \mathcal{C}_1 = \mathbf{e}_j + \mathbf{e}_i + \mathcal{C}_1 = \mathbf{e}_k + \mathcal{C}_1$ for some $k \notin \{i, j\}$, $1 \leq k \leq k$. Hence, $\mathbf{e}_i + \mathbf{e}_j + \mathbf{e}_k + \mathcal{C}_1 = \mathcal{C}_1$, which implies that $\mathbf{e}_i + \mathbf{e}_j + \mathbf{e}_k \in \mathcal{K}_1$ and hence $\mathbf{e}_i + \mathbf{e}_j + \mathbf{e}_k$ is a codeword of $\mathcal{C}_1$ with weight three for each $i \neq j$. Therefore, all $\frac{n-1}{2}$ the codewords of weight three in $\mathcal{C}$ that contain a *one* in the i-th coordinate must be in $\mathcal{K}_1$. This implies that the dimension of $\mathcal{K}_1$ is at least $\frac{n-1}{2}$. This contradicts the assumption that $k_1 = \dim \mathcal{K}_1 < \frac{n-1}{2}$ and, therefore, $x \in \mathcal{K}_1$. Since $x \in \mathcal{K}_1$, it follows that $(x, y^*) + (\mathcal{C}_1 \times \mathcal{C}_2^*) = \mathcal{C}_1 \times \mathcal{C}_2^*$, which implies that y^* must be in the kernel of $\mathcal{C}_2^*$. Thus, $(x, y^*) \in \mathcal{K}_1 \times \mathcal{K}_2^*$ and the proof is completed. □

We will use the Hamming code as one of the codes in Lemma 5.8 and we also assume that there exist 1-perfect codes of length 15 and kernels of dimensions 1, 2, 3, 4, 5, 6, and 7, and also a 1-perfect code of length $2^r - 1$, $r \geq 4$, with kernel of dimension 1. The existence of such codes will be proved later in this section. We infer the following result by applying Lemma 5.8 iteratively.

Corollary 5.5. *For each $n = 2^r - 1$, $r \geq 4$, there exists a 1-perfect code with kernel of dimension k, for any given integer k in the range between 2 and $\frac{n-1}{2}$.*

Recall the definition of the linear sub-code $\mathcal{A}(z)$ of $\mathcal{H}(r)$ from Section 5.5. By abuse of notation, we will also use $A(2^i)$ instead of $A(z)$ if z is the binary representation of the integer 2^i. Recall also that by Theorem 4.1, the codewords of weight three in a 1-perfect code of length $n = 2^r - 1$ form a Steiner system $S(2,3,n)$. Let $b(\ell)$ denote the binary representation of the integer ℓ. A codeword of weight three $\mathbf{e}_i + \mathbf{e}_j + \mathbf{e}_s$ will also be denoted by $\{i,j,s\}$ or $\{b(i), b(j), b(s)\}$.

By Corollary 4.1, the codewords of weight three in a $\mathcal{H}(r)$ form a Steiner system $S(2,3,2^r-1)$. Clearly, for any set S of points in $PG(r-1,2)$ whose rank is k, the linear span of S has $2^k - 1$ points whose triples in $\mathcal{H}(r)$ define a Steiner system $S(2,3,2^k-1)$. A set S of k points in $PG(r-1,2)$ will be called ***independent*** if the smallest system of triples, which forms a Steiner system, and contains the k points of S, has $2^k - 1$ points.

Theorem 5.8. *For any $r \geq 2$ independent points $z_1, z_2, \ldots, z_r$ in $PG(r-1,2)$, the subspace $\mathcal{A}(z_1) \cap \mathcal{A}(z_2) \cap \cdots \cap \mathcal{A}(z_r)$ has dimension* 1.

Proof. The proof is given by induction on r. The basis of the induction, $r = 2$, is trivial and assumes that the claim holds for $r-1$ independent points in $PG(r-2,2)$, where $r \geq 3$. Since the dimension of the intersection is one, it follows that the two codewords in the intersection are $\mathbf{0}$ and $\mathbf{1}$.

For the induction step, consider all the points of $PG(r-1,2)$ as integers between 1 and $2^r - 1$. It can be assumed w.l.o.g. that the r independent points in $PG(r-1,2)$ are represented by the r distinct vectors of length r and weight one. This r independent points form the set $\{b(2^i) : 0 \leq i \leq r-1\}$. For each ℓ, $0 \leq \ell \leq r-1$, the triples that are contained in the generator matrix of $\mathcal{A}(2^\ell)$ are of the form $\{b(2^\ell), b(m), b(m+2^\ell)\}$, where $m \in [2^r-1] \setminus \{2^\ell\}$ and m is taken modulo $2^{\ell+1}$ in the range $[0, 2^\ell - 1]$. Since all the generator matrices of the sub-codes $\mathcal{A}(2^i)$, $0 \leq i \leq r-1$ have codewords of weight three, it follows that any codeword $c \in \mathcal{H}(r)$ in the subspace

$$\mathcal{A}(1) \cap \mathcal{A}(2) \cap \mathcal{A}(4) \cap \ \cdots \ \cap \mathcal{A}(2^{r-2}) \cap \mathcal{A}(2^{r-1})$$

is a linear combination of the same number of rows for each sub-code $\mathcal{A}(2^i)$, $0 \leq i \leq r-1$. Let $c = (c', c'')$, where $c' \in \mathbb{F}_2^{2^{r-1}-1}$, $c'' \in \mathbb{F}_2^{2^{r-1}}$. We distinguish now between two cases, depending on whether the number of rows in the linear combination of c is even or odd.

Case 1. The number of rows in the linear combination is even.

This implies that the weight of c is even, $\text{wt}(c'') = \text{wt}(c')$, and $c'' = 0c'$. This also implies that the entries of c in the positions of $\{2^i : 0 \leq i \leq r-1\}$ are *zeroes* since, in these positions, there are *ones* in all rows of the associated generator matrices of the related $\mathcal{A}(2^i)$, $0 \leq i \leq r-1$. Since $c = (c', c'') = c'0c'$, it follows that c' is formed from linear combinations with the same number of codewords in each of the generator matrices of $\mathcal{A}(1), \mathcal{A}(2), \ldots, \mathcal{A}(2^{r-2})$. By the induction hypothesis, it follows that $c' \in \{\mathbf{0}, \mathbf{1}\}$. Since $c'' = 0c'$ and there is no codeword with weight $2^r - 1$, it follows that c'' is the all-zero codeword.

Case 2. The number of rows in the linear combination is odd.

This implies that the weight of c' is odd, $\text{wt}(c'') = \text{wt}(c') + 1$, and $c'' = 1c'$. Using the same arguments as in Case 1, we have that this also implies that the entries of c in the positions of $\{2^i \ : \ 0 \leq i \leq r-1\}$ are *ones*. Since $c = (c', c'') = c'1c'$, it follows that c' is formed from linear combinations with the same number of codewords in each of the generator matrices of $\mathcal{A}(1), \mathcal{A}(2), \ldots, \mathcal{A}(2^{r-2})$. By the induction hypothesis, it follows that $c' \in \{\mathbf{0}, \mathbf{1}\}$. Since $c'' = 1c'$ and there is no codeword with weight one, it follows that c'' is the all-ones codeword.

The conclusions of Case 1 and Case 2 complete the induction step and hence the subspace $\mathcal{A}(z_1) \cap \mathcal{A}(z_2) \cap \cdots \cap \mathcal{A}(z_r)$ has dimension 1. □

We continue and define the r codewords $c_1, c_2, \ldots, c_r$ from $\mathcal{H}(r)$ as in Section 5.5, where $z_1, z_2, \ldots, z_r$ are independent points in $\text{PG}(r-1, 2)$.

Theorem 5.9. *If* $r \geq 4$, *and*

$$\mathcal{C} \triangleq \left(\mathcal{H}(r) \setminus \left(\bigcup_{i=1}^{r}(c_i + \mathcal{A}(z_i))\right)\right) \bigcup \left(\bigcup_{i=1}^{r}(\mathbf{e}_i + c_i + \mathcal{A}(z_i))\right),$$

where $c_i \in \mathcal{H}(r)$, *then* $\ker\ \mathcal{C} = \bigcap_{i=1}^{r} \mathcal{A}(z_i)$ *and* $\dim \ker\ \mathcal{C} = 1$.

Proof. Recall first that by Lemma 5.5, $\ker\ \mathcal{C}$ is a linear sub-code of $\mathcal{C}$. Let $y \in \ker\ \mathcal{C}$ and distinguish between the cases depending on whether $y \in \mathcal{H}(r) \setminus \bigcup_{i=1}^{r}(c_i + \mathcal{A}(z_i))$ or $y \in \bigcup_{i=1}^{r}(\mathbf{e}_i + c_i + \mathcal{A}(z_i)$.

Case 1. $y \in \mathcal{H}(r) \setminus \bigcup_{i=1}^{r}(c_i + \mathcal{A}(z_i))$.

Since $y \in \ker\ \mathcal{C}$, it follows that for each i, $1 \leq i \leq r$, we have that $y + (\mathbf{e}_i + c_i + \mathcal{A}(z_i)) \subseteq \mathcal{C}$. Since $y \in \mathcal{H}(r)$, $\mathcal{H}(r)$ is a linear code, and $\mathcal{A}(z_i)$ is a linear sub-code of $\mathcal{H}(r)$, it follows that $y + c_i + \mathcal{A}(z_i) \subseteq \mathcal{H}(r)$. This implies that $\mathbf{e}_i + c_i + y + \mathcal{A}(z_i) \subseteq \mathbf{e}_i + \mathcal{H}(r)$. Furthermore, for each j, $(\mathcal{C} \cap (\mathbf{e}_j + \mathcal{H}(r))) \cap \mathcal{H}(r) = \varnothing$ and since $c_i + \mathcal{A}(z_i) \subset \mathcal{H}(r)$, it follows that $\mathcal{C} \cap (\mathbf{e}_i + \mathcal{H}(r)) = \mathbf{e}_i + c_i + \mathcal{A}(z_i)$, for each $i = 1, 2, \ldots, r$. Therefore, for each $i = 1, 2, \ldots, r$, we have that $\mathbf{e}_i + c_i + y + \mathcal{A}(z_i) = \mathbf{e}_i + c_i + \mathcal{A}(z_i)$, which implies that $y \in \mathcal{A}(z_i)$ and thus $y \in \bigcap_{i=1}^{r} \mathcal{A}(z_i)$.

Case 2. $y \in \bigcup_{i=1}^{r}(\mathbf{e}_i + c_i + \mathcal{A}(z_i)$.

Assume that $y \in \mathbf{e}_j + c_j + \mathcal{A}(z_j)$ for some $1 \leq j \leq r$ and since $\mathcal{A}(z_j)$ is a linear sub-code of $\mathcal{H}(r)$, it follows that $y \in \mathbf{e}_j + \mathcal{H}(r)$. Hence, since $\mathbf{e}_i + c_i + \mathcal{A}(z_i) \subseteq \mathbf{e}_i + \mathcal{H}(r)$, it follows that $\mathbf{e}_i + c_i + y + \mathcal{A}(z_i) \subseteq \mathbf{e}_i + y + \mathcal{H}(r) =$

$\mathbf{e}_i + \mathbf{e}_j + \mathcal{H}(r)$. Since $y \in \ker\ \mathcal{C}$, it follows that $y + \mathbf{e}_i + c_i + \mathcal{A}(z_i) \subseteq \mathcal{C}$ and hence $\mathcal{C} \cap (\mathbf{e}_i + \mathbf{e}_j + \mathcal{H}(r)) \neq \varnothing$, which implies that $\mathbf{e}_i + \mathbf{e}_j + \mathcal{H}(r)$ is one of the cosets $\mathbf{e}_s + \mathcal{H}(r)$. Therefore, $\mathbf{e}_i + \mathbf{e}_j + \mathcal{H}(r) = \mathbf{e}_s + \mathcal{H}(r)$, which implies that $\mathbf{e}_i + \mathbf{e}_j + \mathbf{e}_s + \mathcal{H}(r) = \mathcal{H}(r)$, i.e., $\{i, j, s\}$ is a triple in $\mathcal{H}(r)$, contradicting the fact that z_i, z_j, and z_s, are independent points of $\mathrm{PG}(r-1, 2)$. Therefore, there is no $y \in \ker\ \mathcal{C}$ such that $y \in \bigcup_{i=1}^{r} (\mathbf{e}_i + c_i + \mathcal{A}(z_i))$.

From these two cases we have that $y \in \bigcap_{i=1}^{r} \mathcal{A}(z_i)$, which implies, by Theorem 5.8, that $\dim \ker\ \mathcal{C} = 1$. □

Corollary 5.6. *Let $\mathcal{K}$ be a subspace of $\mathcal{H}(r)$ such that $\mathcal{A}(z_i) \subseteq \mathcal{K} \subseteq \mathcal{H}(r)$, for some i, and* $\dim \mathcal{K} \leq (\dim \mathcal{H}(r)) - 2$ *and let $c \in \mathcal{H}(r) \setminus \mathcal{K}$. Then*

$$\mathcal{C} \triangleq (\mathcal{H}(r) \setminus (c + \mathcal{K})) \cup (\mathbf{e}_i + c + \mathcal{K})$$

is a 1-perfect code with kernel $\mathcal{K}$.

Proof. The code $\mathcal{K}$ is a union of cosets of $\mathcal{A}(z_i)$ and hence $\mathcal{K}$ and $\mathbf{e}_i + \mathcal{K}$ are two disjoint perfect coverings with radius one of the same set $\mathcal{U}$. Hence, $c + \mathcal{K}$ and $\mathbf{e}_i + c + \mathcal{K}$ are also two disjoint perfect coverings with radius one of the same set $c + \mathcal{U}$. Therefore, $\mathcal{C}$ is a 1-perfect code. To complete the proof it is suffices to show that $\mathcal{K} = \ker\ \mathcal{C}$.

Let $x \in \mathcal{K}$ and let $c' \in \mathcal{C}$ and consider the word $x + c'$. If $c' \in \mathbf{e}_i + c + \mathcal{K}$, i.e., $c' = \mathbf{e}_i + c + \kappa$, where $\kappa \in \mathcal{K}$, then $x + c' = x + \mathbf{e}_i + c + \kappa \in \mathbf{e}_i + c + \mathcal{K} \subset \mathcal{C}$. If $c' \in \mathcal{H}(r) \setminus (c + \mathcal{K})$, i.e., $c' \notin c + \mathcal{K}$, then since $x \in \mathcal{K} \subset \mathcal{H}(r)$, it follows that $x + c' \in \mathcal{H}(r)$. Since $x \in \mathcal{K}$ and $c' \notin c + \mathcal{K}$, it follow that $x + c' \notin c + \mathcal{K}$ and therefore $x + c' \in \mathcal{C}$. This implies that $\mathcal{K} \subseteq \ker\ \mathcal{C}$.

To complete the proof we have to show that $\ker\ \mathcal{C} \subseteq \mathcal{K}$. For this we will show that a word $x \in \ker\ \mathcal{C}$ is also a word in $\mathcal{K}$. Since by Lemma 5.5 $\ker\ \mathcal{C}$ is a linear sub-code of $\mathcal{C}$, if follows that we can distinguish between two cases depending on whether $x \in \mathcal{H}(r) \setminus (c + \mathcal{K})$ or $x \in \mathbf{e}_i + c + \mathcal{K}$.

Case 1. $x \in \mathcal{H}(r) \setminus (c + \mathcal{K})$.

Since $c, x \in \mathcal{H}(r)$ and $\mathcal{K} \subset \mathcal{H}(r)$, it follows that $\mathbf{e}_i + x + c + \mathcal{K} \subset \mathbf{e}_i + \mathcal{H}(r)$. Nevertheless, $\mathcal{C} \cap (\mathbf{e}_i + \mathcal{H}(r)) = \mathbf{e}_i + c + \mathcal{K}$ since $\mathbf{e}_i + \mathcal{H}(r)$ is a coset of $\mathcal{H}(r)$ and $\mathcal{K} \subset \mathcal{H}(r)$. Therefore, $\mathbf{e}_i + c + x + \mathcal{K} = \mathbf{e}_i + c + \mathcal{K}$, i.e., $x + \mathcal{K} = \mathcal{K}$ and hence $x \in \mathcal{K}$.

Case 2. $x \in \mathbf{e}_i + c + \mathcal{K}$, i.e. $x = \mathbf{e}_i + c + \kappa_1$, where $\kappa_1 \in \mathcal{K}$.

Since $\dim \mathcal{K} \leq (\dim \mathcal{H}(r) - 2)$ and $\mathcal{K}$ is a subspace of $\mathcal{H}(r)$, it follows that there exists a $c' \in \mathcal{H}(r)$, $c' \notin \mathcal{K}$, such that $c' + \mathcal{K} \subseteq \mathcal{H}(r) \setminus (c + \mathcal{K})$ and $c' + \mathcal{K} \neq \mathcal{K}$. Since $c' \notin \mathcal{K}$, it follows that $0 \notin c' + \mathcal{K}$ and hence $x \notin x + c' + \mathcal{K}$ which implies that $x + c' + \mathcal{K} \neq \mathbf{e}_i + c + \mathcal{K}$, i.e., $x + c' + \mathcal{K}$ and $\mathbf{e}_i + c + \mathcal{K}$ are two disjoint cosets of $\mathcal{K}$. Therefore, $x + c' \notin \mathbf{e}_i + c + \mathcal{K}$.

Moreover, $x + c' = \mathbf{e}_i + c + \kappa_1 + c'$ and Since $c, c', \kappa \in \mathcal{H}(r)$, it follow that $x + c' = \mathbf{e}_i + c''$, where $c'' \in \mathcal{H}(r)$ and hence $x + c' \notin \mathcal{H}(r) \setminus (c + \mathcal{K})$. Therefore, $x \notin \ker \mathcal{C}$ (implying that if $x \in \ker \mathcal{C}$, then $x \in \mathcal{H}(r) \setminus (c + \mathcal{K})$.

These two cases imply that if $x \in \ker\ \mathcal{C}$, then $x \in \mathcal{K}$. Thus, $\mathcal{K}$ is the kernel of $\mathcal{C}$. □

Corollary 5.7. *There exists a 1-perfect code $\mathcal{C}$ of length $n = 2^r - 1$ having kernel $\mathcal{K}$, where the dimension of $\mathcal{K}$ is any chosen integer in the set $\{2^{r-1} - 1, 2^{r-1}, \ldots, 2^r - r - 3\}$.*

Proof. Since $\dim \mathcal{A}(z_i) = 2^{r-1} - 1 = \frac{n-1}{2}$, one can choose any subspace $\mathcal{K}$ of $\mathcal{H}(r)$, where $\mathcal{A}(z_i) \subseteq \mathcal{K}$ and $\dim \mathcal{K} \in \{2^{r-1} - 1, 2^{r-1}, \ldots, 2^r - r - 3\}$. Furthermore, let $c \in \mathcal{H}(r) \setminus \mathcal{K}$ and define

$$\mathcal{C} \triangleq (\mathcal{H}(r) \setminus (c + \mathcal{K})) \cup (\mathbf{e}_i + c + \mathcal{K}) \ .$$

As a consequence, the claim follows immediately from Corollary 5.6. □

We now want to establish all the possible dimensions of kernels in nonlinear 1-prefect codes of length $n = 15$. A kernel of dimension one is obtained via Theorem 5.9. A computer search shows that there are 177 codes with kernels of dimension 2 and three codes with kernels of dimension 9.

Consider the Hamming code $\mathcal{H}(4)$ of length 15 whose parity-check matrix is the following 4×15 matrix

$$H = \begin{bmatrix} z_0 & z_1 & z_2 & z_3 & z_4 & z_5 & z_6 & z_7 & z_8 & z_9 & z_{10} & z_{11} & z_{12} & z_{13} & z_{14} \\ \hline 1 & 0 & 1 & 0 & 1 & 0 & 1 & 0 & 1 & 0 & 1 & 0 & 1 & 0 & 1 \\ 1 & 1 & 0 & 0 & 1 & 1 & 0 & 0 & 1 & 1 & 0 & 0 & 1 & 1 & 0 \\ 1 & 1 & 1 & 1 & 0 & 0 & 0 & 0 & 1 & 1 & 1 & 1 & 0 & 0 & 0 \\ 1 & 1 & 1 & 1 & 1 & 1 & 1 & 1 & 0 & 0 & 0 & 0 & 0 & 0 & 0 \end{bmatrix} .$$

Based on the code obtained from H, we present 1-perfect codes of length 15 with dimensions 3, 4, 5, 6, 7, and 8. These are only a small number of many such examples. In both examples, the switching method is applied and the switches made in each row are added to the switches made in the previous rows.

i	coset to switch	codeword c_i	kernel dimension
1	$\mathbf{e}_1 + c_1 + \mathcal{A}(z_1)$	110000000000110	7
2	$\mathbf{e}_2 + c_2 + \mathcal{A}(z_2)$	111111000000110	4
3	$\mathbf{e}_2 + c_3 + \mathcal{A}(z_2)$	000100101101111	5
4	$\mathbf{e}_3 + c_4 + \mathcal{A}(z_3)$	100001000010000	3

i	coset to switch	codeword c_i	kernel dimension
1	$\mathbf{e}_1 + c_1 + \mathcal{A}(z_1)$	111101011010000	7
2	$\mathbf{e}_1 + c_2 + \mathcal{A}(z_1)$	110000000000110	8
3	$\mathbf{e}_2 + c_3 + \mathcal{A}(z_2)$	111101101010110	5
4	$\mathbf{e}_2 + c_4 + \mathcal{A}(z_2)$	011001010000001	6

Theorem 5.10. *For each $r \geq 4$, there exists a nonlinear 1-perfect code of length $n = 2^r - 1$, having a kernel of dimension Δ for each $\Delta \in [2^r - r - 3]$.*

Proof. We have already established the proof for $n = 15$ and $\Delta \geq 2$. By Theorem 5.9, we have a code of length $n = 2^r - 1$, $r \geq 4$, containing the kernel of dimension one. By Corollary 5.7, we can find codes containing kernels of all dimensions from $\frac{n-1}{2}$ up through $2^r - r - 3$, $r \geq 5$. The Hamming code of length $2^r - 1$ has dimension $2^r - r - 1$ and since it is a linear code and its kernel is the code itself, it follows that dimension $2^r - r - 1$ for the kernel is also attained. By Corollary 5.5, we also have codes with dimensions 2 up through $\frac{n-1}{2}$ for $n = 2^r - 1$, $r \geq 5$. $\square$

5.7 Enumeration of Nonequivalent Codes

The parity-check matrix of the q-ary Hamming code, with redundancy r, and length $n_r = \frac{q^r-1}{q-1} = q^{r-1} + q^{r-1} + \cdots + q + 1$ consists of n_r pairwise linearly independent column vectors of length r over $\mathbb{F}_q$. The n_r column vectors, which we choose, will be of the form $(\overbrace{0 \cdots\cdots 0}^{r-1-\ell \text{ times}}\ 1 x_1 \ldots x_\ell)^{\text{tr}}$ for all $0 \leq \ell \leq r-1$, where for $x_i \in \mathbb{F}_q$, $1 \leq i \leq \ell$. Let α be a primitive element in $\mathbb{F}_q$. The $2 \times (q+1)$ parity-check matrix of the q-ary Hamming code with redundancy 2, for the construction, is

$$H_2 = \begin{bmatrix} 0 & 1 & 1 & 1 & \ldots & 1 \\ 1 & 0 & \alpha^0 & \alpha^1 & \ldots & \alpha^{q-2} \end{bmatrix} \tag{5.27}$$

and its $(q-1) \times (q+1)$ generator matrix has the form

$$G_2 = \begin{bmatrix} 1 & \alpha^{q-1} & -\alpha^{q-1} & 0 & \ldots & 0 \\ 1 & \alpha^{q-2} & 0 & -\alpha^{q-2} & \ldots & 0 \\ \vdots & \vdots & \vdots & \vdots & \vdots & \vdots \\ 1 & \alpha & 0 & 0 & \ldots & -\alpha \end{bmatrix} = \begin{bmatrix} 1 & \\ \vdots & X \\ 1 & \end{bmatrix} \tag{5.28}$$

where X is a $(q-1) \times q$ matrix.

Assume that the Hamming code of length $n_r = \frac{q^r-1}{q-1} = q^{r-1}+\cdots+q+1$ with $r \geq 2$ has the $r \times n_r$ parity-check matrix H_r of the form

$$H_r = \begin{bmatrix} 0 & & & & \\ \vdots & S_1 & S_2 & \ldots & S_{n_r-1} \\ 0 & & & & \\ 1 & & & & \end{bmatrix},$$

where the S_i's are column vectors of length r.

Assume also that the $(n_r - r) \times n_r$ generator matrix G_r has the form

$$G_r = \left[\begin{array}{ccccc} 1 & & & & \\ \vdots & X & \mathbf{0} & \cdots\cdots\cdots & \mathbf{0} \\ 1 & & & & \\ 1 & & & & \\ \vdots & \mathbf{0} & X & \cdots\cdots\cdots & \mathbf{0} \\ 1 & & & & \\ \vdots & \vdots & \vdots & \cdots\cdots\cdots & \vdots \\ \vdots & \vdots & \vdots & \cdots\cdots\cdots & \vdots \\ \vdots & \vdots & \vdots & \cdots\cdots\cdots & \vdots \\ 1 & & & & \\ \vdots & \mathbf{0} & \mathbf{0} & \cdots\cdots\cdots & X \\ 1 & & & & \\ \hline & & F & & \end{array}\right],$$

where F is a $t \times n_r$ matrix with $t = (n_r - r) - (q-1)\frac{n_r-1}{q} = n_{r-1} - r + 1$.

Now, we generate the following $(r+1) \times (n_r q + 1)$ parity-check matrix H_{r+1}.

$$H_{r+1} = \begin{bmatrix} 0 & & & & & & & & & 0 & 0 & \cdots & 0 \\ \vdots & S_1 & S_1 & \cdots & S_1 & \cdots & S_{n_r-1} & S_{n_r-1} & \cdots & S_{n_r-1} & \vdots & \vdots & \cdots & \vdots \\ 0 & & & & & & & & & 0 & 0 & \cdots & 0 \\ 0 & & & & & & & & & 1 & 1 & \cdots & 1 \\ 1 & 0 & \alpha^0 & \cdots & \alpha^{q-2} & \cdots & 0 & \alpha^0 & \cdots & \alpha^{q-2} & 0 & \alpha^0 & \cdots & \alpha^{q-2} \end{bmatrix}.$$

Lemma 5.9. *H_{r+1} is a parity-check matrix of the Hamming code of length $n_r q + 1$.*

Proof. Assume that H_r contains all the column vectors of length r of the form $(0 \cdots 01 x_1 \cdots x_\ell)^{\text{tr}}$ for all ℓ, $0 \leq \ell \leq r-1$, where $x_i \in \mathbb{F}_q$, $1 \leq i \leq \ell$. By induction, starting from the parity-check matrix H_2 as a basis, we can easily prove that H_{r+1} is a parity-check matrix of the q-ary Hamming code of length n_{r+1}. □

Now, let G_{r+1} be the following $(n_r q - r) \times (n_r q + 1)$ matrix.

$$G_{r+1} = \left[\begin{array}{cccccc} 1 & & & & & \\ \vdots & X & \mathbf{0} & \cdots & \cdots\cdots & \mathbf{0} \\ 1 & & & & & \\ 1 & & & & & \\ \vdots & \mathbf{0} & X & \cdots & \cdots\cdots & \mathbf{0} \\ 1 & & & & & \\ \vdots & \vdots & \vdots & \cdots & \cdots\cdots & \vdots \\ \vdots & \vdots & \vdots & \cdots & \cdots\cdots & \vdots \\ \vdots & \vdots & \vdots & \cdots & \cdots\cdots & \vdots \\ 1 & & & & & \\ \vdots & \mathbf{0} & \mathbf{0} & \cdots & \cdots\cdots & X \\ 1 & & & & & \\ \hline & & & F' & & \end{array}\right] \tag{5.29}$$

where F' is some $t' \times (n_r q + 1)$ matrix for which $t' = n_r - r$.

Lemma 5.10. *The generator matrix of the Hamming code with parity-check matrix H_{r+1} has the form of G_{r+1}.*

Proof. From the form of H_2 and G_2 given in (5.27) and (5.28), respectively, it follows that any matrix of the form

$$\begin{bmatrix} 0 & & & & & \\ \vdots & S & S & S & \dots & S \\ 0 & & & & & \\ 1 & 0 & \alpha^0 & \alpha^1 & \dots & \alpha^{q-2} \end{bmatrix}$$

is orthogonal to G_2 for any column vector S. These sub-matrices are the building blocks of the matrices G_{r+1} and H_{r+1}. Note now that this immediately implies the claim of the lemma. □

We say that a word v covers the set $\mathcal{U}$ if for any $u \in \mathcal{U}$ we have that $d(v,u) \leq 1$. A code $\mathcal{C}$ covers a set $\mathcal{U}$ if for every element $u \in \mathcal{U}$ there exists a codeword $c \in \mathcal{C}$ such that $d(c,u) \leq 1$.

Lemma 5.11. *If G^1_{r+1} is the generator matrix consisting of the first $n_r(q-1)$ rows of G_{r+1} defined in (5.29), then $\mathcal{C}(G^1_{r+1})$ and $(\alpha^j, 0\cdots 0) + \mathcal{C}(G^1_{r+1})$, $j \geq 0$, perfectly cover the same subset of $\mathbb{F}_q^{n_r q+1} = \mathbb{F}_q^{n_{r+1}}$.*

Proof. Since $G_2 = G^1_2$ and $\mathcal{C}(G_2)$ is a 1-perfect code, it follows that its coset $(\alpha^j, 0\cdots 0) + \mathcal{C}(G_2)$ is also a 1-perfect code and thus $\mathcal{C}(G^1_2)$ and $(\alpha^j, 0\cdots 0) + \mathcal{C}(G^1_2)$ perfectly cover the same subset of $\mathbb{F}_q^{q+1}$. Let $v = (\gamma, u_1, \ldots, u_{n_r}) \in \mathcal{C}(G^1_{r+1})$, where $(\delta_i, u_i) \in \mathcal{C}(G_2)$, $u_i \in \mathbb{F}_q^q$, $\delta_i \in \mathbb{F}_q$, and $\gamma = \sum_{i=1}^{n_r} \delta_i$. This is the form of the codewords from $\mathcal{C}(G^1_{r+1})$ as follows from the form of G_2 and G^1_{r+1}. We will show that every word that is covered by v is also covered by a codeword of $(\alpha^j, 0\cdots 0) + \mathcal{C}(G^1_{r+1})$. Obviously, $v + (\beta, 0\cdots 0)$, $\beta \in \mathbb{F}_q$, is covered by $v + (\alpha^j, 0\cdots 0) \in (\alpha^j, 0\cdots 0) + \mathcal{C}(G^1_{r+1})$.

Accordingly, we only have to show that any word of the form $(\gamma, u_1 \cdots u_{i-1} u'_i u_{i+1} \cdots u_{n_r})$, where u_i and u'_i differ in exactly one position, is covered by a codeword of $(\alpha^j, 0\cdots 0) + \mathcal{C}(G^1_{r+1})$. We know that the word (δ_i, u'_i) is covered by a codeword $v'_i \in (\alpha^j, 0\cdots 0) + \mathcal{C}(G_2)$ because $(\alpha^j, 0\cdots 0) + \mathcal{C}(G_2)$ is a 1-perfect code. Since $(\delta_i, u_i) \in \mathcal{C}(G_2)$ and the minimum distance of $\mathcal{C}(G_2)$ is 3, it follows that any word of the form (ξ, u'_i), where $\xi \in \mathbb{F}_q$, is not in $\mathcal{C}(G_2)$ and hence also not in $(\alpha^j, 0\cdots 0) + \mathcal{C}(G_2)$. Therefore, $v'_i = (\delta_i, u''_i)$, where u''_i differs in exactly one position from u'_i and in exactly two positions from u_i. Since $(\delta_i, u''_i) \in (\alpha^j, 0\cdots 0) + \mathcal{C}(G_2)$, it follows that $(\delta_i - \alpha^j, u''_i) \in \mathcal{C}(G_2)$ and hence $(\gamma - \alpha^j, u_1 \cdots u_{i-1} u''_i u_{i+1} \cdots u_{n_r}) \in \mathcal{C}(G^1_{r+1})$. This implies that the word $(\gamma, u_1 \cdots u_{i-1} u'_i u_{i+1} \cdots u_{n_r})$ is covered by $(\gamma, u_1 \cdots u_{i-1} u''_i u_{i+1} \cdots u_{n_r}) \in (\alpha^j, 0\cdots 0) + \mathcal{C}(G^1_{r+1})$.

Thus, every word that is covered by $\mathcal{C}(G^1_{r+1})$ is also covered by $(\alpha^j, 0\cdots 0) + \mathcal{C}(G^1_{r+1})$ and since $\mathcal{C}(G^1_{r+1})$ and $(\alpha^j, 0\cdots 0) + \mathcal{C}(G^1_{r+1})$ have the same size, the claim in the lemma follows. □

We now write G_r as

$$G_r = \begin{bmatrix} G^1_r \\ \\ F \end{bmatrix}, \quad \text{where} \quad F = \begin{bmatrix} f_1 \\ \vdots \\ f_t \end{bmatrix},$$

and f_i is a $1 \times n_r$ matrix. Let c_j, $1 \leq j \leq q^t$, be the q^t codewords formed from F, where F is considered as a generator matrix. By Lemma 5.11, we have that $c_j + \mathcal{C}(G_r^1)$ and $(\alpha^j, 0 \cdots 0) + c_j + \mathcal{C}(G_r^1)$ perfectly cover the same subset of $\mathbb{F}_q^{n_r}$.

Lemma 5.12. *Given the vector* $(g_1, g_2, \ldots, g_{q^t})$, $g_i \in \mathbb{F}_q$, $1 \le i \le q^t$, *the code*

$$\mathcal{C} \triangleq \bigcup_{i=1}^{q^t} ((g_i, 0 \cdots 0) + c_i + \mathcal{C}(G_r^1))$$

forms a q*-ary 1-perfect code.*

Proof. If $g_i = 0$ for all i, then $\mathcal{C}$ is the Hamming code. The lemma follows from the fact that $c_i + \mathcal{C}(G_r^1)$ and $(g_i, 0 \cdots 0) + c_i + \mathcal{C}(G_r^1)$ perfectly cover the same subset of $\mathbb{F}_q^{n_r}$. □

Let $\Omega(n_r)$ be the set of 1-perfect codes constructed in Lemma 5.12. Obviously, $|\Omega(n_r)| = q^{q^t} = q^{q^{n_r - 1 + 1 - r}} = q^{q^{\frac{n_r - 1}{q} + 1 - \log_q(n_r(q-1)+1)}}$. We say that two codes $\mathcal{C}_1$, $\mathcal{C}_2$ in $\mathbb{F}_q^n$ are equivalent if there exists a word $v \in \mathbb{F}_q^n$ and a permutation π on $[n]$ such that $\mathcal{C}_2 = \{v + \pi(c) \ : \ c \in \mathcal{C}_1\}$. Given a 1-perfect code $\mathcal{C}$ of length n_r, there are at most $q^{n_r} n_r! \le q^{n_r} q^{n_r \log_q n_r} = q^{n_r(1 + \log_q n_r)}$ different 1-perfect codes equivalent to $\mathcal{C}$. Hence we have the following theorem.

Theorem 5.11. *The set* $\Omega(n_r)$ *of codes, constructed in Lemma 5.12, contains at least* $q^{q^{\frac{n_r - 1}{q} + 1 - \log_q(n_r(q-1)+1)} - n_r(1 + \log_q n_r)}$ *nonequivalent 1-perfect codes over* $\mathbb{F}_q$.

A more precise enumeration will improve the result of Theorem 5.11. But this improvement is very minor. Finally, we would like to mention that given a 1-perfect code $\mathcal{C}$, one might permute symbols independently in each position to obtain another 1-perfect code. If we also consider these 1-perfect codes as equivalent, we will have that there are at most $n_r!(q!)^{n_r}$ different 1-perfect codes equivalent to $\mathcal{C}$. This not, however, dramatically different from the asymptotic result of Theorem 5.11.

5.8 On the Nonexistence of Perfect Codes

Are there more nontrivial perfect codes in the Hamming scheme $\mathcal{H}_q(n)$? This problem was completely solved for any prime power q and was almost completely solved for other q's. Two types of polynomials are involved in the solution for this problem when q is a prime power. The first polynomial is the Krawtchouk polynomial $P_e^n(x)$, which is defined by

$$P_e^n(x) \triangleq \sum_{i=0}^{e}(-1)^i\binom{x}{i}\binom{n-x}{e-i}(q-1)^{e-i}\,.$$

The polynomials from which nonexistence results for perfect codes were obtained are called Lloyd polynomials and such a polynomial $L_e^n(x)$ is an instant of the Krawtchouk polynomial defined as follows.

$$L_e^n(x) \triangleq \sum_{i=0}^{e} P_i^n(x) = P_e^{n-1}(x-1) = \sum_{i=0}^{e}(-1)^i(q-1)^{e-i}\binom{x-1}{i}\binom{n-x}{e-i}\,.$$

The following theorem is known as the Lloyd's theorem.

Theorem 5.12. *If there exists an e-perfect code of length n over* $\mathbb{F}_q$*, then the Lloyd polynomial* $L_e^n(x)$ *has e integer* zeroes $x_1, x_2, \ldots, x_e$ *such that* $0 < x_1 < x_2 < \cdots < x_e \leq n$.

The following theorem was obtained in the nineteen seventies of the 20th century.

Theorem 5.13. *Any nontrivial perfect code over* $\mathbb{F}_q$ *in the Hamming scheme has the same parameters as the q-ary Hamming codes, or the binary Golay code, or the ternary Golay code.*

The proof of Theorem 5.13 is complicated and in the following lines the first steps in the proof are presented. An e-perfect code $\mathcal{C}$ must attain the sphere-packing bound with equality, i.e.,

$$|\mathcal{C}| \cdot \sum_{i=0}^{e}\binom{n}{i}(q-1)^i = q^n.$$

Since q is a prime power, i.e., $q = p^r$, where p is a prime, it follows that both $|\mathcal{C}|$ and $\sum_{i=0}^{e}\binom{n}{i}(q-1)^i$ are powers of p. Hence, we have that for some j,

$$\sum_{i=1}^{e}\binom{n}{i}(q-1)^i = p^j - 1.$$

Since $q-1 = p^r - 1$, this implies that $p^r - 1$ divides $p^j - 1$, i.e., r divides j. Hence, there is an integer $t = \frac{j}{r}$ such that

$$\sum_{i=0}^{e}\binom{n}{i}(q-1)^i = p^j = p^{rt} = q^t\,. \qquad (5.30)$$

This is the starting point for the first few nonexistence results. For example, using a computer search and (5.30) one can verify that there is no perfect code for many sets of parameters having large radius and/or large alphabet size, and/or large length. Let $L_e(x) = \sum_{i=0}^{e} \ell_i x^i$. Combining (5.30) with Lloyd's theorem implies that

$$\ell_0 = L_e(0) = \sum_{i=0}^{e} \binom{n}{i} (q-1)^i = q^t \ .$$

Therefore, the coefficient ℓ_e of x^e in $L_e(x)$ shows that the product of the zeroes $x_1, x_2, \ldots, x_e$ of $L_e(x)$ has the form

$$\prod_{i=1}^{e} x_i = \frac{(-1)^e \ell_0}{\ell_e} = \frac{e!}{q^e} \sum_{i=0}^{e} \binom{n}{i} = e! q^{t-e} \ . \tag{5.31}$$

By adding

$$\sum_{i=1}^{e} x_i = \frac{-\ell_{e-1}}{\ell_e} = \frac{e(n-e)(q-1)}{q} + \frac{e(e+1)}{2} \ ,$$

to (5.31), many new nonexistence results can be obtained.

The following theorem cannot be proved with Lloyd's polynomials. To prove the claim of the theorem we will use a completely different method based on orthogonal arrays.

Theorem 5.14. *A 1-perfect code of length 7 over $\mathbb{Z}_6$ does not exist.*

Proof. Assume the contrary, that there exists such a code $\mathcal{C}$. By the sphere-packing bound, the code $\mathcal{C}$ contains $6^5 = \frac{6^7}{1+7\cdot 5}$ codewords. Let A be a $6^5 \times 7$ matrix whose rows are the codewords of $\mathcal{C}$. Since the minimum distance of $\mathcal{C}$ is 3, it follows from Theorem 3.3 that the matrix A is an OA$(5, 7, 6)$. Let $(x, y, z) \in \mathbb{Z}_6^3$ be a fixed 3-tuple. Since A is an OA$(5, 7, 6)$, it follows that (x, y, z) is contained in the projection of the first three columns of A in exactly $6^2 = 36$ codewords. Let B be the 36×4 matrix obtained from the projection of the last four columns, of A, on these 36 codewords. The matrix B is an OA$(2, 4, 6)$, which is equivalent by Theorem 3.4 to a pair of orthogonal Latin squares of order 6. Nevertheless, by Theorem 3.27, such a pair does not exist, and hence such a code $\mathcal{C}$ does not exist. Thus, there is no 1-perfect code of length 7 over $\mathbb{Z}_6$. □

5.9 Playing Games of Hats

One interesting application of 1-perfect codes is with a variant of the well-known combinatorial game of hats. The ***guessing game of hats*** is a game that received lot of attention by mathematicians and computer scientists. The game has a number of versions and some of the solutions involve beautiful concepts of mathematics. The version presented in this section has a beautiful solution based on perfect codes. There are n players receiving n hats with two colors "red" and "blue". Each player can see the color of the hat of each of the other $n-1$ players, but has no idea about the color of his own hat. The colors on the hats are distributed randomly and uniformly to the n players. Each player must guess the color of the hat on his head and all the players must simultaneously announce their guess or say "pass". The players win the game if one of them guesses the right color of his hat and no one guesses a wrong color. The target is to design a strategy that maximizes the probability of winning the game. During the game, the players cannot consult each other, but before the game they can decide on a strategy. For example, if there are three players, their strategy can be that a player who sees that the other two players' hats are of different colors will say "pass" while a player who sees that both players' hats have the same color will say the opposite color. It is easy to verify that the only case in which the players will lose is when the three hats have the same color. These are two configurations out of the eight possible distributions of the hats and hence the probability of success is 3/4. It is interesting to note that in the two losing configurations, all the players will make a wrong guess and, in the winning configurations, exactly one player will make the correct guess.

A ***perfect strategy*** in the hat guessing game is a strategy for which in a winning configuration exactly one player makes the right guess, while in a losing configuration all the players make a wrong guess.

Theorem 5.15. *If there exists a binary 1-perfect binary code of length n in the Hamming scheme, then there exists a perfect strategy for the hat guessing game with probability of $\frac{n}{n+1}$ for success.*

Proof. Given a binary 1-perfect code $\mathcal{C}$ of length n, each player is associated with one of the coordinates. Let *zero* be associated with "red" and *one* be associated with "blue". A player who sees the other colors, translates the configuration into a binary word of length n, where in his position he decides on an assignment of *zero* or *one* if one of the two leads to a codeword in $\mathcal{C}$. If no such assignment exists, then his guess is "pass". If such an assignment exists, then his guess is the opposite one to the color associated with this assignment, i.e., for a *zero* assignment he guesses "blue" and for a *one* assignment he guesses "red". Distinguish now between two cases:

(1) If the configuration of the hats is associated with a codeword $c \in \mathcal{C}$, then each player will be able to have an assignment that will lead to c. But, since he will make the opposite guess to the one associated with the codeword, it follows that each of the n players will make a wrong guess.

(2) If the configuration of the hats is associated with a word x that is not a codeword of $\mathcal{C}$, then since the code is a 1-perfect code, there exists exactly one codeword $c \in \mathcal{C}$ for which $d(x, c) = 1$. If x and c differ in the i-th coordinate, then only the i-th player will make the right guess. Each other player will "pass" since any assignment on its coordinate will not lead to a codeword.

This implies that this strategy is perfect.

Since all the codewords are associated with the losing configurations and the other words are the winning configuration, it follows that the probability of winning in this strategy is $\frac{n}{n+1}$. □

5.10 Notes

Lot of research on nonlinear 1-perfect codes has been done since their introduction in coding theory. As a consequence of the results in Section 5.7, we know that their number is very large. For different applications or different combinatorial and algebraic properties, different codes have to be found. In this chapter we considered only a small fraction of the research done on these codes. For completeness we will now mention some more research, but not all the research, on other topics considered in the literature.

Two papers [Östergård and Pottonen (2009); Östergård, Pottonen, and Phelps (2010)] have classified the 1-perfect codes of length 15. In the first paper, Östergård and Pottonen gave the complete classification and the automorphism groups of these codes. There are 5983 such nonequivalent 1-perfect codes and 2165 such extended 1-perfect codes of length 15 and length 16, respectively. There are 38408 inequivalent shortened codes of length 14. In the follow-up work, Östergård, Pottonen, and Phelps classified these codes, studied them in great detail, and tabulated their main properties. The results include the fact that 33 of the nonisomorphic 80 Steiner triple systems of order 15 occur in such codes. Further understanding is gained on full-rank codes via the switching method, as it turns out that all but two full-rank 1-perfect codes can be obtained through a series of such transformations (switches) from the Hamming code. Other topics studied in this paper include nonsystematic codes and embedded one-error-correcting codes as well as other topics.

It was established by [Best and Brouwer (1977)] that triply shortened 1-perfect codes of length 2^r-1 are optimal. That is, $A(2^r-2,3)=2^{2^r-r-2}$, $A(2^r-3,3)=2^{2^r-r-3}$, and $A(2^r-4,3)=2^{2^r-r-4}$. Referring to tables of the best known codes suggests that shortening any 1-perfect code of length 2^r-1 up to $2^{r-2}-1$ times is likely to produce optimal codes for $r \leq 9$. The result of [Kabatiansky and Panchenko (1988)] however, shows that this is not true in general for large r. Thus we have the following problem.

Problem 5.1. What is the largest integer s_r such that each shortening of a 1-perfect code of length 2^r-1, up to s_r times, produces optimal codes, i.e., $A(2^r-1-s_r,3)=2^{2^r-r-1-s_r}$?

Problem 5.1 is about the optimality of a shortened 1-perfect code. A related question is on the uniqueness of the shortening. Assume $A(2^r-1-j,3)=2^{2^r-r-1-j}$ for some j, $1 \leq j < 2^{r-2}$ and let $\mathcal{C}$ be a code that attains this bound. Can $\mathcal{C}$ be completed to a 1-perfect code of length 2^r-1? The answer to this question is positive for $j=1$ as was proved in [Blackmore (1999)]. Nevertheless, it was proved in [Östergård and Pottonen (2011)] that if $j=2$, then the answer to the question is negative, for $r=4$. This result was generalized for $r>4$ by [Krotov, Östergård, and Pottonen (2011)].

Section 5.1. Constructions of nonequivalent 1-perfect codes have been suggested since the introduction of the first codes in [Hamming (1950)]. Theorem 5.1 was presented first in [Vasil'ev (1962)]. To yield a large number of perfect codes, the construction of this theorem depends on previously obtained perfect codes since it is a recursive construction. Nonetheless, its recursive implementation will yield asymptotically similar number of perfect codes as other direct constructions. Its importance is also in being the first one for nonlinear 1-perfect codes. This construction was generalized in [Mollard (1986)]. The construction in Theorem 5.2 was presented in [Solovieva (1981)] and also independently by [Phelps (1983)]. They have both analyzed the codes obtained and introduced some variants of the construction. The construction was generalized to any field $\mathbb{F}_q$ by [Romanov (2019)]. The construction can be viewed as a variant of the more general and complicated construction introduced in [Heden (1977)]. The general product construction of Theorem 5.3 was introduced in [Phelps (1984a)], which was further generalized in [Phelps (1984b)]. The last general product construction in Theorem 5.4 was presented in [Etzion (2011)]. Other constructions can be found, for example, in [Solovieva (1981, 1989, 1994)].

Section 5.2. The weight distribution of 1-perfect codes, introduced in this section, was obtained in [Etzion and Vardy (1994)].

Section 5.3. The maximum intersection between two 1-perfect codes was obtained in [Etzion and Vardy (1994)] and the minimum intersection was presented in [Etzion and Vardy (1998)]. It was proved in [Avgustinovich, Heden, and Solovieva (2006)] that for each even integer η in the interval $0 \leq \eta \leq 2^{n+1-2\log_2(n+1)}$, there are two 1-perfect codes $\mathcal{C}_1$ and $\mathcal{C}_2$ of length $n = 2^r - 1$, $r \geq 4$, for which $|\mathcal{C}_1 \cap \mathcal{C}_2| = \eta$. A general method to find the possible intersection numbers of isomorphic linear codes was presented in [Bar-Yahalom and Etzion (1997)].

Section 5.4. The intersection numbers between Hamming codes, presented in this section, were found by [Etzion and Vardy (1998)].

Section 5.5. The construction for full-rank 1-perfect codes was presented in [Etzion and Vardy (1994)]. Generalization of the construction for a nonbinary alphabet was done in [Phelps and Villanueva (2002a)]. Full-rank 1-perfect codes play an important role in full-rank tiling [Cohen, Litsyn, Vardy, and Zémor (1996)], where a tiling $(\mathcal{A}, \mathcal{B})$ of $\mathbb{F}_q^n$ is of full-rank if rank $\mathcal{A}$ = rank $\mathcal{B} = n$.

The set of sub-codes, $\mathcal{A}(z)$, $z \in \mathbb{F}_2^r$, plays an important role in various constructions such as 1-perfect codes with various ranks and kernels. One problem in this context is to find the size of the intersection between such various sub-codes and their cosets. The most fundamental problem is to find the size of the intersection of two such sub-codes. This was considered, for example, in [Etzion and Vardy (1994); Phelps and Levan (1995)]. The following lemma is an immediate consequence from Proposition 5.3.

Lemma 5.13. *If $n = 2^r - 1$ and $\mathcal{C}$ is a 1-perfect code of length n, then for each $i \neq j$*

$$|\mathcal{A}(z_i) \cap \mathcal{A}(z_j)| = 2^{2^{r-2}} .$$

Corollary 5.8. *Each coset $x_i + \mathcal{A}(z_i)$ has a nonempty intersection with at most $2^{2^{r-2}-1}$ cosets (in the Hamming code) of $\mathcal{A}(z_j)$, where $i \neq j$.*

Proof. If $x_i + \mathcal{A}(z_i)$ and $x_j + \mathcal{A}(z_j)$, are two distinct cosets, and $y \in \mathcal{A}(z_i) \cap \mathcal{A}(z_j)$, then

$$x_i + \mathcal{A}(z_i) = y + \mathcal{A}(z_i) \text{ and } x_j + \mathcal{A}(z_j) = y + \mathcal{A}(z_j) .$$

But, $(y + \mathcal{A}(z_i)) \cap (y + \mathcal{A}(z_j)) = y + (\mathcal{A}(z_i) \cap \mathcal{A}(z_j))$. Therefore, any coset of $\mathcal{A}(z_j)$ is either disjoint from $x_i + \mathcal{A}(z_i)$ or intersects it in $2^{2^{r-2}}$ words. Since the cosets of $\mathcal{A}(z_j)$ are either equal or disjoint and $x_i + \mathcal{A}(z_i)$ has $2^{2^{r-1}-1}$ codewords, the claim follows. □

Section 5.6. The proof for the possible kernels of 1-perfect codes and the related constructions were presented in [Phelps and Levan (1995)]. Theorem 5.8 in that paper was proved in a completely different way from the proof that we gave for this theorem. The theorem in [Phelps and Levan (1995)] is more general and it is stated as follows.

Theorem 5.16. *For $k \geq 2$ independent points in the projective space associated with the words of weight three in the Hamming code of length $2^r - 1$, the subspace $\mathcal{A}(z_1) \cap \mathcal{A}(z_2) \cap \cdots \cap \mathcal{A}(z_k)$ has dimension 2^{r-k}.*

Theorem 5.16 is also a generalization of Lemma 5.13 (derived from Proposition 5.3) for which we provided a different proof from the one in [Phelps and Levan (1995)]. A tradeoff between the ranks and the kernels of binary 1-perfect codes was obtained in [Phelps and Villanueva (2002b)]. The results were generalized for nonbinary alphabets in [Phelps, Rifà, and Villanueva (2005)].

Another interesting question that was considered is the existence of nonsystematic 1-perfect codes. Nonsystematic 1-perfect codes were considered and constructed in [Avgustinovich and Solovieva (1996a,b)]. Their results were improved later by [Phelps and Levan (1999)], where the switching method was used to construct such codes. They used a construction similar to the one that they used in [Phelps and Levan (1995)] for the construction of 1-perfect codes with different kernels. This means that again the switching method has an important role in the construction of 1-perfect codes with required properties. The problem was considered later also by [Malyugin (2010)]. An interesting connection between systematic and nonsystematic 1-perfect codes on one side and superimposed codes used for group testing and multiple access communication on the other side can be found in [Ericson and Levenshtein (1994)].

Section 5.7. The switching method that was described and used to obtain a lower bound on the number of nonequivalent q-ary 1-perfect codes is taken from [Etzion (1996c)]. Asymptotically, there are no dramatic improvements on the number of codes obtained by this construction, but slight improvements can be obtained and they are increased as the alphabet size q get larger. For example, the constructions mentioned in Section 5.1 can be also used to construct many nonequivalent codes. Analysis of the various constructions can be found in [Cohen, Honkala, Litsyn, and Lobstein (1997), pp. 296–310].

Section 5.8. The final steps in the nonexistence results were proved by [Tietäväinen (1970); Zinoviev and Leontiev (1973); Tietäväinen (1974)] after an extensive work in [van Lint (1970a,b, 1971a,b, 1974)]. van Lint also surveyed all the known results in [van Lint (1975)]. Other results were proved in [Tietäväinen and Perko (1971)]. The results were further generalized to cover non-prime power alphabets in [Best (1983)], where an outline for the proof that there are no nontrivial e-perfect codes with possible exceptions for $e \in \{1, 2, 6, 8\}$, was presented. More detailed proofs and information appear in his Ph.D. thesis [Best (1982)]. It was proved in [Lenstra (1972)] that if there exists a 1-perfect code in $\mathcal{H}_q(n)$, then $n = kq+1$ for some k. Further nonexistence results for 1-perfect codes over non-prime alphabet can be found in [Heden and Roos (2011)], e.g., where the results implied that there is no 1-perfect code in $\mathcal{H}_6(19)$. The proof of Theorem 5.14 for the nonexistence of a 1-perfect code in $\mathcal{H}_6(7)$ is due to Block and Hall and appeared in [Golomb and Posner (1964)].

As noted before, an extended 1-perfect code in $\mathcal{H}_q(q + 2)$, where $q > 1$ is any integer, is also an $\text{OA}(q-1, q+2, q)$. The nonexistence result of [Hill (1978)] (see Theorem 4.10) for the nonexistence of extended 1-perfect codes is only applied for linear codes. For nonlinear codes it was proved in [Gijswijt, Schrijver, and Tanaka (2006)] that there is no extended 1-perfect code in $\mathcal{H}_3(14)$. A more substantial result was given in [Bespalov (2020)] who proved that there is no 1-perfect code in $\mathcal{H}_q(q + 2)$, where q is an odd integer, which also implies that there are no $\text{OA}(q - 1, q + 2, q)$ if q is odd. It was also proved in [Bespalov (2020)] that if q is a power of an odd prime, then there is no extended 1-perfect code of length $\frac{q^r+q-2}{q-1}$ when r is an even integer. It was further proved that if $n > q + 2$, then there is no extended 1-perfect code in $\mathcal{H}_q(n)$ when $q \in \{3, 4\}$. Finally, there is no extended 1-perfect code in $\mathcal{H}_q(n)$ if n is odd for any integer q.

Section 5.9. The version of the hat guessing game introduced in this section is that of [Lenstra and Seroussi (2002)]. There was a lot of research about hat guessing games before and after this paper and a sample of papers includes [Aravamuthan and Lodha (2006); Guo, Kasala, Rao, and Tucker (2007); Butler, Hajiaghayi, Kleinberg, and Leighton (2009); Paterson and Stinson (2010); Feige (2010); Gadouleau and Georgiou (2015); Gadouleau (2018); Alon, Ben-Eliezer, Shangguan, and Tamo (2020)].

Chapter 6

Density and Quasi-Perfect Codes

As we have seen, there are only two families of parameters with nontrivial perfect codes in the Hamming scheme. The Hamming codes over $\mathbb{F}_q$ have length $\frac{q^r-1}{q-1}$ and redundancy r, for each $r \geq 2$. The binary Golay code is a $[23, 12, 7]$ code and the ternary Golay code is a $[11, 6, 5]$ code. This lack of perfect codes has motivated the search for codes that are "almost" perfect. In this chapter we consider such codes only for the binary alphabet. What is an "almost" perfect code whose minimum distance is $2e+1$? One answer can be that in such a code $\mathcal{C}$, beyond the packing radius e, there are only a few words whose distance from the code is $e+1$ (and none that are further apart) and $\mathcal{C}$ is the largest code with these parameters. Such a code belongs to a family of codes called quasi-perfect codes and codes from An upper bound which improves on the sphere-packing bound is developed in Section 6.2. This bound on the size of e-codes takes into account the uncovered words at distance $e+1$ from the code. Codes which meet this bound are called nearly-perfect codes. All prefect codes are nearly perfect and also shortened 1-perfect codes are nearly perfect. The other family of nearly-perfect codes are the punctured Preparata codes, which will be presented in Section 6.3. The punctured Preparata codes are quasi-perfect codes. In a perfect code $\mathcal{C}$, each word in the space is within distance e from exactly one codeword of $\mathcal{C}$. Hence, in a code $\mathcal{C}$ with similar properties to a perfect code, the requirements are that most words will be within distance e from exactly one codeword of $\mathcal{C}$ and only a small fraction of the words in the space will not be at distance e from some codeword of $\mathcal{C}$. This is a property from a point of view of the packing radius. A similar definition can be given from the point of view of the covering radius. From a point of view of the covering radius it is required that each word in the space is within distance e from exactly one codeword, but there are a small number of words in the space which are within distance e from more than one codeword. These requirements lead to the definition of the density of codes, which will be discussed in Section 6.1.

For packing we would like to have dense codes for which most words are covered by the code and for covering we would like to have sparse codes for which most words are covered exactly once by the code. Dense codes with packing radius one or two are the Hamming codes and the Preparata codes, respectively. Hamming codes are also sparse codes with covering radius one. Nearly-perfect codes are dense codes from the packing radius point of view. By the description of nearly-perfect codes, it is clear that if the packing radius of such code is e, then its covering radius is $e + 1$. A code with packing radius e and covering radius $e + 1$ is called a quasi-perfect code. Such codes are discussed in Section 6.4. Are quasi-perfect codes similar to perfect codes? The answer is yes if we examine only the packing radius and the covering radius of these codes. Apparently, however, this is not the most important property of perfect codes and quasi-perfect codes are usually not as dense as perfect codes (as packing codes) for most parameters. Moreover, they are usually not as sparse as perfect codes (as covering codes) for most parameters. Sparse codes with covering radius 2, which are said to be asymptotically perfect, are presented in Section 6.5 and sparse codes with covering radius 3 are presented in Section 6.6. Finally, we remind that in this chapter only binary codes will be considered.

6.1 Density of Codes

Recall that by the sphere-packing bound, a binary code $\mathcal{C}$ of length n with minimum Hamming distance $2e + 1$, is e-perfect if

$$\frac{|\mathcal{C}| \cdot |\mathcal{B}_e(n)|}{2^n} = 1 \; .$$

What about a code for which this fraction is close to 1? Such a code is clearly very dense, i.e., the balls of radius e around codewords are disjoint, and the number of points in the space $\mathbb{F}_2^n$ that are not covered by any ball is very small. To make these observations more formal, we define the notion of density. We distinguish between packing density and covering density.

The ***packing density*** of a code $\mathcal{C}$ of length n with packing radius e is defined by

$$\mu_p(\mathcal{C}) \triangleq \frac{|\mathcal{C}| \cdot |\mathcal{B}_e(n)|}{2^n}.$$

The ***covering density*** of a code $\mathcal{C}$ of length n with covering radius R is defined by

$$\mu_c(\mathcal{C}) \triangleq \frac{|\mathcal{C}| \cdot |\mathcal{B}_R(n)|}{2^n} .$$

Let $\{\mathcal{C}_i\}_{i=1}^{\infty}$ be an infinite family of (n_i, M_i, d) codes with packing radius $e = \left\lfloor \frac{d-1}{2} \right\rfloor$, where $n_i < n_{i+1}$ for each $i \geq 1$. The packing density of this family is defined by

$$\mu_p(\{\mathcal{C}_i\}) \triangleq \lim_{i \to \infty} \mu_p(\mathcal{C}_i),$$

if the limit exists.

Similarly, let $\{\mathcal{C}_i\}_{i=1}^{\infty}$ be an infinite family of (n_i, M_i, d_i) codes, where each code has covering radius R and $n_i < n_{i+1}$ for each $i \geq 1$. The covering density of this family is defined by

$$\mu_c(\{\mathcal{C}_i\}) \triangleq \lim_{i \to \infty} \mu_c(\mathcal{C}_i),$$

if the limit exists.

A family for which the packing (covering) density is 1 is called ***asymptotically perfect***. Clearly, the packing density and the covering density of a perfect code is 1, and if the code is not perfect, then its packing density is smaller than 1 and its covering density is larger than 1. Therefore, we will omit perfect codes in our discussion on the density. Are there other families of codes that are asymptotically perfect? In the following sections such asymptotically perfect codes will be presented both for packing and covering.

Let $K(n, R)$ be the minimum number of codewords in a code of length n with covering radius R, and define the covering density, $\mu_c(n, R)$ with respect to length n and covering radius R by

$$\mu_c(n, R) \triangleq \frac{K(n, R) \cdot |\mathcal{B}_R(n)|}{2^n} .$$

The ***maximum asymptotic covering density*** with respect to covering radius R, $\mu^*(R)$, and the ***minimum asymptotic covering density*** with respect to covering radius R, $\mu_*(R)$, are defined by

$$\mu^*(R) \triangleq \lim_{n \to \infty} \sup \mu_c(n, R)$$

and

$$\mu_*(R) \triangleq \lim_{n \to \infty} \inf \mu_c(n, R) .$$

Similarly, we define the maximum asymptotic packing density and the minimum asymptotic packing density using $A(n, d)$ instead of $K(n, R)$. For example, the packing density, $\mu_p(n,e)$ with respect to length n and packing radius e is defined by

$$\mu_p(n,e) \triangleq \frac{A(n,2e+1)\cdot|\mathcal{B}_e(n)|}{2^n}.$$

These definitions can be further specialized using similar definitions for linear codes. We will not give the other formal definitions, since we will not distinguish between linear codes and nonlinear codes in our exposition. There are many interesting questions related to all these definitions for the density of codes and the density of packing (covering, respectively) with respect to a certain packing radius (covering radius, respectively). Our main concern are those parameters where the density is 1, i.e., the codes are asymptotically perfect, and we also concentrate on the most dense families with small packing radius and the most sparse families with small covering radius. Finally, one can easily verify that unless $\mathcal{C}$ is a perfect code, the code $\mathcal{C}$ can be either dense, with respect to its packing radius, or sparse, with respect to its covering radius, but not both.

6.2 The Johnson Bound

The Johnson bound is an improvement on the sphere-packing bound. It takes into account the words that cannot be covered by the e-balls of the codewords and are at distance $e+1$ from the code. The theorem is proved by considering the words in the balls of radius e around the codewords of a code $\mathcal{C}$ that attains $A(n,2e+1)$ and the words that are at distance $e+1$ from $\mathcal{C}$ that are not covered by $\mathcal{C}$.

Theorem 6.1.

$$A(n,2e+1)\left(\sum_{i=0}^{e}\binom{n}{i}+\frac{\binom{n}{e+1}-\binom{2e+1}{e}A(n,2e+2,2e+1)}{\left\lfloor\frac{n}{e+1}\right\rfloor}\right)\leq 2^n. \quad (6.1)$$

Proof. Let $\mathcal{C}$ be an (n,M,d) binary code with $d = 2e+1$, where $M=A(n,d)$, and $\mathbf{0}\in\mathcal{C}$. Let $\mathcal{D}_0=\mathcal{C}$ and $\mathcal{D}_i$, $i\geq 1$, be the set of words at distance exactly i from $\mathcal{C}$, i.e.,

$$\mathcal{D}_i \triangleq \{x\in\mathbb{F}_2^n \;:\; \mathcal{B}_{i-1}(x)\cap\mathcal{C}=\varnothing,\; \mathcal{B}_i(x)\cap\mathcal{C}\neq\varnothing\}\,.$$

Clearly, $\mathcal{D}_i\cap\mathcal{D}_j=\varnothing$ for $i\neq j$, $\mathcal{D}_d=\varnothing$ since any word at a distance at least d from $\mathcal{C}$ can be added to $\mathcal{C}$ without reducing the minimum distance of $\mathcal{C}$. Hence, $\mathbb{F}_2^n=\bigcup_{i=0}^{d-1}\mathcal{D}_i$. The next step is to estimate $\mathcal{D}_{e+1}$, which is the set of words in the first sphere outside the balls with radius e around the codewords of $\mathcal{C}$. Pick any codeword $c \in \mathcal{C}$ and move it to the origin. Note also that codewords of weight $2e+1$ form a constant weight code with minimum distance $2e+2$, i.e., their number is at most $A(n, 2e+2, 2e+1)$.

Let $\mathcal{W}_{e+1}$ be the set of words in $\mathbb{F}_2^n$ whose weight is $e+1$. Any word in $\mathcal{W}_{e+1}$ is contained in either $\mathcal{D}_e$ or $\mathcal{D}_{e+1}$. With respect to each codeword $c' \in \mathcal{C}$ of weight $2e+1$, there are $\binom{2e+1}{e}$ words of weight $e+1$ at distance e from c'. These words are contained in $\mathcal{W}_{e+1} \cap \mathcal{D}_e$ and they are all distinct. Therefore,

$$|\mathcal{W}_{e+1} \cap \mathcal{D}_{e+1}| = |\mathcal{W}_{e+1}| - |\mathcal{W}_{e+1} \cap \mathcal{D}_e| \geq \binom{n}{e+1} - \binom{2e+1}{e} A(n, 2e+2, 2e+1).$$

A word x in $\mathcal{W}_{e+1} \cap \mathcal{D}_{e+1}$ is at distance $e+1$ from at most $\left\lfloor \frac{n}{e+1} \right\rfloor$ codewords. This can be observed by moving the origin to x and computing the number of codewords that can be at distance $e+1$ from x and are at mutual distance $d+1 = 2e+2$. Such codewords have disjoint sets of *ones* and hence their number is at most $\left\lfloor \frac{n}{e+1} \right\rfloor$. This implies that

$$A(n, 2e+1) \left(\frac{\binom{n}{e+1} - \binom{2e+1}{e} A(n, 2e+2, 2e+1)}{\left\lfloor \frac{n}{e+1} \right\rfloor} \right) \leq |\mathcal{D}_{e+1}| \ .$$

Now let c be computed over all the codewords and count the words in $\bigcup_{i=0}^{e+1} \mathcal{D}_i$, where

$$\left| \bigcup_{i=0}^{e+1} \mathcal{D}_i \right| = A(n, 2e+1) \bigcup_{i=0}^{e} \binom{n}{i} + A(n, 2e+1) \left|\mathcal{D}_{e+1}\right|,$$

to obtain the claim of the Theorem. □

Corollary 6.1.

$$A(n, 2e+1) \left(\sum_{i=0}^{e} \binom{n}{i} + \frac{\binom{n}{e}\left(\frac{n-e}{e+1} - \left\lfloor \frac{n-e}{e+1} \right\rfloor\right)}{\left\lfloor \frac{n}{e+1} \right\rfloor} \right) \leq 2^n \ .$$

Proof. Iterating (2.1) yields

$$A(n, 2e+2, 2e+1) \leq \left\lfloor \frac{n}{2e+1} \left\lfloor \frac{n-1}{2e} \cdots \left\lfloor \frac{n-e}{e+1} \right\rfloor \cdots \right\rfloor \right\rfloor$$

$$\leq \frac{n(n-1)\cdots(n-e+1)}{(2e+1)2e\cdots(e+2)} \left\lfloor \frac{n-e}{e+1} \right\rfloor \ .$$

This equation is substituted in (6.1) to prove the claim. □

The bound of Theorem 6.1 and also the one of Corollary 6.1 is called the Johnson bound. Codes that attain the bound of Corollary 6.1 are called ***nearly-perfect codes***. Perfect codes and shortened 1-perfect codes are nearly-perfect. The other family of nearly-perfect codes are the punctured Preparata codes, which will be discussed in Section 6.3. The bound of Theorem 6.1 can be improved in a few ways. One such way, which considers words at distance $e+1$ or $e+2$ from the code, is presented now.

Let $\mathcal{C}$ be an $(n, M, 2e+1)$ code, for which $M = A(n, 2e+1)$, where $\mathbf{0} \in \mathcal{C}$, and let $\mathcal{C}^*$ be the extended $(n+1, M, 2e+2)$ code. The number of codewords of weight i in $\mathcal{C}$ is denoted by A_i. A word $h \in \mathbb{F}_2^m$, where $m = n$ or $m = n+1$, is called a ***hole*** if $d(h, C) > e$, where C is $\mathcal{C}$ or $\mathcal{C}^*$, repectively. The number of holes of weight i, with respect to C, will be denoted by $NH_i(C)$. Let $NH(C)$ be the total number of holes with respect to C. Finally, we define $NH(c, C, \delta)$ to be the number of holes at distance δ from a codeword $c \in C$, $NC(h, C, \delta)$ to be the number of codewords of C at distance δ from a hole h, and $NH(C, \delta)$ to be the number of holes at distance δ from C.

Lemma 6.1. $NH_{e+1}(\mathcal{C}) = \binom{n}{e+1} - \binom{2e+1}{e+1} A_{2e+1}$.

Proof. The total number of words of weight $e+1$ in $\mathbb{F}_2^n$ is $\binom{n}{e+1}$. Words of weight $e+1$ can be e-covered only by codewords of weight $2e+1$ of $\mathcal{C}$. A codeword of weight $2e+1$ e-covers $\binom{2e+1}{e+1}$ words of weight $e+1$. Finally, two codewords $c_1, c_2 \in \mathcal{C}$ cannot e-cover the same word and hence

$$NH_{e+1}(\mathcal{C}) = \binom{n}{e+1} - \binom{2e+1}{e+1} A_{2e+1}.$$

□

Lemma 6.2. $NH_{e+2}(\mathcal{C}) = \binom{n}{e+2} - \binom{2e+1}{\delta+2} A_{2e+1} - \binom{2e+2}{e+2} A_{2e+2}$.

Proof. The total number of words of weight $e+2$ in $\mathbb{F}_2^n$ is $\binom{n}{e+2}$. Words of weight $e+2$ can be covered either by codewords of weight $2e+2$ or by codewords of weight $2e+1$. A codeword of weight $2e+2$ covers $\binom{2e+2}{e+2}$ words of weight $e+2$. A codeword of weight $2e+1$ e-covers $\binom{2e+1}{e+2}$ words of weight $e+2$. Finally, two codewords $c_1, c_2 \in \mathcal{C}$ cannot e-cover the same word and hence

$$NH_{e+2}(\mathcal{C}) = \binom{n}{e+2} - \binom{2e+1}{e+2} A_{2e+1} - \binom{2e+2}{e+2} A_{2e+2}.$$

□

Lemma 6.3. $NH_{e+2}(\mathcal{C}^*) = NH_{e+1}(\mathcal{C}) + NH_{e+2}(\mathcal{C})$.

Proof. If h_1 is a hole of weight $e+1$ with respect to $\mathcal{C}$, then, clearly, $h_1 1$ is a hole of weight $e+2$ with respect to $\mathcal{C}^*$ and if h_2 is a hole of weight $e+2$ with respect to $\mathcal{C}$, then $h_2 0$ is a hole of weight $e+2$ with respect to $\mathcal{C}^*$. Therefore,

$$NH_{e+2}(\mathcal{C}^*) \geq NH_{e+1}(\mathcal{C}) + NH_{e+2}(\mathcal{C}) \ . \tag{6.2}$$

Let hb be a hole of weight $e+2$ with respect to $\mathcal{C}^*$, where $h \in \mathbb{F}_2^n$ and $b \in \mathbb{F}_2$, i.e., $d(hb, \mathcal{C}^*) > e$. Therefore, $d(h, \mathcal{C}) \geq e$. We claim that h is a hole with respect to $\mathcal{C}$. Assume the contrary, that h is not a hole and distinguish between two cases depending on whether the value of b is 0 or 1.

Case 1. $b = 0$.

This implies that $\text{wt}(h) = e+2$. If h is not a hole, with respect to $\mathcal{C}$, then there exists a codeword $c \in \mathcal{C}$ such that $d(c, h) = e$, and since $d(\mathcal{C}) = 2e+1$ and $\text{wt}(h) = e+2$, it follows that $\text{wt}(c) = 2e+2$. Therefore, $c0 \in \mathcal{C}^*$, $d(c0, hb) = e$ and hb is not a hole with respect to $\mathcal{C}^*$, a contradiction.

Case 2. $b = 1$.

This implies that $\text{wt}(h) = e+1$. If h is not a hole, with respect to $\mathcal{C}$, then there exists a codeword $c \in \mathcal{C}$ such that $d(c, h) = e$, and since $d(\mathcal{C}) = 2e+1$ it follows that $\text{wt}(c) = 2e+1$. Therefore, $c1 \in \mathcal{C}^*$, $d(c1, hb) = e$ and hb is not a hole with respect to $\mathcal{C}^*$, a contradiction.

Both cases imply that h is a hole, with respect to $\mathcal{C}$, and hence

$$NH_{e+2}(\mathcal{C}^*) \leq NH_{e+1}(\mathcal{C}) + NH_{e+2}(\mathcal{C}). \tag{6.3}$$

Equations (6.2) and (6.3) imply that

$$NH_{e+2}(\mathcal{C}^*) = NH_{e+1}(\mathcal{C}) + NH_{e+2}(\mathcal{C}).$$

□

Lemma 6.4. $NH_{e+2}(\mathcal{C}^*) \geq \binom{n+1}{e+2} - \binom{2e+2}{e+2} A(n+1, 2e+2, 2e+2)$.

Proof. By Lemmas 6.1, 6.2, and 6.3,

$$NH_{e+2}(\mathcal{C}^*) = NH_{e+1}(\mathcal{C}) + NH_{e+2}(\mathcal{C})$$

$$= \binom{n}{e+1} - \binom{2e+1}{e+1} A_{2e+1} + \binom{n}{e+2} - \binom{2e+1}{e+2} A_{2e+1} - \binom{2e+2}{e+2} A_{2e+2}$$

and hence

$$NH_{e+2}(\mathcal{C}^*) = \binom{n+1}{e+2} - \binom{2e+2}{e+2}(A_{2e+1} + A_{2e+2}). \tag{6.4}$$

It is easily verified that

$$A_{2e+1} + A_{2e+2} \leq A(n+1, 2e+2, 2e+2), \tag{6.5}$$

and, therefore, by (6.4) and (6.5), we have that

$$NH_{e+2}(\mathcal{C}^*) \geq \binom{n+1}{e+2} - \binom{2e+2}{e+2} A(n+1, 2e+2, 2e+2).$$

□

Corollary 6.2. *If $c \in \mathcal{C}^*$, then*

$$NH(c, \mathcal{C}^*, e+2) \geq \binom{n+1}{e+2} - \binom{2e+2}{e+2} A(n+1, 2e+2, 2e+2).$$

Proof. This is an immediate consequence from Lemma 6.4 by considering the translate $c + \mathcal{C}^*$. □

Lemma 6.5. *If for a hole h with respect to $\mathcal{C}^*$ there exists a codeword $c_1 \in \mathcal{C}^*$ such that $d(h, c_1) = e+2$, then $d(h, \mathcal{C}^*) = e+2$.*

Proof. Clearly $d(h, \mathcal{C}^*) \leq e+2$. To complete the proof it is suffices to show that $d(h, \mathcal{C}^*) \neq e+1$. Assume the contrary, i.e., there exists a codeword $c_2 \in \mathcal{C}^*$ such that $d(h, c_2) = e+1$. This, then, implies that $d(c_1, c_2)$ is odd, a contradiction since all codewords of $\mathcal{C}^*$ have even weights. □

Lemma 6.6.

$$NH(\mathcal{C}^*, e+2) \leq 2^n - M \sum_{i=0}^{e} \binom{n}{i}.$$

Proof. Recall that $M = A(n, 2e+1)$ and, clearly,

$$NH(\mathcal{C}) = 2^n - M \sum_{i=0}^{e} \binom{n}{i}.$$

By Lemma 6.3, each hole of the form hb, where $h \in \mathbb{F}_2^n$ and $b \in \mathbb{F}_2$, with respect to $\mathcal{C}^*$, for which $d(hb, c) = e+2$, for some $c \in \mathcal{C}^*$, is obtained from a hole h of $\mathcal{C}$. To complete the proof it is sufficient to show that if hb is a hole for which $d(hb, c) = e+2$, for some $c \in \mathcal{C}^*$, then $h\bar{b}$, is not a hole for which $d(h\bar{b}, c') = e+2$, for some $c' \in \mathcal{C}^*$. Assume the contrary, i.e., there exist two codewords $c_1, c_2 \in \mathcal{C}^*$ such that $d(hb, c_1) = e+2$ and

$d(h\bar{b}, c_2) = e+2$. This implies that $d(c_1, c_2)$ is odd, a contradiction since all codewords of $\mathcal{C}^*$ have even weights. Thus,

$$NH(\mathcal{C}^*, e+2) \le NH(\mathcal{C}) = 2^n - M \sum_{i=0}^{\delta} \binom{n}{i} .$$

□

Theorem 6.2.

$$A(n, 2e+1)\left(\sum_{i=0}^{e}\binom{n}{i} + \frac{\binom{n+1}{e+2} - \binom{2e+2}{e+2}A(n+1, 2e+2, 2e+2)}{A(n+1, 2e+2, e+2)}\right) \le 2^n.$$

Proof. By Corollary 6.2, we have

$$\sum_{c \in \mathcal{C}^*} NH(c, \mathcal{C}^*, e+2) \ge M\left(\binom{n+1}{e+2} - \binom{2e+2}{e+2}A(n+1, 2e+2, 2e+2)\right). \tag{6.6}$$

For each hole h with respect to $\mathcal{C}^*$, we have that

$$NC(h, \mathcal{C}^*, e+2) \le A(n+1, 2e+2, e+2) . \tag{6.7}$$

By Lemma 6.5, Lemma 6.6, and (6.7), we have that

$$\sum_{h,\ d(h,\mathcal{C}^*)=e+2} NC(h, \mathcal{C}^*, e+2) \le \left(2^n - M\sum_{i=0}^{e}\binom{n}{i}\right) A(n+1, 2e+2, e+2). \tag{6.8}$$

Moreover,

$$\sum_{c \in \mathcal{C}^*} NH(c, \mathcal{C}^*, e+2) = \sum_{h, d(h,\mathcal{C}^*)=e+2} NC(h, \mathcal{C}^*, e+2) \tag{6.9}$$

since each pair (c, h), where $c \in \mathcal{C}^*$, and $d(h, \mathcal{C}^*) = e+2$, is counted exactly once on each side of (6.9).

We substitute (6.6) and (6.8) into (6.9) and use the initial assumption that $M = A(n, 2e+1)$ to obtain

$$A(n, 2e+1)\left(\binom{n+1}{e+2} - \binom{2e+2}{e+2}A(n+1, 2e+2, 2e+2)\right)$$

$$\le \left(2^n - A(n, 2e+1)\sum_{i=0}^{e}\binom{n}{i}\right) A(n+1, 2e+2, e+2)$$

which implies the claim of the theorem. □

It is not difficult to show that the bound of Theorem 6.2 is always better or the same as the bound of Theorem 6.1. One can also verify that for some small values of n and related minimum distance, the new bound coincides with the Plotkin bound, but we still have the following general problem.

Problem 6.1. Characterize all the codes that meet the bound of Theorem 6.2.

The known upper bounds on the sizes of covering code are not as good as the Johnson bound or the bound implied by Theorem 6.2.

Problem 6.2. Is their a bound similar to the Johnson bound for covering codes? Are their codes which attain such a bound with equality?

6.3 The Preparata Code

The ***Preparata code*** C has length 2^m, where m is even and greater than 3. It has a minimum Hamming distance 6 and its number of codewords is 2^{2^m-2m}. The union of 2^{m-1} disjoint translates of the code forms the extended Hamming code $\mathcal{H}^*(m)$. This family of codes is one of the most interesting family of nonlinear codes in coding theory. This section is devoted to the construction of many such codes and analysis of their properties.

In this section, $r \geq 3$ is an odd integer and $n = 2^r - 1$. Let $x \mapsto x^\sigma$ be an automorphism of $\mathbb{F}_{2^r}$, which implies that σ is a power of 2. Assume also that $x \mapsto x^{\sigma-1}$ and $x \mapsto x^{\sigma+1}$ are one-to-one mappings, i.e., $\gcd(\sigma \pm 1, 2^r - 1) = 1$.

For the admissible values of σ, a code $\mathcal{P}(\sigma)$ of length $2n+2 = 2^{r+1}$ is defined. The codewords will be described by pairs (X, Y), where $X, Y \subset \mathbb{F}_{2^r}$ will be interpreted as binary words of length 2^r, which are the characteristic vectors of X and Y. The *zero* element of $\mathbb{F}_{2^r}$ corresponds to the first position in the X-part. For this representation of the codewords by subsets, the addition of codewords X and Y represented by subsets, will be performed by their ***symmetric difference*** $X \triangle Y$, defined by

$$X \triangle Y \triangleq (X \setminus Y) \cup (Y \setminus X) .$$

Definition 6.1. The ***Preparata code*** $\mathcal{P}(\sigma)$ of length 2^{r+1} consists of the codewords described by all the pairs (X, Y) satisfying the following three conditions:

$$|X| \text{ and } |Y| \text{ are even integers} \tag{6.10}$$

$$\sum_{x\in X} x = \sum_{y\in Y} y, \tag{6.11}$$

$$\sum_{x\in X} x^{\sigma+1} + \left(\sum_{x\in X} x\right)^{\sigma+1} = \sum_{y\in Y} y^{\sigma+1}. \tag{6.12}$$

The punctured Preparata code $\mathcal{P}'(\sigma)$ is obtained by deleting the first coordinate of $\mathcal{P}(\sigma)$. For the computations in this section note that

$$(a+b)^{\sigma+1} = a^{\sigma+1} + a^{\sigma}b + ab^{\sigma} + b^{\sigma+1}\,. \tag{6.13}$$

Theorem 6.3. *The code $\mathcal{P}(\sigma)$ is distance invariant.*

Proof. We compare a codeword (X_0, Y_0) with $(\varnothing, \varnothing) = \mathbf{0}$. Let $\alpha = \sum_{x\in X_0} x$. The mapping $(X,Y) \mapsto (U,V)$, where $U = (X\triangle X_0) + \alpha$ and $V = Y\triangle Y_0$, is clearly an one-to-one mapping. We claim that (X,Y) is codeword if and only if (U,V) is a codeword. To verify this claim we have to show that the conditions defined in Definition 6.1 are satisfied. The first two conditions are trivial. For the third condition,

$$\sum_{x\in U} x^{\sigma+1} + \left(\sum_{x\in U} x\right)^{\sigma+1} = \sum_{x\in X\triangle X_0} (x+\alpha)^{\sigma+1} + \left(\sum_{x\in X\triangle X_0} (x+\alpha)\right)^{\sigma+1}$$

$$= \sum_{x\in X} (x+\alpha)^{\sigma+1} + \sum_{x\in X_0} (x+\alpha)^{\sigma+1} + \left(\alpha + \sum_{x\in X} x\right)^{\sigma+1}$$

$$= \sum_{x\in X} x^{\sigma+1} + \sum_{x\in X_0} x^{\sigma+1} + \left(\sum_{x\in X} x\right)^{\sigma+1} + \alpha^{\sigma+1}$$

$$= \sum_{y\in Y} y^{\sigma+1} + \sum_{y\in Y_0} y^{\sigma+1} = \sum_{y\in V} y^{\sigma+1}.$$

□

The proofs of the main properties of these codes become simpler if we find some automorphisms of the codes.

Theorem 6.4. *The group Aut $\mathcal{P}(\sigma)$ contains the permutations*

- $(X,Y) \mapsto (c+X, c+Y)$, $c \in \mathbb{F}_{2^r}$,
- $(X,Y) \mapsto (Y,X)$,
- $(X,Y) \mapsto (\alpha X, \alpha Y)$, $\alpha \in \mathbb{F}_{2^r}^{-}$,
- $(X,Y) \mapsto (X^{\varphi}, Y^{\varphi})$, $\varphi \in Aut\, \mathbb{F}_{2^r}$.

Proof. The first permutation is derived from the third condition of Definition 6.1 using (6.13). The other permutations are readily verified. □

The first two sets of permutations in Theorem 6.4 generate all the translations of the $(r+1)$-dimensional vector space $V = \mathbb{F}_{2^r} \times \mathbb{F}_2$.

Theorem 6.5. *The code $\mathcal{P}(\sigma)$ has a minimum distance of 6.*

Proof. By Theorem 6.3 it is sufficient to show that the minimum weight of a codeword in $\mathcal{P}(\sigma)$ is 6. There are obviously no codewords of weight 2. Hence, we just have to show that there are no codewords of weight 4. Assume the contrary, that a codeword (X, Y) in $\mathcal{P}(\sigma)$ has weight 4, and distinguish between three cases depending on the weights of X and Y.
Case 1. $|X| = |Y| = 2$.

If $(\{x_1, x_2\}, \{y_1, y_2\})$ is a codeword, then we may assume that $x_1 = 0$ by Theorem 6.4. This implies, by condition (6.12) of Definition 6.1, that

$$y_1^{\sigma+1} + y_2^{\sigma+1} = 0,$$

and then the condition on σ implies that $y_1 = y_2$, a contradiction.
Case 2. $|X| = 4$ and $|Y| = 0$.

We can assume that $X = \{0, a, b, c\}$ and hence by Definition 6.1 we have

$$a + b + c = 0,$$

$$a^{\sigma+1} + b^{\sigma+1} + c^{\sigma+1} = 0.$$

Substituting the first equation into the second, using (6.13), yields $ab(a^{\sigma-1} + b^{\sigma-1}) = 0$, i.e., $a = b$, a contradiction (using the fact that $x \mapsto x^{\sigma-1}$ is one-to-one). Finally, we show that there are indeed codewords of weight 6. Given distinct a, b, c, define y by $y^{\sigma+1} = a^{\sigma+1} + b^{\sigma+1} + c^{\sigma+1}$ and let $x \triangleq a + b + c + y$. Then, $(\{0, x\}, \{a, b, c, y\})$ is a codeword.
Case 3. $|X| = 0$ and $|Y| = 4$.

This case is symmetric to Case 2.

The three cases imply that the minimum distance of the code $\mathcal{P}(\sigma)$ is 6. □

Corollary 6.3. *The code $\mathcal{P}'(\sigma)$ has a minimum distance of 5.*

Theorem 6.6. *The size of $\mathcal{P}(\sigma)$ is 2^k, where $k = 2^{r+1} - 2r - 2$.*

Proof. By Definition 6.1, we can choose the set X in 2^n, $n = 2^r - 1$, distinct ways such that the requirement that the size of X and Y be even is satisfied. We count how many sets $Y \subset \mathbb{F}_{2^r}^-$ satisfy the two other conditions of the definition and add the *zero* element to each such set, if necessary to satisfy the condition that the size of X and Y be even. Let α be a primitive element of $\mathbb{F}_{2^r}$ and $m_i(x)$ the minimal polynomial of α^i. The conditions (6.11) and (6.12) of Definition 6.1, for the element y, form equations over $\mathbb{F}_{2^r}$. Considering $\mathbb{F}_{2^r}$ as an r-dimensional space over $\mathbb{F}_2$, these become $2r$ linear equations over $\mathbb{F}_2$. We claim that these equations are independent. The reason is that $\gcd(\sigma+1,n)=1$ and hence $m_{\sigma+1}(x)$ has degree r, i.e., the cyclic code over $\mathbb{F}_2$ of length n and generator $m_1(x)m_{\sigma+1}(x)$ has dimension $n-2r$. This implies that for each choice of X, the last two equations in Definition 6.1 have 2^{n-2r} solutions for Y, where $Y \subset \mathbb{F}_{2^r}^-$. Therefore, we have that

$$|\mathcal{P}(\sigma)| = 2^n \cdot 2^{n-2r} = 2^{2^r-1} \cdot 2^{2^r-1-r} = 2^{2^{r+1}-2r-2} .$$

This completes the proof of the theorem. □

We define the translates of $\mathcal{P}(\sigma)$ as follows. Let $\mathcal{C}_0 \triangleq \mathcal{P}(\sigma)$ and if $\alpha \in \mathbb{F}_{2^r}^-$, then let $\mathcal{C}_\alpha$ be the code obtained by adding the word corresponding to $(\{0,\alpha\},\{0,\alpha\})$ to the codewords of $\mathcal{P}(\sigma)$

Lemma 6.7. *The code $\mathcal{C}_\alpha$ has minimum weight 4.*

Proof. By Theorems 6.3 and 6.5, we only have to show that $\mathcal{C}_\alpha$ does not have a codeword with weight 2. A codeword with weight 2 is possible only if $\mathcal{P}(\sigma)$ contains a codeword of the form $(\{0,\alpha\},\{0,\alpha,\beta,\gamma\})$. By the second condition of Definition 6.1 this is not possible. □

By Theorem 6.5, the codes $\mathcal{C}_\alpha$, where $\alpha \in \mathbb{F}_{2^r}$ (note that $\mathcal{C}_0 = \mathcal{P}(\sigma)$), are pairwise disjoint. We define

$$\mathcal{H} \triangleq \bigcup_{\alpha \in \mathbb{F}_{2^r}} \mathcal{C}_\alpha .$$

By Theorem 6.6, we have that $|\mathcal{H}| = 2^r\,|\mathcal{P}(\sigma)| = 2^{2n-r}$ which is the cardinality of $\mathcal{H}^*(r+1)$ whose of length is $2n+2=2^{r+1}$.

Lemma 6.8. *The code $\mathcal{H}$ is a linear code.*

Proof. Let (X_1,Y_1) and (X_2,Y_2) be codewords in $\mathcal{P}(\sigma)$ and let $\alpha,\beta \in \mathbb{F}_{2^r}$. Define $s_i = \sum_{x \in X_i} x$, where $i=1,2$. For some $\gamma \in \mathbb{F}_{2^r}$, define X and Y by

$$X\triangle\{0,\gamma\} = X_1\triangle X_2\triangle\{\alpha,\beta\},$$

$$Y\triangle\{0,\gamma\} = Y_1\triangle Y_2\triangle\{\alpha,\beta\}.$$

We must show that there is a choice for γ such that $(X, Y) \in \mathcal{P}(\sigma)$. For each choice of γ, the sets X and Y satisfy conditions (6.10) and (6.11) of Definition 6.1. Substitution in the condition (6.12) of the definition yields the equation

$$(s_1 + s_2 + \alpha + \beta + \gamma)^{\sigma+1} = s_1^{\sigma+1} + s_2^{\sigma+1},$$

which has a unique solution γ. □

Corollary 6.4. *The code $\mathcal{H}$ is the extended Hamming code $\mathcal{H}^*(r+1)$.*

Theorem 6.7. *The punctured Preparata code $\mathcal{P}'(\sigma)$ is a nearly-perfect code.*

Proof. By Corollary 6.1,

$$A(n, 2e+1)\left(\sum_{i=0}^{e}\binom{n}{i} + \frac{\binom{n}{e}\left(\frac{n-e}{e+1} - \left\lfloor\frac{n-e}{e+1}\right\rfloor\right)}{\left\lfloor\frac{n}{e+1}\right\rfloor}\right) \le 2^n.$$

When $n = 2^r - 1$, r even, $r \ge 4$, and $e = 2$, by this equation we have that $A(2^r - 1, 5) \le 2^{2^r - 2r}$ and this bound is attained with equality by a punctured Preparata code of length $2^r - 1$. Thus, $\mathcal{P}'(\sigma)$ is a nearly-perfect code. □

Corollary 6.5. *The covering radius of $\mathcal{P}'(\sigma)$ is 3.*

Proof. By the proof of Theorem 6.1, in a nearly-perfect code the difference between the covering radius and packing radius is at most one. Since $\mathcal{P}'(\sigma)$ is a nearly-perfect code, but not a perfect code, and its packing radius is 2, it follows that its covering radius is 3. □

Corollary 6.6. *The covering radius of $\mathcal{P}(\sigma)$ is 4.*

Theorem 6.8. *The family of the punctured Preparata codes is asymptotically 2-perfect (with respect to the packing radius).*

Proof. Recall that the length of $\mathcal{P}'(\sigma)$ is $2^r - 1$ for even r, $r \ge 4$, its size is $2^{2^r - 2r}$, and its packing radius is 2. The size of a ball with radius 2 is

$$\binom{2^r - 1}{0} + \binom{2^r - 1}{1} + \binom{2^r - 1}{2} = 2^{2r-1} - 2^{r-1} + 1.$$

Hence, the density of the punctured Preparata code $\mathcal{P}'(\sigma)$ is

$$\frac{2^{2^r - 2r}(2^{2r-1} - 2^{r-1} + 1)}{2^{2^r - 1}} = 1 - \frac{2^{r-1} - 1}{2^{2r-1}}$$

and hence it is an asymptotically 2-perfect packing code. □

Note that for length 15, the punctured Preparata code has packing density $1 - \frac{7}{2^7}$, which is already very close to one. This code is equivalent to the punctured Nordstorm-Robinson code whose length is 15 and it was constructed in Section 4.2.

Henceforth, we will not distinguish between the various Preparata codes of the same length and only consider the code with the basic properties that have been proved. This is summarized in the following definition.

Definition 6.2. The ***Preparata code*** of order m, $\mathcal{P}(m)$ (or $\mathcal{P}_0(m)$), m an even integer, where $m \geq 4$, is a binary $(2^m, 2^{2^m-2m}, 6)$ code whose covering radius is 4. There exist 2^{m-1} translates of $\mathcal{P}_0(m)$ whose union is the extended Hamming code $\mathcal{H}^*(m)$ (or $\mathcal{H}_0^*(m)$). Let

$$\{\mathcal{P}_i(m) \;:\; 0 \leq i \leq 2^{m-1} - 1\}$$

be the family of codes that consists of these 2^{m-1} translates. Let $\{\mathcal{P}_i(m) \;:\; 0 \leq i \leq 2^{2m-1} - 1\}$ be the family that consists of the 2^{2m-1} translates of the Preparata code, whose union is $\mathbb{E}_2^{2^m}$. Denote

$$\mathcal{H}_j^*(m) = \bigcup_{i=j2^{m-1}}^{(j+1)2^{m-1}-1} \mathcal{P}_i(m)$$

for each j, $0 \leq j \leq 2^m - 1$.

Definition 6.3. The ***punctured Preparata code*** of order m, $\mathcal{P}'(m)$ (or $\mathcal{P}'_0(m)$), m an even integer, where $m \geq 4$, is a binary $(2^m - 1, 2^{2^m-2m}, 5)$ code whose covering radius is 3, which is obtained from $\mathcal{P}_0(m)$ by deleting any coordinate. There exist 2^{m-1} translates of $\mathcal{P}'_0(m)$ whose union is the Hamming code $\mathcal{H}(m)$ (or $\mathcal{H}_0(m)$). Let

$$\{\mathcal{P}'_i(m) \;:\; 0 \leq i \leq 2^{m-1} - 1\}$$

be the family of codes that consists of these 2^{m-1} translates. Let $\{\mathcal{P}'_i(m) \;:\; 0 \leq i \leq 2^{2m-1} - 1\}$ be the family that consists of the 2^{2m-1} translates, of the punctured Preparata code, obtained by deleting the last coordinate in each translate of $\{\mathcal{P}_i(m) \;:\; 0 \leq i \leq 2^{2m-1} - 1\}$. Let

$$\mathcal{H}_j(m) = \bigcup_{i=j2^{m-1}}^{(j+1)2^{m-1}-1} \mathcal{P}'_i(m)$$

for each j, $0 \leq j \leq 2^m - 1$.

Definitions 6.2 and 6.3 will be used for further constructions, especially in the next section.

Henceforth, we denote the family $\{\mathcal{P}_i(m) \ : \ 0 \leq i \leq 2^{m-1}-1\}$ by $\mathbb{P}(m)$ and the family $\{\mathcal{P}'_i(m) \ : \ 0 \leq i \leq 2^{m-1}-1\}$ by $\mathbb{P}'(m)$.

Lemma 6.9. *Each word $x \in \mathbb{F}_2^{2^m-1}$, m even, has one of the following two properties:*

- *x is a word in $\mathcal{P}'_i(m)$ for some $0 \leq i \leq 2^{m-1}-1$.*
- *x is at distance at most two from $\mathcal{P}'_i(m)$ for all $0 \leq i \leq 2^{m-1}-1$ and at distance one from exactly one $\mathcal{P}'_j(m)$, for some $0 \leq j \leq 2^{m-1}-1$.*

Proof. By Corollary 4.3, x is either a codeword of the Hamming code of length 2^m-1, $\mathcal{H}_0(m)$, or at distance one from exactly one of its codewords and at distance two from exactly $2^{m-1}-1$ of its codewords.

If x is a codeword of the Hamming code, then by Definition 6.3 x is a codeword of $\mathcal{H}_0(m) = \bigcup_{i=0}^{2^{m-1}-1} \mathcal{P}'_i(m)$ and hence x is a codeword in $\mathcal{P}'_j(m)$ for some $0 \leq j \leq 2^{m-1}-1$. If x is at distance at most two from exactly 2^{m-1} codewords of $\mathcal{H}_0(m)$, then these codewords are in different translates of $\mathcal{P}'_0(m)$ since otherwise one such translate will contain two of these codewords and its minimum distance will be at most 4, a contradiction. This completes the proof of the claim in the lemma. □

Corollary 6.7. *Each word $x \in \mathbb{F}_2^{2^m}$, m even, has one of the following three properties:*

- *x is a word in $\mathcal{P}_i(m)$ for some $0 \leq i \leq 2^{m-1}-1$.*
- *x is at distance two from $\mathcal{P}_i(m)$ for all $0 \leq i \leq 2^{m-1}-1$.*
- *x is at distance three from $\mathcal{P}_i(m)$ for all $0 \leq i \leq 2^{m-1}-1$, except for exactly one $\mathcal{P}_j(m)$, for some $0 \leq j \leq 2^{m-1}-1$, where its distance is one.*

6.4 Quasi-Perfect Codes

A code C is called ***quasi-perfect*** if its packing radius is e and its covering radius is $e+1$. By the definition of nearly-perfect codes (based on the proof of Theorem 6.1), we have that any nearly-perfect code is also a quasi-perfect code. Unfortunately, there are not many known packing radius parameters with nontrivial quasi-perfect codes, and infinite families of such codes are known only for packing radius 1 and packing radius 2. Nevertheless, for these small number of known radius parameters, there are many interesting open problems.

Problem 6.3. Construct quasi-perfect codes for all possible lengths for a given covering radius and minimum distance.

Problem 6.4. What is the most dense (and the most sparse, respectively) quasi-perfect code for a given length and packing radius (covering radius, respectively)?

Problem 6.5. For a given redundancy, packing radius or covering radius, what is the longest and the shortest quasi-perfect code?

Clearly, however, the most important problem is how to construct quasi-perfect codes with a large radius.

Problem 6.6. Construct a (possibly infinite) family of quasi-perfect codes with packing radius at least 3 for each code in the family.

The length of the shortest code with covering radius R, minimum distance d, and redundancy r, will be denoted by $\ell^*(R,d,r)$. The length of the longest code with covering radius R, minimum distance d, and redundancy r, will be denoted by $n^*(R,d,r)$. The code that attains $n^*(R,d,r)$ with equality is the most dense one, and the code that attains $\ell^*(R,d,r)$ with equality is the most sparse one for the given R, d, and r. It should be noted that while we are given a packing radius e, we would like to consider minimum distance $2e+1$ when the goal is to find a code with a large packing density. If the packing radius is e and the covering radius is $e+1$, we would like to consider distance $2e+2$ when the goal is to find a code with a small covering density.

Definition 6.4. A family of codes $\{\mathcal{C}_1, \mathcal{C}_2, \ldots, \mathcal{C}_\ell\}$, where $\mathcal{C}_i \subseteq \mathbb{F}_2^n$, $1 \leq i \leq \ell$, has *subnorm* t if

$$\min_{1\leq i\leq \ell} d(x,\mathcal{C}_i) + \max_{1\leq i\leq \ell} d(x,\mathcal{C}_i) \leq t$$

holds for all $x \in \mathbb{F}_2^n$.

Remark 6.1. In the family of codes in Definition 6.4 all the codes have the same length. In contrast, in the definition of density of family of codes as done in Section 6.1, the codes in the family have different lengths.

Lemma 6.10. *The subnorm of the family* $\mathbb{P}'(m)$ *is 3 and the subnorm of the family* $\mathbb{P}(m)$ *is 4.*

Proof. Let $x \in \mathbb{F}_2^{2^m-1}$, where m is even and $m \geq 4$, and distinguish between two cases depending on whether x is a codeword of the Hamming code, i.e., x is contained in one of the translates of the punctured Preparata code, or x is not contained in the Hamming code.

Case 1. $x \in \mathcal{H}(m)$, i.e., $x \in \mathcal{P}'_i(m)$ for some i.

This implies that $d(x, \mathcal{P}'_i(m)) = 0$. Since by Corollary 6.5 the covering radius of $\mathcal{P}'_0(m)$ is 3, it follows that for each j, $0 \leq j \leq 2^{m-1} - 1$, $d(x, \mathcal{P}'_j(m) \leq 3$. Therefore,

$$\min_{1 \leq j \leq m} d(x, \mathcal{C}_j) + \max_{1 \leq j \leq m} d(x, \mathcal{C}_j) \leq 3.$$

Case 2. $x \notin \mathcal{H}(m)$, i.e., $x \notin \mathcal{P}'_i(m)$ for each $0 \leq i \leq 2^{m-1} - 1$.

This implies by Lemma 6.9 that there exists one i such that $d(x, \mathcal{P}'_i(m)) = 1$ and for each $j \neq i$, $0 \leq j \leq 2^{m-1} - 1$, $d(x, \mathcal{P}'_i(m)) = 2$. Hence,

$$\min_{1 \leq j \leq m} d(x, \mathcal{C}_j) + \max_{1 \leq j \leq m} d(x, \mathcal{C}_j) \leq 3.$$

Thus, the subnorm of the family $\mathbb{P}'(m)$ is 3.

The proof that the subnorm of the family $\mathbb{P}(m)$ is 4 is done similarly. Consider $x \in \mathbb{F}_2^{2^m}$, where m is even. If $x \in \mathcal{P}_i(m)$ for some i, then we continue as in Case 1. If $x \notin \mathcal{P}_i(m)$ for each $0 \leq i \leq 2^{m-1} - 1$, then distinguish between the cases when the weight of x is even and when it is odd. If the weight of x is even, then by Corollary 6.7 it is at distance two from all the cosets in the family $\mathbb{P}(m)$ and the claim follows. Finally, if the weight of x is odd, then the proof follows from the third property of Corollary 6.7. □

Lemma 6.11. *Let $\mathcal{H}_i^*(m)$, $0 \leq i \leq 2^{m+1} - 1$, be the 2^{m+1} distinct cosets of $\mathcal{H}^*(m)$ and let $\hat{\mathcal{H}}_i^*(m)$, $0 \leq i \leq 2^m - 1$, be the 2^m distinct even cosets of $\mathcal{H}^*(m) = \hat{\mathcal{H}}^*(m)$. The subnorm of $\mathbb{H}^*(m) \triangleq \{\mathcal{H}_i^*(m) : 0 \leq i \leq 2^{m+1} - 1\}$ is 2 and the subnorm of $\hat{\mathbb{H}}^*(m) \triangleq \{\hat{\mathcal{H}}_i^*(m) : 0 \leq i \leq 2^m - 1\}$ is also 2.*

Proof. Consider first the family $\mathbb{H}^*(m)$ and let $x \in \mathbb{F}_2^{2^m}$. Since the union of the cosets of $\mathcal{H}^*(m)$, in this family, is $\mathbb{F}_2^{2^m}$, it follows that $d(x, \mathcal{H}_i^*(m)) = 0$ for some i. Since the covering radius of $\mathcal{H}^*(m)$ is 2 (an immediate consequence of the covering radius of $\mathcal{H}(m)$ which is 1), it follows that for every j, $d(x, \mathcal{H}_j^*(m)) \leq 2$. Thus, the subnorm of this family is 2.

Consider now the family $\hat{\mathbb{H}}^*(m)$, let $x \in \mathbb{F}_2^{2^m}$, and distinguish between two cases depending on whether x has even weight or odd weight.

Case 1. x has even weight.

Since the union of the cosets of $\mathcal{H}^*(m)$, in this family, is $\mathbb{E}_2^{2^m}$, it follows that $d(x, \hat{\mathcal{H}}_i^*(m)) = 0$ for some i. Since the covering radius of the extended Hamming code is 2, it follows that for each j, $1 \le j \le 2^m$, we have that $d(x, \hat{\mathcal{H}}_j^*(m)) \le 2$.

Case 2. x has odd weight.

Since all the words in a coset have even weight and the covering radius of the extended Hamming code is 2, it follows that for each j, $1 \le j \le 2^m$, we have that $d(x, \hat{\mathcal{H}}_j^*(m)) = 1$.

Thus, the arguments and their consequences in these two cases imply that the subnorm of this family is also 2. □

The Blockwise Direct Sum (BDS) Construction

Suppose we are given four codes: an (n_1, M_1, d_1) code $\mathcal{C}^1$ whose covering radius is R_1, an $(n_1, M_2 = \ell M_1, d_2)$ code $\mathcal{C}^2$ whose covering radius is R_2, an (n_3, M_3, d_3) code $\mathcal{C}^3$ whose covering radius is R_3, and an $(n_3, M_4 = \ell M_3, d_4)$ code $\mathcal{C}^4$ whose covering radius is R_4. Assume further that these codes have the following properties:

- The code $\mathcal{C}^2$ is a union of the ℓ disjoint codes $\mathcal{C}_i^1$, $1 \le i \le \ell$, with the parameters of $\mathcal{C}^1$, i.e.,
$$\mathcal{C}^2 = \bigcup_{i=1}^{\ell} \mathcal{C}_i^1 .$$
- The code $\mathcal{C}^4$ is a union of the ℓ disjoint codes $\mathcal{C}_i^3$, $1 \le i \le \ell$, with the parameters of $\mathcal{C}^3$, i.e.,
$$\mathcal{C}^4 = \bigcup_{i=1}^{\ell} \mathcal{C}_i^3 .$$
- The family $\mathbb{C}^2 = \{\mathcal{C}_1^1, \mathcal{C}_2^1, \ldots, \mathcal{C}_\ell^1\}$ has subnorm t_1.
- The family $\mathbb{C}^4 = \{\mathcal{C}_1^3, \mathcal{C}_2^3, \ldots, \mathcal{C}_\ell^3\}$ has subnorm t_3.

Accordingly, the BDS of $\mathcal{C}^2$ and $\mathcal{C}^4$ is the following code $\mathcal{C}$ obtained by the direct product construction, i.e.,
$$\mathcal{C} \triangleq \mathbb{C}^2 \otimes \mathbb{C}^4 \triangleq \bigcup_{i=1}^{\ell} \mathcal{C}_i^1 \times \mathcal{C}_i^3 .$$

Theorem 6.9. *The code $\mathcal{C}$ obtained by the BDS construction is an (n, M, d) with covering radius R and the following parameters:*
$$n = n_1 + n_3, \quad M = \ell M_1 M_3, \quad d \ge \min\{d_1, d_3, d_2 + d_4\}, \quad R \le (t_1 + t_3)/2 .$$

Proof. The length of $\mathcal{C}$ and its size can be readily verified from the definition of $\mathcal{C}$.

We continue to compute the minimum distance d of the code $\mathcal{C}$.

Let $c_1 = (x_1, y_1)$, $c_2 = (x_2, y_2)$ be two distinct codewords of $\mathcal{C}$ and distinguish between three cases related to the relations between x_1 and x_2.

Case 1. $x_1 = x_2 \in \mathcal{C}_i^1$ for some i.

This implies that $y_1, y_2 \in \mathcal{C}_i^3$ and $y_1 \neq y_2$ and since $d(\mathcal{C}^3) = d_3$, it follows that $d(y_1, y_2) \geq d_3$ and hence $d(c_1, c_2) \geq d_3$.

Case 2. $x_1 \neq x_2$ and $x_1, x_2 \in \mathcal{C}_i^1$ for some i.

Since $d(\mathcal{C}^1) = d_1$, it follows that $d(x_1, x_2) \geq d_1$ and hence $d(c_1, c_2) \geq d_1$.

Case 3. $x_1 \in \mathcal{C}_i^1$ and $x_2 \in \mathcal{C}_j^1$ for $j \neq i$.

This implies that $y_1 \in \mathcal{C}_i^3$ and $y_2 \in \mathcal{C}_j^3$. Hence, x_1, x_2 are two distinct codewords in $\mathcal{C}^2$ and y_1, y_2 are two distinct codewords in $\mathcal{C}^4$. Therefore, $d(x_1, x_2) \geq d_2$ and $d(y_1, y_2) \geq d_4$, which implies that $d(c_1, c_2) \geq d_2 + d_4$.

Thus, $d \geq \min\{d_1, d_3, d_2 + d_4\}$.

Finally, we have to compute the covering radius R of the code $\mathcal{C}$.

Let (x, y) be a word in $\mathbb{F}_2^{n_1+n_2}$, where $x \in \mathbb{F}_2^{n_1}$ and $y \in \mathbb{F}_2^{n_2}$, and assume that

$$d(x, \mathcal{C}_\alpha^1) = \min_i d(x, \mathcal{C}_i^1) \quad \text{and} \quad d(x, \mathcal{C}_\beta^3) = \min_i d(x, \mathcal{C}_i^3).$$

Since the family $\mathbb{C}^2 = \{\mathcal{C}_1^1, \mathcal{C}_2^1, \ldots, \mathcal{C}_\ell^1\}$ has subnorm t_1 and $d(x, \mathcal{C}_\alpha^1) = \min_i d(x, \mathcal{C}_i^1)$, it follows that $d(x, \mathcal{C}_i^1) \leq t_1 - d(x, \mathcal{C}_\alpha^1)$ for all i, $1 \leq i \leq \ell$. Similarly, $d(x, \mathcal{C}_i^3) \leq t_3 - d(x, \mathcal{C}_\beta^3)$ for each i, $1 \leq i \leq \ell$, since the family $\mathbb{C}^4 = \{\mathcal{C}_1^3, \mathcal{C}_2^3, \ldots, \mathcal{C}_\ell^3\}$ has subnorm t_3. Therefore,

$$2d((x, y), \mathcal{C}) \leq d((x, y), \mathcal{C}_\alpha^1 \times \mathcal{C}_\alpha^3) + d((x, y), \mathcal{C}_\beta^1 \times \mathcal{C}_\beta^3)$$

$$= d(x, \mathcal{C}_\alpha^1) + d(y, \mathcal{C}_\alpha^3) + d(x, \mathcal{C}_\beta^1) + d(y, \mathcal{C}_\beta^3)$$

$$\leq d(x, \mathcal{C}_\alpha^1) + (t_3 - d(y, \mathcal{C}_\beta^3)) + (t_1 - d(x, \mathcal{C}_\alpha^1))) + d(y, \mathcal{C}_\beta^3) = t_1 + t_3.$$

Thus, we have $R \leq \frac{t_1 + t_3}{2}$.

This completes the proofs of the claims in the theorem. □

If $R = 2$ and $d = 4$, then $n^*(2, 4, r) = 2^{r-1}$. This value is attained by the extended Hamming codes. Next, we give an upper bound on $\ell^*(2, 4, r)$.

Construction 6.1. If $r \equiv 0 \pmod 4$, where $r = 2m$ and m is an even integer, then consider the following code obtained by the BDS construction:

$$\mathcal{C}^1 = \mathcal{P}_0'(m), \ \mathcal{C}^2 = \bigcup_{i=0}^{2^{m-1}-1} \mathcal{P}_i'(m) = \mathcal{H}_0(m),$$

$$\mathcal{C}^3 = \hat{\mathcal{H}}_0^*(m-1),\ \mathcal{C}^4 = \bigcup_{i=0}^{2^{m-1}-1} \hat{\mathcal{H}}_i^*(m-1) = \mathbb{E}_2^{2^{m-1}}.$$

The obtained code by the BDS construction will be denoted by $\Psi(m)$.

Theorem 6.10. *The code $\Psi(m)$, obtained in Construction 6.1, is a $(3 \cdot 2^{m-1} - 1, 2^{3 \cdot 2^{m-1}-1-2m}, 4)$ code with covering radius 2 and covering density $\frac{9}{8} - \frac{3 \cdot 2^{m-2}-1}{2^{2m}}$.*

Proof. The parameters of the codes $\mathcal{C}^1$ and $\mathcal{C}^2$ are given in Definition 6.3 and the subnorm of the related family of codes is proved in Lemma 6.10. The parameters of the codes $\mathcal{C}^3$ and $\mathcal{C}^4$ are readily verified and the subnorm of the related family of codes is proved in Lemma 6.11. Now, the parameter of the code $\Psi(m)$ are implied by the BDS construction and Theorem 6.9.

The size of a ball with radius 2 is

$$\binom{3 \cdot 2^{m-1} - 1}{0} + \binom{3 \cdot 2^{m-1} - 1}{1} + \binom{3 \cdot 2^{m-1} - 1}{2} = 9 \cdot 2^{2m-3} - 3 \cdot 2^{m-2} + 1\,.$$

Hence, the covering density of the code $\Psi(m)$ is

$$\frac{2^{3 \cdot 2^{m-1}-1-2m}(9 \cdot 2^{2m-3} - 3 \cdot 2^{m-2} + 1)}{2^{3 \cdot 2^{m-1}-1}} = \frac{9}{8} - \frac{3 \cdot 2^{m-2} - 1}{2^{2m}}\,.$$

□

Corollary 6.8. *The family of codes $\{\Psi(m) \ :\ m = 2k \geq 4\}$ has covering density $\frac{9}{8}$.*

Corollary 6.9. *If $r \equiv 0 \pmod 4$, where $r = 2m$ and m is an even integer, then $\ell^*(2, 4, 2m) \leq 3 \cdot 2^{m-1} - 1$.*

Construction 6.2. If $r \equiv 1 \pmod 4$, where $r = 2m + 1$, m is an even integer, and $m \geq 4$, then consider the following code obtained by the BDS construction. Let

$$\Psi_j(m) = \bigcup_{i=0}^{2^{m-1}-1} \mathcal{P}_i'(m) \times \hat{\mathcal{H}}_{i+j}^*(m-1),\ 0 \leq j \leq 2^{m-1} - 1,$$

$$\Psi_{2^{m-1}+j}(m) = (10^{3 \cdot 2^{m-1}-3}1) + \Psi_j(m),\ 0 \leq j \leq 2^{m-1} - 1,$$

where the subscript $i + j$ is taken modulo 2^{m-1}.

The codes that are used in the BDS construction are

$$\mathcal{C}^1 = \Psi_0(m),\ \mathcal{C}^2 = \bigcup_{i=0}^{2^m-1} \Psi_i(m),$$

$$\mathcal{C}^3 = \hat{\mathcal{H}}_0^*(m),\ \mathcal{C}^4 = \bigcup_{i=0}^{2^m-1} \hat{\mathcal{H}}_i^*(m) = \mathbb{E}_2^{2^m}.$$

The code obtained by the BDS construction will be denoted by $\Upsilon(m)$.

To analyze the code $\Upsilon(m)$ obtained in Construction 6.2, the following two lemmas are required.

Lemma 6.12. *The following code (which is $\mathcal{C}^2$ in Construction 6.2)*

$$\mathcal{C} \triangleq \bigcup_{i=0}^{2^m-1} \Psi_i(m)$$

is a $(3 \cdot 2^{m-1} - 1, 2^{3 \cdot 2^{m-1}-1-m}, 2)$ code with covering radius one.

Proof. We start be considering the minimum distance of the code $\mathcal{C}$.

Let $c_1 = (x_1, y_1)$, $c_2 = (x_2, y_2)$, where $x_1, x_2 \in \mathbb{F}_2^{2^m-1}$ and $y_1, y_2 \in \mathbb{F}_2^{2^{m-1}}$, be two distinct codewords of $\mathcal{C}$. We distinguish between three cases depending on the relations between x_1, x_2, y_1, and y_2.

Case 1. $x_1 = x_2$.

Since $c_1, c_2 \in \mathcal{C}$, it follows that x_1 is contained either in some $\mathcal{P}_i'(m)$ or in some translate $(10^{2^m-2}) + \mathcal{P}_i'(m)$, where $0 \leq i \leq 2^{m-1} - 1$. If $x_1 \in \mathcal{P}_i'(m)$, then y_1 and y_2 are two distinct words in two even cosets (not necessarily distinct) of the extended Hamming code. This implies that $d(y_1, y_2) \geq 2$, i.e., $d(c_1, c_2) \geq 2$. If $x_1 \in (10^{2^m-2}) + \mathcal{P}_i'(m)$, then y_1 and y_2 are two distinct words in two odd cosets of the extended Hamming code. This implies that $d(y_1, y_2) \geq 2$, i.e., $d(c_1, c_2) \geq 2$.

Case 2. $y_1 = y_2$.

Since $c_1, c_2 \in \mathcal{C}$, it follows that y_1 is contained either in some $\hat{\mathcal{H}}_j^*(m-1)$ or in some $(0^{2^{m-1}-1}1) + \hat{\mathcal{H}}_j^*(m-1)$, where $0 \leq j \leq 2^{m-1} - 1$. If $y_1 \in \hat{\mathcal{H}}_j^*(m-1)$, then x_1 and x_2 are two distinct words in $\mathcal{P}_{i_1}'(m)$ and $\mathcal{P}_{i_2}'(m)$, respectively, where $0 \leq i_1, i_2 \leq 2^{m-1} - 1$. Hence, x_1 and x_2 are two distinct codewords in the Hamming code. This implies that $d(x_1, x_2) \geq 3$, i.e., $d(c_1, c_2) \geq 3$. The same arguments hold if $y_1 \in (0^{2^{m-1}-1}1) + \hat{\mathcal{H}}_j^*(m-1)$.

Case 3. $x_1 \neq x_2$ and $y_1 \neq y_2$.

This implies that $d(x_1, x_2) \geq 1$ and $d(y_1, y_2) \geq 1$ and hence $d(c_1, c_2) \geq 2$.

Thus, these three cases imply that the minimum distance of $\mathcal{C}$ is 2.

We continue to consider the covering radius of the code $\mathcal{C}$. Assume now that $(x, y) \in \mathbb{F}_2^{3 \cdot 2^{m-1}-1}$, where $x \in \mathbb{F}_2^{2^m-1}$ and $y \in \mathbb{F}_2^{2^{m-1}}$. We again distinguish between three cases depending on whether x is a codeword of the Hamming code or in the coset of the Hamming code that contains the word (10^{2^m-2}). If x is not in the Hamming code or in this identified coset which contains the word (10^{2^m-2}), then we distinguish between the case where y is of even weight and the case where y is of odd weight.

Case 1. Assume first that $x \in \mathcal{P}'_i(m)$ or $x \in (10^{2^m-2}) + \mathcal{P}'_i(m)$, for some $0 \leq i \leq 2^{m-1} - 1$ (this implies that $x \in \mathcal{H}(m)$ or $x \in (10^{2^m-2}) + \mathcal{H}(m)$).

If $\text{wt}(y)$ is even, then there exists some j, $0 \leq j \leq 2^{m-1} - 1$, such that $y \in \hat{\mathcal{H}}^*_{i+j}(m-1)$. Therefore, $(x, y) \in \mathcal{P}'_i(m) \times \hat{\mathcal{H}}^*_{i+j}(m-1)$, i.e., $(x, y) \in \Psi_j(m) \subset \mathcal{C}$. If $\text{wt}(y)$ is odd, then using the fact that in each even coset of the extended Hamming code $\mathcal{H}^*(m-1)$ there exists a word z such that $d(y, z) = 1$. Clearly, $(x, z) \in \mathcal{P}'_i(m) \times \mathcal{H}^*(m) \subset \mathcal{C}$ and therefore $d((x, y), \mathcal{C}) \leq 1$ in this case. The same arguments hold if $x \in (10^{2^m-2}) + \mathcal{P}'_i(m)$

Case 2. $x \notin \mathcal{P}'_i(m)$, $x \notin (10^{2^m-2}) + \mathcal{P}'_i(m)$, for all $0 \leq i \leq 2^{m-1} - 1$, and $\text{wt}(y)$ is even.

By Lemma 6.9, we have that x is at distance one from exactly one translate $\mathcal{P}'_i(m)$. Moreover, there exists a j, $0 \leq j \leq 2^{m-1} - 1$, such that $y \in \hat{\mathcal{H}}^*_{i+j}(m-1)$. Therefore, $d((x, y), \mathcal{P}'_i(m) \times \hat{\mathcal{H}}^*_{i+j}(m-1)) = 1$, i.e., $d((x, y), \mathcal{C}) = 1$.

Case 3. $x \notin \mathcal{P}'_i(m)$, $x \notin (10^{2^m-2}) + \mathcal{P}'_i(m)$, for all $0 \leq i \leq 2^{m-1} - 1$, and $\text{wt}(y)$ is odd.

By Lemma 6.9 we have that x is at distance one from exactly one $(10^{2^m-2}) + \mathcal{P}'_i(m)$. Moreover, there exists a j, $0 \leq j \leq 2^{m-1} - 1$, such that $y \in (0^{2^{m-1}-1}1) + \hat{\mathcal{H}}^*_{i+j}(m-1)$. Therefore,

$$d((x, y), ((10^{2^m-2}) + \mathcal{P}'_i(m)) \times ((0^{2^{m-1}-1}1) + \hat{\mathcal{H}}^*_{i+j}(m-1))) = 1,$$

i.e., $d((x, y), \mathcal{C}) = 1$.

This implies that the covering radius of $\mathcal{C}$ is one. □

Lemma 6.13. *The family of codes* $\{\Psi_i(m) : 0 \leq i \leq 2^m - 1\}$ *has sub-norm 3.*

Proof. Let $(x, y) \in \mathbb{F}_2^{3 \cdot 2^{m-1}-1}$, where $x \in \mathbb{F}_2^{2^m-1}$, $y \in \mathbb{F}_2^{2^{m-1}}$. Assume first that $\text{wt}(y)$ is even and distinguish between three cases depending on whether x is a codeword in the Hamming code or which coset of the Hamming code contains x.

Case 1. x is a codeword of $\mathcal{H}(m)$, i.e., $x \in \mathcal{P}'_i(m)$ for some $0 \leq i \leq 2^{m-1}-1$.

Since $\text{wt}(y)$ is even, it follows that there exists a j such that $0 \leq j \leq 2^{m-1}-1$ and $y \in \hat{\mathcal{H}}^*_{i+j}(m-1)$, which implies that (x,y) is a codeword in $\bigcup_{i=0}^{2^{m-1}-1} \Psi_i(m)$. To complete the proof in this case, it suffices to show that $d((x,y),\Psi_j(m)) \leq 3$, for each $0 \leq j \leq 2^m - 1$. For $0 \leq j \leq 2^{m-1}-1$, this follows immediately from the fact that, by Corollary 6.5, the covering radius of $\mathcal{P}'_j(m)$ is 3. For $2^{m-1} \leq j \leq 2^m - 1$, this follows from the fact that x is not a word in $(10^{2^m-2}) + \mathcal{H}(m)$ and hence, by Lemma 6.9, we have that x is at distance at most two from each $\mathcal{P}'_{j-2^{m-1}}(m)$. Moreover, note that for $2^{m-1} \leq j \leq 2^m - 1$, $\Psi_j(m)$ is constructed from the odd cosets of $\mathcal{H}^*(m-1)$ instead of the even cosets used when $0 \leq j \leq 2^{m-1}-1$. Now, since $\text{wt}(y)$ is even, it follows that y is at distance one from each odd coset of $\mathcal{H}^*(m-1)$, which completes the proof in this case.

Case 2. x is a word in $(10^{2^m-2}) + \mathcal{H}(m)$.

Since $\text{wt}(y)$ is even and the covering radius of any coset of the extended Hamming code is two, it follows that the distance between y and any odd coset of the extended Hamming code is one. This implies that for each i, $2^{m-1} \leq i \leq 2^m - 1$, we have that $d((x,y),\Psi_i(m)) = 1$. To complete the proof in this case, it suffices to show that $d((x,y),\Psi_j(m)) \leq 2$, for each $0 \leq j \leq 2^{m-1}-1$. By using similar arguments again, we have that since x is a word in $(10^{2^m-2}) + \mathcal{H}(m)$, it follows by Lemma 6.9 that x is at distance at most two from each $\mathcal{P}'_j(m)$ and hence $d((x,y),\Psi_j(m)) \leq 2$ for each $0 \leq j \leq 2^{m-1}-1$.

Case 3. x is not a codeword in $\mathcal{H}(m)$ and not a word in $(10^{2^m-2}) + \mathcal{H}(m)$.

We apply similar arguments to the ones in Case 2.

These three cases complete the analysis when the weight of y is even. Similar arguments are applied when $\text{wt}(y)$ is odd.

Thus, the subnorm of the family $\{\Psi_i(m) : 0 \leq i \leq 2^m - 1\}$ is 3. □

Theorem 6.11. *The code* $\Upsilon(m)$ *is a* $(5 \cdot 2^{m-1} - 1, 2^{5\cdot 2^{m-1}-2-2m}, 4)$ *code with covering radius 2 and covering density* $\frac{25}{16} - \frac{5\cdot 2^{m-2}-1}{2^{2m+1}}$.

Proof. The parameters of the codes $\mathcal{C}^1$ and $\mathcal{C}^2$ are derived in Theorem 6.10 and Lemma 6.12, and the subnorm of the family of codes is derived in Lemma 6.13. The parameters of the codes $\mathcal{C}^3$ and $\mathcal{C}^4$ are readily verified, and the subnorm of the family of codes is obtained in Lemma 6.11. This implies, by using the BDS construction and Theorem 6.9, that the parameters of the code $\Upsilon(m)$ are exactly as specified by the claim of the theorem.

The size of a ball with radius 2 is

$$\binom{5\cdot 2^{m-1}-1}{0}+\binom{5\cdot 2^{m-1}-1}{1}+\binom{5\cdot 2^{m-1}-1}{2} = 25\cdot 2^{2m-3}-5\cdot 2^{m-2}+1.$$

Hence, the covering density of the code $\Upsilon(m)$ is

$$\frac{2^{5\cdot 2^{m-1}-2-2m}(25\cdot 2^{2m-3}-5\cdot 2^{m-2}+1)}{2^{5\cdot 2^{m-1}-1}} = \frac{25}{16} - \frac{5\cdot 2^{m-2}-1}{2^{2m+1}}.$$

□

Corollary 6.10. *The family of codes* $\{\Upsilon(m) \ : \ m = 2k \geq 4\}$ *has covering density* $\frac{25}{16}$.

Corollary 6.11. *If* $r \equiv 1 \pmod 4$, *where* $r = 2m+1$, m *is an even integer, and* $m \geq 4$, *then* $\ell^*(2,4,2m+1) \leq 5\cdot 2^{m-1}-1$.

Using arguments similar to the ones used in the previous results, one can verify the following simple lemma.

Lemma 6.14. *If* S_1 *is partitioned into* k *subsets* $A_0, A_1, \ldots, A_{k-1}$ *and* S_2 *is partitioned into* t *subsets* $B_0, B_1, \ldots, B_{t-1}$, *then for any* $0 \leq j \leq t-1$, *the code*

$$\bigcup_{i=0}^{k-1} A_i \times B_{i+j}$$

has the space tiling property with respect to the space $S_1 \times S_2$.

Proof. Clearly,

$$S_1 \times S_2 = \bigcup_{i=0}^{k-1}\bigcup_{j=0}^{t-1} A_i \times B_j = \bigcup_{i=0}^{k-1}\bigcup_{j=0}^{t-1} A_i \times B_{i+j},$$

where $i+j$ is taken modulo t. Moreover, for $0 \leq j_1, j_2 \leq t-1$ we have that

$$\left(\bigcup_{i=0}^{k-1} A_i \times B_{i+j_1}\right) \bigcap \left(\bigcup_{i=0}^{k-1} A_i \times B_{i+j_2}\right) = \varnothing \text{ if and only if } j_1 \neq j_2,$$

from which the claim of the lemma follows. □

Lemma 6.15. *The code* $\Upsilon(m)$ *has the space tiling property.*

Proof. By Lemma 6.14 and the properties of the code in the family of $\{\Upsilon(m) \ : \ m = 2k \geq 4\}$, we have that the set $\bigcup_{i=0}^{2^m-1} \Psi_i(m) \times \mathbb{E}_2^{2^m}$ can be partitioned into codes with the parameters of $\Upsilon(m)$, where

$$S_1 \triangleq \bigcup_{i=0}^{2^m-1} \Psi_i(m) \quad \text{and} \quad S_2 \triangleq \mathbb{E}_2^{2^m} = \bigcup_{i=0}^{2^m-1} \hat{\mathcal{H}}_i^*(m)$$

are the related sets in Lemma 6.14. Since for some $j \neq 0$,

$$\bigcup_{i=0}^{2^m-1} \Psi_i(m) = \left(\mathcal{H}_0(m) \times \mathbb{E}_2^{2^{m-1}}\right) \bigcup \left(\mathcal{H}_j(m) \times ((10^{2^{m-1}-1}) + \mathbb{E}_2^{2^{m-1}})\right),$$

it is easy to verify that $\mathbb{F}_2^{5 \cdot 2^{m-1}-1}$ can be partitioned into codes with the parameters of $\bigcup_{i=0}^{2^m-1} \Psi_i(m) \times \mathbb{E}_2^{2^m}$. Hence, $\Upsilon(m)$ has the space tiling property. □

6.5 Asymptotically 2-Perfect Covering Codes

Recall that by Theorem 6.8, each code in the family of the punctured Preparata codes has packing radius 2 and the packing density of this family is 1, i.e., it is asymptotically perfect. The same is true if we shorten this code a constant number of times (and even slightly more), i.e., the shortened Preparata codes are asymptotically perfect codes with packing radius 2. No other such family of codes is known. Moreover, no other family whose packing density is greater than 1/2 is known. After the discussion on codes with packing radius 2, the next step is to discuss codes with covering radius 2. In this section we consider a family of codes with covering radius 2 that has a covering density 1, i.e., this family of codes is asymptotically perfect.

Construction 6.3. Let $r \equiv 3 \pmod 4$, where $r = t \cdot 2^k - 1$, t is odd, and $k \geq 2$. Let $\ell(t,k)$, t odd, $k \geq 1$, $(t,k) \neq (1,1)$, be the length of the shortest code $\mathcal{C}$ with minimum distance 4, covering radius 2, and redundancy r, which has the space tiling property. We claim that for $k \geq 2$

$$\ell(t,k) \leq \sum_{i=0}^{k-1} 2^{t \cdot 2^i} - k + \lfloor 2^{t-2} \rfloor.$$

The BDS construction is used to obtain a code $\mathcal{C}(t,k)$, where

$$\mathcal{C}^1 = \mathcal{P}'(t\cdot 2^{k-1}),\ \mathcal{C}^2 = \bigcup_{i=0}^{2^{t\cdot 2^{k-1}-1}-1} \mathcal{P}'_i(t\cdot 2^{k-1}) = \mathcal{H}(t\cdot 2^{k-1}),$$

$\mathcal{C}^3$ is a code with the space tiling property that meets the value of $\ell(t,k-1)$ and $\mathcal{C}^4 = \mathbb{F}_2^{\ell(t,k-1)}$.

By Lemma 6.14, we can partition $\mathcal{H}(t\cdot 2^{k-1})\times\mathbb{F}_2^{\ell(t,k-1)}$ into codes which have the parameters of the code $\mathcal{C}(t,k)$. The code $\mathcal{C}(t,k-1)$ has redundancy $t\cdot 2^{k-1}-1$ and the code $\mathcal{H}(t\cdot 2^{k-1})$ has redundancy $t\cdot 2^{k-1}$. Hence, the code $\mathcal{C}(t,k)$ obtained by the BDS construction has redundancy $t\cdot 2^k-1$ and the space tiling property. Therefore, for $k\geq 2$ we obtain

$$\ell(t,k)\leq 2^{t\cdot 2^{k-1}}-1+\ell(t,k-1)$$

with the initial conditions $\ell(1,2)=4$ (which is attained by the extended Hamming code of length 4) and $\ell(t,1)=5\cdot 2^{t-2}-1$ for $t\geq 3$ (which is obtained by $\Upsilon(t-1)$ whose redundancy is $2t-1$, Lemma 6.15, and because $\ell(3,1)=9$, a value attained by a linear code, which as any linear code has the linear space tiling property).

It is now easy to verify that

$$\ell(t,k)\leq\sum_{i=0}^{k-1}2^{t\cdot 2^i}-k+\lfloor 2^{t-2}\rfloor$$

and hence

$$\ell^*(2,4,t\cdot 2^k-1)\leq\ell(t,k)\leq\sum_{i=0}^{k-1}2^{t\cdot 2^i}-k+\lfloor 2^{t-2}\rfloor,$$

where t is odd and $k\geq 2$. Therefore, this is a

$$\left(\sum_{i=0}^{k-1}2^{t\cdot 2^i}-k+\lfloor 2^{t-2}\rfloor,\ 2^{\sum_{i=0}^{k-1}2^{t\cdot 2^i}-k+\lfloor 2^{t-2}\rfloor-t\cdot 2^k+1},\ 4\right)$$

code with covering radius 2.

Theorem 6.12. *The family of codes*

$$\{\mathcal{C}(t,k)\ :\ k\geq 1,\ t\ \textit{odd},\ t\geq 1,\ (t,k)\neq(1,1)\}$$

has covering density 1 and hence this family is an asymptotic 2-perfect covering family.

Proof. Define $n(t,k) = \sum_{i=0}^{k-1} 2^{t\cdot 2^i} - k + \lfloor 2^{t-2} \rfloor$, which is the length of the code $\mathcal{C}(t,k)$. This implies that

$$\mu_c\{\mathcal{C}(t,k)\} = \lim_{\substack{k\to\infty \\ t\to\infty}} \frac{2^{n(t,k)-t\cdot 2^k+1}(1+n(t,k)+\binom{n(t,k)}{2})}{2^{n(t,k)}}$$

$$= \lim_{\substack{k\to\infty \\ t\to\infty}} \frac{\binom{2^{t\cdot 2^{k-1}}}{2}}{2^{t\cdot 2^k - 1}} = 1.$$

□

6.6 Dense Covering Codes with Radius Three

In this section our goal is to present a family of codes with covering radius 3 and small covering density, as close to 1 as possible. Our search is for a family of codes with small asymptotic covering density. The best known such family is constructed by the BDS construction.

Construction 6.4. Let $\mathcal{P}_0(m), \mathcal{P}_1(m), \ldots, \mathcal{P}_{2^{m-1}-1}(m)$ be the 2^{m-1} translates of the Preparata code of length 2^m, $m \geq 4$, whose union forms $\mathcal{H}^*(m)$. Let $\mathcal{C}(m)$ be the code over $\mathbb{F}_2^{2^m} \times \mathbb{F}_2^{2^m-1}$ defined by

$$\mathcal{C}(m) \triangleq \{(x,y) \ : \ x \in \mathcal{P}_i(m), \ y \in \mathcal{P}'_i(m), \ 0 \leq i \leq 2^{m-1}-1\} \ .$$

Theorem 6.13. *The code* $\mathcal{C}(m)$ *is a binary* $(2^{m+1}-1, 2^{2^{m+1}-3m-1}, 5)$ *code with covering radius 3 and covering density* $\frac{4}{3} - \frac{1}{2^{2m}}$.

Proof. Recall that by Lemma 6.10, the subnorm of the family $\mathbb{P}(m)$ is 4 and the subnorm of the family $\mathbb{P}'(m)$ is 3. The length of the code $\mathcal{C}(m)$, its size, minimum distance, and covering radius are immediate consequences of Theorem 6.9, the parameters of the Preparata code, and the number of translates of the Preparata code and the punctured Preparata code, which are used in the BDS construction.

The size of a ball with radius 3 is

$$\binom{2^{m+1}-1}{0}+\binom{2^{m+1}-1}{1}+\binom{2^{m+1}-1}{2}+\binom{2^{m+1}-1}{3} = \frac{2^{3m+2}-3\cdot 2^m}{3}.$$

Hence, the covering density of the code $\mathcal{C}(m)$ is

$$\frac{2^{2^{m+1}-3m-1}(2^{3m+2}-3\cdot 2^m)}{3\cdot 2^{2^{m+1}-1}} = \frac{4}{3} - \frac{1}{2^{2m}},$$

which completes the proof. □

Corollary 6.12. *The family of codes $\{\mathcal{C}(m)\}_{m=2k\geq 4}$ has covering density $\frac{4}{3}$.*

Note that for length 31, the code $\mathcal{C}(4)$ has covering density $\frac{4}{3} - \frac{1}{256} = \frac{1023}{768}$, which is already very close to 4/3.

6.7 Notes

Section 6.1. When discussing a "good" error-correcting code $\mathcal{C}$ of length n over $\mathbb{F}_q$, the first measure to be considered is the ***rate*** of the code $\mathcal{C}$ that is defined as $\frac{\log_q |\mathcal{C}|}{n}$. The goal is to obtain codes whose rate is close to 1. Having obtained such codes with a rate close to 1, the next goal is to have a code with small redundancy r, where $r = n - \log_q |\mathcal{C}|$. The smallest redundancy that we can obtain implies the highest density of a code, which is the motivation for the various definitions on the density. A remarkable result by [Kabatiansky and Panchenko (1988)] is that $\mu^*(1) = 1$ for both packing and covering. The reader can refer to a comprehensive analysis of the density in the excellent book of [Cohen, Honkala, Litsyn, and Lobstein (1997)].

Section 6.2. Theorem 6.1 and Corollary 6.1 are contained in the work of [Johnson (1962)]. Codes that attain this bound are nearly-perfect codes and they were first studied by [Goethals and Snover (1972)]. The improvement of Theorem 6.2 and a comparison between the bounds was done in [Mounits, Etzion, and Litsyn (2002)]. All these bounds including the sphere-packing bound can be proved for other metrics that form association schemes as was proved in [Mounits, Etzion and Litsyn (2007)]. Other improvements for Theorem 6.1 can be found in [MacWilliams and Sloane (1977)].

Section 6.3. The Preparata codes were defined in [Preparata (1968)]. The simple representation given in this section of the chapter from Definition 6.1 to Corollary 6.4 is due to [Baker, van Lint, and Wilson (1983)]. It was proved in [Kantor (1983)] that $\mathcal{P}(\sigma_1)$ and $\mathcal{P}(\sigma_2)$ are equivalent if and only if $\sigma_1 = \sigma_2$ or $\sigma_1\sigma_2 = 2^r$. In this paper the group of automorphisms, *Aut* $\mathcal{P}(\sigma)$, was also found.

The properties of the Preparata code presented in the section and some other properties are very intriguing. The code is nonlinear, but has many linear properties (the union of disjoint translates is the linear Hamming codes and many other properties, some of which were discussed in this section). These phenomena were resolved after it was proved in [Hammons, Kumar, Calderbank, Sloane, and Solé (1994)] that by applying a Gray mapping which transfers 00 to 0, 01 to 1, 11 to 2, and 10 to 3, on all the codewords of the Preparata code, the obtained code is a linear code over $\mathbb{Z}_4$. This work motivated an extensive research for codes over rings. For example see [Dinh and López-Permouth (2004)] and the long list of papers quoting this work. Codes over rings are related to codes in the Lee metric and codes in the Manhattan metric which are the topic of Chapters 11 and 12.

The number of codewords in the Preparata code (and also in the punctured Preparata code) is twice as large as the largest linear code with the same length and minimum distance [Brouwer and Tolhuizen (1993)]. Moreover, in this paper an improvement of the Johnson bound for linear codes is given and, as a consequence, it was proved that there is no linear code with the parameters of the shortened Preparata code. The Preparata codes are also used as building blocks for other interesting codes such as the asymptotic 2-perfect covering codes constructed in Section 6.5 or the dense codes with covering radius 3 constructed in Section 6.6. They will also be used for construction of 2-perfect mixed codes in Section 7.3.

The following theorem was proved in [Lindström (1975a)].

Theorem 6.14. *The only binary nearly-perfect codes are the binary perfect codes, the* $[2^r-2, 2^r-r-2, 3]$ *shortened Hamming codes (and other shortened 1-perfect codes), and the punctured Preparata codes.*

Theorem 6.14 was generalized for nonbinary codes in [Lindström (1977)] who proved that the only nonbinary nearly-perfect codes are the nonbinary perfect codes.

Section 6.4. The BDS construction is a generalization of the direct product construction. It was re-introduced along the years, using various variants, in many papers. The packing version can be attributed to [Sloane, Reddy, and Chen (1972)] and the covering version can be attributed to [Honkala (1991)].

The punctured Preparata codes are quasi-perfect codes with packing radius 2 and covering radius 3. These codes are asymptotically 2-perfect codes, i.e., their asymptotic packing density is 1. The redundancy r of these codes equals $r \equiv 3 \pmod 4$. What is the longest length $n^*(3,5,r)$ of codes with other redundancies? For $r \equiv 0 \pmod 8$, codes with asymptotic packing density 1/2 for which $n^*(3,5,r) = 2^{r/2} + 2^{r/4} - 1$ were presented in [Etzion and Mounits (2005)]. For $r \equiv 4 \pmod 8$, linear codes with asymptotic packing density $1/2$ for which $n^*(3,5,r) = 2^{r/2} + 1$ were constructed by Zetterberg and presented in [Moreno (1983)]. For $r \equiv 2 \pmod 4$, linear codes with asymptotic packing density $1/2$ for which $n^*(3,5,r) = 2^{r/2}$ were constructed by Goppa (known as irreducible Goppa codes) and presented in [Moreno (1983)]. Codes with different parameters and especially from covering radius points of view can be found in the survey on covering codes [Cohen, Karpovsky, Mattson, and Schatz (1985)]. It is worth mentioning that in [Wagner (1966)], a search for linear quasi-perfect codes with packing radius 2 was performed. The most notable code that was found is a $[23,14,5]$ code whose packing density is $\frac{277}{512}$. Clearly, there are big gaps in our knowledge on the parameters of quasi-prefect codes with packing radius 2 and covering radius 3. This analysis leads to the following research problem.

Problem 6.7. For each redundancy r, what is the value of $n^*(3,5,r)$?

Consider now some smaller parameters. What is the shortest length, $\ell^*(2,4,r)$ of codes with minimum distance 4 and covering radius 2, and what is the lowest covering density of such a family of codes with a given redundancy r? For $r \equiv 0 \pmod 4$, we constructed the code $\Psi(m)$ whose covering density is $9/8$ (see also [Etzion and Greenberg (1993)]). For $r \equiv 1 \pmod 4$, we constructed the code $\Upsilon(m)$ whose covering density is $25/16$ (see also [Etzion and Mounits (2005)]). Linear codes with this redundancy and covering density $529/256$ were constructed in [Gabidulin, Davydov, and Tombak (1991)]. For $r \equiv 2 \pmod 4$, linear codes with this redundancy and covering density $225/128$ were constructed in [Gabidulin, Davydov, and Tombak (1991)]. For $r \equiv 3 \pmod 4$, we constructed codes with covering density 1 (asymptotically optimal) in Section 6.5. More work on linear quasi-perfect codes with covering radius 2 and minimum distance 4 was done in [Davydov and Tombak (1989a,b)].

Our exposition in this chapter is only on binary codes, but there is some literature on nonbinary quasi-perfect codes. Work in this direction can be found in [Giulietti and Pasticci (2007); Danev and Dodunekov (2008); Danev, Dodunekov, and Radkova (2011); Li and Helleseth (2016)]. In particular, a quasi-perfect $[n, n-r, 4]_q$ code is equivalent to what is called a complete n-cap in $\mathrm{PG}(r-1, q)$. Complete n-caps in $\mathrm{PG}(r-1, q)$ were considered for example in [Hirschfeld and Storme (1998); Giulietti (2000); Bierbrauer, Marcugini, and Pambianco (2006); Giulietti (2007a,b); Davydov, Faina, Marcugini, and Pambianco (2009); Davydov, Giulietti, Marcugini, and Pambianco (2010); Anbar, Bartoli, Giulietti, and Platoni (2014)].

Section 6.5. The first family of asymptotically 2-perfect covering codes was obtained in [Struik (1994), Construction 4.24]. The codes presented in this section were constructed in [Etzion and Mounits (2005)]. For $r = 7$ and $r = 11$, the parameters of the codes presented in this section are the same as those in [Struik (1994)]. For $r \geq 15$, the codes presented in this section are shorter than the codes in [Struik (1994)]. For example, if $r = 15$, the construction in this section produces a code of length 274 whereas the code obtained in [Struik (1994)] has length 276, and for $r \geq 19$, $r = t \cdot 2^k - 1$, t odd, $k \geq 2$, the code of [Struik (1994)] has the following parameters

$$(2^{t\cdot 2^{k-1}} + \frac{23}{16}2^{t\cdot 2^{k-2}} - 4, 2^{2^{t\cdot 2^{k-1}}+\frac{23}{16}2^{t\cdot 2^{k-2}}-3-t\cdot 2^k}, 4)2 \ .$$

Section 6.6. The code $\mathcal{C}(m)$ was constructed in [Etzion and Greenberg (1993)]. The analysis given in the section is contained in the work of [Etzion and Mounits (2005)], where the density of more families of quasi-perfect codes with packing radius 2 or covering radius 3 are analyzed. The code for which $\ell(3,1) = 9$ was presented in [Bruladi, Pless, and Wilson (1989)].

For the covering density, one would like to find the shortest length of such codes, $\ell^*(3,5,r)$. For $r \equiv 0 \pmod 6$, codes of length $2^{(r+3)/3} - 1$ with asymptotic covering density 4/3 were constructed in [Etzion and Greenberg (1993)] and later in [Etzion and Mounits (2005)]. These codes presented in this section are asymptotically the most sparse family known as of 2021. It is not known whether a family of linear codes with these parameters exists (see [Graham and Sloane (1985); Bruladi, Pless, and Wilson (1989)]). For $r \equiv 2 \pmod 6$, codes of length $5 \cdot 2^{(r-2)/3} - 1$ with asymptotic covering density 125/24 were constructed in [Etzion and Mounits (2005)]. For $r \equiv 4 \pmod 6$, codes of length $3 \cdot 2^{(r-1)/3} - 2$ with asymptotic covering density 9/4 were constructed in [Struik (1994)]. For $r \equiv 5 \pmod 6$, codes of length $5 \cdot 2^{(r+4)/3} - 2$ with asymptotic covering density 8/3 were constructed in [Etzion and Mounits (2005)]. A comprehensive analysis of quasi-perfect codes with small radius (packing and covering) is presented in this paper.

Problem 6.8. For each redundancy r, what is the value of $\ell^*(3,5,r)$?

Contrary to dense packing codes with packing radius 2 and covering radius 3 (where dense linear codes are known), families of linear sparse codes with these parameters are not known.

Problem 6.9. Construct sparse linear quasi-perfect codes with packing radius 2 and covering radius 3.

Chapter 7

Codes with Mixed Alphabets

Before moving on to different metrics and different spaces, we continue with the Hamming metric, but with a different space. Instead of words of length n over a finite field $\mathbb{F}_q$ (or any finite alphabet of size q), we consider words of length n, where each coordinate can be taken from an alphabet of a different size. In other words, the code is of length n and the codewords are taken from $\Sigma_1 \times \Sigma_2 \times \cdots \times \Sigma_n$, where Σ_i is an alphabet with q_i symbols, $q_i > 1$, and q_i is not necessarily a power of a prime. For simplicity and w.l.o.g. Σ_i will be taken as $\mathbb{Z}_{q_i}$ or $\mathbb{F}_{q_i}$ if q_i is a power of a prime, $1 \leq i \leq n$. The Hamming distance is the metric in this space, but this space with the Hamming metric is not a scheme (if at least two of the q_i's are different). To prove this claim, assume w.l.o.g. that $q_1 \neq q_2$. Consider the three words $x = (000 \cdots 0)$, $y = (100 \cdots 0)$, and $z = (010 \cdots 0)$. Clearly, $d(x,y) = d(x,z) = 1$ and the number of words which satisfy the equality $d(x,u) = d(y,u) = 1$ is $q_1 - 2$, where $u = (\alpha 00 \cdots 0)$, with $\alpha \notin \{0,1\}$). On the other hand the number of words in the space such that $d(x,u) = d(z,u) = 1$ is $q_2 - 2$. Since $q_1 \neq q_2$, it follows that the related intersection number depends on the chosen words and hence this is not an association scheme. A perfect code in this space will be called a ***perfect mixed code***.

In Section 7.1 such perfect codes with radius one will be discussed. The perfect codes with radius one will be constructed from partitions of the nonzero elements of a group into subgroups (without the identity). The codes derived from these partitions are associated with codes organized in bytes, where the errors are restricted to these bytes. Such partitions and byte-correcting codes will be discussed in Section 7.2. Throughout most of this chapter it will be assumed that in a perfect mixed code, at least two of the coordinates are associated with different alphabet sizes, although most of the results can also be applied to the case where the symbols of all the coordinates are taken from an alphabet of the same size. For byte-correcting codes, there will be no such assumption as all the bytes can also be of the same size. In other words, for a byte-correcting code, we will also consider the case where all the bytes are of the same size, which implies that when it is transferred into a mixed code, all the coordinates are over the same alphabet (which is just a code in the Hamming scheme). In Section 7.3 we present one construction for a family of perfect mixed codes with radius two. This is the only known infinite family of 2-perfect mixed codes with the Hamming distance. These perfect mixed codes will be also extended to diameter perfect mixed codes and a general construction for diameter perfect mixed codes for each possible diameter will be presented. In this section also several nonexistence results for various radii will be presented.

7.1 Perfect Mixed Codes with Radius One

There are several constructions for 1-perfect mixed codes, but as far as parameters for such codes, all of them can be obtained with one simple construction, which is a generalization of the construction for the Hamming codes. The construction is applied only when the alphabet size in each coordinate is a power of the same prime p.

Theorem 7.1. *Let $\{\mathcal{S}_1, \mathcal{S}_2, \ldots, \mathcal{S}_n\}$ be a partition of $\mathbb{F}_{q^m}^-$ into n subsets such that $\mathcal{T}_i \triangleq \mathcal{S}_i \cup \{\mathbf{0}\}$ is a subspace of dimension k_i. The code defined by*

$$\mathcal{C} \triangleq \left\{ (c_1, c_2, \ldots, c_n) \ : \ c_i \in \mathcal{S}_i \cup \{0\}, \ \sum_{i=1}^{n} c_i = 0 \right\},$$

where the sum is performed in $\mathbb{F}_{q^m}$, is a 1-perfect mixed code over $\mathcal{T}_1 \times \mathcal{T}_2 \times \cdots \times \mathcal{T}_n$ which is isomorphic to $\mathbb{F}_{q^{k_1}} \times \mathbb{F}_{q^{k_2}} \times \cdots \times \mathbb{F}_{q^{k_n}}$.

Proof. If c and c' are two codewords of $\mathcal{C}$, then clearly $c + c'$ is also a codeword in $\mathcal{C}$. This implies that $\mathcal{C}$ is a linear code and its minimum distance is the weight of the codeword of minimum weight in the code. It is readily verified that there is no codeword of weight one in $\mathcal{C}$. Therefore, if $(x_1, x_2, \ldots, x_n) \in \mathcal{T}_1 \times \mathcal{T}_2 \times \cdots \times \mathcal{T}_n$ has weight two, then assume w.l.o.g. that $x_i \neq 0$ and $x_j \neq 0$, for some $1 \leq i < j \leq n$. Since $x_i \in \mathcal{T}_i$, $x_j \in \mathcal{T}_j$, $\mathcal{T}_i \cap \mathcal{T}_j = \{\mathbf{0}\}$, and $\mathcal{T}_i$, $\mathcal{T}_j$ are subspaces of $\mathbb{F}_{q^m}^-$ it follows that $x_i + x_j \neq 0$ and hence $\mathcal{C}$ does not have a codeword of weight two. This implies that the minimum distance of $\mathcal{C}$ is three.

Now, to prove the claim of the theorem, it suffices to show that for each word $x = (x_1, x_2, \ldots, x_n) \in \mathcal{T}_1 \times \mathcal{T}_2 \times \cdots \times \mathcal{T}_n$, there exists exactly one codeword $c = (c_1, c_2, \ldots, c_n)$ in $\mathcal{C}$ such that $d(x, c) \leq 1$. Let $s = \sum_{i=1}^{n} x_i$. If $s = 0$, then, clearly, x is a codeword. If $s = \alpha \neq 0$, then let j be the unique integer such that $\alpha \in \mathcal{S}_j$. Clearly, since $\mathcal{S}_j \cup \{\mathbf{0}\}$ is a subspace and also $x_i \in \mathcal{T}_i$, for each $1 \leq i \leq n$, it follows that $x_j - \alpha \in \mathcal{S}_j$. Define $c \triangleq (c_1, c_2, \ldots, c_n)$, where $c_i = x_i$ for $i \neq j$ and $c_j = x_j - \alpha$. This implies that

$$\sum_{i=1}^{n} c_i = \sum_{i=1}^{n} x_i - \alpha = s - \alpha = 0$$

and hence $c \in \mathcal{C}$ and $d(x, c) = 1$.

Assume now that there exists two distinct codewords $c, c' \in \mathcal{C}$ such that $d(x, c) \leq 1$ and $d(x, c') \leq 1$. By the trianle equality we have that

$$d(c, c') \leq d(c, x) + d(x, c') \leq 2,$$

a contradiction to the minimum distance of $\mathcal{C}$.

Thus, c is the unique codeword in $\mathcal{C}$ such that $d(x, c) = 1$. □

If c, c' are two codewords of the code $\mathcal{C}$ constructed in Theorem 7.1, then $c + c' \in \mathcal{C}$. Hence, $\mathcal{C}$ is a linear code. If $k_i = k_j$ for all $1 \leq i < j \leq n$, then $\mathcal{C}$ is a linear code in the Hamming scheme. The q-ary Hamming code is a special case of this code, where each $\mathcal{S}_i \cup \{\mathbf{0}\}$ is a one-subspace of $\mathbb{F}_q^n$. Moreover, the theorem can be generalized and stated in terms of a general group $\mathcal{G}$ and not necessarily $\mathbb{F}_{q^m}$ as follows.

Theorem 7.2. *Let $\{\mathcal{G}_1^-, \mathcal{G}_2^-, \ldots, \mathcal{G}_n^-\}$ be a partition of $\mathcal{G}^-$, where $\mathcal{G}$ is an abelian group, into n nonempty subsets such that $\mathcal{G}_i = \mathcal{G}_i^- \cup \{\mathbf{0}\}$ is a subgroup of $\mathcal{G}$ whose size is k_i. The code defined by*

$$\mathcal{C} \triangleq \left\{ (c_1, c_2, \ldots, c_n) \ : \ c_i \in \mathcal{G}_i \cup \{0\}, \ \sum_{i=1}^{n} c_i = 0 \right\},$$

where the sum is performed in $\mathcal{G}$, is a 1-perfect mixed code over $\mathcal{G}_1 \times \mathcal{G}_2 \times \cdots \times \mathcal{G}_n$.

The proof of Theorem 7.2 is identical to the one for Theorem 7.1. To apply these theorems, we have to find partitions as required by the theorems. Such a partition is called a ***group partition***. There are many such partitions when the group $\mathcal{G}$ of Theorem 7.2 is $\mathbb{F}_{q^m}$ as in Theorem 7.1. Unfortunately, only partitions of this type are known. Some of the known partitions will be discussed in Section 7.2. There are several differences between linear codes in the Hamming scheme and linear mixed codes, where at least two of the coordinates are over alphabets of different sizes. The generator matrix and the parity-check matrix for such linear mixed code is completely different from the representation for linear codes in the Hamming scheme. If $\mathcal{C}$ is such a linear code over $\mathbb{F}_{q^{k_1}} \times \mathbb{F}_{q^{k_2}} \times \cdots \times \mathbb{F}_{q^{k_n}}$, then one might assume that it can be represented by a parity-check matrix $H = [\alpha^{i_1}\alpha^{i_2} \cdots \alpha^{i_n}]$, where α is a primitive element in $\mathbb{F}_{q^m}$. Furthermore, α^{i_r}, $1 \leq r \leq n$, is an element in the subfield $\mathbb{F}_{q^{k_r}}$ in the partition of $\mathbb{F}_{q^m}$. In this case a codeword $c \in \mathcal{C}$ has the form $c = (c_1, c_2, \ldots, c_n) = (\alpha^{j_1}, \alpha^{j_2}, \ldots, \alpha^{j_n})$, where $c_r = \alpha^{j_r}$ is an element in the subfield $\mathbb{F}_{q^{k_r}}$ and $\sum_{r=1}^{n} \alpha^{i_r + j_r} = 0$. Unfortunately, this representation does not lead to a perfect mixed code as in the codes constructed in Theorem 7.1. The proof of Theorem 7.1 is not correct for this representation and this observation if left as an exercise. But, the linear code in Theorem 7.1 has a different representation with a parity-check matrix and a generator matrix. It is equivalent to byte-correcting codes as will be discussed in Section 7.2. The parity-check matrices used for byte-correcting codes will be presented in this section. There are many intriguing questions which remain unsolved in this direction. Two such examples are given in the following problems.

Problem 7.1. Are there nonlinear 1-perfect mixed codes with parameters that cannot be obtained by Theorem 7.2?

Problem 7.2. Are there parameters of 1-perfect codes obtained with Theorem 7.2 that cannot be obtained by Theorem 7.1?

What about nonlinear 1-perfect mixed codes with parameters that can also be obtained by Theorem 7.2? Many types of such codes can be obtained with constructions similar to the ones given in Chapter 5. Two such constructions will be presented. The first construction is a general product construction.

Let $\mathcal{C}$ be a 1-perfect code (mixed or not mixed) over the alphabet $\mathbb{Z}_{q_1} \times \mathbb{Z}_{q_2} \times \ldots \mathbb{Z}_{q_t}$. Let $\mathcal{C}^i$, $1 \leq i \leq t$ be a 1-perfect code (mixed or not mixed) of length n_i in a space $\mathcal{V}_i$ with the space tiling property and q_i be the number of associated translates, where $\mathcal{C}^i_j$, $1 \leq j \leq q_i$ is the j-th translate of the i-th code. In other words, for each $1 \leq i \leq t$, $\mathcal{C}^i_{j_1} \cap \mathcal{C}^i_{j_2} = \varnothing$ for $1 \leq j_1 < j_2 \leq q_i$ and $\cup_{j=1}^{q_i} \mathcal{C}^i_j = \mathcal{V}_i$. Define

$$\mathfrak{C}_1 \triangleq \{(x_{i_1}, x_{i_2}, \ldots, x_{i_t}) \ : \ x_{i_\ell} \in \mathcal{C}^\ell_{i_\ell}, (i_1, i_2, \ldots, i_t) \in \mathcal{C}, \ 1 \leq \ell \leq t\}.$$

Similarly to the proof of Theorem 5.4, one can prove the following theorem.

Theorem 7.3. *If $\mathcal{C}$ and each $\mathcal{C}^i$, $1 \leq i \leq t$, are 1-perfect (mixed) codes, then the code $\mathfrak{C}_1$ is a 1-perfect mixed code of length $n = \sum_{i=1}^{t} n_i$ and size $|\mathcal{C}| \prod_{i=1}^{t} |\mathcal{C}^i|$ over $\mathcal{V}_1 \times \mathcal{V}_2 \times \cdots \times \mathcal{V}_n$.*

The second construction can be viewed as a generalization of the one defined in Theorem 5.1. For this construction, let $\mathcal{C}^1$ be a 1-perfect mixed code of length n_1+1 over $\mathbb{Z}_q \times Q_1$, where $Q_1 = \mathbb{Z}_{q_1} \times \mathbb{Z}_{q_2} \times \cdots \times \mathbb{Z}_{q_{n_1}}$ and let $\mathcal{B}_1^1(n_1+1)$ be the a ball of radius one related to the space $\mathbb{Z}_q \times Q_1$, where the size of a ball with radius one in $\mathbb{Z}_q \times Q_1$ is $|\mathcal{B}_1^1(n_1+1)| = q + \sum_{i=1}^{n_1}(q_i - 1)$.

Let $\mathcal{C}^2$ be a 1-perfect code of length n_2 over $Q_2 = \mathbb{Z}_{q'_1} \times \mathbb{Z}_{q'_2} \times \cdots \times \mathbb{Z}_{q'_{n_2}}$ and let $\mathcal{B}_1^2(n_2)$ be the a ball of radius one related to the space Q_2, where the size of a ball with radius one in Q_2 is q, i.e., $|\mathcal{B}_1^2(n_2)| = q = 1 + \sum_{i=1}^{n_2}(q'_i - 1)$. Let φ be an injective function from $\mathbb{Z}_q$ into Q_2 such that $\varphi(0) = \mathbf{0}$ and the weight of $\varphi(x)$ is one for each $x \in \mathbb{Z}_q^-$. In other words, φ is an one-to-one function which maps $\mathbb{Z}_q$ into the ball $\mathcal{B}_1^2(n_2)$, where $\varphi(0) = \mathbf{0}$. Define the following set of words of length $n = n_2 + n_1$.

$$\mathfrak{C}_2 \triangleq \{(c + \varphi(x), y) \ : \ c \in \mathcal{C}^2, \ (x, y) \in \mathcal{C}^1, \ x \in \mathbb{Z}_q\}.$$

Theorem 7.4. *The code $\mathfrak{C}_2$ is a 1-perfect mixed code of length n_2+n_1 over $Q_2 \times Q_1$.*

Proof. We start and prove that the minimum Hamming distance of the code $\mathfrak{C}_2$ is 3. Let $b_1 = (c_1 + \varphi(x_1), y_1)$, $b_2 = (c_2 + \varphi(x_2), y_2)$, be two distinct codewords of $\mathfrak{C}_2$. Since $x_1, x_2 \in \mathbb{Z}_q$, it follows that $d(x_1, x_2) \in \{0, 1\}$. We distinguish between four cases depending on whether $d(y_1, y_2)$ equals to 0, 1, 2, or 3.

Case 1. If $d(y_1, y_2) = 0$, then $y_1 = y_2$, and hence since $d(\mathcal{C}^1) = 3$ and $x_1, x_2 \in \mathbb{Z}_q$, it follows that $(x_1, y_1) = (x_2, y_2)$, i.e., $x_1 = x_2$, and therefore $\varphi(x_1) = \varphi(x_2)$. As a consequence, we must have that $c_1 \neq c_2$, which implies that $d(b_1, b_2) = d(c_1 + \varphi(x_1), c_2 + \varphi(x_2)) = d(c_1, c_2) \geq 3$.

Case 2. If $d(y_1, y_2) = 1$, then since $d(x_1, x_2) \in \{0, 1\}$, it follows that $d((x_1, y_1), (x_2, y_2)) \leq 2$, contradicting the minimum distance of $\mathcal{C}^1$. This implies that $d(y_1, y_2) \neq 1$ when $(x_1, y_1) \in \mathcal{C}^1$ and $(x_2, y_2) \in \mathcal{C}^1$.

Case 3. If $d(y_1, y_2) = 2$, then first note that $d(c_1, c_2) = 0$ or $d(c_1, c_2) \geq 3$. As a consequence, since $\text{wt}(\varphi(x)) \leq 1$ for $x \in \mathbb{Z}_q$ and $x_1 \neq x_2$ as $(x_1, y_1), (x_2, y_2) \in \mathcal{C}^1$, it follows that $d(\varphi(x_1), \varphi(x_2)) = 2$ and hence $d(c_1 + \varphi(x_1), c_2 + \varphi(x_2)) \geq 1$, which implies that $d(b_1, b_2) \geq 3$.

Case 4. If $d(y_1, y_2) \geq 3$, then obviously, $d(b_1, b_2) \geq 3$.

To complete the proof, it suffices to show that the code attains the sphere-packing bound. An immediate consequence of the definition of $\mathfrak{C}_2$ is that $|\mathfrak{C}_2| = |\mathcal{C}^1| \cdot |\mathcal{C}^2|$. Since $\mathcal{C}^1$ and $\mathcal{C}^2$ are 1-perfect codes, it follows that $|\mathcal{B}_1^1(n_1+1)| \cdot |\mathcal{C}^1| = |\mathbb{Z}_q \times Q_1| = q\,|Q_1|$ and $|\mathcal{B}_1^2(n_2)| \cdot |\mathcal{C}^2| = |Q_2|$. Clearly, the space associated with $\mathfrak{C}_2$ is $Q_2 \times Q_1$ whose size it $|Q_2| \cdot |Q_1|$ and the size of the associated ball $\hat{\mathcal{B}}_1(n_1+n_2)$ is

$$\left|\hat{\mathcal{B}}_1(n_1+n_2)\right| = 1 + \sum_{i=1}^{n_1}(q_i - 1) + \sum_{i=1}^{n_2}(q_i' - 1)$$

$$= \left|\mathcal{B}_1^1(n_1+1)\right| + \left|\mathcal{B}_1^2(n_2)\right| - q = \left|\mathcal{B}_1^1(n_1+1)\right|.$$

Therefore,

$$\left|\hat{\mathcal{B}}_1(n_1+n_2)\right| \cdot |\mathfrak{C}_2| = \left|\mathcal{B}_1^1(n_1+1)\right| \cdot |\mathcal{C}^1| \cdot |\mathcal{C}^2| = q\,|Q_1|\,\frac{|Q_2|}{|\mathcal{B}_1^2(n_2)|} = |Q_1| \cdot |Q_2|$$

and hence $\mathfrak{C}_2$ meets the sphere-packing bound and the proof is completed. □

7.2 Byte-Correcting Codes and Group Partitions

The construction presented in Theorem 7.1 for 1-perfect mixed codes requires a "partition" of $\mathbb{F}_{q^m}$ into subspaces whose intersection is the null space. This problem is exactly related to perfect byte-correcting codes. In most memory and storage systems, the information is stored in bytes. In many of these systems, when an error event occurs it can corrupt a few positions of the same byte. Hence, when we consider error detection and correction in such systems, we want to be able to detect and correct all errors that occur in the same byte. For this purpose, we consider e-byte-correcting codes for these systems, i.e., codes which correct all errors which occur in at most e of the bytes. Usually, all bytes are of the same size, but in various memory systems they can be of different sizes. For 1-perfect byte-correcting codes, the partition into bytes induces a group partition and hence linear 1-perfect byte-correcting codes are equivalent to linear 1-perfect mixed codes. This equivalence holds also for e-perfect mixed codes and e-perfect byte-correcting codes, where $e > 1$, but such codes are not obtained from group partitions.

We distinguish between five types of byte-correcting codes (in other words, five types of group partitions), depending on the different sizes of the bytes in the code. Some of these types are of practical use, and some do not and are provided only for two reasons. First, for their theoretical and mathematical value and also since they form group partitions which can be applied to form perfect mixed codes.

Type 1: All bytes have the same size.
Type 2: One byte is of size b_1 and the other bytes are of size b_2.
Type 3: Each byte is either of size b_1 or of size b_2.
Type 4: The size of each byte is a power of 2 (q if the code is over $\mathbb{F}_q$).
Type 5: All the other cases.

Since a byte-correcting code should be able to correct all errors which occurred within one of its bytes (if all the errors occurred only inside one byte), it follows that all syndromes generated from one byte must be distinct. This implies that a byte of size b can be considered as a subspace of dimension b. All codes considered for this purpose will be linear and will be analyzed by their parity-check matrices. For simplicity, in this section, for most cases only binary codes will be discussed, but the results are generalized in a straightforward way to codes over $\mathbb{F}_q$ for any prime power q. Some of the results will be proved and explained for all alphabets of a finite field $\mathbb{F}_q$.

Note that a specific byte-correcting code can belong to a few different types. For all these types of codes, there is a simple necessary condition for the existence of the corresponding e-perfect byte-correcting codes. Given a code $\mathcal{C}$ of length n, and an integer e, let $\mathcal{B}_e(n)$ be a ball with radius e centered at any codeword. Note, that the ball of radius e centered at a codeword $c \in \mathcal{C}$ contains the set of words that differ in no more than e bytes from c. The size of the ball (as well as its structure) does not depend on the center c. The code $\mathcal{C}$ is an e-perfect byte-correcting code if its minimum distance is $2e+1$ (where the distance between two codewords are the number of bytes in which they differ) and $|\mathcal{C}| \cdot |\mathcal{B}_e(n)| = 2^n$. Assume that $\mathcal{C}$ is a linear code of dimension k. The code $\mathcal{C}$ is an e-perfect byte-correcting code if and only if each syndrome of length $r = n-k$ is produced by exactly one linear combination of columns from no more than e bytes of the parity-check matrix of $\mathcal{C}$. Therefore, a necessary condition for the existence of a linear 1-perfect byte-correcting code with m possible sizes of bytes, i.e., s_i bytes of size b_i, $1 \leq i \leq m$, is that

$$2^r - 1 = \sum_{i=1}^{m} s_i(2^{b_i} - 1). \tag{7.1}$$

The necessary condition of (7.1) implies some other necessary conditions for the existence of some types of 1-perfect byte-correcting codes.

Lemma 7.1. *If $\mathcal{C}$ is a 1-perfect byte-correcting code with redundancy r, one* *byte of size* b_1 *and* s *bytes of size* b_2*, then*

$$s = \frac{2^{b_1}(2^{r-b_1} - 1)}{2^{b_2} - 1}$$

and b_2 *divides* $r - b_1$.

Proof. By the condition of (7.1), we have that

$$2^r - 1 = 2^{b_1} - 1 + s(2^{b_2} - 1)$$

and hence,

$$s = \frac{2^{b_1}(2^{r-b_1} - 1)}{2^{b_2} - 1}. \tag{7.2}$$

Therefore, $2^{b_2} - 1$ divides $2^{r-b_1} - 1$ and hence b_2 divides $r - b_1$. □

Recall that two linear subspaces are called disjoint if their intersection is the null space. Is the necessary condition (7.1) also sufficient for the existence of a 1-perfect byte-correcting code? It is not difficult to realize that this condition is not sufficient. The condition can be satisfied, but the sum of the sizes of the two largest bytes can be greater than the redundancy of the code r. In this case such a code cannot exist since in a space of dimension r there cannot be two disjoint subspaces of dimensions b_1 and b_2, for which $b_1 + b_2 > r$. Therefore, a linear 1-perfect byte-correcting code, with redundancy r and some bytes of size b_1 and some bytes of size b_2, must satisfy

$$b_1 + b_2 \leq r \ .$$

This condition and the condition of (7.1) are two necessary conditions, but they are still not sufficient for the existence of 1-perfect byte-correcting codes of Type 2 as implied by the following theorem.

Theorem 7.5. *A 1-perfect byte-correcting code with one byte of size* b_1 *and all the other bytes of size* b_2*, with* $b_1 < b_2$*, cannot exist.*

Proof. Assume the contrary that H is an $r \times n$ parity-check matrix of a 1-perfect byte-correcting code, with one byte of size b_1, and s bytes of size b_2, where $b_1 < b_2$. The b_1 columns of H in the byte of size b_1 are linearly independent and hence, w.l.o.g. we can assume that the i-th column, $1 \leq i \leq b_1$, is $\mathbf{e}_i$. Let H_1 be the $b_1 \times n$ matrix whose rows are the first b_1 rows of H. Since all the $2^r - 1$ linear combinations of nonempty subsets of columns from the bytes of H consist of all nonzero binary r-tuples, it follows that in all such linear combinations of columns from H_1, each nonzero

b_1-tuple appears the same number of times, and the all-zero b_1-tuple appears one time less. Each nonzero column vector of length r is obtained in exactly one linear combination and hence each nonzero prefix of length b_1 is obtained 2^{r-b_1} times and by (7.2) we have that $1 + s(2^{b_2} - 1)/2^{b_1} = 2^{r-b_1}$. Since, the all-zero column vector of length r is the only r-tuple which is not obtained as such syndrome, it follows that in the linear combinations of columns from H_1 associated with the bytes, the all-zero perfixes of length b_1 is obtained $s(2^{b_2} - 1)/2^{b_1} = 2^{r-b_1} - 1$ times.

Given a $b_1 \times b_2$ matrix A with rank $m \leq b_1$, there are $2^{b_2-m}-1$ nontrivial linear combinations of nonempty subsets of columns from A, which result in the all-zero b_1-tuple. There are also $2^m - 1$ distinct nonzero column vectors of length b_1 in the linear combinations of the columns of A. For each nonzero b_1-tuple v in the subspace spanned by the columns of A, there are exactly 2^{b_2-m} distinct linear combinations of nonempty subsets of columns from A, which result in v. The first b_1 rows of H in each byte of size b_2 can be viewed as such matrix A whose rank is $m \leq b_1$. Since each nonzero b_1-tuple is a result of exactly one linear combination of the byte of size b_1 in H_1, it follows that it should be a result in a total of $2^{r-b_1} - 1$ linear combinations of columns in all the other bytes. However, this is not possible since for any $m \leq b_1$, 2^{b_2-m} is even and greater than 1. Thus, there is no 1-perfect byte-correcting code with one byte of size b_1 and the other bytes of size b_2, with $b_1 < b_2$. □

Example 7.1. Do we have a 1-perfect byte-correcting code of length $2^{b_1}b_2 + b_1$ with 2^{b_1} bytes of size b_2 and one byte of size b_1 where $r = b_1 + b_2$ and $b_1 < b_2$? Condition (7.1) is satisfied since

$$2^{b_1}(2^{b_2} - 1) + 2^{b_1} - 1 = 2^{b_1+b_2} - 1 = 2^r - 1.$$

Nevertheless, by Theorem 7.5 such a code cannot exists.

The connection between byte-correcting codes and mixed codes is very simple. First note that each 1-perfect byte-correcting code form a group partition implied by the subspace spanned by the columns of the parity-check matrix in each associated byte. Let $\mathcal{C}$ be a byte-correcting code over $\mathbb{F}_q$ with m bytes, where the i-th byte has size b_i, $1 \leq i \leq m$ (not necessarily distinct). This code is transformed into a mixed code $\hat{\mathcal{C}}$ over $\mathbb{F}_{q^{b_1}} \times \mathbb{F}_{q^{b_2}} \times \cdots \times \mathbb{F}_{q^{b_m}}$, where a sub-codeword $(c'_1, c'_2, \ldots, c'_{b_i})$, associated with a byte of size b_i, of a codeword $(\cdots\ c'_1, c'_2, \ldots, c'_{b_i}\ \cdots)$, is mapped into the element $\gamma \in \mathbb{F}_q^{b_i}$ whose representation as a vector of length b_i in $\mathbb{F}_q$ is $(c'_1, c'_2, \ldots, c'_{b_i})$. Recall that if $(c^1_1, c^1_2, \ldots, c^1_{b_i})$ is the representation of $\alpha^{j_1} \in \mathbb{F}_q$ and $(c_1^2, c_2^2, \ldots, c_{b_i}^2)$ is the representation of $\alpha^{j_2} \in \mathbb{F}_q$ then $(c_1^1 + c_1^2, c_2^1 + c_2^2, \ldots, c_{b_i}^1 + c_{b_i}^2)$ is the representation of $\alpha^{j_1} + \alpha^{j_2}$. Now, one can verify easily that with this mapping an e-perfect byte-correcting code is mapped to an e-perfect mixed code. Given a perfect mixed code one can use a the inverse mapping to obtain a perfect byte-correcting code. In other words, this mapping transform an e-perfect mixed code into an e-perfect byte-correcting code and vice versa. It is highly possible, and we also conjecture that in all e-perfect mixed codes, $e \geq 1$, the alphabet in all coordinates are powers of the same prime, and if this conjecture is correct, then each mixed code can be translated into a byte-correcting code and the same is true for the related perfect codes.

The constructions in Section 7.1 can be used as constructions for 1-perfect byte-correcting codes, where all the bytes are of the same size b. In general, a basic construction is implied by a direct partition of all vectors of $\mathbb{F}_q^n$ (or equivalently, $\mathbb{F}_{q^n}$) into pairwise disjoint b-subspaces. Such a partition is possible if and only if b divides n. There are many nonequivalent such partitions and some of them will be presented in the rest of this section. We start with one simple construction.

Let α be a primitive element in $\mathbb{F}_{q^n}$, $n = \rho b$, and let $s = (q^{\rho b}-1)/(q^b-1)$. The element α^s is a primitive element in the subfield $\mathbb{F}_{q^b}$ whose elements are

$$\{0, \alpha^0, \alpha^s, \alpha^{2s}, \ldots, \alpha^{(q^b-2)s}\}.$$

This subfield is a subspace of dimension b and the elements $\alpha^0, \alpha^s, \alpha^{2s}, \ldots, \alpha^{(b-1)s}$ form a basis for this subfield. Since we also have that $\alpha^{i+x} + \alpha^{i+y} = \alpha^i(\alpha^x + \alpha^y)$, it follows that for each i, $0 \leq i \leq s-1$, the q^b elements of the set

$$\{0, \alpha^i, \alpha^{i+s}, \alpha^{i+2s}, \ldots, \alpha^{i+(q^b-2)s}\},$$

are closed under addition in $\mathbb{F}_{q^{\rho b}}$ and $\alpha^i, \alpha^{i+s}, \alpha^{i+2s}, \ldots, \alpha^{i+(b-1)s}$ are linearly independent. Therefore, the matrix

$$H = [H_0 \; H_1 \; H_2 \; \cdots \; H_{s-1}],$$

where $s = (q^{\rho b} - 1)/(q^b - 1)$ and

$$H_i = [\alpha^i, \alpha^{i+s}, \alpha^{i+2s}, \ldots, \alpha^{i+(b-1)s}], \quad 0 \leq i \leq s-1,$$

is a parity-check matrix for a 1-perfect byte-correcting code of length sb, redundancy ρb, and s bytes of size b. Clearly, the set

$$\{0, \alpha^i, \alpha^{i+s}, \alpha^{i+2s}, \ldots, \alpha^{i+(q^b-2)s}\}$$

can be further partitioned into disjoint linear subspaces of size $q^{b'}$ if and only if $q^{b'} - 1$ divides $q^b - 1$, i.e., b' divides b. This result can easily be generalized to any b-subspace of any n-space. The idea is summarized in the following results.

Lemma 7.2. *Any b-subspace can be partitioned into $s = (q^b - 1)/(q^{b'} - 1)$ disjoint b'-subspaces, for which b' divides b.*

Lemma 7.2 and condition (7.1) imply the following result.

Theorem 7.6. *Given s_i bytes of size b_i, $1 \leq i \leq m$, such that b_m divides r and for each i, $1 \leq i \leq m-1$, b_i divides b_{i+1}, a 1-perfect byte-correcting code with redundancy r exists if and only if*

$$\sum_{i=1}^{m} s_i(2^{b_i} - 1) = 2^r - 1. \tag{7.3}$$

Proof. If a 1-perfect byte-correcting code exists, then (7.3) is implied by (7.1).

If (7.3) is satisfied, then by Lemma 7.2, we have that $\mathbb{F}_{2^r}$ can be partitioned into $s = (2^r - 1)/(2^{b_m} - 1)$ subspaces of dimension b_m. Consider $s - s_m$ of these b_m-subspaces. Each one can be partitioned into $(2^{b_m} - 1)/(2^{b_{m-1}} - 1)$ subspaces of dimension b_{m-1}. This can be further iterated to obtain the required partition, which is a 1-perfect code since the partition allows us to take a basis for each part and therefore each syndrome is generated by exactly one of the bytes. □

Theorem 7.6 can be applied in many ways to form 1-perfect byte-correcting codes using many types of group partitions. Theorem 7.6 also provides a proof for the fact that the necessary condition (7.1) for the existence of 1-perfect byte-correcting codes of Type 4 is also sufficient if the redundancy of the code is also a power of 2. Next, we continue to consider the case when one byte is of size b_1 and the other bytes are of size b_2. We have already proved in Lemma 7.1 and Theorem 7.5 that a necessary condition for the existence of such a code with redundancy r is that b_2 divides $r - b_1$ and $b_1 > b_2$.

Construction 7.1. Let $b_1 > b_2$, α be a primitive element in $\mathbb{F}_{2^{\rho b_2}}$, β a primitive element in $\mathbb{F}_{2^{b_1}}$, and $s = (2^{\rho b_2} - 1)/(2^{b_2} - 1)$. Let H be the matrix defined by

$$H \triangleq [A\ B\ C],$$

where

(1) A is a $(\rho b_2 + b_1) \times b_1$ matrix defined by

$$A \triangleq \begin{bmatrix} I_{b_1} \\ \mathbf{0} \end{bmatrix},$$

where $\mathbf{0}$ is a $\rho b_2 \times b_1$ all-zero matrix.

(2) B is a $(\rho b_2 + b_1) \times (sb_2)$ matrix of the form

$$B \triangleq [B_0 \; B_1 \; \cdots \; B_{s-1}],$$

where B_i, $0 \le i \le s-1$, is a $(\rho b_2 + b_1) \times b_2$ matrix defined by

$$B_i \triangleq \begin{bmatrix} \mathbf{0} & \mathbf{0} & \cdots & \mathbf{0} \\ \alpha^i & \alpha^{i+s} & \cdots & \alpha^{i+(b_2-1)s} \end{bmatrix},$$

where $\mathbf{0}$ is the column vector of length b_1.

(3) C is a $(\rho b_2 + b_1) \times s(2^{b_1}-1)b_2$ matrix of the form

$$C \triangleq [C_0 \; C_1 \; \cdots \; C_{s(2^{b_1}-1)-1}],$$

where C_k, $k = js + i$, $0 \le j \le 2^{b_1} - 2$, and $0 \le i \le s-1$, is a $(\rho b_2 + b_1) \times b_2$ matrix defined by

$$C_k \triangleq \begin{bmatrix} \beta^j & \beta^{j+1} & \cdots & \beta^{j+b_2-1} \\ \alpha^i & \alpha^{i+s} & \cdots & \alpha^{i+(b_2-1)s} \end{bmatrix}.$$

Theorem 7.7. *Construction 7.1 generates a parity-check matrix for a 1-perfect byte-correcting code with redundancy $\rho b_2 + b_1$, one byte of size b_1 and the other bytes of size b_2.*

Proof. We have to show that each nonzero syndrome of length $\rho b_2 + b_1$ is produced by exactly one linear combination of columns from one byte of H. The syndromes produced from A are exactly all those vectors whose last ρb_2 entries are *zeros*. As described before, the linear span of the B_i's, without the zero element, form a partition of $\mathbb{F}_{2^{\rho b_2}}^{-}$. Hence, the syndromes produced from the linear combinations of the columns inside the bytes of B are distinct and are exactly all those column vectors with zeros in the first b_1 entries. It remains to show that each syndrome that is nonzero in the first b_1 entries and also nonzero in the last ρb_2 entries is produced by exactly one linear combination of the C_k's. Let v be such a syndrome, where $v^{\text{tr}} = ((\beta^{\ell_1})^{\text{tr}}, (\alpha^{\ell_2})^{\text{tr}})$. There exists a unique i, $0 \le i \le s-1$, and a unique linear combination such that

$$\alpha^{\ell_2} = \sum_{m=0}^{b_2-1} c_m \alpha^{i+ms}, \qquad c_m \in \{0,1\}.$$

Since $b_1 > b_2$, it follows that the set of elements $\{\beta^m : 0 \leq m \leq b_2 - 1\}$ are also linearly independent. Therefore,

$$\gamma = \sum_{m=0}^{b_2-1} c_m \beta^m \neq 0$$

is obtained by this unique linear combination of the elements in $\{\beta^m : 0 \leq m \leq b_2 - 1\}$. Clearly, there exists a unique j, $0 \leq j \leq 2^{b_1} - 2$, such that $\beta^{\ell_1} = \beta^j \gamma$, and hence v is obtained by a unique linear combination from C_{js+i}. □

By Lemma 7.1 and Theorems 7.5 and 7.7, we can classify the set of 1-perfect byte-correcting codes of Type 2.

Corollary 7.1. *A 1-perfect byte-correcting code with redundancy r, one byte of size b_1 and the other bytes of size b_2, exists if and only if b_2 divides $r - b_1$ and $b_1 > b_2$.*

Construction 7.2. Let H_1 be an $r \times n$ parity-check matrix of a 1-perfect byte-correcting code $\mathcal{C}_1$ with s_i bytes of size b_i, $1 \leq i \leq m$. Further, let α be a primitive element in $\mathbb{F}_{2^r}$ and b_{m+1} be a positive integer less than or equal to r (b_{m+1} is not necessarily distinct from the other b_i's). We define a matrix H_2 as follows:

$$H_2 \triangleq \begin{bmatrix} H_1 & \mathbf{0} & A_0 & A_1 & \cdots & A_{2^r-2} \\ \mathbf{0} & I_{b_{m+1}} & I_{b_{m+1}} & I_{b_{m+1}} & \cdots & I_{b_{m+1}} \end{bmatrix},$$

where $\mathbf{0}$ are all-zeroes matrices of the appropriate sizes and A_i, $0 \leq i \leq 2^r - 2$, is an $r \times b_{m+1}$ matrix defined by

$$A_i \triangleq [\alpha^i \; \alpha^{i+1} \; \cdots \; \alpha^{i+b_{m+1}-1}].$$

The following theorem is proved in very similar way to that done for Theorem 7.7.

Theorem 7.8. *The parity-check matrix of the codes obtained in Construction 7.2 is an $(r + b_{m+1}) \times (n + b_{m+1}2^r)$, $b_{m+1} \leq r$, parity-check matrix for a 1-perfect byte-correcting code with s_i bytes of size b_i, $1 \leq i \leq m$, and 2^r bytes of size b_{m+1}.*

Construction 7.2 can be further applied to obtain 1-perfect byte-correcting codes with various parameters. The parameters of the 1-perfect byte-correcting codes obtained in Construction 7.1 can also be obtained via Construction 7.2, but the presentation in Construction 7.1 is simpler and can be implemented easily. To obtain more 1-perfect byte-correcting codes, Lemma 7.2 can be used to replace a byte of size b by $(2^b-1)/(2^{b'}-1)$ bytes of size b' for any b' that divides b. Several other methods in which several bytes of size b_1 are replaced by several bytes of size b_2 to obtain 1-perfect byte-correcting codes with other parameters can be obtained in a similar way.

We continue to consider nonisomorphic linear 1-perfect byte-correcting codes. Two linear codes $\mathcal{C}_1$ and $\mathcal{C}_2$ are *isomorphic* if there exists a permutation π such that $\mathcal{C}_1 = \{\pi(c) \ : \ c \in \mathcal{C}_2\}$. As mentioned in Chapter 4, the linear 1-perfect code, i.e., the Hamming code, is a unique code (i.e., any two such linear 1-perfect codes of the same length are isomorphic). When 1-perfect byte-correcting codes are considered, some slight changes should be made in the definition of isomorphic codes. Such a permutation π for isomorphic codes can permute elements only within the same byte and can also permute between the bytes. For simplicity, we only give the formal definition for 1-perfect byte-correcting codes with bytes of size b.

Two linear 1-perfect byte-correcting codes $\mathcal{C}_1$ and $\mathcal{C}_2$ of length mb, with m bytes of size b, are called ***isomorphic*** if there exists such a permutation π for which $\mathcal{C}_1 = \{\pi(c) \ : \ c \in \mathcal{C}_2\}$ and $\pi = (\pi_0, \pi_1, \dots, \pi_{mb-1})$, where for each i, $0 \leq i \leq m-1$,

$$\{\pi_{ib}, \pi_{ib+1}, \dots, \pi_{ib+b-1}\} = \{jb, jb+1, \cdots, jb+b-1\}$$

for some j, $0 \leq j \leq m-1$. For the other types of byte-correcting codes the exact formal definition is slightly more complicated, but the idea is the same. We will now present for each redundancy ρb, $\rho \geq 3$, $b \geq 2$, parity-check matrices for two nonisomorphic 1-perfect byte-correcting codes $\mathcal{C}_1$ and $\mathcal{C}_2$ where all the bytes are of size b, and the redundancy of the codes is ρb.

Construction 7.3. For a primitive element $\alpha \in \mathbb{F}_{2^{(\rho-1)b}}$, construct the following four sets of $(\rho b) \times b$ matrices.

(1) The first set consists of one $\rho b \times b$ matrix

$$A \triangleq \begin{bmatrix} I_b \\ \mathbf{0} \end{bmatrix},$$

where $\mathbf{0}$ is a $(\rho-1)b \times b$ all-zero matrix.

(2) The second set consists of $s = (2^{(\rho-1)b} - 1)/(2^b - 1)$ matrices of the form

$$B_i = \begin{bmatrix} \mathbf{0} & \mathbf{0} & \cdots & \mathbf{0} \\ \alpha^i & \alpha^{i+s} & \cdots & \alpha^{i+(b-1)s} \end{bmatrix}, \qquad 0 \leq i \leq s-1 ,$$

where $\mathbf{0}$ is the all-zero column vector of length b.

(3) The third set consists of $2^{(\rho-1)b}-1$ matrices of the form

$$C_k = \begin{bmatrix} 1 & 0 & \cdots & 0 \\ 0 & 1 & \cdots & 0 \\ \vdots & \vdots & \vdots & \vdots \\ 0 & 0 & \cdots & 1 \\ \alpha^k & \alpha^{k+1} & \cdots & \alpha^{k+b-1} \end{bmatrix}, \quad 0 \le k \le 2^{(\rho-1)b}-2\,.$$

(4) The fourth set consists of $2^{(\rho-1)b}-1$ matrices of the form

$$D_k = \begin{bmatrix} 1 & 0 & \cdots & 0 \\ 0 & 1 & \cdots & 0 \\ \vdots & \vdots & \vdots & \vdots \\ 0 & 0 & \cdots & 1 \\ \alpha^{i+js} & \alpha^{i+(j+1)s} & \cdots & \alpha^{i+(j+b-1)s} \end{bmatrix},$$

where $0 \le i \le s-1$, $0 \le j \le 2^b-2$, and $k = js+i$.

Define $\mathcal{C}_1$ as the code whose parity-check matrix is

$$H_1 \triangleq \left[A\ B_0\ \cdots\ B_{s-1}\ C_0\ \cdots\ C_{2^{(\rho-1)b}-1}\right]$$

and $\mathcal{C}_2$ as the code whose parity-check matrix is

$$H_2 \triangleq \left[A\ B_0\ \cdots\ B_{s-1}\ D_0\ \cdots\ D_{2^{(\rho-1)b}-1}\right],$$

where each $\rho b \times b$ matrix defined in H_1 and H_2 corresponds to a distinct byte of size b.

Theorem 7.9. *The codes $\mathcal{C}_1$ and $\mathcal{C}_2$ defined in Construction 7.3 are nonisomorphic 1-perfect byte-correcting codes of length $b(2^{\rho b}-1)/(2^b-1)$, where all the bytes are of size b.*

We have constructed many types of 1-perfect mixed codes and 1-perfect byte-correcting codes. If the codes are linear, then e-perfect mixed codes are equivalent to e-perfect byte-correcting codes for any $e \geq 1$. Unfortunately, we are not aware on any linear e-perfect byte-correcting code, where $e > 1$ and the size of at least one byte is larger than one.

7.3 Codes with a Larger Radius and Mixed Steiner Systems

When the radius of the perfect code is larger than one, the only nontrivial perfect codes over $\mathbb{F}_q$, in the Hamming scheme, are the two Golay codes. The situation is slightly better for perfect mixed codes with radius larger than one. One infinite family of perfect codes with radius two is known. We start this section by describing this family of such perfect mixed codes whose radius is greater than one.

Recall that the Preparata code $\mathcal{P}(m)$ has length 2^m, where m is even and greater than 3. It has a minimum Hamming distance of 6 and its size is 2^{2^m-2m}. We now define a simple direct product construction for a 2-perfect mixed code over $\mathbb{F}_2^{2^m} \times \mathbb{Z}_{2^{m-1}}$.

Construction 7.4. Let $\mathcal{P}_0(m), \mathcal{P}_1(m), \ldots, \mathcal{P}_{2^{m-1}-1}(m)$ be the 2^{m-1} translates of the Preparata code of length 2^m whose union forms the extended Hamming code $\mathcal{H}^*(m)$ of length 2^m (see Definition 6.2). Let $\mathcal{M}(m)$ be the code over $\mathbb{F}_2^{2^m} \times \mathbb{Z}_{2^{m-1}}$ defined by

$$\mathcal{M}(m) \triangleq \{(x,i) \ : \ x \in \mathcal{P}_i(m), \ i \in \mathbb{Z}_{2^{m-1}}\} \ .$$

Theorem 7.10. *The code $\mathcal{M}(m)$ is a 2-perfect mixed code over $\mathbb{F}_2^{2^m} \times \mathbb{Z}_{2^{m-1}}$.*

Proof. We first claim that $\mathcal{M}(m)$ has a minimum Hamming distance of 5. Let (x,i) and (y,j) be two distinct codewords in $\mathcal{M}(m)$. If $i=j$, then x and y are two distinct words in the same translate $\mathcal{P}_i(m)$ of the Preparata code and hence $d((x,i),(y,j)) = d(x,y) \geq 6$. If $i \neq j$, then $x \neq y$ and since both x and y are codewords of $\mathcal{H}^*(m)$, it follows that $d((x,i),(y,j)) = d(x,y) + d(i,j) \geq 4+1 = 5$. This completes the proof of the first claim.

Since the number of codewords in $\mathcal{H}^*(m)$ is 2^{2^m-m-1}, it follows that also $\mathcal{M}(m)$ has 2^{2^m-m-1} codewords.

The size of a ball $\mathcal{B}_2(2^m+1)$ with radius 2 of a word over $\mathbb{F}_2^{2^m} \times \mathbb{Z}_{2^{m-1}}$ is

$$|\mathcal{B}_2(2^m+1)| = 1 + 2^m + 2^{m-1} - 1 + \binom{2^m}{2} + 2^m(2^{m-1}-1) = 2^{2m} \ .$$

The number of words in $\mathbb{F}_2^{2^m} \times Z_{2^{m-1}}$ is $2^{2^m}2^{m-1} = 2^{2^m+m-1}$ and hence

$$|\mathcal{M}(m)| \cdot |\mathcal{B}_2(2^m+1)| = 2^{2^m-m-1} \cdot 2^{2m} = 2^{2^m+m-1} = \left|\mathbb{F}_2^{2^m} \times \mathbb{Z}_{2^{m-1}}\right| ,$$

i.e., $\mathcal{M}(m)$ meets the sphere-packing bound.

Thus, $\mathcal{M}(m)$ is a 2-perfect mixed code. □

Are there any other e-perfect mixed codes with radius $e > 1$? No such code is known, but there are a few nonexistence results. The following lemma suggests one necessary condition for the existence of such a code. and a lower bound on its length.

Lemma 7.3. *If $\mathcal{C}$ is an e-perfect mixed over $\mathbb{Z}_{q_1} \times \mathbb{Z}_{q_2} \times \cdots \times \mathbb{Z}_{q_n}$, where $q_n > 2$, $q_{i+1} \geq q_i$, $1 \leq i \leq n-1$, then $n \geq e \cdot q_n + 1$.*

Proof. W.l.o.g. assume that the all-zero word is a codeword. This implies that the codewords of minimum weight have weight $2e+1$. Consider the word $u = (\overbrace{1 \cdots\cdots 1}^{e \text{ times}} \overbrace{0 \cdots\cdots 0}^{n-e-1 \text{ times}} \alpha)$, where $\alpha \in \mathbb{Z}_{q_n}$. This word of weight $e+1$ must be covered by a codeword whose weight is $2e+1$ that starts with e *ones* and ends with α. The other e nonzero entries of this codeword are in the middle of the codeword, in positions where u has *zeros*. Since $\mathcal{C}$ is an e-perfect code, it follows that for the two words $(\overbrace{1 \cdots\cdots 1}^{e \text{ times}} \overbrace{0 \cdots\cdots 0}^{n-e-1 \text{ times}} \alpha_1)$ and $(\overbrace{1 \cdots\cdots 1}^{e \text{ times}} \overbrace{0 \cdots\cdots 0}^{n-e-1 \text{ times}} \alpha_2)$, where $\alpha_1, \alpha_2 \in \mathbb{Z}_{q_n}$, $\alpha_1 \neq \alpha_2$, which are covered by two codewords c_1 and c_2, respectively, these $2e$ nonzero entries in the middle (e for each codeword) are in distinct positions. Since the last position in any word with the structure of u can be chosen in $q_n - 1$ distinct ways, it follows that $n - e - 1 \geq (q_n - 1)e$ and, therefore, $n \geq e \cdot q_n + 1$. $\square$

Corollary 7.2. *The code $\mathcal{M}(m)$ meets the lower bound of Lemma 7.3 on the length of a perfect mixed code.*

Are there more perfect mixed codes that meet the lower bound of Lemma 7.3 on the length of a perfect mixed code? Consider the following code. Let $\hat{\mathcal{H}}_i^*(m)$, $0 \leq i \leq 2^m - 1$, be the even cosets of the extended Hamming code $\mathcal{H}^*(m)$ and define the following code $\mathcal{K}(m)$ over $\mathbb{F}_2^{2^m} \times \mathbb{Z}_{2^m}$.

$$\mathcal{K}(m) \triangleq \{(x, i) \ : \ x \in \hat{\mathcal{H}}_i^*(m), \ i \in \mathbb{Z}_{2^m}\} .$$

Theorem 7.11. *The code $\mathcal{K}(m)$ is a 1-perfect mixed code over $\mathbb{F}_2^{2^m} \times \mathbb{Z}_{2^m}$ whose length is $2^m + 1$. It meets the lower bound of Lemma 7.3 on the length of a perfect mixed code.*

Proof. We first claim that $\mathcal{K}(m)$ has a minimum Hamming distance of 3. Let (x, i) and (y, j) be two distinct codewords in $\mathcal{K}(m)$. If $i = j$, then x and y are two distinct words in the same even cosets of $\mathcal{H}^*(m)$ and hence $d((x,i),(y,j)) = d(x,y) \geq 4$. If $i \neq j$, then $x \neq y$

and since both x and y are codewords of even weight, it follows that $d((x,i),(y,j)) = d(x,y) + d(i,j) \geq 2 + 1 = 3$. This completes the proof of the first claim.

Since the number of words with even weight in $\mathbb{F}_2^{2^m}$ is 2^{2^m-1}, it follows that also $\mathcal{K}(m)$ has 2^{2^m-1} codewords.

The size of a ball $\mathcal{B}_1(2^m+1)$ with radius one of a word over $\mathbb{F}_2^{2^m} \times \mathbb{Z}_{2^m}$ is

$$|\mathcal{B}_1(2^m+1)| = 1 + 2^m + 2^m - 1 = 2^{m+1} .$$

The number of words in $\mathbb{F}_2^{2^m} \times \mathbb{Z}_{2^m}$ is $2^{2^m} 2^m = 2^{2^m+m}$ and hence

$$|\mathcal{K}(m)| \cdot |\mathcal{B}_1(2^m+1)| = 2^{2^m-1} \cdot 2^{m+1} = 2^{2^m+m} = \left|\mathbb{F}_2^{2^m} \times \mathbb{Z}_{2^m}\right| ,$$

i.e., $\mathcal{K}(m)$ meets the sphere-packing bound.

Thus, $\mathcal{K}(m)$ is a 1-perfect mixed code. Finally, it is readily verified that the length of $\mathcal{K}(m)$ meets the lower bound of Lemma 7.3. □

The code-anticode bound (Corollary 2.15) can be applied to the Hamming metric on words over a mixed alphabet, as it does for the Hamming scheme, since the metric is distance invariant with addition as the binary operation. It is not difficult to find D-diameter perfect mixed codes, for each positive integer D. When increasing the length of a code $\mathcal{C}$, it is more difficult to construct such codes without considerably increasing the alphabet size in the new coordinates compared to the alphabet size of the coordinates in $\mathcal{C}$. We start with a construction for which the alphabet size is not dramatically increased. It is easily verified that the following extended code of $\mathcal{M}(m)$, $\mathcal{M}^*(m)$, over $\mathbb{F}_2^{2^m} \times \mathbb{Z}_{2^{m-1}}^2$,

$$\mathcal{M}^*(m) \triangleq \{(x,i,i) \ : \ x \in \mathcal{P}_i(m), \ i \in \mathbb{Z}_{2^{m-1}}\} ,$$

has a minimum Hamming distance of 6. Note that $\mathcal{B}_2(2^m+1)$, the ball with radius 2 in $\mathbb{F}_2^{2^m} \times \mathbb{Z}_{2^{m-1}}$, is also an anticode with diameter 4 and hence

$$\mathcal{A}_5(2^m+2) \triangleq \{(x,i) \ : \ x \in \mathcal{B}_2(2^m+1), \ i \in \mathbb{Z}_{2^{m-1}}\}$$

is an anticode with diameter 5 in $\mathbb{F}_2^{2^m} \times \mathbb{Z}_{2^{m-1}} \times \mathbb{Z}_{2^{m-1}}$. Hence, we have that

$$|\mathcal{M}^*(m)| \cdot |\mathcal{A}_5(2^m+2)| = 2^{2^m-m-1} 2^{2m} 2^{m-1}$$
$$= 2^{2^m+2m-2} = \left|\mathbb{F}_2^{2^m} \times \mathbb{Z}_{2^{m-1}} \times \mathbb{Z}_{2^{m-1}}\right| .$$

Thus, by the code-anticode bound, we have the following theorem.

Theorem 7.12. *The code $\mathcal{M}^*(m)$ is a 5-diameter perfect mixed code over $\mathbb{F}_2^{2^m} \times \mathbb{Z}_{2^{m-1}} \times \mathbb{Z}_{2^{m-1}}$.*

Do other diameter mixed perfect codes similar to $\mathcal{M}^*(m)$ whose minimum distance is 4 exist? The answer is that there exists at least one family of such codes which is the extended code of $\mathcal{K}(m)$, i.e.,

$$\mathcal{K}^*(m) \triangleq \{(x,i,i) \ : \ x \in \hat{\mathcal{H}}_i^*(m), \ i \in \mathbb{Z}_{2^m}\} \ .$$

Theorem 7.13. *The code $\mathcal{K}^*(m)$ is a 3-diameter perfect mixed code over $\mathbb{F}_2^{2^m} \times \mathbb{Z}_{2^m} \times \mathbb{Z}_{2^m}$.*

Proof. The proof that $\mathcal{K}^*(m)$ has minimum distance 4 is similar to the proof of Theorem 7.10. Note that $\mathcal{B}_1(2^m+1)$, the ball with radius 1 in $\mathbb{F}_2^{2^m} \times \mathbb{Z}_{2^m}$, is also an anticode with diameter 2 and hence

$$\mathcal{A}_3(2^m+2) \triangleq \{(x,i) \ : \ x \in \mathcal{B}_1(2^m+1), \ i \in \mathbb{Z}_{2^m}\}$$

is an anticode with diameter 3 in $\mathbb{F}_2^{2^m} \times \mathbb{Z}_{2^m} \times \mathbb{Z}_{2^m}$. Hence, we have that

$$|\mathcal{K}^*(m)| \cdot |\mathcal{A}_3(2^m+2)| = 2^{2^m-1}2^{m+1}2^m = 2^{2^m+2m} = \left|\mathbb{F}_2^{2^m} \times \mathbb{Z}_{2^m} \times \mathbb{Z}_{2^m}\right| .$$

Thus, by the code-anticode bound, the claim in the theorem follows. □

The following construction can be used to construct a D-diameter perfect mixed code for each positive integer D.

Construction 7.5. Let $\mathcal{C}$ be a D-diameter perfect mixed code over $Q = \mathbb{Z}_{q_1} \times \mathbb{Z}_{q_2} \times \cdots \times \mathbb{Z}_{q_n}$, with ℓ distinct codeword $c_0, c_1, \ldots, c_{\ell-1}$. Define the following code over $Q \times \mathbb{Z}_\ell$:

$$\mathcal{C}^* \triangleq \{(c_i,i) \ : \ c_i \in \mathcal{C}, \ i \in \mathbb{Z}_\ell\}.$$

Theorem 7.14. *The code $\mathcal{C}^*$ defined in Construction 7.5 is a $(D+1)$-diameter perfect mixed code over $Q \times \mathbb{Z}_\ell$.*

Proof. Let $\mathcal{A}$ be a maximum size anticode over Q whose diameter is D. Since $\mathcal{C}$ is a D-diameter perfect mixed code over Q, it follows that

$$|\mathcal{C}| \cdot |\mathcal{A}| = |Q| \ .$$

Define $\mathcal{A}^* \triangleq \{(a,i) \ : \ a \in \mathcal{A}, \ i \in \mathbb{Z}_\ell\}$, where $\ell = |\mathcal{C}|$. This implies that the diameter of $\mathcal{A}^*$ is $D+1$ and its size is $\ell \cdot |\mathcal{A}|$. Since $\mathcal{C}$ is a D-diameter perfect code, it follows that the minimum distance of $\mathcal{C}$ is $D+1$ and since each codeword of $\mathcal{C}$ was extended with a different symbol of $\mathbb{Z}_\ell$ to obtain $\mathcal{C}^*$, it follows that the minimum distance of $\mathcal{C}^*$ is $D+2$. Moreover, $|\mathcal{C}^*| = |\mathcal{C}|$, and hence

$$|\mathcal{C}^*| \cdot |\mathcal{A}^*| = |\mathcal{C}| \cdot \ell \cdot |\mathcal{A}| = \ell \cdot |Q| = |Q \times \mathbb{Z}_\ell| \ .$$

Thus, by the code-anticode bound we have that $\mathcal{C}^*$ is a $(D+1)$-diameter perfect mixed code over $Q \times \mathbb{Z}_\ell$. □

The idea in the constructions of the codes $\mathcal{M}^*(m)$ and $\mathcal{K}^*(m)$ can be used to generalize Construction 7.5. as follows.

Construction 7.6. Let $\mathcal{C}$ be a D-diameter perfect mixed code over $Q = \mathbb{Z}_{q_1} \times \mathbb{Z}_{q_2} \times \cdots \times \mathbb{Z}_{q_n}$, with ℓ distinct codeword $c_0, c_1, \ldots, c_{\ell-1}$. Define the following code over $Q \times \mathbb{Z}_\ell^\delta$, where $\delta \geq 1$.

$$\mathcal{C}^\delta \triangleq \{(c_i, \overbrace{i, i, \ldots, i}^{\delta \text{ times}}) \ : \ c_i \in \mathcal{C}, \ i \in \mathbb{Z}_\ell\}.$$

Theorem 7.15. *The code $\mathcal{C}^\delta$ which is defined in Construction 7.6 is a $(D+\delta)$-diameter perfect mixed code over $Q \times \mathbb{Z}_\ell^\delta$.*

Proof. Let $\mathcal{A}$ be a maximum size anticode over Q whose diameter is D. Since $\mathcal{C}$ is a D-diameter perfect mixed code over Q, it follows that

$$|\mathcal{C}| \cdot |\mathcal{A}| = |Q| \ .$$

Define $\mathcal{A}^\delta \triangleq \{(a, i_1, i_2, \ldots, i_\delta) \ : \ a \in \mathcal{A}, \ i_j \in \mathbb{Z}_\ell, \ 1 \leq j \leq \delta\}$, where $\ell = |\mathcal{C}|$. This implies that the diameter of $\mathcal{A}^*$ is $D+\delta$ and its size is $\ell^\delta \cdot |\mathcal{A}|$. Since $\mathcal{C}$ is a D-diameter perfect code, it follows that the minimum distance of $\mathcal{C}$ is $D+1$ and it is readily verified that the minimum distance of $\mathcal{C}^\delta$ is $D+1+\delta$. Moreover, $\left|\mathcal{C}^\delta\right| = |\mathcal{C}|$, and hence

$$\left|\mathcal{C}^\delta\right| \cdot \left|\mathcal{A}^\delta\right| = |\mathcal{C}| \cdot \ell^\delta \cdot |\mathcal{A}| = \ell^\delta \cdot |Q| = \left|Q \times \mathbb{Z}_\ell^\delta\right| \ .$$

Thus, by the code-anticode bound we have that $\mathcal{C}^\delta$ is a $(D+\delta)$-diameter perfect mixed code over $Q \times \mathbb{Z}_\ell^\delta$. □

All the diameter perfect codes that we have constructed are based on extensions of perfect codes and diameter perfect codes. This leads to the following research problem.

Problem 7.3. Construct diameter perfect mixed codes, with new parameters, which are not based on extensions of perfect codes and diameter perfect codes.

We continue with some nonexistence theorems for perfect mixed codes. The first two theorems can be obtained by using a generalization of Lloyd's polynomials.

Theorem 7.16. *If $\mathcal{C}$ is an e-perfect mixed code and the prime p divides the alphabet size in at least one of the coordinates, then p divides the size of the ball with radius e.*

Theorem 7.17. *If $\mathcal{C}$ is an e-perfect mixed code, of length n and the prime p divides the alphabet size of exactly t of the coordinates, then $e > n - t$.*

Corollary 7.3. *If $\mathcal{C}$ is a 1-perfect mixed code and the prime p divides the alphabet size of at least one coordinate, then p divides the alphabet size in all the coordinates.*

Proof. To satisfy the bound $e > n - t$ of Theorem 7.17, t must be equal to n and the claim follows. □

Corollary 7.4. *If $\mathcal{C}$ is an e-perfect mixed code and the prime p divides the alphabet size of at least one coordinate, then p divides the alphabet size in at least $n - e + 1$ coordinates.*

It is important to note that these results do not have any implications for the case where all coordinates are of divisible by the same primes. Theorems 7.16 and 7.17 imply many restrictions on the different alphabet sizes. For example, if $\mathcal{C}$ is a 1-perfect mixed code, then each prime p that divides the alphabet size on one of the coordinates also divides the alphabet size on each coordinate. For a 2-perfect mixed code, such a prime p must divide the size of each coordinate except maybe one. This considerably restricts the alphabet size for the coordinates in a perfect mixed code. There are other results that are implied by Lloyd's polynomials.

Theorem 7.18. *There is no 2-perfect mixed code, of length n, with alphabet of size q_i at coordinate i, $1 \leq i \leq n$, where q_i divides 6.*

Theorem 7.19. *There is no 3-perfect mixed code, of length n, with q_i the alphabet at coordinate i, $1 \leq i \leq n$, where $q_j = q > 2$ for $1 \leq j \leq n - 1$.*

For an e-perfect code over $\mathbb{Z}_q$, where the all-zero word is a codeword, the codewords of minimum weight form a (generalized) Steiner system. There is a similar property for perfect mixed codes.

Definition 7.1. A *mixed Steiner system* $\text{MS}(t, k, Q)$, over the mixed alphabet $Q = \mathbb{Z}_{q_1} \times \mathbb{Z}_{q_2} \times \cdots \times \mathbb{Z}_{q_n}$, is a pair $(Q, \mathcal{C})$, where $\mathcal{C}$ is a set of codewords of weight k, over Q, and for each word x of weight t over Q, there exists exactly one codeword $c \in \mathcal{C}$, such that c covers x, i.e., $d(x, c) = k - t$.

Lemma 7.4. *The number of codewords in a mixed Steiner system*

MS(t,k,Q), *over* $Q = \mathbb{Z}_{q_1} \times \mathbb{Z}_{q_2} \times \cdots \times \mathbb{Z}_{q_n}$, *is*

$$\left(\sum_{\substack{Y \subseteq [n] \\ |Y|=t}} \prod_{j \in Y} (q_j - 1) \right) \Big/ \binom{k}{t} .$$

Proof. Let $S = (Q, \mathcal{C})$ be a mixed Steiner system MS(t,k,Q). Each word of weight t must be covered by exactly one codeword of $\mathcal{C}$. For each t coordinates $i_1, i_2, \ldots, i_t$, $i_1 < i_2 < \cdots < i_t$ there are $\prod_{j=1}^{t}(q_{i_j} - 1)$ words of weight t over Q whose support is $\{i_1, i_2, \ldots, i_t\}$. Each of these $\prod_{j=1}^{t}(q_{i_j} - 1)$ words must be covered by exactly one codeword of $\mathcal{C}$. A codeword X of weight k covers exactly $\binom{k}{t}$ words of weight t and hence the claim of the lemma follows. □

Similarly to Lemma 3.4 we have the following result.

Lemma 7.5. *If* $(Q \times \mathbb{Z}_q, \mathcal{C})$ *is a mixed Steiner system* MS$(t,k,Q \times \mathbb{Z}_q)$, *where* $Q = \mathbb{Z}_{q_1} \times \mathbb{Z}_{q_2} \times \cdots \times \mathbb{Z}_{q_n}$, *then for each* $\alpha \in \mathbb{Z}_q^-$, *the set*

$$\mathcal{C}_\alpha \triangleq \{c \ : \ (c, \alpha) \in \mathcal{C}\}$$

is a mixed Steiner system MS$(t-1, k-1, Q)$.

Corollary 7.5. *A necessary condition for the existence of a mixed Steiner system* MS(t,k,Q), *where* $Q = \mathbb{Z}_{q_1} \times \mathbb{Z}_{q_2} \times \cdots \times \mathbb{Z}_{q_n}$, *is that for each* i, $0 \leq i \leq t-1$, *and each subset* X *of* $[n]$ *whose size is* $n - i$, *we have*

$$\sum_{\substack{Y \subseteq X \\ |Y|=t-i}} \prod_{j \in Y} (q_j - 1) \equiv 0 \left(\text{mod } \binom{k-i}{t-i} \right) .$$

Theorem 7.20. *The codewords of weight* $2e+1$ *of an* e*-perfect code* $\mathcal{C}$ *(which contains the all-zero word), over a mixed alphabet* $Q = \mathbb{Z}_{q_1} \times \mathbb{Z}_{q_2} \times \cdots \times \mathbb{Z}_{q_n}$ *form a mixed Steiner system* MS$(e+1, 2e+1, Q)$.

Proof. Clearly, the all-zero codeword covers all the words of weight at most e, over Q, and no word of weight larger than e. Hence, since the code $\mathcal{C}$ contains the all-zero codeword, it follows that it does not contain any codeword of weight between one and $2e$. Therefore, the words of weight $e+1$, over Q, must be covered by codewords from $\mathcal{C}$ of weight $2e+1$. Each of these words of weight $e+1$ must be covered by exactly one codeword of weight $2e+1$. This implies that the codewords of weight $2e+1$ in $\mathcal{C}$ form a mixed Steiner system MS$(e+1, 2e+1, Q)$. □

Corollary 7.6. *If $\mathcal{C}$ is an e-perfect mixed code of length n, over $Q = \mathbb{Z}_{q_1} \times \mathbb{Z}_{q_2} \times \cdots \times \mathbb{Z}_{q_n}$, where q_i, $1 \le i \le n$, is the alphabet size at coordinate i, then for each t, $1 \le t \le e+1$ and each subset X of $[n]$ whose size is $n-e+t-1$, we have*

$$\sum_{\substack{Y \subseteq X \\ |Y|=t}} \prod_{j \in Y} (q_j - 1) \equiv 0 \left(\text{mod } \binom{e+t}{t} \right).$$

Corollary 7.7. *If $\mathcal{C}$ is an e-perfect mixed code of length n, where q_i is the alphabet size at coordinate i and $q_r \le q_s$ for each $1 \le r < s \le n$, then $e+1$ divides $q_r - q_s$ for each $1 \le r < s \le n$.*

Proof. Choose $t = 1$ in Corollary 7.6. Since $0 < e < n$, it follows that there exist two subsets X_1 and X_2 of $\{1, 2, \ldots, n\}$ whose size is $n - e$ such that $X_1 \setminus X_2 = \{r\}$ and $X_2 \setminus X_1 = \{s\}$. By Corollary 7.6, we have that

$$\sum_{j \in X_1} (q_j - 1) \equiv 0 \text{ (mod } e+1), \tag{7.4}$$

$$\sum_{j \in X_2} (q_j - 1) \equiv 0 \text{ (mod } e+1). \tag{7.5}$$

By subtracting (7.4) from (7.5), the claim is obtained. □

Corollary 7.8. *If $\mathcal{C}$ is an e-perfect mixed code of length n, where q_i is the alphabet size at coordinate i, then $e \le \min\{q_t - q_s \ : \ s < t,\ q_s < q_t\} - 1$.*

The next theorem asserts that similar 3-perfect mixed codes, such as $\mathcal{M}(m)$, do not exist.

Theorem 7.21. *There is no 3-perfect mixed code over $\mathbb{F}_2^{n-1} \times \mathbb{Z}_q$, where $n > 3$ and $q > 2$.*

Proof. Assume $\mathcal{C}$ is a 3-perfect mixed code over $\mathbb{F}_2^{n-1} \times \mathbb{Z}_q$, $q > 2$. By Corollary 7.4, we have that if p is a prime which divides q, then p divides at least $n-2$ of the coordinates. Since $n > 3$ and only 2 divides the alphabet size in $n-1$ of the coordinates, it follows that q must be a power of 2, i.e., $q = 2^\ell$ for $\ell > 1$. Corollary 7.7 implies that 4 divides $2^\ell - 2$, which is clearly impossible. □

Theorem 7.21 can be generalized in a naive way for other e-perfect mixed codes over $\mathbb{F}_2^{n-1} \times \mathbb{Z}_q$.

Problem 7.4. Prove that there is no e-perfect mixed code over $\mathbb{F}_2^{n-1} \times \mathbb{Z}_q$, for $n > 2$ and $q > 2$.

All the necessary conditions that are implied by the results obtained so far rule out most parameters for e-perfect mixed codes, where $e > 1$. There, however, are still some open problems.

Problem 7.5. Find parameters where the necessary conditions (given in this section) for the existence of perfect mixed codes are satisfied. For which of these parameters it can be proved that there is no associated perfect mixed code?

Corollary 7.6 can serve as a motivation to consider mixed Steiner systems for certain alphabet sizes. To complete our discussion in this section we add a few words on mixed Steiner systems. Clearly, each 1-perfect mixed code, over a mixed alphabet Q, constructed in Section 7.1 yields a mixed Steiner system $\text{MS}(2,3,Q)$. These systems are less interesting and they can be constructed easily from 1-perfect mixed codes. The more interesting systems are mixed Steiner systems $\text{MS}(t,k,Q)$, where $t > 2$ and especially those for which $k-t > 1$ or $t > 3$. One such family is derived from the codewords having a minimum weight of 5 in the mixed perfect code $\mathcal{M}(m)$, which form a mixed Steiner system $\text{MS}(3,5,\mathbb{F}_2^{2^m} \times Z_{2^m-1})$. It is interesting to note that in this mixed Steiner system there is no codeword which is an element of $\mathbb{F}_2^{2^m} \times \{0\}$. The following construction can be used to form such systems and also other mixed Steiner systems.

Construction 7.7. Let $\{S_1, S_2, \ldots, S_r\}$ be a set of r pairwise disjoint mixed Steiner systems $\text{MS}(t-1,k,Q)$, where $\bigcup_{i=1}^{r} S_i$ is a mixed Steiner system $\text{MS}(t,k,Q)$. Let $\hat{\mathcal{M}}$ be the system defined on $Q \times [r]$ as follows

$$\hat{\mathcal{M}} \triangleq \{(c,i) \ : \ c \in S_i, \ 1 \leq i \leq r\}.$$

The claim of the following theorem can be verified easily.

Theorem 7.22. *The set $\hat{\mathcal{M}}$ obtained in Construction 7.7 is a mixed Steiner system* $\text{MS}(t,k+1,Q \times \mathbb{Z}_{r+1})$.

The codewords of minimum weight in the code $\mathcal{M}(m)$, obtained in Construction 7.4, can also be obtained via Construction 7.7 by considering the codewords of weight four in the even translate of the Preparata code of length 2^m whose union forms the Hamming code. In each such translate, which is not the Preparata code itself (in which there are no codewords of weight four), the codewords of weight four form a Steiner system $\text{S}(2,4,2^m)$, and the union of these codewords forms the Steiner system $\text{S}(3,4,2^m)$. By

Theorem 7.22, we have that in this case Construction 7.7 yields a mixed Steiner system $\text{MS}(3, 5, \mathbb{F}_2^{2^m} \times \mathbb{Z}_{2^m-1})$.

Problem 7.6. Construct new mixed Steiner systems $\text{MS}(t, k, Q)$, where $t > 2$ and especially when $k - t > 1$ or $t > 3$.

Problem 7.7. Make a comprehensive exposition on properties and bounds of mixed Steiner systems.

7.4 Notes

Codes with mixed binary and ternary alphabets are important in the context of football pools, already discussed in Section 4.4. ***Football pools*** are a practical game that is played around the world for the purpose of winning money by correctly guessing of results of football games in advance. It can be readily generalized to any two-team (player) game and also to games with more participants.

Assume that there are n football matches and in each one two teams A and B play. The outcome of the match can be a win for A, or a win for B, or a draw. Therefore, the total number of possible outcomes of the n matches is 3^n. In the football pools, one has to guess the results of the n matches. One makes a prediction about the results of the n matches and each such prediction, of the results of the n matches, is a bet that costs a certain amount of money. Each such bet can be represented by a ternary word of length n. If in one of the bets there are at least $n - e$ correct guesses, where e is usually a very small integer, then there is a prize (in real life, money) for this bet. Of course, by having 3^n bets one can correctly guess all the results of the n matches, but this will cost a large amount of money that will not be covered by the prize given for such a correct bet. Therefore, one wants to guarantee at least $n - e$ correct guesses. This is a covering problem in which we want to find the smallest subset S of the set of 3^n ternary words, $\mathbb{F}_3^n$, such that each element of $\mathbb{F}_3^n$ is within Hamming distance e from at least one element of S. If S is not an e-perfect code, then there are some possible results that are within distance e from more than one element of S. This implies that there are some redundant bets if the target is only to guarantee at least $n - e$ correct guesses. To avoid this redundancy, i.e., to save some money, one has to use a perfect code to make the bet.

If the target is to guarantee at least $n - 1$ correct guesses in one of the bets, then the solution will be to use a ternary 1-perfect code. Such

1-perfect code exists for $n = \frac{3^r-1}{2}$, $r > 1$, i.e., 4 matches, 13 matches, 40 matches, 121 matches, and so on. Of course, no one bets on 121 matches and even 40 matches seems to be unlikely. If the target will be to guarantee at least $n-2$ correct guesses, then a ternary 2-perfect code should be used. The ternary Golay code $\mathcal{G}_{11}$ is such a code with 729 codewords. The total number of words in $\mathbb{F}_3^{11}$ is 177,147, so 729 is a small fraction (one bet for each 243 words of the space; one bet for each ball with radius 2).

This idea can be further generalized if, for example, we consider only two possible results for each game. This can occur if we are betting on games that have only two possible results or in football pools if we are confident that one outcome is impossible. In such cases (there are others too), the 1-perfect code in the Hamming scheme is the solution and the feasible parameters might be for $n = 7$, $n = 15$, or $n = 31$. For example, if there are 15 matches, then there are 32,768 possible bets, from which we can use only 2048 bets to guarantee 14 correct guesses, and hopefully the last one will also be in our favor. A straightforward generalization is to have a larger alphabet when each game has more than three possible outcomes. The last possible generalization is when in some matches we are confident that one of the outcomes is impossible, while in the other matches we still consider the three possible outcomes. In this case the solution is based on a mixed alphabet, where some coordinates have a binary alphabet and some have a ternary alphabet. Unfortunately, by Corollary 7.7, such a 1-perfect code does not exist in this case and hence one has to settle for a good covering code for this purpose.

Section 7.1. The construction of Theorem 7.4 was presented in [Heden (1977)]. The other constructions in this section are straightforward generalizations of the constructions of nonlinear 1-perfect codes. This representation of perfect codes as suggested in Theorem 7.1 and Theorem 7.2 has been reproduced several times, e.g., [Zaremba (1950, 1952)]. Some other constructions for perfect mixed codes and group partitions can be found in [Schönheim (1970); Herzog and Schönheim (1971); Herzog and Schöönheim (1972); Lindström (1975b)].

Section 7.2. Information on error-correcting and error-detecting codes in memories where the information is organized in bytes can be found in [Chen (1983, 1986)] and in the book by [Rao and Fujiwara (1989)]. The results in this section are contained in [Etzion (1998)]. There are many other results on space partitioning. Construction 7.1 is similar to the one in [Hong and Patel (1972)], but the representation given in this section of Chapter 7 is much simpler. Construction 7.3 was presented in [Etzion (1998)] and the omitted proof of Theorem 7.9 was also presented in [Etzion (1998)].

A ***partial k-spread*** of $\mathbb{F}_q^n$ is a set of pairwise disjoint k-subspaces of $\mathbb{F}_q^n$. A ***k-spread*** of $\mathbb{F}_q^n$ is a partial k-spread in which each element of $\mathbb{F}_q^n$ is contained in exactly one of the k-subspaces. A construction of a k-spread of $\mathbb{F}_q^n$ is equivalent to a construction of a 1-perfect byte-correcting code, where the size of a byte is equal to k. A partial k-spread can always be completed to form a 1-perfect byte-correcting code, with bytes of size k for all the k-subspaces of the partial k-spread. The other elements that are not contained in these k-subspaces are partitioned into subspaces of various sizes (in the worse case, all of them are of dimension one). k-spreads have many applications in various problems in coding theory, some of which will be further discussed in Chapter 10. Spreads together with other concepts in projective geometry are used to construct optimal codes, e.g. [Hamada and Tamari (1982)]. Spreads are also considered as combinatorial designs and they have also applications in modern technologies of the 21st century, e.g. [Chee, Etzion, Kiah, and Vardy (2018); Zhang, Etzion, and Yaakobi (2020)].

This problem of partitioning $\mathbb{F}_q^n$ into subspaces has been the subject of extensive research. Some of the work in this direction was done in [El-Zanati, Seelinger, Sissokho, Spence and Vanden Eynden (2007); Blinco, El-Zanati, Seelinger, Sissokho, Spence and Vanden Eynden (2008); El-Zanati, Jordon, Seelinger, Sissokho, Spence (2008); El-Zanati, Seelinger, Sissokho, Spence and Vanden-Eynden (2009); Khare (2009); Heden (2009a,b); El-Zanati, Heden, Seelinger, Sissokho, Spence and Vanden Eynden (2010); Seelinger, Sissokho, Spence and Vanden Eynden (2012a,b)].

Section 7.3. The perfect mixed code with radius two was introduced in [Etzion and Greenberg (1993)]. Theorems 7.16 and 7.17 were proved in [Heden (1975)]. Theorems 7.18 and 7.19 were proved by [Reuvers (1977)]. Corollary 7.6 was proved by [van Wee (1991)]. It was generalized later to give a more expanded version for necessary conditions for the existence of perfect mixed codes in [Perkins, Sakhnovich, and Smith (2006)]. This paper contains also references to other papers that contain bounds on the sizes of error-correcting mixed codes. The new proof, for Corollary 7.6, using the new concept of mixed Steiner systems that we give in this section of Chapter 7 is much simpler. Corollaries 7.7 and 7.8 were proved in [van Wee (1991)]. Theorem 7.21 was proved in [Reuvers (1977)] and the proof was shortened by [van Wee (1991)].

In [Teirlinck (1994)], a construction for Steiner systems $S(3,4,n)$, which can be partitioned into Steiner systems $S(2,4,n)$, was presented. Such a construction was presented for each $n = 2 \cdot 7^m + 2$ and each $n = 2 \cdot 31^m + 2$, whenever $m \geq 1$. This Steiner system $S(3,4,n)$ is called 2-resolvable. Resolutions in block design, and in particular for Steiner systems, are interesting and intriguing.

Problem 7.8. Find new constructions for new parameters of 2-resolvable $S(3,4,n)$.

Chapter 8

Binary Constant-Weight Codes

Constant-weight codes have drawn lot of interest in coding theory during the years. The reason is that they are used to construct general codes and also since upper bounds on the size of constant-weight codes imply upper bounds on the size of general codes. These codes have also found applications in modern technologies and hence the interest in them has become more intensive. This chapter is concerned with the existence and constructions of binary perfect constant-weight codes. These codes are related to the Johnson scheme. It is conjectured that there are no nontrivial perfect codes in this scheme, but this conjecture is far from settled. A large part of this chapter is devoted to various techniques to prove this conjecture. We will use the Johnson scheme for a comprehensive demonstration how to rule out the possible existence of perfect codes, something which was done and will be done sporadically for other metrics.

There are three different directions in the nonexistence proofs – excluding graphs in which there are no perfect codes, excluding radii for which such codes cannot exist, and finding a tradeoff between the various parameters of possible perfect codes. These directions are presented in Sections 8.1 through 8.7. Concepts and techniques from block design and in particular Steiner systems play an important role in these directions. In Section 8.8, we discuss diameter perfect codes in the Johnson scheme and again Steiner systems have an important role in this discussion.

8.1 The Johnson Scheme

Constant-weight codes are related to the Johnson scheme. The Johnson scheme $\mathrm{J}(n, w)$ is the most important scheme after the Hamming scheme. In a Johnson scheme $\mathrm{J}(n, w)$, we are given two integers, n and w, such that $0 \leq w \leq n$. In a code $\mathcal{C}$, all the codewords are binary words having length n and constant weight w. Two words u and v are at ***Johnson distance*** (J-distance in short) d apart if there are exactly d positions in which u has *ones* and v has *zeroes*. Obviously, there are exactly d other positions in which u has *zeroes* and v has *ones*. This implies the following simple result.

Lemma 8.1. *The J-distance of two words is exactly half of their Hamming distance.*

Corollary 8.1. *A code $\mathcal{C}$ in the Johnson scheme has minimum J-distance δ if and only if its minimum Hamming distance is 2δ.*

Lemma 8.1 and Corollary 8.1 induce a close connection between the Hamming scheme and the Johnson scheme, which is a subset of the Hamming scheme. Nevertheless, the definition of the graph $\mathrm{J}(n, w)$ must be done using the Johnson distance, since the graph will not be connected when using the Hamming distance (there are no words whose Hamming distance is one). A binary word of length n and weight w can be represented by its support and, by using this representation, the elements of the Johnson scheme are w-subsets of an n-set instead of binary words of length n and weight w. We associate the Johnson graph $\mathrm{J}(n, w)$ with the Johnson scheme. The vertex set, V_w^n, of the Johnson graph consists of all w-subsets of a fixed n-set. Two such w-subsets are adjacent (connected by an edge) if and only if their intersection has size $w-1$. With this representation in hand, the J-distance between the two words x and y of weight w becomes

$$d_J(x, y) \triangleq |\mathrm{supp}(x) \setminus \mathrm{supp}(y)| = |\mathrm{supp}(y) \setminus \mathrm{supp}(x)|\ ,$$

where $\mathrm{supp}(x)$ and $\mathrm{supp}(y)$ are the related w-subsets. Throughout this chapter we will use both representations for the Johnson scheme. In some cases the proof will have a mixed language of vector notation and set notation. The representation that will be used should be understood from the context and the translation between the two representations should be clear. It is easy to verify that the Johnson distance defines an association scheme. It is not trivial to compute the intersection numbers $p_{i,j}^{\ell}$, but it is trivial to verify that they do not depend on the pair of vertices x, y for which $d(x, y) = \ell$. A code $\mathcal{C}$ of such w-subsets is called an e-perfect code in $\mathrm{J}(n, w)$ (or in the Johnson scheme) if the e-balls of all the codewords of $\mathcal{C}$ form a partition of V_w^n. In other words, $\mathcal{C}$ is an e-perfect code if for each element $v \in V_w^n$, there exists a unique element $c \in \mathcal{C}$, such that the J-distance between v and c is less than or equal to e. Except for the usual trivial perfect codes, there is another family of trivial perfect codes in $\mathrm{J}(n,w)$. If $w = 2e+1$ and $n = 2w$, then any two complement words of weight w form an e-perfect code in $\mathrm{J}(2w,w)$. In this chapter we will try to figure out if there are other perfect codes in the Johnson scheme. Given Lemma 8.1 and Corollary 8.1, we have the following result.

Corollary 8.2. *The Hamming distance of an e-perfect code in the Johnson scheme is* $4e+2$.

We start by noting that generalizations of Lloyd polynomials mentioned in Section 5.8 do not lead to significant nonexistence results and hence other techniques will be used. The Johnson's ball of radius e is given in the following lemma whose proof is straightforward.

Lemma 8.2. *The size of an e-ball in* $\mathrm{J}(n,w)$, $\mathcal{B}_e(n,w)$, *is*

$$|\mathcal{B}_e(n,w)| = \sum_{i=0}^{e} \binom{w}{i}\binom{n-w}{i}.$$

Hence, by the sphere-packing bound, the number of codewords of an e-perfect code $\mathcal{C}$ in $\mathrm{J}(n,w)$ is

$$|\mathcal{C}| = \frac{\binom{n}{w}}{|\mathcal{B}_e(n,w)|}$$

and, therefore,

$$|\mathcal{B}_e(n,w)| \;\Big|\; \binom{n}{w}. \tag{8.1}$$

We may, however, do much better than this, as will be demonstrated in Section 8.5.

Since $d_H(\bar{x},\bar{y}) = d_H(x,y)$ for any two words $x,y \in \mathbb{F}_2^n$, the following theorem follows from Corollary 8.1.

Theorem 8.1. *The code* $\mathcal{C}$ *is an e-perfect code in* $\mathrm{J}(n,w)$ *if and only if* $\bar{\mathcal{C}}$ *is an e-perfect code in* $\mathrm{J}(n,n-w)$.

In view of Theorem 8.1, we only have to consider perfect codes in $\mathrm{J}(n,w)$, where $2w \leq n$. Hence, in the following sections, when the nonexistence of perfect codes in $\mathrm{J}(n,w)$ will be considered, we will assume that $n \geq 2w$ in $\mathrm{J}(n,w)$, unless otherwise is stated.

For an e-perfect code $\mathcal{C}$ in $\mathrm{J}(n,w)$, we say that $u \in \mathcal{C}$ *J-cover* $v \in V_w^n$ if the J-distance between u and v is less than or equal to e. For a given two subsets u and v, we say that u *C-cover* v if v is a subset of u (containment).

8.2 Configuration Distribution

The weight distribution of a code $\mathcal{C}$ is an important tool in the Hamming scheme to obtain many interesting results on the code $\mathcal{C}$. Related to the weight distribution is the distance distribution. These concepts were considered for perfect codes in the Hamming scheme (see Section 5.2). If the code is a linear code or a 1-perfect code in the Hamming scheme, then these concepts coincide (see Theorem 2.7 and Theorem 5.5). For the Johnson scheme, all the codewords have the same weight. When one codeword is considered to be the zero codeword, a definition for the weight distribution is an immediate consequence of the distance of the codewords from the zero codeword. On the other hand, it is straightforward to define the distance distribution of the code. The definition of the weight distribution by these observations can be generalized to the concept of configuration distribution, which is defined next. This definition will enable us to obtain many nonexistence results on e-perfect codes in the Johnson scheme.

Definition 8.1. Assume $\mathcal{C}$ is a code in $\mathrm{J}(n,w)$ and let $\mathcal{N}$ be the set of n coordinates of the Johnson scheme. Assume further that $\mathcal{N}$ is partitioned into two subsets (called ***parts***) $\mathcal{A}$ and $\mathcal{B}$ such that $|\mathcal{A}| = r$ and $|\mathcal{B}| = n-r$. A word x of $\mathrm{J}(n,w)$ is in ***configuration*** (i,j), where $i+j=w$ if $|x \cap \mathcal{A}| = i$ and $|x \cap \mathcal{B}| = j$.

Definition 8.2. Assume $\mathcal{C}$ is a code in $\mathrm{J}(n,w)$ and let $\mathcal{N}$ be the set of n coordinates of the Johnson scheme. Assume further that $\mathcal{N}$ is partitioned into two subsets (called ***parts***) $\mathcal{A}$ and $\mathcal{B}$ such that $|\mathcal{A}| = w$ and $|\mathcal{B}| = n-w$. Let $\{D_{(i,j)} \ : \ 0 \leq i,j, \ i+j=w\}$ denote the ***configuration distribution*** of the code, i.e., $D_{(i,j)}$ denote the number of codewords from configuration (i,j).

The distinction between Definition 8.1 and Definition 8.2 is that only for the partition of the coordinate set into $\mathcal{A}$ and $\mathcal{B}$, such that $|\mathcal{A}| = w$ and $|\mathcal{B}| = n-w$, a configuration distribution is defined in this section. Indeed, such a definition can be given to any partition as defined in Definition 8.1, but such a configuration (distance) distribution is not used in this chapter (although different partitions are defined in the chapter). Other configuration distributions will be discussed in Section 8.5. Moreover, one might infer some interesting results based on different distance distributions associated with partition of the coordinate set into parts of different sizes as defined in Definition 8.1. Different partitions can lead to different results.

Examples for this claim are demonstrated in Theorems 8.6 and 8.7 which follow.

Theorem 8.2. *There are exactly* $e+1$ *possible different configuration distributions for an e-perfect code in* $\mathrm{J}(n,w)$*. If* CD_k*,* $0 \leq k \leq e$*, is the set of the k-th configuration distribution, then* CD_k *contains* $D_{(w-k,k)}$ *and* $D_{(w-2e-1+k,2e+1-k)}$ *as the only nonzero elements among* $D_{(w-i,i)}$*,* $0 \leq i \leq 2e+1-k$*.*

Proof. Assume $\mathcal{C}$ is an e-perfect code in $\mathrm{J}(n,w)$ and the coordinate set $\mathcal{N}$ is partitioned into two parts $\mathcal{A}$ and $\mathcal{B}$ such that $|\mathcal{A}| = w$ and $|\mathcal{B}| = n - w$. Let k be the smallest integer such that $\mathcal{C}$ has a codeword from configuration $(w-k,k)$. Since the word from configuration $(w,0)$ must be J-covered by $\mathcal{C}$, it follows that $0 \leq k \leq e$. Since by Corollary 8.1 the minimum Hamming distance of $\mathcal{C}$ is $4e+2$, it follows that there is exactly one codeword from configuration $(w-k,k)$ and no codeword from any configuration $(w-j,j)$, $k+1 \leq j \leq 2e-k$. The codeword from configuration $(w-k,k)$ J-covers all words from configurations $(w-i,i)$ for all i, $0 \leq i \leq e-k$. Some words from configuration $(w-e+k-1, e-k+1)$, $k>0$, are J-covered by the codewords from configuration $(w-k,k)$ and the other words, which are the majority, can be J-covered only by codewords from configuration $(w-2e-1+k, 2e+1-k)$. Note that we can always partition the coordinate set $\mathcal{N}$ into two subsets $\mathcal{A}$ and $\mathcal{B}$ such that the first codeword will have $w-k$ *ones* in the $\mathcal{A}$-part and k *ones* in the $\mathcal{B}$-part, and hence w.l.o.g. $\mathcal{C}$ contains a codeword from configuration $(w-k,k)$. To complete the proof, it is sufficient to show that once we are given k, $0 \leq k \leq e$, such that a word from configuration $(w-k,k)$ is a codeword, i.e., $D_{(w-k,k)} = 1$, $D_{(w-i,i)} = 0$, $0 \leq i \leq 2e-k$, $i \neq k$, then the configuration distribution is determined. The proof is by induction; assume we have determined all the values $D_{(w-i,i)}$, $0 \leq i \leq r$, for some r, $r \geq 2e-k$, and all words from configurations $(w-j,j)$, $0 \leq j \leq r-e$, are J-covered by codewords from configurations $(w-i,i)$, $0 \leq i \leq r$. To evaluate $D_{(w-r-1,r+1)}$, notice that by considering how words from configuration $(w-r+e-1, r-e+1)$ are J-covered, we have

$$\binom{w}{r-e+1}\binom{n-w}{r-e+1} = \sum_{i=r-2e+1}^{r+1} C_{(w-i,i)}^{(w-r+e-1,r-e+1)} \cdot D_{(w-i,i)}$$

where $C_{(x_1,y_1)}^{(x_2,y_2)}$ is the number of words from configuration (x_2, y_2) that are

J-covered by a codeword from configuration (x_1, y_1). Hence we have

$$D_{(w-r-1,r+1)} = \frac{\left[\binom{w}{r-e+1}\binom{n-w}{r-e+1} - \sum_{i=r-2e+1}^{r} C_{(w-i,i)}^{(w-r+e-1,r-e+1)} \cdot D_{(w-i,i)}\right]}{C_{(w-r-1,r+1)}^{(w-r+e-1,r-e+1)}}$$

and, therefore, $D_{(w-r-1,r+1)}$ is determined and all words from configurations $(w-j, j)$, $0 \leq j \leq r-e+1$, are J-covered by codewords from configurations $(w-i, i)$, $0 \leq i \leq r+1$.

Thus, since k has exactly one of the values between 0 and e, it follows that there are exactly $e+1$ possible different configuration distributions for e-perfect codes. □

Remark 8.1. To convert the configuration distribution into a distance distribution we have to require that the word from configuration $(w, 0)$ is a codeword. This can easily be done by permuting coordinates between the $\mathcal{A}$-part and the $\mathcal{B}$-part. By Theorem 8.2 there will be exactly one configuration distribution in this case. This implies that similarly to the Hamming scheme, we have the following property on the the distance distribution of an e-perfect code in $\mathrm{J}(n, w)$.

Corollary 8.3. *The distance distribution $\{D_i : 0 \leq i \leq w\}$ of an e-perfect code in $\mathrm{J}(n, w)$ does not depend on the e-perfect code. It is derived from CD_0 as defined in Theorem 8.2, where $D_i = D_{(w-i,i)}$ for $0 \leq i \leq w$.*

Proof. The proof is derived from the unique configuration distribution when $D_{(w,0)} = 1$, as proved in Theorem 8.2, and the simple observation that each codeword can be chosen as the unique codeword from configuration $(w, 0)$. □

Lemma 8.3. *If $\mathcal{C}$ is an e-perfect code in $\mathrm{J}(2w+e+1, w)$, then the intersection between any two codewords of $\mathcal{C}$ is at least of size e.*

Proof. Assume $\mathcal{C}$ is an e-perfect code in $\mathrm{J}(2w+e+1, w)$ and the coordinate set $\mathcal{N}$ is partitioned into two parts $\mathcal{A}$ and $\mathcal{B}$ such that $|\mathcal{A}| = w$ and $|\mathcal{B}| = w+e+1$, and $\mathcal{C}$ contains the word c, from configuration $(0, w)$, which ends with $e+1$ *zeroes* (in the $\mathcal{B}$-part), as a codeword. The only words from configuration $(0, w)$ that are not J-covered by c are the $\binom{w}{e+1}$ words ending with $e+1$ *ones*. Since $\mathcal{C}$ is an e-perfect code and hence any codeword that J-covers some of these $\binom{w}{e+1}$ words should have J-distance at least $2e+1$ from c, it follows that these words are J-covered by codewords from configuration $(e, w-e)$, which end with $e+1$ *ones*. Moreover, note that since $D_{(0,w)} = 1$, it follows that the entire configuration distribution of $\mathcal{C}$ is determined (as was proved in Theorem 8.2). Now, by Theorem 8.2, for this configuration distribution we have that $D_{(w-k,k)} = 1$ for exactly one k, $0 \le k \le e$, and for $i \ne k$, $0 \le i \le 2e-k$, $D_{(w-i,i)} = 0$. If $k < e$, we can exchange one point from $\mathcal{A}$, with a *one* in the codeword from configuration $(w-k,k)$, with a point from $\mathcal{B}$, with *zeroes* in the codeword from configurations $(w-k,k)$ and the codeword from configuration $(0,w)$. This is possible since the codeword c from configuration $(0,w)$ has $e+1$ *zeroes* in the $\mathcal{B}$-part and the codeword from configuration $(w-k,k)$ has at most k *ones*, where $k < e$, in the $e+1$ positions of the *zeroes*. The obtained e-perfect code $\mathcal{C}'$ has $D_{(0,w)} = 1$ and $D_{(w-k-1,k+1)} = 1$, $k+1 \le e$, contradicting the fact that by Theorem 8.2 we have that $D_{(0,w)} = 1$ determines all the configuration distribution of such an e-perfect code, while $\mathcal{C}$ and $\mathcal{C}'$ have two distinct such configuration distributions. Thus, $D_{(0,w)} = 1$, $D_{(w-e,e)} = 1$, and $D_{(w-i,i)} = 0$ for $0 \le i \le e-1$ (note that $k = e$ does not yield any such contradiction).

Now assume the contrary, that there exists a codeword from configuration $(w-k-r,k+r)$, $k < e$, $r > 0$, that intersects the codeword from configuration $(0,w)$ in exactly k positions. This intersection in exactly k positions implies that we can exchange r points from the $\mathcal{A}$-part, with *zeroes* in the codeword from configuration $(w-k-r,k+r)$, with r points from the $\mathcal{B}$-part with *ones* in this codeword and *zeroes* in the codeword from configuration $(0,w)$. The obtained e-perfect code $\mathcal{C}'$ has $D_{(0,w)} = 1$ and $D_{(w-k,k)} = 1$, for $k < e$, a contradiction as before, since in $\mathcal{C}$, $D_{(0,w)} = 1$ and $D_{(w-e,e)} = 1$. Therefore, the intersection of each codeword with the codeword from configuration $(0,w)$ is at least e. Since each codeword can be chosen as the codeword from configuration $(0,w)$ by using an appropriate permutation on the coordinate set, it follows that the intersection of any two codewords is at least e. □

Theorem 8.3. *There is no e-perfect code in* $\mathrm{J}(2w+e+1,w)$.

Proof. Assume the contrary, that $\mathcal{C}$ is an e-perfect code in $\mathrm{J}(2w+e+1,w)$ and the coordinate set $\mathcal{N}$ is partitioned into two parts $\mathcal{A}$ and $\mathcal{B}$ such that $|\mathcal{A}| = w$ and $|\mathcal{B}| = w+e+1$, and the unique word from configuration $(w,0)$ is a codeword in $\mathcal{C}$. By Lemma 8.3 the intersection between any two codewords is at least e and hence $\mathcal{C}$ cannot contain a codeword from any configuration $(i,w-i)$, $0 \le i \le e-1$. Therefore, the $\binom{w+e+1}{w}$ words from configuration $(0,w)$ are J-covered only by codewords from configuration $(e,w-e)$. Moreover, every $e+1$ positions in the $\mathcal{B}$-part must be C-covered by the $2e+1$ *zeroes* of exactly one codeword from configuration $(e, w-e)$. Let $\mathcal{C}_1$ be the set of codewords of $\mathcal{C}$ from configuration $(e, w-e)$. Thus, the complements in the projection of that $\mathcal{B}$-part on $\mathcal{C}_1$ form a Steiner system $\mathrm{S}(e+1, 2e+1, w+e+1)$.

This Steiner system $\mathrm{S}(e+1, 2e+1, w+e+1)$ implies that each e points of the $\mathcal{B}$-part of $\mathcal{C}_1$ have exactly $\frac{w+1}{e+1}$ codewords with e *zeroes*. By exchanging any e points from the $\mathcal{A}$-part that contain e *ones* in a codeword $c' \in \mathcal{C}_1$ with e points of the $\mathcal{B}$-part that contain e *zeroes* in c', we obtain an e-perfect code $\mathcal{C}'$. Clearly, $\mathcal{C}'$ contains a codeword from configuration $(0, w)$. Since $\mathcal{C}$ has a codeword from configuration $(w, 0)$, it follows that $\mathcal{C}'$ has a codeword from configuration $(w-e, e)$.

Consider now the $\frac{w+1}{e+1}$ codewords of $\mathcal{C}_1$ with e *zeroes* in the first e coordinates of the $\mathcal{B}$-part. Clearly, not all $\binom{w}{e}$ subsets of e coordinates of the $\mathcal{A}$-part have e *ones* in these codewords of $\mathcal{C}_1$. If we exchange any such e points of the $\mathcal{A}$-part with the first e points of the $\mathcal{B}$-part we obtain a code $\mathcal{C}''$ with a codeword from configuration $(w-e, e)$ but no codeword from configuration $(0, w)$. This is in contradiction to Theorem 8.2, which states that all such codes with a codeword from configuration $(w-e, e)$ have the same configuration distribution, while $\mathcal{C}'$ and $\mathcal{C}''$ have different configuration distributions.

Thus, there is no e-perfect code in $\mathrm{J}(2w+e+1, w)$. □

Another interesting consequence of Theorem 8.2 is on the structure of e-perfect codes in $\mathrm{J}(2w, w)$.

Theorem 8.4. *An e-perfect code in* $\mathrm{J}(2w, w)$ *is a self-complement code.*

Proof. Let $\mathcal{C}$ be an e-perfect code in $\mathrm{J}(2w, w)$, and assume $\mathcal{N}$ is partitioned into two parts $\mathcal{A}$ and $\mathcal{B}$ such that $|\mathcal{A}| = |\mathcal{B}| = w$ and the unique word from configuration $(w, 0)$ is a codeword in $\mathcal{C}$. By Theorem 8.2 for exactly one k, $0 \leq k \leq e$, we have that $D_{(k,w-k)} = 1$ and for $i \neq k$, $0 \leq i \leq 2e-k$, we have that $D_{(i,w-i)} = 0$. If $k > 0$, then we can exchange one point from the $\mathcal{A}$-part with a *zero* in the codeword from configuration $(k, w-k)$ with a point from the $\mathcal{B}$-part with a *zero* in the codeword from configuration $(k, w-k)$ to obtain a new e-perfect code $\mathcal{C}'$. In $\mathcal{C}'$ we have $D_{(w-1,1)} = 1$ and $D_{(k,w-k)} = 1$ in contradiction to the unique configuration distribution when $D_{(k,w-k)} = 1$, $0 \leq k \leq e$, obtained in Theorem 8.2.

Thus, $k = 0$ and $\mathcal{C}$ is a self-complement code. □

8.3 Steiner Systems Embedded in a Perfect Code

It appears that if an e-perfect code exists in $\mathrm{J}(n, w)$, then there are many Steiner systems embedded in it. One such example of a Steiner system was presented in the proof of Theorem 8.3. These Steiner systems imply several necessary conditions, derived in Corollary 3.1, which must be satisfied. Hence, these Steiner systems also yield necessary conditions for the existence of the related e-perfect codes.

Theorem 8.5. *If there exists an e-perfect code in* $\mathrm{J}(n, w)$, *then there exists a Steiner system* $\mathrm{S}(e+1, 2e+1, w)$.

Proof. Assume $\mathcal{C}$ is an e-perfect code in $\mathrm{J}(n, w)$. Partition the coordinate set $\mathcal{N}$ into two subsets $\mathcal{A}$ and $\mathcal{B}$, such that $|\mathcal{A}| = w$ and $|\mathcal{B}| = n - w$, and the unique word from configuration $(w, 0)$ is a codeword. This codeword J-covers exactly all the words of all configurations $(w-i, i)$, where $0 \leq i \leq e$. Since $\mathcal{C}$ is an e-perfect code and all words from all configurations of the form $(w-i, i)$, where $0 \leq i \leq e$, are J-covered, it follows that $\mathcal{C}$ does not contain any codeword of any configurations of the form $(w-i, i)$, where $1 \leq i \leq 2e$. Therefore, all words of configuration $(w-e-1, e+1)$ must be J-covered by codewords from configuration $(w-2e-1, 2e+1)$. Consider now all the $\binom{w}{e+1}$ words in configuration $(w-e-1, e+1)$ with $e+1$ *ones* in $e+1$ fixed positions of the $\mathcal{B}$-part. These words are J-covered by codewords from configuration $(w-2e-1, 2e+1)$ with $2e+1$ *ones* in positions of the $\mathcal{B}$-part that C-cover these $e+1$ fixed positions. Let $\mathcal{C}_1$ be this set of codewords. Each subset of $e+1$ *zeroes* in the $\mathcal{A}$-part with these $e+1$ fixed positions in the $\mathcal{B}$-part must be C-covered and no such subset can be C-covered twice (since the code is e-perfect). Hence, the complements of these codewords formed by the projection of the coordinates in the $\mathcal{A}$-part on $\mathcal{C}_1$ form a Steiner system $\mathrm{S}(e+1, 2e+1, w)$. □

Corollary 8.4. *If there exists an e-perfect code in* $\mathrm{J}(n, w)$, *then there exists a Steiner system* $\mathrm{S}(e+1, 2e+1, n-w)$.

Corollary 8.5. *If there exists an e-perfect code in* $\mathrm{J}(n, w)$, *then* $n - w \equiv w \equiv e \pmod{e+1}$ *and hence* $e+1$ *divides* $n - 2w$.

Corollary 8.6. *If there exists an e-perfect code in* $\mathrm{J}(n, w)$, *then there exists a Steiner system* $\mathrm{S}(2, e+2, w-e+1)$ *and there exists a Steiner system* $\mathrm{S}(2, e+2, n-w-e+1)$.

Theorem 8.6. *If there exists an e-perfect code in* $\mathrm{J}(n, w)$, *then*

$$n \leq (w-1)(2e+1)/e .$$

Proof. Assume $\mathcal{C}$ is an e-perfect code in $\mathrm{J}(n, w)$. Partition the coordinate set $\mathcal{N}$ into two subsets $\mathcal{A}$ and $\mathcal{B}$, such that $|\mathcal{A}| = n - w + 1$ and $|\mathcal{B}| = w - 1$, and a word from configuration$(e+1, w-e-1)$ is a codeword. The J-distance between a word from configuration $(e+1, w-e-1)$ and a word from configuration $(e+1-i, w-e-1+i)$, where $i > 0$, is less than $2e+1$ and hence $\mathcal{C}$ does not have any codeword from configuration $(e+1-i, w-e-1+i)$, where $i > 0$. Therefore, all the words from configuration $(1, w-1)$ are J-covered by codewords from configuration $(e+1, w-e-1)$. Since $|\mathcal{B}| = w - 1$, it follows that to J-cover each word from configuration $(1, w-1)$ exactly once, there must be exactly $\frac{n-w+1}{e+1}$ codewords from configuration $(e+1, w-e-1)$. Since the minimum J-distance of $\mathcal{C}$ is $2e+1$, it follows that in any two codewords from configuration $(e+1, w-e-1)$, the subsets of coordinates with the e *zeroes* in the $\mathcal{B}$-part are disjoint. Hence, $w - 1 \geq \frac{n-w+1}{e+1}e$, which is equivalent to

$$n \leq (w-1)(2e+1)/e.$$

□

Theorem 8.7. *If there exists an e-perfect code in* $\mathrm{J}(n, w)$, *where* $n < (w-1)\frac{2e+1}{e}$, *then there exists a Steiner system* $\mathrm{S}(2, e+2, n-w+2)$.

Proof. Assume $\mathcal{C}$ is an e-perfect code in $J(n, w)$ and partition the coordinate set $\mathcal{N}$ into two subsets $\mathcal{A}$ and $\mathcal{B}$, such that $|\mathcal{A}| = n - w + 1$ and $|\mathcal{B}| = w - 1$, and there are $\frac{n-w+1}{e+1}$ codewords from configuration $(e+1, w-e-1)$ (see the proof of Theorem 8.6). Since $n < (w-1)(2e+1)/e$, i.e., $\frac{n-w+1}{e+1}e < w - 1$, it follows that there exists at least one coordinate in the $\mathcal{B}$-part that has *ones* in all the codewords from configuration $(e+1, w-e-1)$. Remove this coordinate from the $\mathcal{B}$-part to obtain a new subset $\mathcal{B}_1$ and join it to the $\mathcal{A}$-part to obtain a new subset $\mathcal{A}_1$. It this new partition we have that $|\mathcal{A}_1| = n - w + 2$, $|\mathcal{B}_1| = w - 2$, and $\mathcal{C}$ does not have any codeword from configuration $(e+2-i, w-e-2+i)$, $i > 0$, with respect to this new partition. Therefore, in this new partition, all the words from configuration $(2, w-2)$ are J-covered by codewords from configuration $(e+2, w-e-2)$. Since each word from configuration $(2, w-2)$ must be J-covered by exactly one codeword from configuration $(e+2, w-e-2)$, it follows that the codewords obtained from the $\mathcal{A}_1$-part of the codewords from configuration $(e+2, w-e-2)$ form a Steiner system $\mathrm{S}(2, e+2, n-w+2)$. □

Corollary 8.7. *If there exists a nontrivial e-perfect code in* $\mathrm{J}(n, w)$ *and* $w \leq n - w$*, then there exists a Steiner system* $\mathrm{S}(2, e+2, w+2)$.

Proof. If $w < n - w$, then Theorem 8.6 implies that

$$n \leq (w-1)(2e+1)/e < (n-w-1)(2e+1)/e$$

and the claim follows from Theorem 8.7.

If $n = 2w$, then Theorem 8.6 implies that $n \leq (w-1)(2e+1)/e$, where equality holds if and only if $w = 2e+1$, i.e., for a trivial perfect code. Hence, $n < (w-1)(2e+1)/e$ the claim follows from Theorem 8.7. □

Finally, we would like to find new nonexistence results by using other distance distribution for other partitions of the coordinate set.

Problem 8.1. Define the configuration distribution for other partitions of the coordinate set and use it to obtain new nonexistence results.

The theorems obtained in the previous sections make it possible to reduce the range in which perfect codes in the Johnson scheme can exist. If there exists an e-perfect code in $\mathrm{J}(n, w)$, then by Theorem 8.5 and Corollary 8.4 there exist a Steiner systems $\mathrm{S}(e+1, 2e+1, w)$ and a Steiner system $\mathrm{S}(e+1, 2e+1, n-w)$. By the divisibility conditions of Corollary 3.1 we have that $e+1$ should divide $w-e$ and $n-w-e$ and hence we have $n - w \equiv w \equiv e \pmod{e+1}$. This condition itself limits the range in which e-perfect codes can exist. Combining this condition with the nonexistence of e-perfect codes in $\mathrm{J}(2w+e+1, w)$ obtained in Theorem 8.3, we have the following theorem.

Theorem 8.8. *There are no perfect codes in* $\mathrm{J}(2w+p, w)$*, where* p *is a prime.*

Proof. Assume the contrary, that there exists an e-perfect code in $\mathrm{J}(2w+p, w)$, where p is a prime. By Theorem 8.5, Corollary 8.4, and Corollary 3.1, we have that $e+1$ divides $w-e$ and $e+1$ divides $w+p-e$, and hence $e+1$ divides p, which implies that $e+1 = p$. Nevertheless, by Theorem 8.3, there is no perfect code in $\mathrm{J}(2w+e+1, w)$, which completes the proof. □

The following theorem will be given without its proof.

Theorem 8.9. *There are no perfect codes in* $\mathrm{J}(2w+1, w)$.

In fact, we can obtain many more results similar to Theorem 8.8, e.g., there are no perfect codes in $J(2w+2p, w)$, p prime, $p \neq 3$ or there are no perfect codes in $J(2w+3p, w)$, p prime, $p \neq 2$, $p \neq 3$, and $p \neq 5$, and other similar theorems. The proofs involve careful examination of the divisibility conditions of Corollary 3.1 for a Steiner system $S(e+1, 2e+1, w)$ and for a Steiner system $S(e+1, 2e+1, n-w)$, and using Theorem 8.3. Checking all the necessary conditions we obtain that for $e = 1$, we must have that $n - w \equiv w \equiv 1 \pmod{6}$, and for $e = 2$, we have that $n - w \equiv w \equiv 2$, 17, 26, 41 or 50 (mod 60) and so on. Compiling all this data, we have that there are no nontrivial perfect codes in $J(2w-r, w)$ and $J(2w+r, w)$ for all $1 \leq r \leq 14$ with possible exceptions for $r = 6$, 9, and 12. This comes together with modulo conditions imposed on w and $n-w$ for any e-perfect code in $J(n, w)$. Assume the coordinate set $\mathcal{N}$ is partitioned into two subsets $\mathcal{A}$ and $\mathcal{B}$, such that $|\mathcal{A}| = w$ and $|\mathcal{B}| = n - w$, and the unique word from the configuration $(w, 0)$ is a codeword. By considering the way in which the words of the configuration $(w-e-2, e+2)$ are J-covered, we can get some more divisibility conditions that rule out perfect codes in some more graphs. Similarly, other results can be obtained from the configuration distributions, but the outcome is less significant.

From the proof of Theorem 8.5 we can see the complicated structure of e-perfect codes in $J(n, w)$. The coordinate set $\mathcal{N}$ is partitioned into two subsets $\mathcal{A}$ and $\mathcal{B}$, such that $|\mathcal{A}| = w$ and $|\mathcal{B}| = n - w$, and the unique word of the configuration $(w, 0)$ is a codeword. In the codewords of configuration $(w-2e-1, 2e+1)$, the projection of the coordinates in the $\mathcal{A}$-part forms a complement of a Steiner system $S(e+1, 2e+1, w)$ for each set of codewords that C-cover any fixed $e+1$ positions of the $\mathcal{B}$-part. Similarly, we can obtain (and this can be an alternative proof for Corollary 8.4) that in the $\mathcal{B}$-part there exists a Steiner system $S(e+1, 2e+1, n-w)$ for each set of codewords for which the complements C-cover any fixed $e+1$ positions in the $\mathcal{A}$-part. This complicated structure, together with other embedded Steiner systems that can be obtained in a perfect code, looks to be impossible to exist.

8.4 Tradeoff between Length, Weight, and Radius

The three parameters for an e-perfect code in $J(n, w)$ are n, w, and e. The tradeoff between these parameters can exclude other possible e-perfect codes. We first show that no trivial e-perfect code achieves the bound of Theorem 8.6 with equality.

Theorem 8.10. *If there exists an e-perfect code in* $\mathrm{J}(n,w)$ *then*

$$n < (w-1)\frac{2e+1}{e} .$$

Proof. Assume the contrary, that $\mathcal{C}$ is an e-perfect code in $\mathrm{J}(n,w)$, where $n = 2w+\alpha$ and $n = (w-1)\frac{2e+1}{e}$. If $\alpha = 0$, then $n = 2w$ and $n = (w-1)\frac{2e+1}{e}$ imply that $w = 2e+1$, i.e., $\mathcal{C}$ is a trivial perfect code. By Theorem 8.9 we have that there are no perfect codes in $\mathrm{J}(2w+1,w)$ and, therefore, we assume that $\alpha \geq 2$.

Let $\beta = e+1$; by Corollary 8.5, we have that $e+1$ divides $n-w$, i.e., β divides α, and hence $2 \leq \beta \leq \alpha$. Substituting $\beta = e+1$ and $n = 2w+\alpha$, in $n = (w-1)\frac{2e+1}{e}$, we obtain $w = \alpha\beta - \alpha + 2\beta - 1$. By the analysis done so far, we have the existence of several Steiner systems.

- By Corollary 8.6 there exists a Steiner system $\mathrm{S}(2,\beta+1,\alpha\beta-\alpha+\beta+1)$. Thus, by Corollary 3.1 we have that $\frac{\binom{\alpha\beta-\alpha+\beta+1}{2}}{\binom{\beta+1}{2}}$ must be an integer.
- By Corollary 8.6 there also exists a Steiner system $\mathrm{S}(2,\beta+1,\alpha\beta+\beta+1)$. Thus, by Corollary 3.1 we have that $\frac{\binom{\alpha\beta+\beta+1}{2}}{\binom{\beta+1}{2}}$ must be an integer.
- By Corollary 8.7 we have that there also exists a Steiner system $\mathrm{S}(2,\beta+1,\alpha\beta-\alpha+2\beta+1)$. Thus, by Corollary 3.1 we have that $\frac{\binom{\alpha\beta-\alpha+2\beta+1}{2}}{\binom{\beta+1}{2}}$ must be an integer.

Therefore,

$$\frac{\binom{\alpha\beta+\beta+1}{2}}{\binom{\beta+1}{2}} - \frac{\binom{\alpha\beta-\alpha+\beta+1}{2}}{\binom{\beta+1}{2}} = \frac{2\alpha^2 - \frac{\alpha^2}{\beta} + 2\alpha + \frac{\alpha}{\beta}}{\beta+1}$$

is an integer, and hence

$$2\alpha^2 - \frac{\alpha^2}{\beta} + 2\alpha + \frac{\alpha}{\beta} \equiv 0 \pmod{\beta+1}.$$

But, since $\beta \equiv -1 \pmod{\beta+1}$, we have that

$$3\alpha^2 + \alpha \equiv 0 \pmod{\beta+1}. \tag{8.2}$$

We also have

$$\frac{\binom{\alpha\beta-\alpha+2\beta+1}{2}}{\binom{\beta+1}{2}} - \frac{\binom{\alpha\beta-\alpha+\beta+1}{2}}{\binom{\beta+1}{2}} = \frac{2\alpha\beta - 2\alpha + 3\beta + 1}{\beta+1}$$

is an integer, and hence

$$2\alpha\beta - 2\alpha + 3\beta + 1 \equiv 0 \pmod{\beta+1}.$$

Again, we have that $\beta \equiv -1 \pmod{\beta+1}$, which implies that

$$4\alpha + 2 \equiv 0 \pmod{\beta+1}. \tag{8.3}$$

By (8.2) and (8.3) we have that

$$8(3\alpha^2 + \alpha) - (6\alpha - 1)(4\alpha + 2) \equiv 0 \pmod{\beta+1}. \tag{8.4}$$

Nevertheless, $8(3\alpha^2+\alpha)-(6\alpha-1)(4\alpha+2) = 2$, and since $\beta \geq 2$, it follows that 2 is not divisible by $\beta+1$, a contradiction. Hence, $n < (w-1)\frac{2e+1}{e}$. □

By combining Theorem 8.7, Corollary 8.7, and Theorem 8.10 we obtain the the following result.

Corollary 8.8. *If there exists an e-perfect code in* $\mathrm{J}(n, w)$, *then there exist a Steiner system* $\mathrm{S}(2, e+2, w+2)$ *and a Steiner system* $\mathrm{S}(2, e+2, n-w+2)$.

Assume again that there exists an e-perfect code in $\mathrm{J}(n, w)$. By Corollaries 8.6 and 8.8, we have that the following Steiner systems exist:

$$\mathrm{S}(2, e+2, w+2) \qquad \mathrm{S}(2, e+2, n-w+2)$$

$$\mathrm{S}(2, e+2, w-e+1) \qquad \mathrm{S}(2, e+2, n-w-e+1).$$

Hence, by Corollary 3.1, we have that

- $(e+1)(e+2)$ divides $(w+1)(w+2)$.
- $(e+1)(e+2)$ divides $(n-w+1)(n-w+2)$.
- $(e+1)(e+2)$ divides $(w-e)(w-e+1)$.
- $(e+1)(e+2)$ divides $(n-w-e)(n-w-e+1)$.

Since $(n-w+1)(n-w+2) - (w+1)(w+2) = (n+3)(n-2w)$, it follows that

$$(e+1)(e+2) \text{ divides } (n+3)(n-2w). \tag{8.5}$$

Since $(n-w-e)(n-w-e+1)-(w-e)(w-e+1) = (n-2e+1)(n-2w)$, it follows that

$$(e+1)(e+2) \text{ divides } (n-2e+1)(n-2w). \tag{8.6}$$

By Corollary 8.5 we have that $e+1$ divides $n-2w$ and, therefore, $(e+1)(e+2)$ divides $(e+2)(n-w)$. Hence, by (8.6) we have that

$$(e+1)(e+2) \text{ divides } (n+5)(n-2w). \tag{8.7}$$

Thus, from (8.5) and (8.7) we have that

$$(e+1)(e+2) \text{ divides } 2(n-2w). \tag{8.8}$$

Therefore, by Corollary 8.5, (8.5), and (8.8), we obtain the following theorem.

Theorem 8.11. *Assume $\mathcal{C}$ is an e-perfect code in* $\mathrm{J}(n,w)$.

- *If e is odd, then n is even and $(e+1)(e+2)$ divides $n-2w$.*
- *If e is even and n is even, then $(e+1)(e+2)$ divides $n-2w$.*
- *If e is even and n is odd, then $e \equiv 0 \pmod 4$ and $\frac{(e+1)(e+2)}{2}$ divides $n-2w$.*

Corollary 8.9. *Assume $\mathcal{C}$ is an e-perfect code in* $\mathrm{J}(n,w)$.

- *If n is even, then $(e+1)(e+2)$ divides $n-2w$.*
- *If n is odd, then $e \equiv 0 \pmod 4$ and $(e+1)(e+2)/2$ divides $n-2w$.*

Corollary 8.10. *There are no perfect codes in*

- $\mathrm{J}(2w+p^i, w)$, *p is a prime and $i \geq 1$.*
- $\mathrm{J}(2w+pq, w)$, *p and q primes, $q<p$, and $p \neq 2q-1$.*

Now we present a lower bound on w if there exists an e-perfect code $\mathcal{C}$ in $\mathrm{J}(n,w)$.

Theorem 8.12. *Assume there exists an e-perfect code in* $\mathrm{J}(n,w)$, $w<n-w$. *If n is odd, then* $w > \frac{e(e+1)(e+2)}{2} + 2e + 1$. *If n is even, then* $w > e(e+1)(e+2) + 2e + 1$.

Proof. Assume first that n is odd. By Corollary 8.9 we have that $\frac{(e+1)(e+2)}{2}$ divides $n-2w$ and hence $\frac{(e+1)(e+2)}{2} \leq n-2w$. By Theorem 8.10 we have that $n-2w < \frac{w-2e-1}{e}$ and hence $\frac{(e+1)(e+2)}{2} < \frac{w-2e-1}{e}$. Thus, $w > \frac{e(e+1)(e+2)}{2} + 2e + 1$.

The case where n is even is proved similarly. □

We now handle the case of $n = 2w$. We denote $w = 2e + 1 + \epsilon$, and $n = 4e+2+2\epsilon$, where $\epsilon \geq 0$ (since, clearly, $w \geq 2e+1$ and $n-w \geq 2e+1$). We partition the set of coordinates $\mathcal{N}$ into two subsets, $\mathcal{A}$ and $\mathcal{B}$, such that $|\mathcal{A}| = |\mathcal{B}| = w$, and the unique word from configuration $(w, 0)$ is a codeword. Recall that $D_{(i,w-i)}$ denote the number of codewords with i ones in the positions of $\mathcal{A}$. One can easily verify that

$$D_{(w-2e-1,2e+1)} = \left[\frac{(2e+1+\epsilon)!e!}{(2e+1)!(e+\epsilon)!}\right]^2$$

$$D_{(w-2e-2,2e+2)} = D_{(w-2e-1,2e+1)}\frac{\epsilon^2 - 2e(e+1)\epsilon}{(2e+2)^2}.$$

Since $D_{(w-2e-2,2e+2)}$ is obviously nonnegative, we have that

$$\epsilon^2 \geq 2e(e+1)\epsilon.$$

We note that $\epsilon > 0$ or else the e-perfect code is trivial. Therefore,

$$\epsilon \geq 2e(e+1).$$

Therefore, we have the following theorem.

Theorem 8.13. *If there exists an e-perfect code in* $\mathrm{J}(n, w)$*, where* $n = 2w$*, then*

$$w \geq 2e^2 + 4e + 1.$$

8.5 Regularity of Codes

In this section we present a different approach to rule out the existence of e-perfect codes in $\mathrm{J}(n, w)$. We note that so far all the divisibility conditions that rule out perfect codes are derived from Steiner systems. In this section and the following two sections we investigate the divisibility conditions that are derived from the size of the code as given by the sphere-packing bound. For this, we introduce the notion of k-regular codes.

Definition 8.3. Let $\mathcal{C}$ be a code in $\mathrm{J}(n, w)$ and let $\mathcal{A}$ be a subset of the coordinate set $\mathcal{N} = \{1, \ldots, n\}$. For all $0 \leq i \leq |\mathcal{A}|$, we define

$$\mathcal{C}_{\mathcal{A}}(i) \triangleq |\{c \in \mathcal{C} \ : \ |c \cap \mathcal{A}| = i\}|.$$

Also, for each $I \subseteq \mathcal{A}$, we define

$$\mathcal{C}_{\mathcal{A}}(I) \triangleq |\{c \in \mathcal{C} \ : \ c \cap \mathcal{A} = I\}|.$$

Note that for a given partition of the coordinate set $\mathcal{N}$ into two parts $\mathcal{A}$ and $\mathcal{B}$ such that $|\mathcal{A}| = k$ and $|\mathcal{B}| = n-k$, $\mathcal{C}_{\mathcal{A}}(i)$ represents a configuration distribution for this partition.

Definition 8.4. A code $\mathcal{C}$ in $\mathrm{J}(n, w)$ is said to be k-***regular***, if the following two conditions hold:

(c.1) There exist integers $\alpha(0), \ldots, \alpha(k)$ such that if $\mathcal{A} \subseteq \mathcal{N}$, $|\mathcal{A}| = k$, then $\mathcal{C}_{\mathcal{A}}(i) = \alpha(i)$ for all $0 \leq i \leq k$.

(c.2) For any given k-subset $\mathcal{A}$ of $\mathcal{N}$, there exist integers $\beta_{\mathcal{A}}(0), \ldots, \beta_{\mathcal{A}}(k)$ such that if $I \subseteq \mathcal{A}$, then $\mathcal{C}_{\mathcal{A}}(I) = \beta_{\mathcal{A}}(|I|)$.

Note that if a code is k-regular, $k \geq 1$, then it is also $(k-1)$-regular. Equation (8.1) is a simple consequence from the following theorem and the fact that all codes are trivially 0-regular.

Theorem 8.14. *If an e-perfect code $\mathcal{C}$ in $\mathrm{J}(n, w)$ is k-regular, then*

$$|\mathcal{B}_e(n, w)| \;\Big|\; \binom{n-i}{w-i},$$

for all $0 \leq i \leq k$.

Proof. Let $\mathcal{C}$ be an e-perfect code in $\mathrm{J}(n, w)$, which is k-regular. Let $0 \leq i \leq k$, and by condition (c.1), let Φ denote the number of all-one words of length i appearing in a projection of $\mathcal{C}$ onto i coordinates. We, therefore, may write the following equation, which counts in two different ways the total number of all-one words of length i appearing in all the projections of $\mathcal{C}$ onto i coordinates:

$$\frac{\binom{n}{w}}{|\mathcal{B}_e(n, w)|}\binom{w}{i} = \binom{n}{i}\Phi.$$

Therefore,

$$\Phi = \frac{\binom{n-i}{w-i}}{|\mathcal{B}_e(n, w)|} \tag{8.9}$$

for each i, $0 \leq i \leq k$. □

For the rest of this section and for Sections 8.6 and 8.7, we examine e-perfect codes in $\mathrm{J}(2w + a, w)$. We define the following polynomial that plays a crucial role in our examination:

$$\sigma_e(w, a, k) \triangleq \sum_{j=0}^{e}(-1)^j\binom{k}{j}\sum_{i=0}^{e-j}\binom{w-j}{i}\binom{w+a-k+j}{i+j}.$$

Theorem 8.15. *Let $\mathcal{C}$ be an e-perfect code in* $\mathrm{J}(2w+a,w)$, *and let* $1 \le k \le w$. *If* $\sigma_e(w,a,m) \neq 0$ *for all the integers* $1 \le m \le k$, *then* $\mathcal{C}$ *is a k-regular code.*

Proof. We prove the theorem by induction on k. Let $\mathcal{C}$ be an e-perfect code in $\mathrm{J}(2w+a,w)$. We partition the coordinate set into two subsets $\mathcal{A}$ and $\mathcal{B}$ such that $|\mathcal{A}| = k$ and $|\mathcal{B}| = 2w+a-k$.

The basis for the induction is $k = 1$. We obtain the following two equations:

$$\mathcal{C}_{\mathcal{A}}(0)\sum_{i=0}^{e}\binom{w}{i}\binom{w+a-1}{i}+\mathcal{C}_{\mathcal{A}}(1)\sum_{i=0}^{e-1}\binom{w-1}{i}\binom{w+a}{i+1}=\binom{2w+a-1}{w}$$

$$\mathcal{C}_{\mathcal{A}}(0)+\mathcal{C}_{\mathcal{A}}(1)=\frac{\binom{2w+a}{w}}{|\mathcal{B}_e(2w+a,w)|}.$$

The first equation describes the way codewords of configuration $(0,w)$ and $(1,w-1)$ J-cover words of configuration $(0,w)$. The second equation simply relates $\mathcal{C}_{\mathcal{A}}(0)$ and $\mathcal{C}_{\mathcal{A}}(1)$ to the total number of codewords. To see that this equation set has exactly one solution we have to show that the determinant

$$\begin{vmatrix} \sum_{i=0}^{e}\binom{w}{i}\binom{w+a-1}{i} & \sum_{i=0}^{e-1}\binom{w-1}{i}\binom{w+a}{i+1} \\ 1 & 1 \end{vmatrix} \tag{8.10}$$

is nonzero. The determinant is simply $\sigma_e(w,a,1)$, which is nonzero. Since our solution does not depend on the partition, we see immediately that the conditions of Definition 8.4 are satisfied. Therefore, the basis is proved.

For the induction hypothesis, assume that $\mathcal{C}$ is a $(k-1)$-regular code. Hence, there exist integers $\alpha'(0),\dots,\alpha'(k-1)$, such that for each $(k-1)$-subset $\mathcal{A}'$ of $\mathcal{N}$, we have that $\mathcal{C}_{\mathcal{A}'}(i) = \alpha'(i)$, for all $0 \le i \le k-1$. We now prove the induction step, i.e., that $\mathcal{C}$ is also a k-regular code. Again, let $\mathcal{A}$ and $\mathcal{B}$ be a partition of the coordinate set $\mathcal{N}$ into two subsets, with $|\mathcal{A}| = k$ and $|\mathcal{B}| = 2w+a-k$. We start by showing that condition (c.2) in Definition 8.4 for the regularity is satisfied. This is done by induction on the weight of the $\mathcal{A}$-part. For weight 0, the claim is obvious. Now assume the claim holds for weight i, i.e., each word of length k and weight i appears in the $\mathcal{A}$-part of the codewords the same amount of times. We prove that the claim holds for weight $i+1$.

Let $\mathcal{A}' \subseteq \mathcal{A}$, where $|\mathcal{A}'| = k-1$, and let $\mathcal{B}' \supseteq \mathcal{B}$, where $|\mathcal{B}'| = 2w+a-k+1$, be a partition of the coordinates that is obtained from the $\mathcal{A}$-part and the $\mathcal{B}$-part by moving one coordinate η from the $\mathcal{A}$-part to the $\mathcal{B}$-part. With these two partitions, fix a word ω of length $k-1$ and weight i in the $\mathcal{A}'$-part. The number of codewords having this word in their $\mathcal{A}'$-part is given by $\alpha'(i)/\binom{k-1}{i}$ since the code is $(k-1)$-regular. By our last induction assumption concerning weight i, the number of codewords containing ω in the $\mathcal{A}'$-part and a *zero* in coordinate η is given by $\mathcal{C}_{\mathcal{A}}(i)/\binom{k}{i}$. Hence, the number of codewords containing ω in their $\mathcal{A}'$-part and a *one* in coordinate η is the difference,

$$\frac{\alpha'(i)}{\binom{k-1}{i}} - \frac{\mathcal{C}_{\mathcal{A}}(i)}{\binom{k}{i}}.$$

We now note that the choice of coordinate η has no bearing on the last arguments, i.e., we can use any coordinate of $\mathcal{A}'$ instead of η. Therefore, the number of codewords containing a word of a given weight $i+1$ in the $\mathcal{A}$-part is $\mathcal{C}_{\mathcal{A}}(i+1)/\binom{k}{i+1}$. Hence, condition (c.2) for the regularity is satisfied. Again, note that (c.2) may hold while (c.1) is not satisfied. In fact, we have proved that if (c.1) and (c.2) hold for k, then (c.2) also holds for $k+1$.

Therefore, we have k equations in $k+1$ variables:

$$\frac{\mathcal{C}_{\mathcal{A}}(i)}{\binom{k}{i}} + \frac{\mathcal{C}_{\mathcal{A}}(i+1)}{\binom{k}{i+1}} = \frac{\alpha'(i)}{\binom{k-1}{i}} \qquad \text{for all } 0 \le i \le k-1. \tag{8.11}$$

Just like in the induction basis, to prove condition (c.1) for regularity we add the following equation,

$$\sum_{j=0}^{\min(k,e)} \mathcal{C}_{\mathcal{A}}(j) \sum_{i=0}^{e-j} \binom{w-j}{i}\binom{w+a-k+j}{i+j} = \binom{2w+a-k}{w}. \tag{8.12}$$

This set of equations has exactly one solution if and only if its determinant is nonzero. This determinant is easily seen to be equal to

$$\prod_{i=0}^{k-1} \frac{1}{\binom{k}{i}} \cdot \left[\sum_{j=0}^{e} (-1)^j \binom{k}{j} \sum_{i=0}^{e-j} \binom{w-j}{i}\binom{w+a-k+j}{i+j} \right] = \sigma_e(w,a,k) \cdot \prod_{i=0}^{k-1} \frac{1}{\binom{k}{i}}.$$

By our assumption on σ_e we have a unique solution to the set of equations in (8.11) and (8.12). Since the partition does not affect the above arguments, it follows that condition (c.1) for regularity holds and $\mathcal{C}$ is a k-regular code. □

8.6 Regularity of Codes with Radius One

The next step is to focus on 1-perfect codes and show that they are k-regular for a relatively wide range of values of k.

Theorem 8.16. *If a code* $\mathcal{C}$ *is a 1-perfect code in* $\mathrm{J}(2w+a, w)$*, then it is also a* k*-regular code for all*

$$0 \leq k < \frac{2w+a+1-\sqrt{(a+1)^2+4(w-1)}}{2}.$$

Proof. By Theorem 8.15, a 1-perfect code is also a k-regular code in $\mathrm{J}(2w+a, w)$ when

$$\sigma_1(w,a,k) = k^2 - (2w+a+1)k + w(w+a) + 1$$

has no integer roots in the range $[1, k]$. Considered as a polynomial in k, the smaller of the two possible roots is $\frac{2w+a+1-\sqrt{(a+1)^2+4(w-1)}}{2}$, so the range of k described in the theorem contains no integer roots. □

Corollary 8.11. *If a 1-perfect code exists in* $\mathrm{J}(n, w)$*,* $n = 2w + a$*, then*

$$|\mathcal{B}_1(n,w)| = 1 + w(n-w) \;\Big|\; \binom{n-i}{w-i},$$

for all $0 \leq i < \frac{2w+a+1-\sqrt{(a+1)^2+4(w-1)}}{2}$.

The following theorem by the 19th century mathematician Ernst Kummer on binomial coefficients will be used later to determine the non-divisibility of binomial coefficients by powers of primes.

Theorem 8.17. *Let* p *be a prime. The number of times* p *appears in the factorization of* $\binom{a}{b}$ *equals the number of carries when adding* b *to* $a-b$ *in base* p.

Theorem 8.18. *There are no 1-perfect codes in* $\mathrm{J}(n, w)$*, when*

$$|\mathcal{B}_1(n,w)| = 1 + w(n-w) \equiv 0 \pmod{4}.$$

Proof. Assume there exists a 1-perfect code in $\mathrm{J}(n, w)$, $n = 2w + a$ for $2^m \leq n \leq 2^{m+1} - 1$. We distinguish between two cases depending on whether $w \leq 2^{m-1} - 1$ or $2^{m-1} \leq w$.

Case 1. $2^{m-1} \leq w \leq n/2$.

In this case,

$$w - 2^{m-1} \leq \frac{w}{2} < \frac{2w+a+1-\sqrt{(a+1)^2+4(w-1)}}{2},$$

and hence by Corollary 8.11, we have that

$$1 + w(n - w) \;\Big|\; \binom{n - w + 2^{m-1}}{2^{m-1}}.$$

Theorem 8.17 implies that

$$\binom{n - w + 2^{m-1}}{2^{m-1}} \not\equiv 0 \pmod 4,$$

and hence

$$1 + w(n - w) \not\equiv 0 \pmod 4.$$

Case 2. $w \leq 2^{m-1} - 1$.

Note that by Theorem 8.10 we also have that $a < w - 3$. If we want to use Corollary 8.11, we have to show that

$$n - (2^m - 1) < \frac{2w + a + 1 - \sqrt{(a+1)^2 + 4(w-1)}}{2}, \tag{8.13}$$

but, after rearranging, this is equivalent to showing that

$$2w + a + \sqrt{(a+1)^2 + 4(w-1)} < 2^{m+1} - 1.$$

We now notice the following,

$$\begin{aligned} &2w + a + \sqrt{(a+1)^2 + 4(w-1)} \\ &< 3w - 3 + \sqrt{(w-2)^2 + 4(w-1)} && \text{since } a < w - 3 \\ &\leq 2^{m+1} - 7 && \text{since } w \leq 2^{m-1} - 1 \\ &< 2^{m+1} - 1, \end{aligned}$$

which we wanted to show. Hence (8.13) holds, and then by Corollary 8.11, we have that

$$1 + w(n - w) \;\Big|\; \binom{2^m - 1}{w - n + 2^m - 1}.$$

Theorem 8.17 implies that

$$\binom{2^m - 1}{w - n + 2^m - 1} \not\equiv 0 \pmod 4,$$

and hence

$$1 + w(n - w) \not\equiv 0 \pmod 4.$$

□

Corollary 8.12. *If there exists a 1-perfect code in* $\mathrm{J}(n, w)$ *then either* $w \equiv n - w \equiv 1 \pmod{12}$ *or* $w \equiv n - w \equiv 7 \pmod{12}$.

8.7 Regularity of Codes with Larger Radius

We now discuss nontrivial e-perfect codes when $e \geq 2$. We show that if such a code exists, it must be k-regular for a wide range of values of k. We start by giving two simple lemmas, which can be proved by basic combinatorial techniques.

Lemma 8.4. *Vandermonde's convolution:*

$$\binom{n}{m} = \sum_{k=0}^{p} \binom{n-p}{m-k}\binom{p}{k}.$$

Lemma 8.5.

$$\binom{n-p}{m} = \sum_{k=0}^{p} (-1)^k \binom{n-k}{m-k}\binom{p}{k}.$$

Theorem 8.19. *If an e-perfect code, $e \geq 2$, exists in $\mathrm{J}(2w+a, w)$, then it is a k-regular code for all $0 \leq k < \frac{w}{e} - e$.*

Proof. Our aim is to show that $\sigma_e(w, a, k) \neq 0$ for all $0 \leq k < \frac{w}{e} - e$ for the required range of parameters (w, a, and k). We actually show a stronger claim. We show that σ_e is strictly positive in the required range of parameters. We start by noting that the polynomial may be rewritten in the following manner by summing in a different order:

$$\sigma_e(w, a, k) = \sum_{i=0}^{e}\sum_{j=0}^{i} (-1)^j \binom{k}{j}\binom{w-j}{i-j}\binom{w+a-k+j}{i}.$$

We continue and show that in the inner sum, each of the positive summands is greater than its following negative summand in absolute value. This is equivalent to showing that

$$\frac{\binom{k}{j+1}\binom{w-j-1}{i-j-1}\binom{w+a-k+j+1}{i}}{\binom{k}{j}\binom{w-j}{i-j}\binom{w+a-k+j}{i}} < 1.$$

Since, $j \geq 0$, $i \leq e$, $a \geq 0$, and $k < w/e - e$,

$$\frac{\binom{k}{j+1}\binom{w-j-1}{i-j-1}\binom{w+a-k+j+1}{i}}{\binom{k}{j}\binom{w-j}{i-j}\binom{w+a-k+j}{i}} < \frac{(w-e^2)(we-w+e^2+e)}{w(we-w+e)}.$$

So it suffices to show that

$$\frac{(w-e^2)(we-w+e^2+e)}{w(we-w+e)} \leq 1,$$

but this is equivalent to

$$w(e-2) + e(e+1) \geq 0,$$

which always holds. □

Corollary 8.13. *An e-perfect code in* $\mathrm{J}(n, w)$ *is an e-regular code.*

Proof. Assume there exists an e-perfect code in $\mathrm{J}(n, w)$. By Theorems 8.12 and 8.13, we have that $w > 2e^2$ and by Theorem 8.19 such a code is k-regular for all $k < \frac{w}{e} - e$, and hence the code is e-regular. □

In the next theorem we extend the range of regularity given in Theorem 8.19. We use Corollary 8.13 as the starting point for the proof. The method used in the proof of Theorem 8.19 no longer works for the extended range, so an asymptotic approach is used.

Theorem 8.20. *For all* $e \geq 2$*, there exists* $W_e > 0$ *such that for all* $w \geq W_e$*, all e-perfect codes in* $\mathrm{J}(2w + a, w)$ *are* $\lfloor \frac{w}{2} \rfloor$*-regular.*

Proof. Our proof starts essentially the same way as the proof of Theorem 8.19. We actually want to show that for a large enough w, with $a \geq 0$ and $k \leq w/2$,

$$\sigma_e(w, a, k) = \sum_{i=0}^{e} \sum_{j=0}^{i} (-1)^j \binom{k}{j} \binom{w-j}{i-j} \binom{w+a-k+j}{i} > 0.$$

By Corollary 8.13, we may consider $k \geq e$, so we have to show that

$$\sum_{i=0}^{e} \sum_{j=0}^{i} (-1)^j \binom{k}{j} \binom{w-j}{i-j} \binom{w+a-k+j}{i} > 0.$$

The left side of the equation can be rewritten as

$$\sum_{i=0}^{e} \frac{\binom{w}{i}}{\binom{w}{k}} \binom{w+a-k}{i} \sum_{j=0}^{i} (-1)^j \binom{i}{j} \binom{w-j}{k-j} \frac{\binom{w+a-k+j}{i}}{\binom{w+a-k}{i}}.$$

We continue by proving that for all $0 \leq i \leq e$, the inner sum is positive, i.e.,

$$\sum_{j=0}^{i} (-1)^j \binom{i}{j} \binom{w-j}{k-j} \frac{\binom{w+a-k+j}{i}}{\binom{w+a-k}{i}} > 0.$$

Now,

$$\begin{aligned}
&\sum_{j=0}^{i}(-1)^j\binom{i}{j}\binom{w-j}{k-j}\frac{\binom{w+a-k+j}{i}}{\binom{w+a-k}{i}}\\
&\geq \sum_{\substack{j=0\\ j\text{ even}}}^{i}\binom{i}{j}\binom{w-j}{k-j}-\frac{\binom{w+a-k+i}{i}}{\binom{w+a-k}{i}}\sum_{\substack{j=0\\ j\text{ odd}}}^{i}\binom{i}{j}\binom{w-j}{k-j}\\
&=\sum_{j=0}^{i}(-1)^j\binom{i}{j}\binom{w-j}{k-j}-\left(\frac{\binom{w+a-k+i}{i}}{\binom{w+a-k}{i}}-1\right)\sum_{\substack{j=0\\ j\text{ odd}}}^{i}\binom{i}{j}\binom{w-j}{k-j}\\
&=\binom{w-i}{k}-\left(\frac{\binom{w+a-k+i}{i}}{\binom{w+a-k}{i}}-1\right)\sum_{\substack{j=0\\ j\text{ odd}}}^{i}\binom{i}{j}\binom{w-j}{k-j},
\end{aligned}$$

where the last step is taken by using Lemma 8.5. It is now sufficient to prove that

$$\left(\frac{\binom{w+a-k+i}{i}}{\binom{w+a-k}{i}}-1\right)\sum_{\substack{j=0\\ j\text{ odd}}}^{i}\binom{i}{j}\binom{w-j}{k-j}<\binom{w-i}{k}. \tag{8.14}$$

We note that the sum may be rewritten in the following manner:

$$\begin{aligned}
\sum_{\substack{j=0\\ j\text{ odd}}}^{i}\binom{i}{j}\binom{w-j}{k-j}&=\frac{1}{2}\left(\sum_{j=0}^{i}\binom{i}{j}\binom{w-j}{k-j}-\sum_{j=0}^{i}(-1)^j\binom{i}{j}\binom{w-j}{k-j}\right)\\
&=\frac{1}{2}\left(\sum_{j=0}^{i}\binom{i}{j}\binom{w-j}{k-j}-\binom{w-i}{k}\right)\quad\text{by Lemma 8.5.}
\end{aligned}$$

Plugging this into (8.14) we have to prove that,

$$\left(\frac{\binom{w+a-k+i}{i}}{\binom{w+a-k}{i}}-1\right)\left(\frac{1}{\binom{w-i}{k}}\sum_{j=0}^{i}\binom{i}{j}\binom{w-j}{k-j}-1\right)<2. \tag{8.15}$$

Finally, we have the following chain of inequalities:

$$\left(\frac{\binom{w+a-k+i}{i}}{\binom{w+a-k}{i}}-1\right)\left(\frac{1}{\binom{w-i}{k}}\sum_{j=0}^{i}\binom{i}{j}\binom{w-j}{k-j}-1\right)$$
$$\leq\left(\left(\frac{w+a-k+1}{w+a-k-i+1}\right)^{i}-1\right)\left(\frac{1}{\binom{w-i}{k}}\sum_{j=0}^{i}\binom{i}{j}\binom{w-j}{k-j}-1\right)$$
$$=\left(\left(\frac{w+a-k+1}{w+a-k-i+1}\right)^{i}-1\right)\left(\sum_{j=0}^{i}\binom{i}{j}\sum_{\ell=0}^{i-j}\binom{i-j}{\ell}\frac{\binom{w-i}{k-j-\ell}}{\binom{w-i}{k}}-1\right)\text{ by Lemma 8.4}$$
$$\leq\left(\left(\frac{w+a-k+1}{w+a-k-i+1}\right)^{i}-1\right)\left(\sum_{j=0}^{i}\binom{i}{j}\sum_{\ell=0}^{i-j}\binom{i-j}{\ell}\left(\frac{k}{w-i-k+1}\right)^{j+\ell}-1\right)$$
$$=\left(\left(\frac{w+a-k+1}{w+a-k-i+1}\right)^{i}-1\right)\left(\sum_{j=0}^{i}\binom{i}{j}\left(\frac{k}{w-i-k+1}\right)^{j}\left(\frac{w-i+1}{w-i-k+1}\right)^{i-j}-1\right)$$
$$=\left(\left(\frac{w+a-k+1}{w+a-k-i+1}\right)^{i}-1\right)\left(\left(\frac{w-i+k+1}{w-i-k+1}\right)^{i}-1\right)\text{ by Newton's binomial}$$
$$\leq\left(\left(\frac{w/2+1}{w/2-e+1}\right)^{e}-1\right)\left(\left(\frac{3w/2-e+1}{w/2-e+1}\right)^{e}-1\right)\text{ since } a\geq 0,\ i\leq e,\ k\leq w/2.$$

Therefore, it is enough that we show that

$$\left(\left(\frac{w/2+1}{w/2-e+1}\right)^{e}-1\right)\left(\left(\frac{3w/2-e+1}{w/2-e+1}\right)^{e}-1\right)<2. \tag{8.16}$$

For a fixed value of e,

$$\lim_{w\to\infty}\left(\left(\frac{w/2+1}{w/2-e+1}\right)^{e}-1\right)\left(\left(\frac{3w/2-e+1}{w/2-e+1}\right)^{e}-1\right)=0,$$

and hence, a W_e exists as required. □

Theorem 8.21. *There are no e-perfect codes in* $\mathrm{J}(n,w)$, $e\geq 2$, *which are also* $\lfloor w/2\rfloor$*-regular, when* $|\mathcal{B}_e(n,w)|\equiv 0 \pmod 4$.

Proof. Let $\mathcal{C}$ be a $\lfloor w/2\rfloor$-regular e-perfect code in $\mathrm{J}(n,w)$, $n=2w+a$, for $2^m\leq n\leq 2^{m+1}-1$. We distinguish between two cases depending on whether $w\leq 2^{m-1}-1$ or $2^{m-1}\leq w$.

Case 1: $2^{m-1}\leq w\leq n/2$.

In this case,

$$w-2^{m-1}\leq\frac{w}{2}.$$

Since the code is a $\lfloor w/2 \rfloor$-regular code, it follows by Theorem 8.14 that

$$|\mathcal{B}_e(n,w)| \;\Big|\; \binom{n-w+2^{m-1}}{2^{m-1}}.$$

Theorem 8.17 implies that

$$\binom{n-w+2^{m-1}}{2^{m-1}} \not\equiv 0 \pmod 4,$$

and hence

$$|\mathcal{B}_e(n,w)| \not\equiv 0 \pmod 4.$$

Case 2: $w \leq 2^{m-1} - 1$.

Note that by Theorem 8.10 we also have that $a < \frac{w-(2e+1)}{e} < \frac{w}{2}$. If we want to use Theorem 8.14, we have to show that

$$n - (2^m - 1) \leq \frac{w}{2}. \tag{8.17}$$

But now,

$$n - \frac{w}{2} = 2w + a - \frac{w}{2} < 2w < 2^m - 1.$$

Hence (8.17) holds, and then by Theorem 8.14 we have that

$$|\mathcal{B}_e(n,w)| \;\Big|\; \binom{2^m - 1}{w - n + 2^m - 1}.$$

Theorem 8.17 implies that

$$\binom{2^m - 1}{w - n + 2^m - 1} \not\equiv 0 \pmod 4,$$

and hence

$$|\mathcal{B}_e(n,w)| \not\equiv 0 \pmod 4.$$

□

Theorem 8.22. *There are no e-perfect codes in* $\mathrm{J}(n,w)$, $e \geq 2$, *which are also* $\lfloor w/2 \lfloor$*-regular, when* $\mathcal{B}_e(n,w) \equiv 0 \pmod{p^2}$, $p \geq 3$ *a prime.*

Proof. Let $\mathcal{C}$ be an e-perfect code in $\mathrm{J}(n,w)$, for $p^m \leq n \leq p^{m+1} - 1$. Now, if $w \leq p^{m-1} - 1$, we have that $w < n/p$, which is impossible for $p \geq 3$ by Theorem 8.6. Hence, let $kp^{m-1} \leq w \leq (k+1)p^{m-1} - 1$, for some $1 \leq k \leq p^2 - 1$. In this case,

$$w - kp^{m-1} \leq \frac{w}{2}.$$

Since the code is $\lfloor w/2 \rfloor$-regular, it follows by Theorem 8.14 that

$$|\mathcal{B}_e(n,w)| \; \Big| \; \binom{n-w+kp^{m-1}}{kp^{m-1}}.$$

Theorem 8.17 implies that

$$\binom{n-w+kp^{m-1}}{kp^{m-1}} \not\equiv 0 \pmod{p^2},$$

and hence

$$|\mathcal{B}_e(n,w)| \not\equiv 0 \pmod{p^2}.$$

$\square$

Corollary 8.14. *There are no e-perfect codes in* $\mathrm{J}(n,w)$*, where* $e \geq 2$*, which are also* $\lfloor w/2 \rfloor$*-regular, where* $|\mathcal{B}_e(n,w)| \equiv 0 \pmod{p^2}$ *and* p *is a prime.*

For the next theorem we draw on another interesting theorem on binomial coefficients. This theorem was developed by another 19th century mathematician Edouard Lucas. Let $a \geq 0$ be some integer. We then denote by $\mathsf{d}_p(a,i)$, the i-th digit of a when written in base p. Hence,

$$a = \sum_{i=0}^{\infty} \mathsf{d}_p(a,i) p^i.$$

Theorem 8.23. *Let* p *be a prime, and* $n \geq m \geq 0$ *two integers, then*

$$\binom{n}{m} \equiv \prod_{i=0}^{\infty} \binom{\mathsf{d}_p(n,i)}{\mathsf{d}_p(m,i)} \pmod{p}.$$

Theorem 8.24. *Let* p *be a prime, and* $e \equiv -1 \pmod{p^2}$*. If there exists an e-perfect in* $\mathrm{J}(n,w)$*, then*

$$|\mathcal{B}_e(n,w)| \equiv 0 \pmod{p^2}.$$

Proof. Let $\mathcal{C}$ be an e-perfect code in $\mathrm{J}(n,w)$. By Corollary 8.8, there exist a Steiner system $\mathrm{S}(2, e+2, w+2)$ and a Steiner system $\mathrm{S}(2, e+2, n-w+2)$. Hence by Corollary 3.1, $\frac{w+1}{e+1}$ and $\frac{n-w+1}{e+1}$ must be an integer, and hence $w+1 \equiv 0 \pmod{p^2}$ and $n-w+1 \equiv 0 \pmod{p^2}$. In other words, the two least significant digits in the representation in base p of e, w, and $n-w$, are both $p-1$, i.e.,

$$\begin{aligned} &\mathsf{d}_p(w,0) = \mathsf{d}_p(w,1) = \mathsf{d}_p(n-w,0) \\ &= \mathsf{d}_p(n-w,1) = \mathsf{d}_p(e,0) = \mathsf{d}_p(e,1) = p-1. \end{aligned} \tag{8.18}$$

Let $0 \leq j < e$ be some integer such that $j \equiv 0 \pmod{p}$. Now,

$$\binom{w}{j+1}\binom{n-w}{j+1} = \binom{w}{j}\binom{n-w}{j}\frac{(w-j)(n-w-j)}{(j+1)^2}.$$

Note, however that $w-j$, $n-w-j$, and $j+1$ are co-prime to p^2. Furthermore, $w-j \equiv n-w-j \equiv -(j+1) \pmod{p^2}$. Hence,

$$\binom{w}{j}\binom{n-w}{j} \equiv \binom{w}{j+1}\binom{n-w}{j+1} \pmod{p^2}.$$

This may be repeated to get,

$$\binom{w}{j}\binom{n-w}{j} \equiv \binom{w}{j+1}\binom{n-w}{j+1} \equiv \cdots \equiv \binom{w}{j+p-1}\binom{n-w}{j+p-1} \pmod{p^2}. \tag{8.19}$$

Now let $0 \leq j < e$ be some integer such that $j \equiv 0 \pmod{p^2}$. Note that in all the integers of the form $j+ip$, where $0 \leq i \leq p-1$, only the second digit in base p changes while the first digit is always zero. We examine the following sum modulo p using Theorem 8.23:

$$\begin{aligned}\sum_{i=0}^{p-1}\binom{w}{j+ip}\binom{n-w}{j+ip} &\equiv \sum_{i=0}^{p-1}\prod_{\ell=0}^{\infty}\left[\binom{\mathsf{d}_p(w,\ell)}{\mathsf{d}_p(j+ip,\ell)}\binom{\mathsf{d}_p(n-w,\ell)}{\mathsf{d}_p(j+ip,\ell)}\right]\\ &\equiv \left(\sum_{i=0}^{p-1}\binom{p-1}{i}^2\right)\left(\prod_{\ell=2}^{\infty}\left[\binom{\mathsf{d}_p(w,\ell)}{\mathsf{d}_p(j+ip,\ell)}\binom{\mathsf{d}_p(n-w,\ell)}{\mathsf{d}_p(j+ip,\ell)}\right]\right)\\ &\equiv \binom{2(p-1)}{p-1}\prod_{\ell=2}^{\infty}\left[\binom{\mathsf{d}_p(w,\ell)}{\mathsf{d}_p(j+ip,\ell)}\binom{\mathsf{d}_p(n-w,\ell)}{\mathsf{d}_p(j+ip,\ell)}\right] \pmod{p}.\end{aligned}$$

We also have that $\mathsf{d}_p(2(p-1),0) = p-2 < p-1 = \mathsf{d}_p(p-1,0)$, and, therefore, by Theorem 8.17, $\binom{2(p-1)}{p-1} \equiv 0 \pmod{p}$. Hence, the previous sum is congruent to 0 modulo p. Now, for some integer k we have

$$\sum_{i=0}^{p-1}\binom{w}{j+ip}\binom{n-w}{j+ip} = kp. \tag{8.20}$$

We continue by examining the following sum modulo p^2:

$$\begin{aligned}\sum_{i=0}^{p^2-1}\binom{w}{j+i}\binom{n-w}{j+i} &\equiv \sum_{\ell=0}^{p-1}\sum_{i=0}^{p-1}\binom{w}{j+ip+\ell}\binom{n-w}{j+ip+\ell}\\ &\equiv p\sum_{i=0}^{p-1}\binom{w}{j+ip}\binom{n-w}{j+ip} && \text{by (8.19)}\\ &\equiv kp^2 && \text{by (8.20)}\\ &\equiv 0 \pmod{p^2}.\end{aligned}$$

Finally, the sphere size modulo p^2 equals,

$$\begin{aligned}
|\mathcal{B}_e(n,w)| &\equiv \sum_{i=0}^{e} \binom{w}{i}\binom{n-w}{i} \\
&\equiv \sum_{\substack{0\le j<e \\ j\equiv 0 \pmod{p^2}}} \sum_{i=0}^{p^2-1} \binom{w}{j+i}\binom{n-w}{j+i} \quad \text{since } e \equiv -1 \pmod{p^2} \\
&\equiv 0 \pmod{p^2}.
\end{aligned}$$

□

Corollary 8.15. *For any given $e \geq 2$, $e \equiv -1 \pmod{p^2}$, p prime, there are finitely many nontrivial e-perfect codes in the Johnson scheme.*

A simple observation is that the left side of (8.16) is a monotonously decreasing function in w. Hence, a simple computer search can find the value of W_e of Theorem 8.20 and validate that $\sigma_e(w,a,k)$ has no integer roots for $k \leq w/2$ and $w \leq W_e$. Such a computer search was done for $e = 3, 7, 8$ and indeed no such roots were found. Therefore, we conclude with the following result.

Proposition 8.1. *There are no nontrivial* 3*-perfect,* 7*-perfect, and* 8*-perfect codes in the Johnson graph.*

Another computer search was conducted to test the divisibility conditions of Theoren 8.14. The computer search was used to prove the following result.

Proposition 8.2. *There are no* 2*-perfect codes in* $\mathrm{J}(n,w)$ *for all* $n \leq 40000$.

Proving the conjecture that there are no perfect codes in $\mathrm{J}(n,w)$ appears to be impossible to prove in the near future. We suggest the following four problems as the main targets for future research in this direction.

Problem 8.2. Prove that there are no 1-perfect codes in the Johnson scheme.

Problem 8.3. Prove that there are no perfect codes in $\mathrm{J}(2w,w)$.

Problem 8.4. Prove that for each $e > 0$ there are finitely many nontrivial e-perfect codes in the Johnson graph $\mathrm{J}(n,w)$.

Problem 8.5. Prove that for each $n > 2w > 0$ there are finitely many nontrivial perfect codes in the Johnson graph $\mathrm{J}(n,w)$.

8.8 Diameter Perfect Codes

The proof of Lemma 2.14 does not hold for the Johnson scheme since there is no binary operation that makes the metric right or left distance invariant. Fortunately, we can provide an alternative proof for the Johnson scheme.

Lemma 8.6. *Let $\mathcal{C}_\mathcal{D}$ be a code in* $\mathrm{J}(n,w)$ *with distances between the codewords of $\mathcal{C}_\mathcal{D}$ taken from a subset $\mathcal{D}$. Let $\mathcal{A}$ be a subset of* $\mathrm{J}(n,w)$ *and let $\mathcal{C}'_\mathcal{D} \subseteq \mathcal{A}$ be the largest code in $\mathcal{A}$ with distances taken from $\mathcal{D}$. Then*

$$\frac{|\mathcal{C}_\mathcal{D}|}{\binom{n}{w}} \leq \frac{|\mathcal{C}'_\mathcal{D}|}{|\mathcal{A}|} \ . \tag{8.21}$$

Proof. Consider the set of pairs,

$$\mathcal{P} = \{(c,\pi) \ : \ c \in \mathcal{C}_\mathcal{D}, \ \pi \in S_n, \ \pi(c) \in \mathcal{A}\}.$$

For a fixed $c \in \mathcal{C}_\mathcal{D}$ and a fixed $a \in \mathcal{A}$, there are exactly $w!(n-w)!$ choices for π such that $a = \pi(c)$. Hence, the number of pairs in $\mathcal{P}$ equals $|\mathcal{C}_\mathcal{D}| \cdot |\mathcal{A}| \cdot w! \cdot (n-w)!$.

Note that for each permutation π and two elements $x, y \in \mathrm{J}(n,w)$, we have that $d(\pi(x),\pi(y)) = d(x,y)$. This implies that a fixed permutation $\pi \in S_n$ can transfer the elements of $\mathcal{C}_\mathcal{D}$ into at most $|\mathcal{C}'_\mathcal{D}|$ elements of $\mathcal{A}$. Therefore, each permutation π contributes at most $|\mathcal{C}'_\mathcal{D}|$ pairs to $\mathcal{P}$, and hence the number of pairs in $\mathcal{P}$ is at most $|\mathcal{C}'_\mathcal{D}|\, n!$, which implies that

$$|\mathcal{C}_\mathcal{D}| \cdot |\mathcal{A}| \cdot w! \cdot (n-w)! \leq |\mathcal{C}'_\mathcal{D}|\, n! \ ,$$

and the claim of the lemma follows. □

Lemma 8.6 implies that the code-anticode bound is satisfied for the Johnson scheme.

Corollary 8.16. *Let $\mathcal{C}_\mathcal{D}$ be a code in* $\mathrm{J}(n,w)$ *with distances between the codewords of $\mathcal{C}_\mathcal{D}$ taken from the range $[2\delta, n]$. Let $\mathcal{A}$ be a subset of* $\mathrm{J}(n,w)$ *and let $\mathcal{C}'_\mathcal{D} \subseteq \mathcal{A}$ be the largest code in $\mathcal{A}$ with distances from $[2\delta, n]$. Then*

$$A(n, 2\delta, w) \leq \frac{\binom{n}{w} |\mathcal{C}'_\mathcal{D}|}{|\mathcal{A}|} \tag{8.22}$$

As a consequence of Corollary 8.16, the two Johnson bounds (Lemma 2.3 and Lemma 2.4) can be obtained using proofs, which are completely different from the ones given in Chapter 2.

Corollary 8.17.

$$A(n, 2\delta, w) \leq \left\lfloor \frac{n}{w} A(n-1, 2\delta, w-1) \right\rfloor .$$

Proof. In (8.22) take $\mathcal{A}$ to be the set of all the words in V_w^n with a *one* in a fixed coordinate. □

Corollary 8.18.

$$A(n, 2\delta, w) \leq \left\lfloor \frac{n}{n-w} A(n-1, 2\delta, w) \right\rfloor .$$

Proof. In (8.22) take $\mathcal{A}$ to be the set of all the words in V_w^n with a *zero* in a fixed coordinate. □

The following two lemmas are readily verified, with the proof of the first one being trivial.

Lemma 8.7. *The set $\mathcal{A}_2(n, w, t)$, where $0 \leq t \leq w \leq \frac{n}{2}$, defined by*

$$\{(\overbrace{1 \cdots\cdots 1}^{t \text{ times}}, c_1, \ldots, c_{n-t}) \ : \ c_j \in \mathbb{F}_2, \ 1 \leq j \leq n-t, \ wt(c_1, \ldots, c_{n-t}) = w-t\},$$

is an anticode in $\mathrm{J}(n, w)$ *whose diameter is* $w - t$ *and size is* $\binom{n-t}{w-t}$.

Lemma 8.8. *The set $\bar{\mathcal{A}}_2(n, w, t)$, where $0 \leq t \leq w \leq \frac{n}{2}$, defined by*

$$\{(\overbrace{0 \cdots\cdots 0}^{t \text{ times}}, c_1, \ldots, c_{n-t}) \ : \ c_j \in \mathbb{F}_2, \ wt(c_1, \ldots, c_{n-t}) = n-w\},$$

is an anticode in $\mathrm{J}(n, n-w)$ *whose diameter is* $w - t$ *and size is* $\binom{n-t}{w-t}$.

Proof. Clearly, $\bar{\mathcal{A}}_2(n, w, t)$ is the set of complements of the elements from $\mathcal{A}_2(n, w, t)$ and hence $|\bar{\mathcal{A}}_2(n, w, t)| = \mathcal{A}_2(n, w, t) = \binom{n-t}{w-t}$. Moreover, for each two words $x, y \in \mathrm{J}(n, w)$, we have that $d(x, y) = d(\bar{x}, \bar{y})$ and hence the diameter of $\bar{\mathcal{A}}_2(n, w, t)$ equals the diameter of $\mathcal{A}_2(n, w, t)$. The claim follows now directly from Lemma 8.7. □

Theorem 8.25. *Any Steiner system* $\mathrm{S}(t, w, n)$ *forms a* $(w - t)$*-diameter perfect code.*

Proof. If $\mathcal{C}$ is an $(n, 2(w-t+1), w)$ code constructed from a Steiner system $\mathrm{S}(t, w, n)$, then its Johnson distance is $w - t + 1$, and

$$|\mathcal{C}| = \frac{\binom{n}{t}}{\binom{w}{t}} = \frac{\binom{n}{w}}{\binom{n-t}{w-t}} .$$

On the other hand, by the code-anticode bound, $|\mathcal{C}| \leq \frac{\binom{n}{w}}{\mathcal{A}}$, where $\mathcal{A}$ is any anticode in $\mathrm{J}(n, w)$ whose diameter is $w - t$, and, therefore, $\mathcal{A} \leq \binom{n-t}{w-t}$.

Since by Lemma 8.7 $\mathcal{A}_2(n, w, t)$ is an anticode in $\mathrm{J}(n, w)$ of size $\binom{n-t}{w-t}$ whose diameter is $w - t$, the claim of the theorem follows. □

Theorem 8.26. *Any complement of a Steiner system* $\mathrm{S}(t, w, n)$ *forms a* $(w-t)$*-diameter perfect code.*

Proof. For each two words $x, y \in \mathrm{J}(n, w)$, $d(x, y) = d(\bar{x}, \bar{y})$ and hence the complement of a Steiner system $\mathrm{S}(t, w, n)$ has minimum Hamming distance $2(w-t+1)$, i.e., minimum J-distance $w-t+1$. By Lemma 8.8, $\bar{\mathcal{A}}_2(n, w, t)$ and $\mathcal{A}_2(n, w, t)$ have the same diameter and the same size. Moreover, the weight of anticodewords in $\bar{\mathcal{A}}_2(n, w, t)$ is $n-w$ and since $\binom{n}{n-w} = \binom{n}{w}$, it follows by Theorem 8.25 that the complement of a Steiner system $\mathrm{S}(t, w, n)$ forms a $(w-t)$-diameter perfect code in $\mathrm{J}(n, n-w)$. $\square$

Corollary 8.19. *Any Steiner system* $\mathrm{S}(t, w, n)$ *and any complement of a Steiner system* $\mathrm{S}(t, w, n)$ *forms a* $(w-t)$*-diameter perfect code in* $\mathrm{J}(n, w)$ *and* $\mathrm{J}(n, n-w)$*, respectively.*

Problem 8.6. Are there diameter perfect codes in $\mathrm{J}(n, w)$, asides from Steiner systems and their complements?

Conjecture 8.1. *There are no nontrivial diameter perfect codes in* $\mathrm{J}(n, w)$*, asides from Steiner systems and their complements.*

Problem 8.7. Are there perfect sets in $\mathrm{J}(n, w)$, asides from Steiner systems? (see the definition in Section 2.4).

8.9 Notes

Constant-weight have found many applications throughout the years. Optical orthogonal codes [Chung, Salehi, and Wei (1989)] also known as cyclically permutable codes [Moreno, Zhang, Kumar, and Zionviev (1995)] are two such applications. They have also found applications in storage devices like flash memories [En Gad, Langberg, Schwartz, and Bruck (2011)]. Constant-weight codes in $\mathrm{J}(2w, w)$, called ***balanced codes*** are of special interest and have drawn lot of attention [Knuth (1986)]. They are important in the context of constrained codes, a family of code which found lot of applications in the 20th and the 21st centuries.

Section 8.1. The Johnson scheme was given its name by Delsartė [Delsarte (1973)] based on Johnson bounds on the sizes of constant-weight codes [Johnson (1972)]. These bounds are presented in Lemmas 2.3 and 2.4. The definition of the Johnson distance as half of the Hamming distance was given in [Delsarte (1973)]. Delsarte was the first to consider perfect codes in the Johnson scheme. In his work, [Delsarte (1973)] wrote in page 55:

> "After having recalled that there are "very few" perfect codes in the Hamming schemes, one must say that, for $1 < \delta < n$, there is not a single one known in the Johnson schemes. It is tempting to risk the conjecture that such codes do not exist. Certain results contained in the present work could be useful to attack this problem; especially the generalized Lloyd theorem of sec. 5.2.2 and theorem 4.7 about t-designs."

It was proved in [Bannai (1977)] that there are no e-perfect codes in $\mathrm{J}(2w+1, w)$ for $e \geq 2$. The proof is based on a generalization of Lloyd's theorem for the Johnson scheme. In [Hammond (1982)] a different approach was used to prove that a class of completely regular codes in certain distance-regular graphs does not exist. The results by this approach imply that there are no perfect codes in $\mathrm{J}(2w+1, w)$ and $\mathrm{J}(2w+2, w)$ (and hence Theorem 8.9 is true).

Section 8.2. The concept of configuration distribution and some discussions regarding it were presented in [Etzion (1996a)]. A comprehensive work on configuration distributions was done in [Etzion (2007)].

Section 8.3. The idea of looking at Steiner systems embedded in perfect codes was suggested by [Etzion (1996a)] and further developed in [Etzion (2001b)]. Enumeration for the number of Steiner systems embedded in a perfect code was done in [Etzion (2007)]. Theorem 8.6 was proved for first time in [Roos (1983)] using the code-anticode bound. The proof presented in this section is the result work by [Etzion (2001b)]. The improvement of the inequality for a strict inequality as proved in Theorem 8.10 was presented in [Etzion and Schwartz (2004)].

The bound was further improved in [Bannai and Noda (2016)] as follows.

Theorem 8.27. *If there exists an e-perfect code in $\mathrm{J}(n, w)$, where $e \geq 2$, then*

$$n \leq \frac{2we}{e-1} - \frac{7e+1}{2(e-1)} - \frac{\sqrt{D(e)}}{2e(e-1)},$$

where

$$D(e) = 8e(e+1)\left(w - \frac{e+3}{2}\right)^2 - e(e+2)(e-1)^2.$$

The analysis of graphs with no perfect codes is from [Etzion (1996a)], where the nonexistence proof of perfect codes in $\mathrm{J}(2w+1, w)$, which was presented in [Hammond (1982)] (see Theorem 8.9), is used.

Section 8.4. Tradeoff between the various parameters (length, weight, radius) for the nonexistence of perfect codes was done first in [Etzion (1996a)], and improved later in [Etzion (2001b)]. The other results that appear in this section were proved in [Etzion and Schwartz (2004)].

Section 8.5. The concept of regularity of codes in the Johnson scheme was presented first in [Etzion and Schwartz (2004)] and the analysis in this section is taken from this paper. It was further investigated in [Etzion (2007)], where some proofs were given using a different approach. Based on the analysis of the regularity of codes, the nonexistence of perfect codes for some radii was proved in [Etzion and Schwartz (2004)]. Computer search was also used for elimination of some radii up to a certain length of codewords. In particular, Corollary 8.15 was proved in [Etzion and Schwartz (2004)].

Section 8.6. The analysis in this section is taken from [Etzion and Schwartz (2004)]. It was further proved in [Gordon (2006)] that there are no 1-perfect codes in $\mathrm{J}(n, w)$ when $n \leq 2^{250}$. The proof used computer search after a proof of some related results from number theory.

Section 8.7. The analysis in this section is also taken from [Etzion and Schwartz (2004)]. Finally, for a given e and a, it was examined by [Etzion and Schwartz (2004)] in which graph $\mathrm{J}(2w+a, w)$ the existence of e-perfect codes was not ruled out. The results of the previous sections and careful analysis show the following:

Theorem 8.28. *For* $1 \leq a \leq 35$ *there are no e-perfect codes in* $\mathrm{J}(2w+a, w)$ *with the following possible exceptions: 1-perfect codes and 2-perfect codes in* $\mathrm{J}(2w+12, w)$ *and* $\mathrm{J}(2w+24, w)$, *and 4-perfect codes in* $\mathrm{J}(2w+15, w)$ *and* $\mathrm{J}(2w+30, w)$.

It is worth mentioning that, as proved in [Shimabukuro (2005)], there are also no perfect codes in $\mathrm{J}(2w+p^2, w)$, p prime, and in $\mathrm{J}(2w+5p, w)$, p prime different from 3.

Two older results in number theory were used in the proofs of this section. Theorem 8.23 is due to Edouard Lucas in his book "Théorie des Nombres" and a short proof was given in [Fine (1947)]. Theorem 8.17 was proved by Ernst Kummer and can be found in [Graham, Knuth, and Patashnik (1994), p. 245].

There are many other properties which perfect codes (if exist) in the Johnson scheme must satisfy. They can be found in [Martin (1992); Etzion (2007); Silberstein (2007); Silberstein and Etzion (2010)]

Section 8.8. The results of this section were taken from [Ahlswede, Aydinian, and Khachatrian (2001)] and [Etzion (2021)]. [Ahlswede, Aydinian, and Khachatrian (2001)] have pointed out on the connection between the diametric problem in the Johnson scheme and the intersection problem for systems of finite sets. This problem was completely solved in [Ahlswede and Khachatrian (1997)] as follows.

A system of subsets $\mathcal{A} \subset \mathrm{J}(n, w)$ is called ***t-intersecting*** if

$$|A_1 \cap A_2| \geq t \text{ for all } A_1,\ A_2 \subset \mathcal{A}\ .$$

Define the function

$$M(n, w, t) \triangleq \max \left\{ |\mathcal{A}| \ : \ \mathcal{A} \text{ is a} t\text{-intersecting system, } \mathcal{A} \subset \binom{[n]}{w} \right\},$$

where $1 \leq t \leq w \leq n$. It was proved in [Ahlswede and Khachatrian (1997)] that $M(n, w, t)$ is the size of the maximum size anticode with Johnson distance $w - t$. Define

$$\mathcal{F}_i \triangleq \left\{ A \in \binom{[n]}{w} \ : \ |A \cap [t + 2i]| \geq t + i \right\}$$

for $0 \leq i \leq \frac{n-t}{2}$.

As a consequence, the following theorem was proved in [Ahlswede and Khachatrian (1997)].

Theorem 8.29. *Let, t, w, and n be integers such that $1 \leq t \leq w \leq n$.*

- *If $(w-t+1)(2+\frac{t-1}{r+1}) < n < (w-t+1)(2+\frac{t-1}{r})$ for some $r \in \mathbb{N} \cup \{0\}$, then $M(n, w, t) = |\mathcal{F}_r|$. The set $\mathcal{F}_r$ is up to permutation the unique optimum set for $M(n, w, t)$, where by convention $\frac{\alpha}{\beta} = \infty$ for $\alpha \neq 0$ and $\beta = 0$.*
- *If $(w - t + 1)(2 + \frac{t-1}{r+1}) = n$ for $r \in \mathbb{N} \cup \{0\}$, then $M(n, w, t) = |\mathcal{F}_r| = |\mathcal{F}_{r+1}|$. An optimal system equals up to permutations either to $\mathcal{F}_r$ or to $\mathcal{F}_{r+1}$.*

If the maximum cardinality of an anticode in $\mathrm{J}(n, w)$ whose diameter is D is denoted by $\mathcal{A}(n, w, D)$, then

$$\mathcal{A}(n, w, D) = M(n, w, t), \text{ if } D = w - t.$$

The parameters of a maximum anticode of diameter $\delta - 1$ can be obtained using Theorem 8.29 from the following inequalities

$$\delta \left(2 + \frac{w - \delta}{r + 1} \right) \leq n < \delta \left(2 + \frac{w - \delta}{r} \right) . \tag{8.23}$$

We must have that $w - \delta + 1 + 2r = w$ and, therefore, $r = \frac{\delta-1}{2} = e$. Hence, the Roos bound (Theorem 8.6)

$$n \leq (2e+1)(w-1)/w$$

is proved again.

These results on the maximum size anticode are important in our context since by the code-anticode bound we have that

$$A(n, 2\delta, w) \cdot |\mathcal{F}_r| \leq \binom{n}{w} .$$

In their paper [Ahlswede, Aydinian, and Khachatrian (2001)] also proved a result that is analogous to Theorem 8.5 and to Corollary 8.5. The proof of this result is also based on the structure of the maximum size anticodes, i.e., the largest t-intersecting families.

Theorem 8.30. *If there exists an $(\delta-1)$-diameter perfect code $\mathcal{C}$ in $\mathrm{J}(n, w)$ and r is a nonnegative integer obtained as the parameter in (8.23), then there exist a Steiner system $\mathrm{S}(\delta-r, \delta, w)$ and a Steiner system $\mathrm{S}(r+1, \delta, w)$.*

For example, there exists a Steiner system $\mathrm{S}(5, 8, 24)$ (embedded in the extended Golay codes), which is a $(24, 8, 8)$ diameter perfect code. Since $(k-t+1)(t+1) = 24$, it follows by Theorem 8.29 that there are two choices for r, i.e., $r = 0$ or $r = 1$. Therefore, by Theorem 8.30 (with $r = 1$) we obtain two Steiner systems, $\mathrm{S}(3, 4, 8)$ and $\mathrm{S}(2, 4, 16)$.

In view of Theorem 8.30, more necessary conditions on the existence of diameter perfect codes in the Johnson scheme can be obtained from the necessary conditions on the existence of the related Steiner systems.

Finally, [Ahlswede, Aydinian, and Khachatrian (2001)] have also presented the following bound on the weight of a diameter perfect code.

Theorem 8.31. *A $(\delta-1)$-diameter perfect code exists in $\mathrm{J}(n, w)$ only if $w \geq (r+1)(\delta-r+1)$.*

Chapter 9

NonBinary Constant-Weight Codes

All the words in the Johnson scheme $\mathrm{J}(n, w)$ are binary words of length n and constant weight w. The Johnson distance between two words is exactly half of their Hamming distance. The definition of a perfect code is based on the Johnson graph and the Johnson distance. Using the Hamming distance on binary words will give us the obvious result that an e-perfect code in $\mathrm{J}(n, w)$ is also an $(2e)$-perfect code when the space is the set of all binary words of length n and constant weight w and the metric used is the Hamming metric. The definition, however, will not be based on the graph, which is not a connected graph. The reason is that all the Hamming distances are even and the edges in the graph correspond to words for which the distance is one. To define the same perfect code using the Hamming distance we say that in an e-perfect code $\mathcal{C}$ in $\mathrm{J}(n, w)$, the $(2e)$-balls (using the Hamming distance) around the codewords of $\mathcal{C}$ form a partition of $\mathrm{J}(n, w)$. By Lemma 8.1 these $(2e)$-balls are exactly the e-balls using the J-distance. The same definition can be done using the Hamming graph $\mathcal{H}_2(n)$ with the Hamming distance and considering a code which consists only of words with the same weight w. The $(2e)$-balls around a code $\mathcal{C}$ with codewords of weight w are disjoint. Hence, this definition is exactly the same one as the one defined on $\mathrm{J}(n, w)$.

The next natural problem is to consider nonbinary perfect constant-weight codes, which is done as follows. Let $\mathrm{J}_q(n, w)$ be the set of words of length n and weight w over an alphabet with q symbols. This set of words is our space $\mathcal{V}$ and the metric defined on $\mathrm{J}_q(n, w)$ is the Hamming distance, i.e., for two words $x, y \in \mathrm{J}_q(n, w)$, $d(x, y)$ is the number of coordinates in which x and y differ. In other words, if $x = (x_1, x_2, \ldots, x_n)$ and $y = (y_1, y_2, \ldots, y_n)$, then

$$d(x, y) \triangleq |\{i \ : \ x_i \neq y_i\}| \ .$$

The distinction between the binary and the nonbinary cases is the graph that represents the space with its distance. The graph based on $J_q(n, w)$, which represents nonbinary constant-weight words, is not a connected graph, unless $w = n$, and hence the definition of an e-perfect code cannot be based on the graph $J_q(n, w)$ in the same way that it is used for the binary case, when the Johnson distance is used. Nevertheless, there is one exception – when $w = n$, which will be discussed later. For the definition of an e-perfect codes, we simply use the usual definition that considers the balls with radius e by using the Hamming distance. Now, we can ask whether there exist e-perfect codes for $J_q(n, w)$ with the Hamming distance. Can we use the same definition as the one we used for $J(n, w)$, i.e., taking the e-balls around codewords of a code $\mathcal{C}$ in $J_q(n, w)$ to defined a code with minimum distance $2e + 1$ in $J_q(n, w)$. Unfortunately, this definition does not work since these e-balls can be disjoint in $J_q(n, w)$, but the minimum distance of the code can be smaller than $2e + 1$. For example, if $e = 1$ and the two codewords are $(01 \cdots 11)$ and $(11 \cdots 10)$, then the 1-balls centered at these codewords are disjoint, but their distance is two and hence they cannot be contained in a 1-perfect code. Hence, the distance between the codewords should be taken into account. To overcome this problem we can use the Hamming graph $\mathcal{H}_q(n)$, consider only codewords with weight w and require that the e-balls centered at the codewords are disjoint and contain all the words of weight w (but not all the vertices of the graph are contained in these e-balls). This is equivalent to require minimum distance $2e + 1$ for disjoint e-balls in $J_q(n, w)$.

The next question is whether the set of words in $J_q(n, w)$ with the Hamming distance a scheme? As noted before, the graph is not connected, but there is a representation of e-codes with a connected graph. Hence, we would like to know whether related intersection numbers are independent of the given words in $J_q(n, w)$. Consider $n \geq 3$ and $w = 2$, and the three words $x = (110 \cdots 0)$, $y = (220 \cdots 0)$, $z = (1010 \cdots 0)$. Clearly, $d(x, y) = d(x, z) = 2$, and a word u, in the space, for which $d(x, u) = d(y, u) = 1$ is in the set $\{(120 \cdots 0), (210 \cdots 0)\}$, while there is no word u such that $d(x, u) = d(z, u) = 1$. Hence, this space with the given metric is not an association scheme. A similar example can be given for any weight $w < n$. This is our starting point to consider nonbinary perfect constant-weight codes.

The rest of this chapter is organized as follows. Nonbinary perfect constant-weight codes are considered in Section 9.1. The code-anticode bound is proved for nonbinary constant-weight codes in Section 9.2 and six families of diameter perfect constant-weight codes are identified. These families of codes are discussed in Section 9.3 through Section 9.8. Maximum size anticodes for these families of codes are also defined in these sections. Four families of maximum size anticodes are identified in Section 9.9 and they are analyzed to find anticodes, related to diameter perfect constant-weight codes, with the same parameters and size, but a different structure.

9.1 Nonbinary Perfect Constant-Weight Codes

When $w = n$, the graph $\mathrm{J}_q(n, w)$ is the same as the graph $\mathcal{H}_{q-1}(n)$ since words in $\mathrm{J}_q(n, n)$ do not have *zeroes* and hence only the $q - 1$ nonzero symbols are used. Accordingly, the only difference between the two graphs is the symbols used (and hence we say that it is the same graph and not just isomorphic graphs). In this case we have one family of important e-perfect constant-weight codes that can be constructed easily based on the following theorem.

Theorem 9.1. *There exists an e-perfect code of length n over an alphabet of size q in the Hamming space, if and only if there exists an e-perfect constant-weight code of length n and weight $w = n$ over an alphabet of size $q + 1$.*

Proof. Let $\mathcal{C}$ be an e-perfect code of length n over an alphabet Q of size q in the Hamming space. Let $Q = \{1, 2, \ldots, q\}$ be the alphabet for $\mathcal{C}$. The code $\mathcal{C}' \triangleq \mathcal{C}$ defined over the alphabet $Q \cup \{0\}$ is an e-perfect constant-weight code of length n and weight $w = n$ over an alphabet of size $q + 1$.

If $\mathcal{C}$ is an e-perfect constant-weight code of length n and weight $w = n$ over an alphabet Q of size $q + 1$, where $0 \in Q$, then the same code defined over Q^- is an e-perfect code, of length n over Q^-, in the Hamming space. □

By Theorem 9.1 we have that if $w = n$, then e-perfect constant-weight codes over an alphabet of size $q + 1$ are equivalent to e-perfect codes over an alphabet of size q in the Hamming scheme. We continue to consider the sphere-packing bound for nonbinary constant-weight codes in $\mathrm{J}_q(n, w)$. For the reminder of this section we will consider only 1-perfect constant-weight codes.

Lemma 9.1. *If $\mathcal{C}$ is a $(q+1)$-ary constant-weight code of length n, weight w,* *and minimum Hamming distance 3, then*

$$|\mathcal{C}| \leq \frac{\binom{n}{w}q^w}{(q-1)w+1} .$$

Proof. First, we enumerate the total number of words of length n and weight w, which form the whole space. The number of possible w-subsets of the n coordinates is $\binom{n}{w}$ and on each such w coordinates we can form q^w words of weight w, where in the other $n-w$ coordinates there are *zeroes.* Therefore, the number of words of length n and weight w over an alphabet of size $q+1$ is $\binom{n}{w}q^w$.

Since the minimum Hamming distance of the code is 3, it follows that the balls with radius one around the codeword of $\mathcal{C}$ must be disjoint. The size of a ball with radius one is $(q-1)w+1$ since we can change at most one coordinate. Moreover, to remain with weight w, this change should be made from a nonzero value to another nonzero value in the same coordinate. There are w coordinates that have nonzero alphabet symbols. There are q nonzero alphabet symbols and hence there are $q-1$ possible changes to each such coordinate.

Therefore, the sphere-packing bound implies that $|\mathcal{C}| \leq \frac{\binom{n}{w}q^w}{(q-1)w+1}$. □

Corollary 9.1. *Let $\mathcal{C}$ be a $(q+1)$-ary constant-weight code of length n, weight w, and minimum distance 3. $\mathcal{C}$ is a 1-perfect code if and only if*

$$|\mathcal{C}| = \frac{\binom{n}{w}q^w}{(q-1)w+1} .$$

Let $\mathcal{C}$ be a $(q+1)$-ary 1-perfect constant-weight code of length n, weight w, and minimum Hamming distance 3. Define the following n sub-codes of $\mathcal{C}$.

$$\mathcal{C}_i \triangleq \{c \ : \ c = (c_1, c_2, \ldots, c_n) \in \mathcal{C}, \ c_i = 0\}, \quad 1 \leq i \leq n.$$

Each codeword in $\mathcal{C}$ is contained in exactly $n-w$ of these sub-codes and hence

$$\sum_{i=1}^{n} |\mathcal{C}_i| = (n-w)|\mathcal{C}|$$

and, therefore, by Corollary 9.1, we have that

$$\frac{1}{n}\sum_{i=1}^{n} |\mathcal{C}_i| = \frac{n-w}{n}|\mathcal{C}| = \frac{n-w}{n}\frac{\binom{n}{w}q^w}{(q-1)w+1} = \frac{\binom{n-1}{w}q^w}{(q-1)w+1}. \tag{9.1}$$

Let $\mathcal{C}'_i$ be the shortening of $\mathcal{C}$ with respect to the i-th coordinate, i.e., the code obtained from $\mathcal{C}_i$ by puncturing with respect to the i-th coordinate.

Clearly, $\mathcal{C}_i'$ is a $(q+1)$-ary constant-weight code of length $n-1$, weight w, and minimum Hamming distance 3, for which $|\mathcal{C}_i'| = |\mathcal{C}_i|$. Therefore, by (9.1) we have that $\sum_{i=1}^{n} |\mathcal{C}_i'| = n\frac{\binom{n-1}{w}q^w}{(q-1)w+1}$ and since, by Lemma 9.1, $|\mathcal{C}_i'| \leq \frac{\binom{n-1}{w}q^w}{(q-1)w+1}$, it follows by Corollary 9.1 that $\mathcal{C}_i'$ is a $(q+1)$-ary 1-perfect constant-weight code of length $n-1$ and weight w. These arguments work as long as $w < n$ and hence we have the following lemma.

Lemma 9.2. *If $\mathcal{C}$ is a $(q+1)$-ary 1-perfect constant-weight code of length n and weight w, where $w < n$, then its shortened code, with respect to any coordinate, is a $(q+1)$-ary 1-perfect constant-weight code of length $n-1$ and weight w.*

Corollary 9.2. *If there exists a $(q+1)$-ary 1-perfect constant-weight code of length n, weight w, then $(q-1)w+1$ divides q^w.*

Proof. Lemma 9.2 is applied iteratively $n-w$ times on a $(q+1)$-ary 1-perfect constant-weight code of length n and weight w, to obtain a $(q+1)$-ary 1-perfect constant-weight code of length $n = w$, weight w, minimum Hamming distance 3, and $\frac{q^w}{(q-1)w+1}$ codewords. □

Consider the $(q+1)$-ary 1-perfect constant-weight code of length n and weight $w = n$, obtained via the proof of Corollary 9.2. The code does not have any codeword with *zeroes* and hence it is a q-ary 1-perfect code of length n in the Hamming scheme. Consequently, its existence is equivalent to the existence of 1-perfect codes in the Hamming scheme, as was proved in Theorem 9.1.

We distinguish now between a ternary alphabet and an alphabet of size greater than three. If the alphabet size is three, i.e., $q = 2$, then by Corollary 9.2 we have that $w+1$ is a power of two.

Let $\mathcal{H}^*(r)$ be the $[2^r, 2^r-1-r, 4]$ extended Hamming code whose parity-check matrix is $H_r = [h_0, h_1, \ldots, h_{2^r-2}, h_\infty]$, where $h_i = (\alpha^i, 1)^{\text{tr}}$, $0 \leq i \leq 2^r-2$, α is a primitive element in GF(2^r), and α^i is represented by a binary vector of length r. The code has 2^{r+1} cosets (see Section 4.1), 2^r cosets whose words have even weight and 2^r cosets whose words have odd weight. Consider the cosets with words of odd weight. These cosets will be denoted by $\mathcal{H}_i^o$, $0 \leq i \leq 2^r-2$, and $\mathcal{H}_\infty^o$, where the coset leader in $\mathcal{H}_i^o$ is $\mathbf{e}_i$. Similarly, $\mathcal{H}_\infty^o$ is the coset with a coset leader of weight one whose unique *one* is in the last coordinate. Let $\mathcal{T}_i$, $0 \leq i \leq 2^r-2$, be the code obtained from $\mathcal{H}_i^o$ by replacing the symbol in the $(i+1)$-th position by the symbol *two*, where indices are taken modulo 2^r-1. Similarly, let $\mathcal{T}_\infty$ be the code obtained from $\mathcal{H}_\infty^o$ by replacing the symbol in the last coordinate of each codeword by the symbol *two*.

The following three lemmas will lead to the main claim that the union of all the $\mathcal{T}_i$'s has minimum distance 3, from which a ternary 1-perfect constant-weight code will be constructed. The first lemma is a simple observation from the definition.

Lemma 9.3. *Each codeword in $\mathcal{T}_i$, $0 \leq i \leq 2^r - 2$, and each codeword in $\mathcal{T}_\infty$ has a unique position with the symbol* two.

Lemma 9.4. *The minimum distance of each $\mathcal{T}_i$, $0 \leq i \leq 2^r - 1$ is 3, and the minimum distance of $\mathcal{T}_\infty$ is 3.*

Proof. The minimum distance of $\mathcal{H}^*(r)$, and also of each of its cosets, is 4. Replacing the symbol in the same position of all the codewords by the symbol *two* can reduce the minimum distance at most by one, i.e., to minimum distance 3. □

Lemma 9.5. *In $\mathcal{H}^*(r)$ there is no codeword of weight four with the* ones *in four distinct positions i, $i+1$, j, $j+1$, where these positions are taken modulo $2^r - 1$, and they do not include the last position.*

Proof. Assume the contrary, that there exists such a codeword of weight four with the *ones* in distinct positions i, $i+1$, j, $j+1$. By the structure of the parity-check matrix of $\mathcal{H}^*(r)$, this implies that $\alpha^i + \alpha^{i+1} + \alpha^j + \alpha^{j+1} = 0$. Since, the characteristic of the field is 2, it follows that $\alpha^i + \alpha^{i+1} = \alpha^j + \alpha^{j+1}$, i.e., $\alpha^i(\alpha+1) = \alpha^j(\alpha+1)$ or $\alpha^i = \alpha^j$, which implies that $i = j$, a contradiction of the fact that the four positions are distinct. □

Theorem 9.2. *The ternary code*

$$\mathcal{T} \triangleq \mathcal{T}_\infty \cup \bigcup_{i=0}^{2^r-2} \mathcal{T}_i \ ,$$

has minimum distance 3.

Proof. By Lemma 9.4, each $\mathcal{T}_i$ (including $\mathcal{T}_\infty$) has minimum Hamming distance 3. Hence, to complete the proof it suffices to show that the Hamming distance between two codewords c_1 and c_2 from two distinct $\mathcal{T}_i$'s is at least 3. By Lemma 9.3, each codeword in $\mathcal{T}_i$ has exactly one coordinate with the symbol *two*. This symbol appears in a different coordinate for each $\mathcal{T}_i$ and, therefore, the distance between any two codewords from two distinct $\mathcal{T}_i$'s is at least two. Note that by the definition of $\mathcal{T}_i$, $0 \leq i \leq 2^r - 2$, each codeword of $\mathcal{T}_i$ differs in exactly two coordinates from a codeword in $\mathcal{H}^*(r)$, one related to the coset leader and one coordinate with the symbol *two*. On the other hand, a codeword in $\mathcal{T}_\infty$ differs exactly in the last coordinate from a codeword of $\mathcal{H}^*(r)$. Moreover, note that any two codewords from distinct $\mathcal{T}_i$'s (including $\mathcal{T}_\infty$) that were formed from the same codeword of $\mathcal{C}$ differ in at least three coordinates, two related to the coset leaders and at least one that was changed to *two*. Let c_1 and c_2 be two words from distinct $\mathcal{T}_i$'s (including $\mathcal{T}_\infty$), formed from two distinct codewords of $\mathcal{H}^*(r)$, and assume the contrary that $d(c_1, c_2) = 2$. Assume further that c_i, $i = 1, 2$, was obtained from the codeword $c'_i \in \mathcal{H}^*(r)$. Distinguish between three cases depending on which $\mathcal{T}_i$ contains c_1 and which $\mathcal{T}_j$ contains c_2.

Case 1. $c_1 \in \mathcal{T}_i$, $0 \leq i \leq 2^r - 2$ and $c_2 \in \mathcal{T}_\infty$.

This case implies that the only other position (except for position $i+1$ modulo $2^r - 1$ and the last position, in which the symbol was changed to *two*) in which c'_1 and c'_2 might differ is the i-th position. Hence, $d(c'_1, c'_2) \leq 3$, which contradicts the fact that $d(\mathcal{H}^*(r)) = 4$.

Case 2. $c_1 \in \mathcal{T}_i$, $0 \leq i \leq 2^r - 2$ and $c_2 \in \mathcal{T}_{i+1}$, where $i + 1$ is taken modulo $2^r - 1$.

This case implies that the only other position (except for positions $i+1$ and $i+2$ modulo $2^r - 1$) in which c'_1 and c'_2 might differ is the i-th position. Hence, $d(c'_1, c'_2) \leq 3$, which contradicts the fact that $d(\mathcal{H}^*(r)) = 4$.

Case 3. $c_1 \in \mathcal{T}_i$, $c_2 \in \mathcal{T}_j$ $0 \leq i < j \leq 2^r - 2$, $i \neq j + 1$ and $j \neq i + 1$, where these additions are taken modulo $2^r - 1$.

This case implies that the only other positions (except for positions $i+1$ modulo $2^r - 1$ and $j + 1$ modulo $2^r - 1$) in which c'_1 and c'_2 might differ are i and j. Hence, this implies that $d(c'_1, c'_2) \leq 4$ and since $d(\mathcal{H}^*(r)) = 4$, it follows that $d(c'_1, c'_2) = 4$. But, $d(c'_1, c'_2) = 4$ if and only if $c'_1 + c'_2$ is a codeword of weight four in $\mathcal{H}^*(r)$ with *ones* in positions i, $i + 1$, j, $j + 1$ modulo $2^r - 1$, a contradiction to Lemma 9.5.

Thus, $d(\mathcal{T}) \geq 3$ and the theorem is proved. □

Theorem 9.3. *The code*

$$\mathcal{T}^* \triangleq (\overbrace{1 \cdots 1}^{n\ times}) + \mathcal{T} = (\overbrace{1 \cdots 1}^{n\ times}) + \left(\mathcal{T}_\infty \cup \bigcup_{i=0}^{2^r-2} \mathcal{T}_i \right)$$

is a ternary 1-perfect constant-weight code of length $n = 2^r$ and weight $w = 2^r - 1$.

Proof. The code $\mathcal{T}^*$ is a translate of the code $\mathcal{T}$ and hence it has, by Theorem 9.2, minimum distance 3. Moreover, by Lemma 9.3 and Theorem 9.2, each codeword of $\mathcal{T}$ has exactly one position with the symbol *two*, which implies that in each codeword of its translate $\mathcal{T}^*$, there is a unique *zero* and hence the weight of each codeword of $\mathcal{T}^*$ is $2^r - 1$. The total number of ternary words of length $n = 2^r$ and weight $w = 2^r - 1$ is $2^r \cdot 2^{2^r-1}$. The size of a ball with radius one is 2^r, and the size of the code $\mathcal{T}^*$ is $2^r \cdot 2^{2^r-1-r}$ and hence, by Corollary 9.1, the code $\mathcal{T}^*$ is a ternary 1-perfect constant-weight code of length $n = 2^r$ and weight $w = 2^r - 1$. □

Theorem 9.4. *If $\mathcal{C}$ is a ternary 1-perfect constant-weight code, then either $\mathcal{C}$ is essentially a binary 1-perfect code of length $2^r - 1$ (and hence $n = w = 2^r - 1$), or $w = 2^r - 1$ and $n = w + 1 = 2^r$ for some $r \geq 2$.*

Proof. In view of Theorem 9.3 and Lemma 9.2, a ternary 1-perfect constant-weight code $\mathcal{C}$ with weight w can be shortened to obtain a shorter 1-perfect constant-weight code with the same weight w. By iteratively shortening the obtained code, we end up with a ternary constant-weight code of weight w and length $n = w$. By Theorem 9.1 this code is equivalent to a binary 1-perfect code and hence its length is $2^r - 1$ (which implies that $n = w = 2^r - 1$). Since by Theorem 9.3 there exists a ternary 1-perfect constant-weight code of length 2^r and weight $2^r - 1$, it follows by Lemma 9.2 that to complete the proof it suffices to prove that a ternary 1-perfect constant-weight code with parameters $w = 2^r - 1$ and $n = w + 2$ does not exist.

Assume the contrary, that $\mathcal{C}$ is a ternary constant-weight code of weight $w = 2^r - 1$, length $n = w + 2$, and minimum Hamming distance 3, for some $r \geq 2$. For

$$S \triangleq \{(c, x) \ : \ c \in \mathcal{C}, \ x \in \{1, 2\}^n, \ x_i = c_i \text{ if } c_i \neq 0, \ 1 \leq i \leq n\},$$

we have that $|S| = 4|\mathcal{C}|$ since each codeword of $\mathcal{C}$ has two *zeroes* and there are four assignments in S for these two positions in x to obtain a pair $(c, x) \in S$ for a given $c \in \mathcal{C}$. Let $x \in \{1, 2\}^n$, c and c' be two distinct codewords of $\mathcal{C}$ for which the two pairs (c, x) and (c', x) are in S. This implies that $x_i = c_i = c'_i$, whenever $c_i \neq 0$ and $c'_i \neq 0$. Accordingly, since c and c' have exactly two positions with *zeroes* and also $d(c, c') \geq 3$, it follows that there is no position j for which $c_j = c'_j = 0$. Therefore, since n is odd, it follows that for each $x \in \{1, 2\}^n$, the size of the set $\{(c, x) \ : \ c \in \mathcal{C}\}$ is at most $\frac{n-1}{2}$, i.e.,

$$|S| \leq 2^n \frac{n-1}{2} \ .$$

Hence,

$$|\mathcal{C}| = \frac{|\mathcal{S}|}{4} \leq 2^{n-3}(n-1) = 2^{w-1}(w+1) < 2^{w-1}(w+2) = \frac{2^w \binom{w+2}{w}}{w+1},$$

which contradicts Corollary 9.1. Thus, the proofs of the claims in the theorem are completed. □

Theorem 9.4 settles the existence question about ternary 1-perfect constant-weight codes. It is natural to ask whether there exist e-perfect constant-weight codes, except for the ones obtained Theorem 9.1 and the ones obtained by Theorem 9.3. The answer is that there exists at least one more such family of such codes that generalizes the ternary code of length 4. This family is based on the extended 1-perfect Hamming codes of length $q+2$ over $\mathbb{F}_q$, where $q = 2^r$, that was presented in Theorem 4.4. The new 1-perfect constant-weight codes are of length $n = 2^r + 2$, weight $w = 2^r + 1$, over an alphabet with $2^r + 1$ symbols.

Let $Q \triangleq \mathbb{F}_q \cup \{\infty\}$, where $q = 2^r$ and $r \geq 2$, be the alphabet. The symbol ∞ will play the role of the symbol *two* for the construction of the ternary code when $q = 2$. As in the ternary case, at the end of the construction we interchange the symbols *zero* and ∞ (in the ternary case this was done by adding the all-one word to all the codewords), which will imply that all the codewords will have weight w. For $r > 1$, the construction works only for $w = q+1$ and $n = w+1 = 2^r + 2$.

Consider the cyclic q-ary Hamming code $\mathcal{C}$ of length w. Clearly, $\mathcal{C}$ has minimum Hamming distance 3 and the extended code $\mathcal{C}^*$ has minimum Hamming distance 4 (see Theorem 4.4).

For $0 \leq i < w$ we define $\mathcal{C}_i^*$ to be the coset of $\mathcal{C}^*$ that contains $\mathbf{e}_i$. The code $\mathcal{T}_\infty$ is obtained from $\mathcal{C}^*$ by replacing the symbol in the last position by the symbol ∞. Similarly, the code $\mathcal{T}_i$ is obtained from $\mathcal{C}_i^*$ by replacing, in each codeword of $\mathcal{C}^*$, the symbol in position $i+1$ modulo w by the symbol ∞. Clearly, $\mathcal{T}_\infty$ and each $\mathcal{T}_i$ has minimum Hamming distance 3.

Define

$$\mathcal{T} \triangleq \mathcal{T}_\infty \cup \bigcup_{i=0}^{w-1} \mathcal{T}_i.$$

This construction leads to the following theorem.

Theorem 9.5. *The code $\mathcal{T}$ is a 1-perfect constant-weight code. For $q = 2^r$, $r \geq 2$, there exists a 1-perfect constant-weight code of length $q+2$ and weight $q+1$, over an alphabet with $q+1$ symbols.*

By Theorem 4.10, for length greater than $q+1$, there is no extended Hamming code with minimum Hamming distance 4. Therefore, the idea of the construction does not work (except of course if $r=1$ which yields the ternary alphabet). Perfect codes with parameters corresponding to Corollary 9.1 and Lemma 9.2, however, could still exist.

Problem 9.1. Prove or disprove that the only nonbinary 1-perfect constant-weight codes are those introduced in this section, i.e., the ones with $n=w$ obtained from the Hamming scheme, the ternary ones of length 2^r, where $r \geq 2$, and the ones of length $q+2$ over an alphabet of size $q=2^r$, where $r \geq 2$.

Problem 9.2. Are there e-perfect constant-weight codes with $e>1$ except for the ones implied by Theorem 9.1? Develop the theory for such codes over any alphabet with $q+1$ symbols, $q>1$.

9.2 Nonbinary Diameter Perfect Constant-Weight Codes

We will now consider nonbinary diameter perfect constant-weight codes. It appears that there are a few families of such codes. We will distinguish between six families of such codes and four families of related maximum size anticodes.

We already saw that Steiner systems are diameter perfect codes in the Johnson scheme. Steiner systems are binary constant-weight codes. It is quite natural to ask whether the generalized Steiner systems defined in Section 3.1 are also diameter perfect codes? Recall that $\text{GS}(t,w,n,q)$ is a constant-weight code $\mathcal{C}$ of length n, weight w for each codeword, over an alphabet Q of size q such that the minimum Hamming distance of $\mathcal{C}$ is $2(w-t)+1$ and each word of length n and weight t over Q is covered by exactly one codeword of $\mathcal{C}$. By Lemma 3.3, the number of codewords in a generalized Steiner system $\text{GS}(t,w,n,q)$ is

$$\frac{\binom{n}{t}}{\binom{w}{t}}(q-1)^t \ .$$

To verify whether a $\text{GS}(t,w,n,q)$ is a diameter perfect code, we have to prove the code-anticode bound for nonbinary constant-weight codes and find the maximum size of an anticode with codewords of length n and weight w, over Q, such that the maximum Hamming distance of the anticode is $2(w-t)$. To prove the code-anticode bound we have to prove the local inequality lemma, which was proved in Lemma 8.6 for the Johnson bound. This is required since the code-anticode theorem of Delsarte (see Section 2.5) cannot be applied because the metric is not an association scheme. Moreover, our metric does not satisfy the conditions of Lemma 2.14 since we do not have a binary operation that will make the metric right or left distance invariant. The same technique that was used in the proof of Lemma 8.6 will not work for the nonbinary case and the proof technique should be amended. For our proof of such a lemma for nonbinary constant-weight codes, it is required to prove the following simple lemma.

Lemma 9.6. *For each $q \geq 2$ and any given pair (t, n), where $1 \leq t \leq n$, there exists some $\lambda \geq 1$ for which there exists an $\text{OA}_\lambda(t, n, q)$.*

Proof. Consider a matrix $\mathcal{M}$ whose rows are all the q^n distinct words of length n over $\mathbb{Z}_q$. Clearly, in each projection of t coordinates from $\mathcal{M}$ each t-tuple is contained in $\frac{q^n}{q^t} = q^{n-t}$ distinct rows (codewords). Thus, the $q^n \times n$ matrix $\mathcal{M}$ forms an $\text{OA}_\lambda(t, n, q)$, where $\lambda = q^{n-t}$. $\square$

Lemma 9.7. *Let $\mathcal{C}_\mathcal{D}$ be a constant-weight code of length n and weight w over $\mathbb{Z}_q$, $q > 2$, with distances between the codewords of $\mathcal{C}_\mathcal{D}$ taken from a subset $\mathcal{D}$. Let $\mathcal{A}$ be a subset of $\text{J}_q(n, w)$ and let $\mathcal{C}'_\mathcal{D} \subseteq \mathcal{A}$ be the largest code in $\mathcal{A}$ with distances taken from $\mathcal{D}$. Then*

$$\frac{|\mathcal{C}_\mathcal{D}|}{\binom{n}{w}(q-1)^w} \leq \frac{|\mathcal{C}'_\mathcal{D}|}{|\mathcal{A}|} . \tag{9.2}$$

Proof. Consider the set of pairs

$$\mathcal{P} = \{(c, \pi) \ : \ c \in \mathcal{C}_\mathcal{D}, \ \text{supp}(\pi(c)) = \text{supp}(a), \ \pi \in S_n, \ a \in \mathcal{A}\}.$$

For a fixed $c \in \mathcal{C}_\mathcal{D}$ and a fixed $a \in \mathcal{A}$ there are exactly $w!(n-w)!$ choices for π, for which $\text{supp}(\pi(c)) = \text{supp}(a)$. Hence, the number of pairs in $\mathcal{P}$ equals to $|\mathcal{C}_\mathcal{D}| \cdot |\mathcal{A}| \cdot w! \cdot (n-w)!$.

For the word $v = (v_1, v_2, \ldots, v_n) \in \mathbb{Z}_{q-1}^n$, we form a subset $\mathcal{A}_v$ of $\text{J}_q(n, w)$ as follows. Given a word $x = (x_1, x_2, \ldots, x_n)$ of $\mathcal{A}$, the word $a_v = (a_1, a_2, \ldots, a_n)$ is constructed in $\mathcal{A}_v$ as follows.

(1) If $x_i = 0$, then $a_i = x_i = 0$.
(2) If $x_i \neq 0$, then $a_i = x_i + v_i$ when $x_i + v_i < q$ and $a_i = x_i + v_i - (q-1)$ when $x_i + v_i \geq q$. In other words, if $j = v_i$, then a_i takes the j-th nonzero value of $\mathbb{Z}_q$ after the value of x_i, where 1 follows $q-1$.

Using this definition, we have that $\text{supp}(a_v) = \text{supp}(x)$.

Clearly, $\mathcal{A}_v$ is obtained from $\mathcal{A}$ by permuting the nonzero elements in each one of the w nonzero coordinates, of the words in $\mathcal{A}$, by some w cyclic permutations (a permutation for each coordinate) on the $q-1$ nonzero symbols of $\mathbb{Z}_q$ (which can be different for each coordinate) and hence $|\mathcal{A}_v| = |\mathcal{A}|$. Moreover, $\mathcal{A}_v$ and $\mathcal{A}$ are isomorphic subsets of $\mathrm{J}_q(n,w)$.

Now, let $\mathcal{M}$ be any orthogonal array $\mathrm{OA}_\lambda(w,n,q-1)$, for some $\lambda \geq 1$, whose existence is implied by Lemma 9.6. The number of rows of M is $\lambda(q-1)^w$.

Consider now the set of triples

$$\mathcal{T} = \{(c,\pi,v) \;:\; c \in \mathcal{C}_\mathcal{D},\ \pi \in S_n,\ v \in \mathcal{M},\ \pi(c) \in \mathcal{A}_v\}.$$

Let (c,π) be a pair in $\mathcal{P}$, i.e., $c \in \mathcal{C}_\mathcal{D}$, $\pi \in S_n$, and $\mathrm{supp}(\pi(c)) = \mathrm{supp}(a)$ for some $a \in \mathcal{A}$. Let $X = \mathrm{supp}(a)$ and let $u = (u_1, u_2, \ldots, u_n)$ be a word in $\mathbb{Z}_{q-1}^n$ such that $a = \pi(c)_u$. It is easy to verify that for each word $v \in \mathbb{Z}_{q-1}^n$, for which the projection of the coordinates in X on u and the projection of the coordinates in X on v are equal, we have that $\pi(c)_v = a = \pi(c)_u$. Since $\mathcal{M}$ contains λ rows for which these projections are equal, it follows that $|\mathcal{T}| = \lambda\,|\mathcal{P}|$.

Note, that for each permutation $\pi \in S_n$ and two elements $x, y \in \mathrm{J}_q(n,w)$, we have that $d(\pi(x),\pi(y)) = d(x,y)$. This implies that a fixed permutation π with a fixed row $v \in \mathcal{M}$ can transfer the elements of $\mathcal{C}_\mathcal{D}$ into at most $|\mathcal{C}'_\mathcal{D}|$ elements of $\mathcal{A}_v$. Therefore, the number of triples in $\mathcal{T}$ is at most $\lambda \cdot |\mathcal{C}'_\mathcal{D}| \cdot n! \cdot (q-1)^w$ which implies that

$$\lambda \cdot |\mathcal{C}_\mathcal{D}| \cdot |\mathcal{A}| \cdot w! \cdot (n-w)! = \lambda \cdot |\mathcal{P}| = |\mathcal{T}| \leq \lambda \cdot |\mathcal{C}'_\mathcal{D}| \cdot n! \cdot (q-1)^w \;,$$

and hence the claim of the lemma is proved. □

Lemma 9.7 implies that the code-anticode bound holds for $\mathrm{J}_q(n,w)$. Therefore, we can use this bound to search for nonbinary diameter perfect constant-weight codes and related maximum size anticodes. The bound will also yield interesting maximum size t-intersecting families that will be found based on coding theory rather than through extremal combinatorics. We distinguish between six families of such diameter perfect codes in $\mathrm{J}_q(n,w)$, where $q > 2$.

[**F1**] Nonbinary diameter perfect constant-weight codes for which $w = n$.

[**F2**] Diameter perfect constant-weight codes over an alphabet of size $2^k + 1$ for which $w = n - 1$.

[**F3**] Nonbinary diameter perfect constant-weight codes which are generalized Steiner systems.

[**F4**] Nonbinary diameter perfect constant-weight codes for which $d = w$. These codes are called maximum distance separable constant-weight codes. Each such code has $\binom{n}{w}(q-1)$ codewords.

[**F5**] Nonbinary diameter perfect constant-weight codes for which $d = w+1$. Such a code has $\binom{n}{w}$ codewords.

[**F6**] Nonbinary diameter perfect constant-weight codes for which $d < w$. These codes are called multiple orthogonal arrays constant-weight codes. Each such code has $\binom{n}{w}(q-1)^{w-d+1}$ codewords.

Remark 9.1. The number of codewords in the $(n, d, w)_q$ codes of the families [**F4**], [**F5**], and [**F6**] is $\binom{n}{w}(q-1)^{w-d+1}$. But, each has different properties and constructions and hence they are separated.

Problem 9.3. Are there more families of diameter perfect constant-weight codes, except for these six families. We believe that these six families contain all such codes, but a proof for such a claim can be very challenging. One possible direction is to show sets of parameters with tradeoff between n, w, and d, where such codes cannot exist.

9.3 Diameter Perfect Codes for which $w = n$

In the first family, [**F1**], of diameter perfect codes in $\mathrm{J}_q(n, w)$ we have that $w = n$. Theorem 9.1 is adapted for this case.

Theorem 9.6. *There exists a D-diameter perfect code of length n over an alphabet of size $q-1$ in the Hamming scheme, if and only if there exists a D-diameter perfect constant-weight code of length n and weight $w = n$ over an alphabet of size q.*

Proof. Let $\mathcal{C}$ be a D-diameter perfect code of length n over the alphabet $\{1, 2, \ldots, q-1\}$ in the Hamming scheme. Let $\mathcal{A}$ be the related maximum size anticode with diameter D for which $|\mathcal{C}| \cdot |\mathcal{A}| = (q-1)^n$. We define the same code $\mathcal{C}' \triangleq \mathcal{C}$ and the same anticode $\mathcal{A}' \triangleq \mathcal{A}$ over the extended alphabet $Q \triangleq \{0, 1, 2, \ldots, q-1\}$. We claim that $\mathcal{C}'$ is a D-diameter perfect constant-weight code of length n, weight $w = n$, and minimum Hamming distance $D + 1$, over Q. We also claim that $\mathcal{A}'$ a maximum size anticode of length n, weight $w = n$, and maximum Hamming diatance D, over Q, respectively. Clearly, the minimum Hamming distance of $\mathcal{C}'$ is equal to the minimum Hamming distance of $\mathcal{C}$, i.e., $D + 1$. Similarly, the maximum Hamming distance of $\mathcal{A}'$ is equal to the maximum Hamming distance of $\mathcal{A}$, i.e., D. Moreover,

$$|\mathcal{C}'| \cdot |\mathcal{A}'| = |\mathcal{C}| \cdot |\mathcal{A}| = (q-1)^n = |\mathrm{J}_q(n,n)|\,,$$

which completes the proof of our claim.

Let $\mathcal{C}$ be a D-diameter perfect constant-weight code of length n and weight $w = n$ over an alphabet $Q = \{0, 1, 2, \ldots, q-1\}$. Using similar arguments, in reverse order, the same code defined over Q^- is a D-diameter perfect code, of length n over Q^-, in the Hamming scheme. Similarly, if $\mathcal{A}$ is a maximum size anticode, of length n and weight $w = n$, with diameter D, over Q, then the same anticode defined over Q^- is a maximum size anticode over Q^-. □

In other words, Theorem 9.6 implies that when $w = n$, in each word all the coordinates are nonzero. Hence, the words in $\mathrm{J}_q(n,n)$ are over an alphabet with only $q-1$ symbols. This implies that any code in $\mathrm{J}_q(n,n)$ can be considered as a code in the Hamming scheme over an alphabet with $q-1$ symbols. Therefore, any D-diameter perfect code of length n over an alphabet with $q-1$ symbols is also a D-diameter perfect code in $\mathrm{J}_q(n,n)$. Similarly, each maximum size anticode of length n and diameter D over an alphabet with $q-1$ symbols (with no *zeroes*) is also a maximum size anticode with diameter D in $\mathrm{J}_q(n,n)$. All the linear diameter perfect codes in the Hamming scheme over an alphabet whose size is a prime power were characterized in Theorem 4.9. Accordingly, the family implied by Theorem 9.6 also includes codes derived from the extended Hamming codes, extended Golay codes, and MDS codes. Other nonlinear codes include codes with the same parameters as the Hamming codes and also orthogonal arrays with index unity over any alphabet and not necessary a prime power alphabet. Each such nonbinary diameter perfect constant-weight code with $w = n$ imply some maximum size nonbinary constant-weight anticodes for which $w = n$.

9.4 Codes with Alphabet Size $2^k + 1$ for which $w = n - 1$

By Theorem 2.11 an e-perfect code in $\mathrm{J}_q(n,w)$ is also a $(2e)$-diameter perfect code in $\mathrm{J}_q(n,w)$. All known nontrivial nonbinary perfect constant-weight codes of length n have weight $w = n-1$ and they form an important class of the second family [**F2**] of nonbinary diameter perfect constant-weight codes. Two classes of such perfect codes are known and constructed in Section 9.1. The first class consists of ternary codes of length 2^r, weight $2^r - 1$, and minimum Hamming distance 3. The second class consists of codes over an alphabet with 2^r symbols, length $2^r + 2$, weight $2^r + 1$, and minimum Hamming distance 3. For these two classes, balls are the maximum size anticodes, but there are other maximum size anticodes.

Lemma 9.8. *If $n \geq 4$, then the size of an anticode with diameter 2 in* $\text{J}_3(n, n-1)$ *is at most n.*

Proof. Assume that $\mathcal{A}$ is an anticode with diameter 2 in $\text{J}_3(n, n-1)$ having at least n anticodewords.

Assume first that there are two anticodewords where the unique *zero* is in the same position.

Assume further that the distance between these two anticodewords is two and w.l.o.g. these two anticodewords are $0111 \cdots 1$ and $0221 \cdots 1$. It is easy to verify that there is no other anticodeword whose *zero* is not in the first position. This implies that the only two words that can be added to the anticode are $0121 \cdots 1$ and $0211 \cdots 1$, and hence the anticode has size four.

Assume now that the distance between these two anticodewords is one and w.l.o.g. these two anticodewords are $0111 \cdots 1$ and $0211 \cdots 1$. It is now easy to verify that any anticodeword without a *zero* in the first position must have its *zero* in the second position and there are at most two such anticodewords and in this case (one or two anticodewords with a *zero* in the second position) there are no more anticodewords with a *zero* in the first position. Thus, in this case their are either four anticodewords or n anticodewords which have their *zero* in the same position.

If there are no two anticodewords with *zeroes* in the same position, then clearly size of $\mathcal{A}$ is at most n and the claim of the lemma follows. □

Lemma 9.8 implies that if $n \in \{2, 3\}$, then the maximum size anticode in $\text{J}_3(n, n-1)$ has four anticodewords and if $n \geq 4$, then the maximum size anticode in $\text{J}_3(n, n-1)$ has n anticodewords. There are two nonisomorphic anticodes with n anticodewords when $n \geq 5$ (there are four nonisomorphic when $n = 4$). The first anticode of zise n is a ball and the second anticode consists of n binary words in $\text{J}_3(n, n-1)$. Perfect diameter constant-weight codes in this case are perfect constant-weight codes since they meet the sphere-packing bound which is the same as the code-anticode bound in this case.

Example 9.1. If $n = 3$ then the four anticodewords $\{011,\ 012,\ 110,\ 210\}$ form an anticode in $\text{J}_3(3, 2)$ whose diameter is two. Another nonisomorphic anticode of the same size and diameter is $\{011,\ 012,\ 021,\ 022\}$.

If $n = 4$ then there are four nonisomorphic anticodes of size four and diameter two in $J_3(4, 3)$. These four anticodes can be taken as:

(1) {1101, 1102, 1110, 1120}.
(2) {1011, 1012, 1021, 1022}.
(3) {0111, 0211, 0121, 0112}.
(4) {0111, 1011, 1101, 1110}.

If $n = 7$ then there are two nonisomorphic anticodes of size 7 with diameter two in $J_3(7, 6)$. These two anticodes can be taken as:

(1) {0111111, 0211111, 0121111, 0112111, 0111211, 0111121, 0111112}.
(2) {0111111, 1011111, 1101111, 1110111, 1111011, 1111101, 1111110}.

Are there ternary perfect 3-diameter codes in $J_3(n, n-1)$? To answer this question we first find the size of the maximum size anticode with diameter 3.

Lemma 9.9. *If* $n \geq 5$, *then the size of an anticode with diameter 3 in* $J_3(n, n-1)$ *is at most* $3n - 2$.

Proof. Assume that $\mathcal{A}$ is an anticode with diameter 3 in $J_3(n, n-1)$ having at least $2n + 1$ anticodewords. This implies that there are at least 3 anticodewords x, y, z with a *zero* in a common position. It is easy to verify that the distance between two of these three anticodewords is at least two. We distinguish between two cases depending on whether this distance is two or three.

Case 1. Assume that two of the anticodewords, say x and y, have distance 3. W.l.o.g. we assume that

$$\begin{aligned} x &= 011111\cdots 1, \\ y &= 022211\cdots 1, \\ z &= 0122\phi 1\cdots 1, \end{aligned}$$

where ϕ can be either 1 or 2. By the structure of x ans y we have that there is no anticodeword with a *zero* in one of the last $n - 4$ positions. Moreowver, if the *zero* is in position 2, 3, or 4, then the last $n-4$ positions are *ones*. Hence, there are only the following 8 (or less if ϕ is 1) possible anticodewords at distance at most 3 from each of x, y, z:

$$\begin{array}{cccc} 1012\,1\cdots 1, & 2012\,1\cdots 1, & 1021\,1\cdots 1, & 2021\,1\cdots 1, \\ 2120\,1\cdots 1, & 1120\,1\cdots 1, & 2102\,1\cdots 1, & 1102\,1\cdots 1. \end{array}$$

Since the distance between a pair of words in a column (of these 8 possible anticodewords) is 4, it follows that at most four of these words are contained in $\mathcal{A}$. The set of anticodewords in $\mathcal{A}$ with a *zero* in the first position form a binary anticode of length $n-1$ and diameter 3 in the Hamming scheme. Such an anticode has at most $2(n-1)$ anticodewords. Hence, in this case $\mathcal{A}$ has at most $2n+2$ anticodewords and since $3n-2 \geq 2n+2$ if $n \geq 4$, the claim of the lemma follows,

Case 2. Assume now that x and y have distance 2. W.l.o.g. we assume that

$$x = 01211 \cdots 1,$$
$$y = 02111 \cdots 1,$$
$$z = 011\phi 1 \cdots 1.$$

Note, that if $n \geq 4$, then there are at most n anticodewords with a *zero* in the first position since these anticodewords form a binary anticode of length $n-1$ and diameter 2 in the Hamming scheme.

Consider now the number of anticodewords with a *zero* not in the first position. There are $2(n-3)$ words with a *zero* in the last $n-3$ positions that could be anticodewords in $\mathcal{A}$. These words are

$$11101\cdots 1, \quad 111101\cdots 1, \quad \ldots, \quad 111\cdots 10, \quad 111\cdots 101, \tag{9.3}$$
$$21101\cdots 1, \quad 211101\cdots 1, \quad \ldots, \quad 211\cdots 10, \quad 211\cdots 101. \tag{9.4}$$

There are only eight words with a *zero* in the second or third position that could be anticodewords in $\mathcal{A}$. The first four words are

$$10111\cdots 1, \quad 20111\cdots 1, \quad 11011\cdots 1, \quad 21011\cdots 1.$$

The other four words are

$$10211\cdots 1, \quad 20211\cdots 1, \quad 12011\cdots 1, \quad 22011\cdots 1.$$

If $n = 4$ these eight words and the four anticodewords with a *zero* in the first position form and anticode of size 12. But, each one of the last four words is at distance four from each one of the words either in (9.3) or in (9.4). Thus, if $n \geq 5$, then there are at most $n + 2(n-3) + 4 = 3n-2$ anticodewords in $\mathcal{A}$. □

Lemma 9.9 implies that if $n \geq 5$, then the maximum size anticode in $\mathrm{J}_3(n, n-1)$ has at most $3n-2$ anticodewords. Fortunately, such an anticode with $3n-2$ anticodewords always exists. Let x be a word in $\mathrm{J}_3(n, n-1)$ and let x_1, x_2 be the two words in $\mathrm{J}_3(n, n)$ whose distance from x is exactly one, where x_1 has a *one* instead of the *zero* in x and x_2 has a *two* instead of the *zero* in x. One can easily verify that the set

$$\{y \ : \ d(x,y) \leq 1 \text{ or } d(x_1,y) \leq 1 \text{ or } d(x_2,y) \leq 1, \ y \in \mathrm{J}_3(n,n-1)\},$$

is an anticode with diameter 3 in $\mathrm{J}_3(n,n-1)$, whose size is $3n-2$ and hence it is a maximum size anticode.

Example 9.2. When $n = 6$ the following anticode in $\mathrm{J}_3(6,5)$ has diameter 3 and 16 anticodewords:

$$\begin{array}{cccc} 011111, & 021111, & 012111, & 011211, \\ 011121, & 011112, & 111011, & 111101, \\ 111110, & 211011, & 211101, & 211110, \\ 101111, & 201111, & 110111, & 210111. \end{array}$$

If a code $\mathcal{C}$ in $\mathrm{J}_3(n,n-1)$ is a 3-diameter perfect code then by the code-anticode bound we have that $3n-2$ divides $|\mathrm{J}_3(n,n-1)| = n2^{n-1}$. Since $\gcd(3n-2,n) = 1$, it follows that $3n-2$ divides 2^{n-1}, i.e., $3n-2 = 2^{2m}$ for some m. When $m = 2$ we have that $n = 6$ and there exists such a code of cardinality 12 whose codewords are

$$\begin{array}{cccc} 011221, & 022112, & 102121, & 201212, \\ 120211, & 210122, & 212011, & 121022, \\ 221101, & 112202, & 111110, & 222220. \end{array}$$

The next value of m is 3 which implies that $n = 22$. Do we have a 3-diameter perfect code in $\mathrm{J}_3(n,n-1)$ for $n = 22$ or for a larger n? To continue this discussion we will mention the concept of a perfect coloring. A coloring of a graph is called a ***perfect coloring*** if the multiset of colors of all neighbours of a vertex depends only on its own color. In particular for a perfect 2-coloring there are two colors μ_1 and μ_2, for the vertices of the graph, and the neighbours of the colors can be represented by an 2×2 ***partition array***

$$\begin{bmatrix} a_{11} & a_{12} \\ a_{21} & a_{22} \end{bmatrix},$$

where a_{ij} is the number of neighbours with the color μ_j of a vertex with the color μ_i. We are interested in the 2-coloring of the n-dimensional hypercube (the vertices are the elements of $\mathbb{F}_2^n$ and $x, y \in \mathbb{F}_2^n$ are connected by an edge if and only if $d_H(x,y) = 1$) which is isomorphic to $\mathcal{H}_2(n)$. The following tradeoff between the entries of the partition array of a perfect 2-coloring is known.

Theorem 9.7. *If $\{A, B\}$ is a partition of the colors associated with a perfect 2-coloring of the graph $\mathcal{H}_2(n)$ with the partition array*

$$\begin{bmatrix} a_{11} & a_{12} \\ a_{21} & a_{22} \end{bmatrix},$$

where $a_{12} \neq a_{21}$, then $a_{11} \geq \frac{3a_{21}-a_{12}}{4}$.

Assume now that $\mathcal{C}$ is a perfect 3-diameter code in $\mathrm{J}_3(n, n-1)$. By the code-anticode bound we have that $|\mathcal{C}| = \frac{n2^{n-1}}{3n-2}$.

Let $\mathcal{C}_1 \triangleq \{x \in \mathbb{F}_2^n : d(x, \mathcal{C}) = 1\}$ and $\mathcal{C}_2 \triangleq \mathbb{F}_2^n \setminus \mathcal{C}_1$. Clearly, $|\mathcal{C}_1| = 2|\mathcal{C}| = \frac{n2^n}{3n-2}$ and $|\mathcal{C}_2| = 2^n - |\mathcal{C}_1| = \frac{(n-1)2^{n+1}}{3n-2}$.

Since $\mathcal{C}$ is a code in $\mathrm{J}_3(n, n-1)$, while $\mathcal{C}_1$ contain words only from $\mathbb{F}_2^n$ which are at distance one from $\mathcal{C}$, it follows that each codeword of $\mathcal{C}_1$ has exactly one neighbour from $\mathcal{C}_1$ and hence $n-1$ neighbours in $\mathcal{C}_2$. Consider now the number of neighbours from $\mathcal{C}_1$ that a word from $\mathcal{C}_2$ has. Assume $(b_1, b_2, b_3, \ldots, b_n)$ is a word in $\mathcal{C}_2$ and $(\beta_1, b_2, b_3, \ldots, b_n)$, $(b_1, \beta_2, b_3, \ldots, b_n)$ are words in $\mathcal{C}_1$, where $\beta_i = 1$ if $b_i = 2$ and $\beta_i = 2$ if $b_i = 1$, where $i \in \{1, 2\}$. The two associated codewords of $\mathcal{C}$ which are at distance one to these two words in $\mathcal{C}_1$ must have their *zeroes* in different positions since the minimum distance in $\mathcal{C}$ is 4. This implies that each word from $\mathcal{C}_2$ has at most $n/2$ neighbours from $\mathcal{C}_1$.

Each word in $\mathcal{C}_1$ is a neighbour of $n-1$ words in $\mathcal{C}_2$ and hence the total number of pairs in the set $\mathcal{P} \triangleq \{(c_1, c_2) : c_1 \in \mathcal{C}_1,\ c_2 \in \mathcal{C}_2\}$ is $\frac{2^n}{3n-2}(n-1)$. Since $|\mathcal{C}_2| = \frac{(n-1)2^{n+1}}{3n-2}$, it follows that in average a word in $\mathcal{C}_2$ is contained in $n/2$ pairs of $\mathcal{P}$. Since each word in $\mathcal{C}_2$ has at most $n/2$ neighbours from $\mathcal{C}_1$, it follows that it has exactly $n/2$ neighbours from $\mathcal{C}_1$. As a consequence each word in $\mathcal{C}_2$ has exactly $n/2$ neighbours also from $\mathcal{C}_2$. If all the words of $\mathcal{C}_1$ are colored by μ_1 and all the words of $\mathcal{C}_2$ are colored by μ_2, then this 2-coloring is a perfect 2-coloring with the partition array

$$\begin{bmatrix} 1 & n-1 \\ n/2 & n/2 \end{bmatrix}.$$

But, by Theorem 9.7 such a perfect 2-coloring cannot exist if $n > 10$. On the other hand we saw a code which yield such a perfect 2-coloring for $n = 6$. Since such a perfect 2-coloring of $\mathbb{F}_2^n$ exists only for $n = 6$, it follows that a perfect 3-diameter code in $\mathrm{J}_3(n, n-1)$ exists only for $n = 6$.

Finally, to end this section we will mention that 4-diameter perfect code in $J_3(n, n-1)$ is known to exist for each $n = 2^m$, where $m \geq 3$ is and m is an odd integer and also for $n = 64$.

9.5 Generalized Steiner Systems

We continue with the third family, [**F3**], of nonbinary diameter constant-weight codes. We already saw that a Steiner system $S(t, w, n)$ is a binary $(w-t)$-diameter perfect constant-weight code. For a nonbinary alphabet, we can use the definition of generalized Steiner system that was introduced in Section 3.1.

By Lemma 3.3 we have that the number of codewords in a generalized Steiner system $GS(t, w, n, q)$ is

$$\frac{\binom{n}{t}}{\binom{w}{t}}(q-1)^t .$$

By definition, the minimum Hamming distance ofa generalized Steiner system $GS(t, w, n, q)$ is $2(w-t)+1$.

Let $\mathcal{A}^s(n, w, t)$ be the anticode defined by

$$\mathcal{A}^s(n, w, t) \triangleq \{(\overbrace{1 \cdots\cdots 1}^{t \text{ times}}, a_1, \ldots, a_{n-t}) \;:\; a_i \in \mathbb{Z}_q, \; \text{wt}(a_1 \cdots a_{n-t}) = w-t\} .$$

Note that when $q = 2$, we have that $\mathcal{A}^s(n, w, t)$ is identical to $\mathcal{A}_2(n, w, t)$ defined in Lemma 8.7.

The following lemma can be readily verified.

Lemma 9.10. *The anticode $\mathcal{A}^s(n, w, t)$, where $2w - t \leq n$, over $\mathbb{Z}_q$, has maximum distance $2(w-t)$ and $\binom{n-t}{w-t}(q-1)^{w-t}$ anticodewords of length n and weight w.*

Lemma 9.11. *If there exists a generalized Steiner system $S(t, w, n, q)$, then the anticode $\mathcal{A}^s(n, w, t)$ is a maximum size anticode of length n, weight w, and maximum distance $2(w-t)$, over $\mathbb{Z}_q$.*

Proof. Let $\mathcal{C}$ be a generalized Steiner system $GS(t, w, n, q)$ and let $\mathcal{A}$ be the anticode $\mathcal{A}^s(n, w, t)$. By the definition of a generalized Steiner system and by Lemma 3.3, $\mathcal{C}$ has minimum Hamming distance $2(w-t)+1$ and its size is $\frac{\binom{n}{t}}{\binom{w}{t}}(q-1)^t$. By Lemma 9.10, the anticode $\mathcal{A}^s(n, w, t)$ has maximum distance $2(w-t)$ and its size is $\binom{n-t}{w-t}(q-1)^{w-t}$. Since

$$|\mathcal{C}| \cdot |\mathcal{A}| = \frac{\binom{n}{t}}{\binom{w}{t}}(q-1)^t \cdot \binom{n-t}{w-t}(q-1)^{w-t} = \binom{n}{w}(q-1)^w = |J_q(n, w)| ,$$

it follows by the code-anticode bound that $\mathcal{A}^{s}(n,w,t)$ is a maximum size anticode of length n and weight w, over $\mathbb{Z}_q$, whose maximum distance is $2(w-t)$. $\square$

Corollary 9.3. *A generalized Steiner system* $\mathrm{GS}(t,w,n,q)$ *is a* $2(w-t)$-*diameter perfect code.*

As mentioned in the Chapter 2, an anticode in some metrics is equivalent to a t-intersecting family and an anticode of maximum size is equivalent to a t-intersecting family of the largest size. In our context we have by Lemma 9.11 that $\mathcal{A}^{s}(n,w,t))$ is such a family of maximum size. It is readily verified that all the codewords in the anticode $\mathcal{A}^{s}(n,w,t)$ have the same entries in the first t coordinates and hence they form a t-intersecting family in $\mathrm{J}_q(n,w)$. This is a t-intersecting family of maximum size when the related generalized Steiner system exists (but also for some other parameters), i.e., the largest set in $\mathrm{J}_q(n,w)$ in which each pair of words have at least t coordinates with the same nonzero entries.

9.6 Maximum Distance Separable Constant-Weight Codes

The next family, [**F4**], of nonbinary diameter perfect constant-weight codes in $\mathrm{J}_q(n,w)$ contains codes for which some can be derived from codewords of minimum weight in an MDS code (or an orthogonal array with the all-zero codeword). In this family, however, there are also nonbinary codes whose parameters are not associated with minimum weight codewords of MDS codes or orthogonal arrays with index unity (since MDS codes or orthogonal arrays with these parameters do not exist).

Assume we are given a constant-weight code of length n, and weight w over an alphabet with q symbols and $\binom{n}{w}(q-1)$ codewords, where each w coordinates are the support of exactly $q-1$ codewords. If $q=2$, then the minimum distance of the code is 2. If $q>2$, then there are at least two codewords on each projection of w coordinates and hence the minimum distance of the code should satisfy $d \leq w$.

Definition 9.1. An (n,w,q) **MDS *constant-weight code*** (MDS-CW code in short) is a constant-weight code of length n, weight w, minimum distance $d=w$, over an alphabet with q symbols, and $\binom{n}{w}(q-1)$ codewords. In other words, each w coordinates form a support for exactly $q-1$ codewords.

Note, that an (n, w, q) MDS-CW code is an $(n, w, w)_q$ code. If $q = 2$, then the minimum distance of an MDS-CW code must be 2, and hence a binary MDS-CW code exists if and only if $n \geq 2$ and $w = 2$. If $q > 2$, then codewords of minimum weight $d = w$ in an MDS code over $\mathbb{F}_q$ form an (n, w, q) MDS-CW code. Are there any other MDS-CW codes? The answer is clearly positive when orthogonal arrays, which are nonlinear codes, take the role of the MDS codes. W.l.o.g. we can assume that the all-zero row is a codeword in such an orthogonal array. With this assumption, the nonzero rows with minimum weight, in such an orthogonal array, define an MDS-CW code as will be proved in Theorem 9.10. There are also more MDS-CW codes that are not derived directly from MDS codes or orthogonal arrays.

MDS codes exist for all parameters that are specified in Theorem 3.29 and there are also some other infinite families of orthogonal arrays. Finally, by Theorem 3.4, OA$(2, n, q)$ is equivalent to a set of $n - 2$ pairwise disjoint orthogonal Latin squares of order q. From all these facts, one can construct MDS-CW codes with various parameters. We are interested in introducing constructions that are not derived from the codewords of minimum weight in an orthogonal array. Hence, we turn our discussion first to bounds and constructions for MDS-CW codes and their connection with orthogonal arrays.

Theorem 9.8. *If there exists an (n, w, q) MDS-CW code, then there exists an $(n - 1, w, q)$ MDS-CW code.*

Proof. Let $\mathcal{C}$ be an (n, w, q) MDS-CW code. Shortening $\mathcal{C}$, with respect to any coordinate, yields an $(n - 1, w, q)$ MDS-CW code. □

Theorem 9.9. *If there exists an (n, w, q) MDS-CW code, then there exists an $(n - 1, w - 1, q)$ MDS-CW code.*

Proof. Let $\mathcal{C}$ be an (n, w, q) MDS-CW code and define

$$\mathcal{C}_1 \triangleq \{(c_1, c_2, \ldots, c_{n-1}) \ : \ (c_1, c_2, \ldots, c_{n-1}, c_n) \in \mathcal{C}, \ c_n \neq 0\}.$$

It is readily verified that $\mathcal{C}_1$ is an $(n - 1, w - 1, q)$ MDS-CW code. □

Theorem 9.10. *The rows of weight $w = n - t + 1$ in an orthogonal array* OA(t, n, q), *which contains the all-zero row, form an (n, w, q) MDS-CW code.*

Proof. Assume that A is an OA(t, n, q) over an alphabet Q with q symbols that contains an all-zero row. The first step is to prove that any

$w = n - t + 1$ coordinates are supports of exactly $q - 1$ codewords. This will also prove the weight w of the code. Assume that the orthogonal array A is on the set of coordinates $\mathbb{Z}_n$. Given any set W of w coordinates, if $x \in W$, then $\mathbb{Z}_n \setminus W \cup \{x\}$ contains $n - (n - t + 1) + 1 = t$ coordinates, which must include all possible t-tuples since in an $\mathrm{OA}(t, n, q)$, each t-tuple, over Q, is contained exactly once in any projection on t coordinates of A. By taking the $q - 1$ words with $t - 1$ zeroes in the $t - 1$ coordinates of $\mathbb{Z}_n \setminus W$ and each of the $q - 1$ nonzero symbols of the alphabet Q in the coordinate of x, we must have that the corresponding $q - 1$ rows in A have weight w, i.e., all the other coordinates are nonzero. Otherwise the all-zero t-tuple will be contained twice in some t coordinates, one in these $q - 1$ rows and one in the all-zero row. This completes the proof of the first step as the set W can be taken as any set of w coordinates. Finally, the minimum Hamming distance of the code is also an immediate result since by Theorem 3.3, the minimum distance of the code derived from an $\mathrm{OA}(t = n - w + 1, n, q)$ is w. □

Corollary 9.4. *If there exists an* $\mathrm{OA}(t, n, q)$, *then there exists an* $(n, n - t + 1, q)$ *MDS-CW code.*

Trivial MDS-CW codes are derived similarly to (or from) trivial orthogonal arrays. For $w = 1$ all possible words of weight one and length n over an alphabet of size q form an $(n, 1, q)$ MDS-CW code related to an $\mathrm{OA}(n, n, q)$. For $w = 2$, the set of all $\binom{n}{2}(q - 1)$ possible words with weight two with two equal nonzero entries on the two nonzero coordinates forms an $(n, 2, q)$ MDS-CW code. This code is of size $\binom{n}{2}(q - 1)$ and is related to an $\mathrm{OA}(n - 1, n, q)$. For $w = n$, the set $\{(\alpha, \alpha, \ldots, \alpha) \ : \ \alpha \in [q - 1]\}$ forms an (n, n, q) MDS-CW code related to an $\mathrm{OA}(1, n, q)$. The construction of MDS-CW codes from orthogonal arrays is simple, but the main question is whether there exist MDS-CW codes that cannot be obtained from orthogonal arrays. This will be the next goal in our exposition.

Theorem 9.11. *If there exists an* (n, w, q_1) *MDS-CW code and there exists an* (n, w, q_2) *MDS-CW code, then there exists an* $(n, w, q_1 + q_2 - 1)$ *MDS-CW code.*

Proof. Assume that there exists an (n, w, q_i) MDS-CW code $\mathcal{C}_i$ over Q_i, $i = 1, 2$, where $Q_1 \cap Q_2 = \{0\}$. It is easy to verify that $\mathcal{C}_1 \cup \mathcal{C}_2$ is an $(n, w, q_1 + q_2 - 1)$ MDS-CW code over $Q_1 \cup Q_2$. □

Corollary 9.5. *If there exists an (n, w, q) MDS-CW code, then there exists an $(n, w, r(q-1)+1)$ MDS-CW code for each $r > 0$.*

By Corollary 9.4, from an $\mathrm{OA}(2, n, q)$ we can obtain an $(n, n-1, q)$ MDS-CW code. Now, we will show another type of an MDS-CW code obtained from an $\mathrm{OA}(2, n, q)$. Assume A is an $\mathrm{OA}(2, n, q)$ over $[q]$, which implies that there are no *zeroes* in the array, such that the first q symbols in the first column of A are *ones*, the next q symbols in the first column are *twos*, and so on. Delete the first column of A to obtain an array B. In B, replace the first symbol in the first q rows with *zeroes*, the second symbol in the next q rows will be replaced with *zeroes*, the third symbol in the next q rows will be replaced with *zeroes*, and so on. Rows for which no symbol was replaced are removed and the new array obtained is $\mathcal{M}$. Clearly, the array $\mathcal{M}$ has $n-1$ columns and, therefore, $(n-1)q$ rows (note that in an $\mathrm{OA}(2, n, q)$ there are q^2 rows and by Corollary 3.5 we have that $n \leq q+1$, and hence there are at least $(n-1)q$ rows in the array A and in the array B). The constructed array $\mathcal{M}$ is an $(n-1, n-2, q+1)$ MDS-CW code that implies the following theorem.

Theorem 9.12. *If there exists an* $\mathrm{OA}(2, n, q)$, *then there exists an* $(n-1, n-2, q+1)$ *MDS-CW code.*

An $(n-1, n-2, q+1)$ MDS-CW code obtained by Theorem 9.12 cannot be always extended into an $\mathrm{OA}(2, n, q+1)$. For example, from an $\mathrm{OA}(2, 6, 5)$ (which is equivalent to four orthogonal Latin squares of order 5 by Theorem 3.4), we obtain by Theorem 9.12 a $(5, 4, 6)$ MDS-CW code. If this $(5, 4, 6)$ MDS-CW code forms the $5 \cdot 4^4$ rows of minimum weight of an orthogonal array, then this array will be an $\mathrm{OA}(2, 5, 6)$, which is equivalent to three orthogonal Latin squares of order 6. But, by Theorem 3.27, there is no pair of orthogonal Latin squares of order 6, and hence there is no $\mathrm{OA}(2, 4, 6)$ and, of course, no $\mathrm{OA}(2, 5, 6)$. More generally, from an $\mathrm{OA}(2, q+1, q)$ (which is equivalent to $q-1$ orthogonal Latin squares of order q by Theorem 3.4), where q is a power of a prime, we obtain by Theorem 9.12 a $(q, q-1, q+1)$ MDS-CW code. If this code forms the minimum weight codewords of an orthogonal array, then this orthogonal array will be an $\mathrm{OA}(2, q, q+1)$. If $q+1$ is not a power of a prime, no such orthogonal array is known. The cases when both q and $q+1$ are powers of primes are quite rare. These cases are related to Mersenne primes when $q+1$ is a power of two or Fermat primes when q is a power of 2. Combinations of the arrays obtained from Theorem 9.12, other known MDS-CW codes obtained from orthogonal arrays, Theorem 9.11, and Corollary 9.5, would result in other MDS-CW codes that cannot be obtained from the known parameters of orthogonal arrays.

The construction that led to Theorem 9.12 will be generalized in Section 9.8 to obtain other MDS-CW codes and also other types of nonbinary diameter constant-weight codes.

We now want to derive bounds on the size of the alphabet, q, of an (n, w, q) MDS-CW code for $3 \leq w \leq n-1$. If $w \leq n-1$, we know that on each support of size w there are $q-1$ codewords. These codewords have distinct nonzero symbols in each coordinate and in each coordinate each of the nonzero $q-1$ symbols appears. Let S be such a set of codewords whose support is the first w coordinates. Another codeword c that shares exactly $w-1$ coordinates with the $q-1$ codewords of S cannot have more than one common symbol with each of these $q-1$ codewords (if they share the same nonzero symbols in two coordinates, then their distance will be less than w). Each entry of these $w-1$ entries in c must share a nonzero symbol with a different codeword of S. Therefore, we must have $q-1 \geq w-1$, i.e., $q \geq w$.

If $w \geq 3$, we consider a set S of codewords with nonzero symbols in the first $w-1$ coordinates and the same nonzero symbol in the first coordinate of all these codewords. This implies that $|S| \leq q-1$. Each codeword of S must have its last nonzero symbol in a distinct coordinate from the last $n-w+1$ coordinates. Therefore, $|S| \leq n-w+1$. Moreover, each of the last $n-w+1$ coordinates must have a nonzero symbol for one of the codewords in S. Therefore, $|S| \geq n-w+1$ and hence $|S| = n-w+1$. Finally, $|S| \leq q-1$ and hence $q-1 \geq n-w+1$, i.e., $q \geq n-w+2$.

Thus, we have proved the following theorem.

Theorem 9.13. *Let C be an (n, w, q) MDS-CW code.*

- *If $w \leq n-1$, then $q \geq w$.*
- *If $w \geq 3$, then $q \geq n-w+2$.*

So far, we have constructed MDS-CW codes from orthogonal arrays. In the next result we start with an MDS-CW code to obtain an orthogonal array.

Theorem 9.14. *If there exists an (n, w, w) MDS-CW code, where $n > w$, then there exists an* $\mathrm{OA}(2, w+1, w)$.

Proof. Let C be an $(n, w, w-1)$ MDS-CW code. We apply Theorem 9.8, by considering all the codewords with nonzero entries only in the first $w+1$ coordinates. These codewords form an $(w+1,w,w)$ MDS-CW code. If we add the all-zero word to this code we obtain a code of length $w+1$ over an alphabet with w symbols, $(w-1)(w+1)+1=w^2$ codewords, and minimum distance w. Hence, this code is an $\mathrm{OA}(2,w+1,w)$ and the claim of the theorem follows. □

Theorems 9.10 and 9.14 imply the following result.

Corollary 9.6. *There exists an* $\mathrm{OA}(2,w+1,w)$ *if and only if there exists an* $(w+1,w,w)$ *MDS-CW code.*

Given w and n, where $1 \leq w \leq n$, there exists some alphabet of size q for which there exists an (n,w,q) MDS-CW code since there exists an $[n,n-w+1,w]_q$ MDS code for each power of a prime $q \geq n-1$. Given n and w, is there an q' such that for each $q \geq q'$ there exists an (n,w,q) MDS-CW code? Such a result can be obtained based on the well-known partition theorem of Frobenius.

Theorem 9.15. *Every integer* q*, such that* $q > 2^{2m+1} - 3 \cdot 2^{m+1} + 3$ *can be represented as* $q = r_1(2^m - 1) + r_2(2^{m+1} - 1)$*, for some* $r_l, r_2 \geq 0$.

Theorem 9.16. *For each* n *and* w *there exists a* q_0 *such that for each* $q \geq q_0$ *there exists an* (n,w,q) *MDS-CW code.*

Proof. Let m be the smallest integer such that $2^m \geq n-1$. By Theorem 3.29, there exists an $\mathrm{OA}(n-w+1,n,2^m)$ and hence there exists an $(n,w,2^m)$ MDS-CW code. Similarly, there exists an $(n,w,2^{m+1})$ MDS-CW code. By Theorem 9.15 we have that every integer q, such that $q > 2^{2m+1} - 3 \cdot 2^{m+1} + 3$, can be represented as $q = r_1(2^m - 1) + r_2(2^{m+1} - 1)$, for some $r_l, r_2 \geq 0$. Therefore, by Theorem 9.11 and Corollary 9.5, for each n and w there exists a q_0 such that for each $q \geq q_0$ there exists an (n,w,q) MDS-CW code. □

Let $\mathrm{QMDS}(n,w)$ be the smallest integer such that for each $q \geq \mathrm{QMDS}(n,w)$ there exists an (n,w,q) MDS-CW code. The upper bounds on $\mathrm{QMDS}(n,w)$ implied by Theorems 9.15 and 9.16 might be weak, while the lower bounds implied by Theorem 9.13 might be impossible to attain. This leads to the natural research problem.

Problem 9.4. Find better lower and upper bounds on $\mathrm{QMDS}(n,w)$.

Usually, the bound implied by Theorems 9.15 and 9.16 can be improved by the same technique used in Theorem 9.16, if we find a prime power q, $n-1 \leq q < 2^{m+1}$ such that $q-1$ and $2^m - 1$ are relatively primes.

We continue with our goal to find diameter perfect codes in $\mathrm{J}_q(n,w)$ and examine if an (n,w,q) MDS-CW code whose size is $\binom{n}{w}(q-1)$ forms a diameter perfect constant-weight code in $\mathrm{J}_q(n,w)$. Let $\mathcal{A}^{\mathrm{m}}(n,w,\delta)$, $1 \leq \delta \leq w$, be the anticode defined as follows

$$\mathcal{A}^{\mathrm{m}}(n,w,\delta) \triangleq \{(a_1,a_2,\ldots,a_\delta,\overbrace{1\cdots\cdots1}^{w-\delta \text{ times}},\overbrace{0\cdots\cdots0}^{n-w \text{ times}}) \ : \ a_i \in \mathbb{Z}_q^-, 1 \leq i \leq \delta\} \ .$$

The following lemma can be readily verified.

Lemma 9.12. *The anticode $\mathcal{A}^{\mathrm{m}}(n,w,\delta)$ has anticodewords of length n, weight w, with maximum distance δ. The number of anticodewords in $\mathcal{A}^{\mathrm{m}}(n,w,\delta)$ is $(q-1)^\delta$.*

Lemma 9.13. *If there exists an (n,w,q) MDS-CW code, then the anticode $\mathcal{A}^{\mathrm{m}}(n,w,w-1)$ is a maximum size anticode of length n, weight w, and maximum distance $w-1$, over $\mathbb{Z}_q$.*

Proof. Let $\mathcal{C}$ be an (n,w,q) MDS-CW code and $\mathcal{A}$ be the anticode $\mathcal{A}^{\mathrm{m}}(n,w,w-1)$. By the definition, an (n,w,q) MDS-CW code, has minimum distance w and size $\binom{n}{w}(q-1)$. By Lemma 9.12, the anticode $\mathcal{A}^{\mathrm{m}}(n,w,w-1)$ has maximum distance $w-1$ and size $(q-1)^{w-1}$. Since

$$|\mathcal{C}| \cdot |\mathcal{A}| = \binom{n}{w}(q-1)\cdot(q-1)^{w-1} = \binom{n}{w}(q-1)^w = |\mathrm{J}_q(n,w)| \ ,$$

it follows by the code-anticode bound that the anticode $\mathcal{A}^{\mathrm{m}}(n,w,w-1)$ is a maximum size anticode of length n, weight w, and maximum distance $w-1$, over $\mathbb{Z}_q$. □

Corollary 9.7. *An (n,w,q) MDS-CW code is a $(w-1)$-diameter perfect code.*

Lemma 9.13 will be generalized later to show that $\mathcal{A}^{\mathrm{m}}(n,w,\delta)$ is a maximum size anticode for other parameters too.

9.7 Codes for which $d = w + 1$

The fifth family, [**F5**], of nonbinary diameter perfect constant-weight codes in $\mathrm{J}_q(n,w)$ is for $d = w+1$. When $d = w+1$ we are looking for an $(n,w+1,w)_q$ code, i.e., a constant-weight code of length n, weight w, and minimum Hamming distance $w + 1$, over $\mathbb{Z}_q$. In such a code, each subset of w coordinates will be the support exactly one codeword, which implies that the number of codewords is $\binom{n}{w}$. It is rather easy to verify that such a code is a w-diameter perfect constant-weight code and it exists for any given n and w as proved in the following theorem.

Theorem 9.17. *If n and w are integers such that $1 \leq w \leq n-1$, then there exists a $q_0(w, n)$ such that for each $q \geq q_0(w, n)$ there exist an $(n, w+1, w)_q$ w-diameter perfect code $\mathcal{C}$.*

Proof. First, note that since the minimum distance of an $(n, w + 1, w)_q$ code $\mathcal{C}$ is $w+1$, it follows that each subset of w coordinates can be a support for at most one codeword. If each such subset of w coordinates supports exactly one codeword, then the total number of codewords in $\mathcal{C}$ will be $\binom{n}{w}$. Assume further that in $\mathcal{C}$ for each coordinate all the nonzero elements in the codewords of $\mathcal{C}$ have distinct symbols. This implies that in each coordinate there are $\binom{n}{w}\frac{w}{n} = \binom{n-1}{w-1}$ nonzero symbols. Let $q' \triangleq 1 + \binom{n-1}{w-1}$ and let Q be an alphabet with $q = q' + \epsilon$ symbols, where $\epsilon \geq 0$. Assign now for each coordinate a different nonzero symbols from the $q' + \epsilon - 1$ nonzero symbols of Q^- to each codeword that has a nonzero symbol in this coordinate. Clearly, $\mathcal{C}$ in an $(n, w + 1, w)_q$ code with $\binom{n}{w}$ codewords.

Let $\mathcal{A}$ be the anticode $\mathcal{A}^{\mathrm{m}}(n, w, w)$ over Q. By Lemma 9.12, the anticode $\mathcal{A}$, has diameter w and $(q-1)^w$ anticodewords. Clearly,

$$|\mathcal{C}| \cdot |\mathcal{A}| = \binom{n}{w}(q-1)^w = |\mathrm{J}_q(n, w)|$$

and hence by the code-anticode bound, $\mathcal{C}$ is an $(n, w + 1, w)_q$ w-diameter perfect constant-weight code over the alphabet Q of size q. □

Corollary 9.8. *If there exists an $(n, w + 1, w)_q$ code with $\binom{n}{w}$ codewords, then $\mathcal{A}^{\mathrm{m}}(n, w, w)$ is a maximum size anticode of length n, weight w, and diameter w, over an alphabet with q symbols.*

The proof of Theorem 9.17 implies that indeed an $(n, w + 1, w)_q$ w-diameter constant-weight perfect code $\mathcal{C}$ has $\binom{n}{w}$ codewords, where each w-subset of w coordinates of $\mathcal{C}$ is the support of exactly one codeword of $\mathcal{C}$. In view of Theorem 9.17 our goal now is to find $q_0(w, n)$ which is the smallest size alphabet q for such an $(n, w + 1, w)_q$ code exists.

Corollary 9.9. *For each alphabet Q of size q, where $q \geq 1 + \binom{n-1}{w-1}$, there exists an $(n, w + 1, w)_q$ code, i.e., $q_0(w, n) \leq 1 + \binom{n-1}{w-1}$.*

Lemma 9.14. *For each $w \geq 1$ there exists an $(w+1, w+1, w)_{w+1}$ code which is a w-diameter perfect code.*

Proof. Follows immediately from the fact that if there is a codeword on each subset of w coordinates, then there are exactly w codewords with nonzero symbols on each coordinates. □

Corollary 9.10. *If $w \geq 1$, then $q_0(w, w+1) = w+1$.*

Theorem 9.17 implies the existence of an $(n, w+1, w)_q$ code for each $q \geq q_0(w, n)$, but the upper bound $1 + \binom{n-1}{w-1}$ on $q_0(w, n)$, inferred in Corollary 9.9, is quite large. Can we find a better upper bound on $q_0(w, n)$? The answer is definitely positive and for this purpose we have the following results.

Lemma 9.15. *If $n > w+1$, then $q_0(w, n) \geq q_0(w, n-1)$.*

Proof. Assume that $\mathcal{C}$ is an $(n, w+1, w)_q$ w-diameter constant-weight perfect code and let S be any subset of $n-1$ coordinates. By definition, the set codewords whose supports are subsets of S form a w-diameter perfect $(n-1, w+1, w)_q$ constant-weight code. Thus, the claim of the lemma follows. □

Corollary 9.11. *If $n > w+1$ then $q_0(w, n) \geq w+1$.*

Lemma 9.16. *If $n > w+1$, then $q_0(w, n) \geq n - w + 2$.*

Proof. Let $\mathcal{C}$ be a w-diameter perfect $(n, w+1, w)_q$ code $\mathcal{C}$. Consider the sub-code $\mathcal{C}'$ of codewords from $\mathcal{C}$ for which there is no *zero* in the first $w-1$ coordinates. Since, each one of the other $n-w+1$ coordinates must have a nonzero symbol with exactly one of these codewords, it follows that the sub-code $\mathcal{C}'$ contains $n-w+1$ codewords. Since the distance of $\mathcal{C}'$ is $w+1$, each pair of codewords of $\mathcal{C}'$ have only two distinct coordinates in their supports, and each pair of codewords of $\mathcal{C}'$ have $w-1$ joint coordinates with nonzero symbols, it follows that in each given coordinate of the first $w-1$ coordinates the codewords of $\mathcal{C}'$ have distinct nonzero symbols. Since $|\mathcal{C}'| = n - w + 1$, it follows that $\mathcal{C}$ has at least $n - w + 1$ nonzero symbols and hence $q \geq n - w + 2$. □

Corollary 9.12. *If $n > 2$, then $q_0(2, n) = n$.*

Proof. By Lemma 9.16 we have that $q_0(2, n) \geq n$ and by Corollary 9.9 we have that $q_0(2, n) \leq n$. Thus, $q_0(2, n) = n$. □

The proof of the next theorem requires two more concepts, a one-factorization and a near-one-factorization (see also Section 3.1). A *one-factorization* of the complete graph K_n, n even, is a partition of the edges of K_n (or all the pairs on an n-set) into perfect matchings. In other words, the set

$$\mathcal{F} = \{\mathcal{F}_1, \mathcal{F}_2, \ldots, \mathcal{F}_{n-1}\}$$

is a one-factorization of K_n if each $\mathcal{F}_i$, $1 \leq i \leq n-1$, is a perfect matching (called a *one-factor*), and the $\mathcal{F}_i$'s are pairwise disjoint.

If n is odd, then there is no perfect matching in K_n and we define a *near-one-factorization*

$$\mathcal{F} = \{\mathcal{F}_1, \mathcal{F}_2, \ldots, \mathcal{F}_n\}$$

to be a partition of the edges in K_n into sets of $\frac{n-1}{2}$ pairwise disjoint edges, where each $\mathcal{F}_i$ has one isolated vertex. Each $\mathcal{F}_i$ is called a *near-one-factor*.

Example 9.3. Let n be an odd integer and define the set

$$\mathcal{F} = \{\mathcal{F}_0, \mathcal{F}_1, \ldots, \mathcal{F}_{n-1}\},$$

where

$$\mathcal{F}_i\{\{i,j\} \ : \ i+j \equiv i \pmod{n}\}$$

for each i, $0 \leq i \leq n-1$. For each such i, the only integer which is not contained in a pair of $\mathcal{F}_i$ is $i/2$ modulo n. $\mathcal{F}$ is a near-one-factorization on $\mathbb{Z}_n$ from which a one-factorization on $\mathbb{Z}_{n+1}$ was constructed in Section 3.1.

Theorem 9.18. *If n is odd, then $q_0(3,n) = n-1$, and if n is even, then $q_0(3,n) = n$.*

Proof. By Lemma 9.16 we have that $q_0(3,n) \geq n-1$ and this bound is applied when n is odd.

Assume now that n is even and let $\mathcal{C}$ be a related code. Let $\mathcal{C}_1$ be the set of codewords in $\mathcal{C}$ with a nonzero symbol in the first coordinate. By the definition of this family of codes, it follows that $|\mathcal{C}_1| = \binom{n-1}{2} = \frac{(n-1)(n-2)}{2}$. Since the minimum distance of $\mathcal{C}_1$ is four and n is even, it follows that the number of codewords in $\mathcal{C}_1$ with a given nonzero symbol σ in the first coordinate is at most $\frac{n-2}{2}$. Since $|\mathcal{C}_1| = \frac{(n-1)(n-2)}{2}$, it follows that there are at least $n-1$ nonzero symbols in the first coordinate. Therefore, $q_0(3,n) \geq n$ if n is even.

Regarding the upper bound on $q_0(3,n)$ we distinguish again between two cases, depending on whether n is odd or n is even.

Case 1. n is odd.

Let $\mathcal{N}$ be the set of n coordinates and let $Q \triangleq \{0, \sigma_1, \ldots, \sigma_{n-2}\}$ be an alphabet of size $n - 1$. Let $\mathcal{C}$ be a code of length n and weight 3 with $\binom{n}{3}$ codewords, a codeword for each support of size 3. Consider the i-th coordinate, $i \in \mathcal{N}$ and let $\mathcal{F} = \{\mathcal{F}_1, \mathcal{F}_2, \ldots, \mathcal{F}_{n-2}\}$ be a one-factorization on the $n - 1$ points of $\mathcal{N} \setminus \{i\}$. Given a triple $\{i, j, k\}$, where $\{j, k\} \in \mathcal{F}_r$, we assign σ_r to the symbol in coordinate i of the codeword $\{i, j, k\}$, where the nonzero symbols are in coordinates i, j, k. It is readily verified that we have constructed a code of length n and weight 3, over an alphabet Q of size $n - 1$. Clearly, if two codewords share at most one coordinate, then their Hamming distance is at least 4. Now, assume that two codewords c_1 and c_2 share nonzero symbols in two coordinates i and j. If the symbols in the i-th coordinate of c_1 and c_2 are distinct and the symbols in the j-th coordinate of c_1 and c_2 are distinct, then clearly $d(c_1, c_2) = 4$. Now, assume for the contrary that in one coordinate, say i, c_1 and c_2 have the same symbol. By the construction, we have that the two other pairs of nonzero coordinates in c_1 and c_2 must be disjoint (they belong to the same one-factor), a contradiction. Therefore, the minimum distance of $\mathcal{C}$ is 4 and hence $q_0(3, n) \leq n - 1$.

Case 2. n is even.

Let $\mathcal{N}$ be the set of n coordinates and let $Q \triangleq \{0, \sigma_1, \ldots, \sigma_{n-1}\}$ be an alphabet of size n. Let $\mathcal{C}$ be a code of length n and weight 3 with $\binom{n}{3}$ codewords, a codeword for each support of size 3. Consider the i-th coordinate, $i \in \mathcal{N}$ and let $\mathcal{F} = \{\mathcal{F}_1, \mathcal{F}_2, \ldots, \mathcal{F}_{n-1}\}$ be a near-one-factorization on the $n - 1$ points of $\mathcal{N} \setminus \{i\}$. Given a triple $\{i, j, k\}$, where $\{j, k\} \in \mathcal{F}_r$, we assign σ_r to the symbol in coordinate i of the codeword $\{i, j, k\}$. It is readily verified that we have constructed a code of length n and weight 3, over an alphabet Q of size n. As in Case 1 the minimum distance of $\mathcal{C}$ is 4 and therefore, $q_0(3, n) \leq n$.

Thus, these two cases complete the proof of the theorem. □

Similarly to the technique used in the proof of Theorem 9.18 one can construct w-diameter perfect $(n, w + 1, w)_q$ codes, for relatively small q, when w is small using techniques coming from combinatorial designs. The same is true for w-diameter perfect $(n, w + 1, w)_q$ codes, when n is not much larger than w. Such constructions are left for future research. Moreover, the technique used in Theorem 9.18 to obtain the upper bound on $q_0(3, n)$ can be used to obtain better upper bounds on $q_0(w+1, n+1)$ than the trivial one, i.e., $q_0(w + 1, n + 1) \leq 1 + \binom{n}{w}$. The idea is to partition the set of all binary words of length n and weight w into pairwise disjoint constant-weight codes of length n, weight w, and minimum Hamming distance $w+2$. Let $\chi(n,w)$ be the minimum number of codes in such a partition. With an identical proof as the one in Theorem 9.18 we can prove that $q_0(w+1,n+1) \leq \chi(n,w)+1$. Note, that the minimum distance of a constant-weight code is always even and hence the proof will be more effective for even w, i.e., for bounds on $q_0(w+1,n+1)$ when $w+1$ is odd. This kind of partitions lead to the following result.

Theorem 9.19. *If w is even and p is the smallest prime power for which $p \geq n$, then*

$$q_0(w+1,n+1) \leq 1+p^{w/2} .$$

Although the bound in Theorem 9.19 is much better than the trivial bound $q_0(w+1,n+1) \leq 1+\binom{n}{w}$, it is reasonable to assume that this bound can considerably be improved.

Problem 9.5. Improve the upper bound on $q_0(w,n)$.

9.8 Multiple Orthogonal Arrays Constant-Weight Codes

In the last family of diameter perfect constant-weight codes we have similarly as in the families [**F4**] and [**F5**] that each projection of any w coordinates is the support of some specified number of codewords. The distinction from the families [**F4**] and [**F5**] is that the minimum distance of the code in this family, [**F6**], is strictly smaller than the weight of the codewords. More precisely we have the following definition.

Definition 9.2. An $(n,d,w)_q$ ***multiple orthogonal arrays constant weight*** (MOA-CW in short) code is a code of length n, constant weight w, minimum distance $d < w$, where each subset of w coordinates is the support of exactly $(q-1)^{w-d+1}$ codewords, i.e., these codewords form an $\text{OA}(w-d+1,w,q-1)$.

One might asks why this family does not include the MDS-CW codes, where $d=w$. The two families of codes, [**F4**] and [**F6**] share some properties such as similar expression of their size (which is also shared by the family [**F5**]), they are both related to orthogonal arrays (MDS codes) in a way that codewords with no *zeroes* in the same w coordinates form an orthogonal array. The main reason for the distinction between the two families is that an MDS-CW code either forms the codewords of minimum weight in an orthogonal array or has the same parameters as it would have had, if such an orthogonal array have been possible. There is no similar property for an MOA-CW code. The codewords of an MOA-CW code are not associated with codewords of some weights in MDS codes or orthogonal arrays. Another important distinction is in the simple union construction of Theorem 9.11 which cannot be applied to MOA-CW codes. The similarity of the two families will also be demonstrated in one construction of such codes which is a joint construction for both families of codes. Similarly, some bounds on the tradeoff between the parameters of these codes are joint bounds for the two families of codes.

Theorem 9.20. *An $(n,d,w)_q$ MOA-CW code is a $(d-1)$-diameter perfect constant-weight code.*

Proof. By definition, the size of an $(n,d,w)_q$ MOA-CW code $\mathcal{C}$ is $\binom{n}{w}(q-1)^{w-d+1}$ and by Lemma 9.12 the related anticode $\mathcal{A}$ with diameter $d-1$ in $\mathrm{J}_q(n,w)$, $\mathcal{A}^{\mathrm{m}}(n,w,d-1)$, has size $(q-1)^{d-1}$. Therefore we have that,

$$|\mathcal{C}|\cdot|\mathcal{A}| = \binom{n}{w}(q-1)^{w-d+1}(q-1)^{d-1} = \binom{n}{w}(q-1)^{w} = |J_q(n,w)| \ ,$$

which by the code-anticode bound implies that $\mathcal{C}$ is a $(d-1)$-diameter perfect constant-weight code. □

Corollary 9.13. *If there exists an $(n,d,w)_q$ MOA-CW code, then $\mathcal{A}^{m}(n,w,d-1)$ is a maximum size anticode of length n, weight w, and diameter $d-1$, over an alphabet with q symbols.*

Next, we present a construction for MOA-CW codes which can also serve as a construction for MDS-CW codes. The construction is a generalization and a modification of a construction related to Theorem 9.12 to obtain MDS-CW codes. Let $\mathcal{M}$ be an OA(t,n,q) over Q, where $Q=\{1,2,\ldots,q\}$. Assume further that the symbols in the last coordinate of the first q^{t-1} codewords of $\mathcal{M}$ are *ones*, the symbols in the last coordinate of the next q^{t-1} codewords of $\mathcal{M}$ are *twos*, and so on, where the symbols in the last coordinate of the last q^{t-1} codewords of $\mathcal{M}$ are q's. Assume further that $q \geq \binom{n-1}{\ell}$ for a given ℓ, $1\leq \ell \leq n-1$. Let $S_1,S_2,\ldots,S_r$, where $r=\binom{n-1}{\ell}$, be a sequence containing all the ℓ-subsets of $\{1,2,\ldots,n-1\}$. Let $\mathcal{M}'$ be the $(\binom{n-1}{\ell}q^{t-1})\times(n-1)$ array constructed from $\mathcal{M}$ as follows.

(1) If $S_1=\{i_1,i_2,\ldots,i_\ell\}$, then replace all the symbols in the first q^{t-1} rows of column i_j in $\mathcal{M}$, for each $1\leq j\leq \ell$, with *zeroes*.

(2) If $S_2 = \{i_1, i_2, \ldots, i_\ell\}$, then replace all the symbols in the next q^{t-1} rows of column i_j in $\mathcal{M}$, for each $1 \leq j \leq \ell$, with *zeroes*.
(3) Continue the same process with S_3, S_4, and so on until S_r.
(4) Remove the last column of $\mathcal{M}$.
(5) Remove the last $q^t - (\binom{n-1}{\ell} q^{t-1})$ rows of $\mathcal{M}$.

Theorem 9.21. *The rows of the array $\mathcal{M}'$, obtained in the construction, form an $(n-1, n-t-\ell+1, n-\ell-1)_{q+1}$ code $\mathcal{C}$ that is an $(n-t-\ell)$-diameter perfect constant-weight code over $Q \cup \{0\}$.*

Proof. Exactly one column was deleted from $\mathcal{M}$ to obtain $\mathcal{M}'$ and hence the length of the code $\mathcal{M}'$ is $n-1$. In each codeword of length $n-1$ exactly ℓ *zeroes* were inserted instead of nonzero symbols and hence the weight of each codeword is $n-1-\ell$. Since $\mathcal{M}$ is an $\text{OA}(t, n, q)$, it follows that $|\mathcal{M}| = q^t$, and since $q \geq \binom{n-1}{\ell}$, it follows that $q^t \geq \binom{n-1}{\ell} q^{t-1}$ and hence $\mathcal{M}$ has at least $\binom{n-1}{\ell} q^{t-1}$ rows as required by the construction. Furthermore, note that the minimum distance of the code defined by $\mathcal{M}$ is $n-t+1$. Let c_1 and c_2 be two codewords in $\mathcal{M}'$. If the *zeroes* in c_1 and c_2 are on the same ℓ coordinates, then c_1 and c_2 were derived from two rows $c_1'\alpha$ and $c_2'\alpha$ of $\mathcal{M}$, where $\alpha \in \{1, 2, \ldots, q\}$, and $d(c_1'\alpha, c_2'\alpha) \geq n-t+1$. Since the same ℓ coordinates were changed in c_1' and c_2', respectively, to obtain c_1 and c_2, respectively, it follows that $d(c_1, c_2) \geq n-t+1-\ell$. If the *zeroes* in c_1 and c_2 are not on the same coordinates, then c_1 and c_2 were derived from two rows $c_1'\alpha$ and $c_2'\beta$, where $\alpha, \beta \in \{1, 2, \ldots, q\}$, $\alpha \neq \beta$, and $d(c_1'\alpha, c_2'\beta) \geq n-t+1$, which implies that $d(c_1', c_2') \geq n-t$. The number of coordinates in which both c_1 and c_2 have *zeroes* is at most $\ell - 1$ and hence $d(c_1, c_2) \geq d(c_1', c_2') - (\ell - 1) \geq n - t - (\ell - 1) = n - t - \ell + 1$. Thus, $d(\mathcal{C}) \geq n - t - \ell + 1$.

As an immediate consequence from the construction, the number of rows in the array $\mathcal{M}'$ is $\binom{n-1}{\ell} q^{t-1}$ and its alphabet $\{0, 1, 2, \ldots, q\}$ is of size $q+1$. Let $\mathcal{A}$ be a related anticode of length $n-1$ and diameter $n-t-\ell$. By Lemma 9.12 there exists such a constant-weight anticode $\mathcal{A}^{\text{m}}(n-1, n-1-\ell, n-t-\ell)$, over $\mathbb{Z}_{q+1}$, whose size is $q^{n-t-\ell}$. Therefore,

$$|\mathcal{C}| \cdot |\mathcal{A}| = |\mathcal{C}| \cdot |\mathcal{A}^{\text{m}}(n-1, n-1-\ell, n-t-\ell)| = \binom{n-1}{\ell} q^{t-1} \cdot q^{n-t-\ell}$$

$$= \binom{n-1}{n-1-\ell} q^{n-\ell-1} = |\text{J}_{q+1}(n-1, n-\ell-1)| ,$$

which implies by the code-anticode bound that $\mathcal{M}'$ is an $(n-t-\ell)$-diameter perfect code. □

Corollary 9.14. *When $t = 2$ the code $\mathcal{M}'$, obtained in the construction, is an $(n-1, n-\ell-1, q+1)$ MDS-CW code.*

Corollary 9.15. *When $t > 2$ the code $\mathcal{M}'$, obtained in the construction, is an $(n-1, n-t-\ell+1, n-\ell-1)_{q+1}$ MOA-CW code.*

Theorem 9.22.

(1) *If there exists an $(n, d, w)_q$ MOA-CW code, then there exists an $(n-1, d, w)_q$ MOA-CW code.*
(2) *If there exists an $(n, d, w)_q$ MOA-CW code, then there exists an $(n-1, d-1, w-1)_q$ MOA-CW code.*

Proof. Let $\mathcal{C}$ be $(n, d, w)_q$ MOA-CW code and define the following two code

$$\mathcal{C}_1 \triangleq \{(x_2, x_3, \dots, x_n) \; : \; (0, x_2, x_3, \dots, x_n) \in \mathcal{C}\}$$

and

$$\mathcal{C}_2 \triangleq \{(x_2, x_3, \dots, x_n) \; : \; (x_1, x_2, x_3, \dots, x_n) \in \mathcal{C}, \; x_1 \neq 0\} \, .$$

One can easily verify that $\mathcal{C}_1$ is an $(n-1, d, w)_q$ MOA-CW code and $\mathcal{C}_2$ is an $(n-1, d-1, w-1)_q$ MOA-CW code. □

After constructing $(d-1)$-diameter constant-weight perfect codes for $d < w$, where each w coordinates are the support of exactly $(q-1)^{w-d+1}$ codewords we would like to have some lower bounds on the alphabet size of such codes and upper bounds on their length and their weight. Since each w coordinates are the support of exactly $(q-1)^{w-d+1}$ codewords, it follows that the projection on each w coordinates on these codewords forms an orthogonal array OA$(w-d+1, w, q-1)$ and the related bounds on orthogonal arrays in Corollary 3.5, Theorem 3.5, and Theorem 3.6, can be applied. This implies the following theorem which present a tradeoff between the alphabet size and the minimum distance of the code.

Theorem 9.23.

(1) *If there exists an $(n, w-1, w)_q$ MOA-CW code, then $w \leq q$.*
(2) *If there exists an $(n, w-\delta, w)_q$ MOA-CW code where $2 \leq \delta \leq w-1$ and q is even, then $w \leq q + \delta$.*
(3) *If there exists an $(n, w-\delta, w)_q$ MOA-CW code, where $2 \leq \delta \leq w-1$ and q is odd, then $w \leq q + \delta - 1$.*
(4) *If there exists an $(n, w-\delta, w)_q$ MOA-CW code, where $q - 1 \leq \delta + 1$, then $w \leq \delta + 2$.*

Proof. All the claims are direct consequences of Corollary 3.5, Theorem 3.5, and Theorem 3.6, where the length n in the $\mathrm{OA}(t,n,q)$ is restricted to w, the alphabet size is $q-1$, and $w-\delta = n-t+1$. □

Theorem 9.23 implies upper bounds on w as a function of the alphabet size q and the minimum distance d of the code and, similarly, lower bounds on q as a function of the weight w and the minimum distance of the code. Since $d = w-\delta$, it follows that these bounds can be written as bounds only on the tradeoff between d and q.

Corollary 9.16.

(1) *If there exists an $(n,d,w)_q$ MOA-CW code, where $1 \leq d \leq w-2$ and q is even, then $d \leq q$.*
(2) *If there exists an $(n,d,w)_q$ MOA-CW code, where $1 \leq d \leq w-2$ and q is odd, then $d+1 \leq q$.*

The next bound presents a tradeoff between the length, the weight, and the alphabet size, of the code. It is interesting to note that the minimum distance has no influence on the bound.

Theorem 9.24. *If there exists an $(n,d,w)_q$ MOA-CW code, then $n \leq q+w-2$ (equivalently, $q \geq n-w+2$).*

Proof. Let $\mathcal{C}$ be an $(n,d,w)_q$ MOA-CW code and consider a set S of codewords in $\mathcal{C}$ which have only nonzero symbols in the first $w-1$ coordinates and in these $w-1$ coordinates of S, all the codewords of S share the same prefix, say v, of length $w-d+1$. Each two such codewords of S can differ in at most two coordinates out of the last $n-w+1$ coordinates and in the first $d-2$ coordinates. Hence, since the minimum distance of $\mathcal{C}$ is d, it follows that two such codewords of S differ exactly in these d coordinates. This implies the following observations:

(1) Consider the first $w-1$ coordinates and one coordinate from the last $n-w+1$ coordinates. The codewords whose supports are these coordinates form an $\mathrm{OA}(w-d+1,w,q-1)$ with exactly one codewords with the prefix v in the first $w-d+1$ coordinates. This implies that S contains exactly one codeword with a nonzero symbol in each one of the last $n-w+1$ coordinates and hence $|S| = n-w+1$.
(2) Each two codewords of S differ in all the symbols of their last $d-2$ coordinates out of the first $w-1$ coordinates. This implies that $q-1 \geq |S|$.

Thus, $q-1 \geq n-w+1$, which completes the proof of the theorem. □

There are many intriguing question on MDS-CW codes and MOA-CW codes. Some example are the following problems.

Problem 9.6. Present new constructions for MDS-CW codes (and also for MOA-CW codes). The only construction, which is not derived directly from an orthogonal array, which is known was analyzed in Theorem 9.21. Another construction is the union construction for MDS-CW codes as mentioned in Theorem 9.11. Is there a related construction for MOA-CW codes? We would like to see new different constructions as well as amendments to the construction which was given in this section.

Problem 9.7. Present new bounds on the tradeoff between the parameters of MDS-CW codes (and also for MOA-CW codes). This is especially important to see some new directions which will enable to conjecture for which parameters such codes exist.

Problem 9.8. Given $1 < d < w < n$, does there exist a $q_0(n, d, w)$ for which there exists an $(n, d, w)_q$ MOA-CW for all $q \geq q_0(n, d, w)$? Recall that for MDS-CW codes such a value called QMDS(w, n) exists as was proved in Theorem 9.16.

9.9 Comparison Between Maximum Size Anticodes

So far we have characterized the families of diameter perfect constant-weight codes. Each family is associated with some maximum size anticodes. In this section we will characterize these families of maximum size anticodes and compare some of them.

The first family of maximum size anticodes is associated with the family [**F1**] of nonbinary diameter perfect constant-weight codes for which $w = n$. Clearly, for these anticodes the length of a codeword is n and the weight of each codeword is $w = n$. Moreover, the anticodes are derived from the associated anticodes in the Hamming scheme, by replacing the *zeroes* in the anticodes of the Hamming scheme with the additional nonzero symbol of the constant-weight code.

The second family of maximum size anticodes is associated with the family [**F2**] of diameter perfect constant-weight codes over an alphabet of size $2^k + 1$ for which $w = n - 1$. Clearly, the related anticodes also have length n and weight $w = n - 1$. If the nonbinary diameter perfect code is in fact a nonbinary perfect code with minimum distance 3, then the related anticode is a ball, but another anticode with the same paramter as a ball can be of the same size (see the discussion after the proof of Lemma 9.8). If the nonbinary diameter perfect code is not a nonbinary perfect code, then the related anticode is not a ball and it should be computed for each set of parameters. If the minimum distance of the code is 4, i.e., the diameter of the anticode is 3, then a maximum size anticode in $\mathrm{J}_3(n, n-1)$ has $3n-2$ anticodewords (see Lemma 9.9 and the discussion which follows it). For diameter 4 the maximum size anticode in $\mathrm{J}_3(n, n-1)$ has n^2 codewords if $n \geq 8$ is a power of 2.

The third family of maximum size anticodes is associated with the family of generalized Steiner system. The related anticode $\mathcal{A}^{\mathrm{s}}(n, w, t)$ was defined by

$$\mathcal{A}^{\mathrm{s}}(n,w,t) \triangleq \{(\overbrace{1\cdots\cdots 1}^{t \text{ times}}, a_1, \ldots, a_{n-t}) \;:\; a_i \in \mathbb{Z}_q,\ \mathrm{wt}(a_1 \cdots a_{n-t}) = w-t\}\,.$$

This anticode has diameter $2(w-t)$ (when $n-t \geq 2(w-t)$) and its size is $\binom{n-t}{w-t}(q-1)^{w-t}$.

The last family of maximum size anticodes is associated with the families [**F4**], [**F5**], and [**F6**] of the diameter perfect constant-weight codes. The related anticode $\mathcal{A}^{\mathrm{m}}(n, w, \delta)$ was defined by

$$\mathcal{A}^{\mathrm{m}}(n,w,\delta) \triangleq \{(a_1, \ldots, a_\delta, \overbrace{1\cdots\cdots 1}^{w-\delta \text{ times}}, \overbrace{0\cdots\cdots 0}^{n-w \text{ times}}) \;:\; a_i \in \mathbb{Z}_q^-\}\,.$$

This anticode has diameter δ and its size is $(q-1)^\delta$.

One can be easily observe that nontrivial anticodes of the first two families cannot have the same parameters since they have different weights compared to their lengths. Moreover, it can be observed that nontrivial anticodes from these two families cannot have the same parameters as the anticodes from the last two families. In the first family of anticodes we have that $w = n$. It should be noted that most anticodes in the second family are over ternary alphabet and the other anticodes in this family have alphabet smaller by one than the weight. Hence, we will compare the anticodes from the last two families.

Unless the two anticodes represent one of two trivial cases ($w = n$ and $\delta = w-t$; or $w = t$) they cannot be isomorphic. This can be observed from the fact that the *zeroes* of $\mathcal{A}^{\mathrm{m}}(n, w, \delta)$ are in $n-w$ fixed coordinates, while the *zeroes* of $\mathcal{A}^{\mathrm{s}}(n, w, t)$ are in any combination of $n-w$ coordinates in the last $n-t$ coordinates.

Do there exist two different maximum size anticodes, related to two diameter perfect codes of different families (when the length, weight, and diameter are the same)? Note first that this implies that the related code from the family [**F4**], or the family [**F5**], or the family [**F6**] must be also a generalized Steiner system since the two codes will have the same parameters. Since $|\mathcal{A}^{\text{s}}(n,w,t)| = \binom{n-t}{w-t}(q-1)^{w-t}$ and $|\mathcal{A}^{\text{m}}(n,w,\delta)| = (q-1)^{\delta}$, it follows that the two anticodes are of equal size if and only if $\binom{n-t}{w-t} = (q-1)^{\ell}$ for some nonnegative integer $\ell = \delta + t - w$. If $\ell = 0$, then either $n = w$ (the first trivial case) or $w = t$ (the second trivial case). We distinguish now between two cases depending on whether $\ell = 1$ or $\ell > 1$.

(1) If $\ell = 1$, then $\binom{n-t}{w-t} = q - 1$ and we distinguish between three cases, depending on whether the related code is from the family [**F4**], the family [**F5**], or the family [**F6**].
Case 1.1. The diameter perfect code is from the family [**F4**] which implies that $\delta = w - 1$ and hence $t = \ell + 1 = 2$.
For the generalized Steiner system $\text{GS}(2, w, n, q)$ and the (n, w, q) MDS-CW code to be equal they must have the same minimum distance and hence $2(w-2)+1 = w$, i.e., $w = 3$. Since also $\binom{n-2}{w-2} = q-1$, it follows that $n = q + 1$. Two codes are considered in this case. The first one is a generalized Steiner system $\text{GS}(2, 3, q + 1, q)$ derived from a 1-perfect Hamming code over $\mathbb{F}_q$. The second one is an $(q + 1, 3, q)$ MDS-CW code derived from a $[q + 1, q - 1, 3]_q$ MDS code. For these parameters the 1-perfect Hamming code is also an MDS code and hence both constant-weight codes are the same code. There might be other such constant-weight codes for q which is not a power of a prime, but no such code is known.
Case 1.2. The diameter perfect code is from the family [**F5**] which implies that $\delta = w$ and hence $t = \ell = 1$.
Since $t = 1$, it follows that $\binom{n-1}{w-1} = q - 1$ and hence by Corollary 9.9, there exists an $(n, w + 1, w)_q$ code of the family [**F5**]. If $w = 2$, then $q = n$ and the two related codes form a generalized Steiner system $\text{GS}(1, 2, n, n)$ which is also an $(n, 3, 2)_n$ code from the family [**F5**]. Such a code exists by Corollary 9.12. If $2 < w < n$, then a related code with $\binom{n}{w}$ codewords cannot be a generalized Steiner system $\text{GS}(1, w, n, q)$.
Case 1.3. The diameter perfect code is from the family [**F6**], for which the MOA-CW code has minimum Hamming distance $d < w$, which implies that $\delta = d - 1$ and hence $t = \ell + 1 + w - d = w - d + 2$. Hence, the related codes are generalized Steiner systems $\text{GS}(t, w, n, q)$

and an $(n, d, w)_q$ MOA-CW code. The codes have the same minimum Hamming distance and hence $d = 2(w-t)+1 = 2d-3$, i.e., $d = 3$, which implies that $w = t+1$. Since $\binom{n-t}{w-t} = q-1$, it follows that $n-t = q-1$, i.e., $n = q+t-1$, and hence one of our codes is a generalized Steiner system $\text{GS}(t, t+1, q+t-1, q)$. By iteratively applying Lemma 3.4, we obtain a generalized Steiner system $\text{GS}(2, 3, q+1, q)$ which is the code in the previous case. Unfortunately, no generalized Steiner system $\text{GS}(t, t+1, q+t-1, q)$ is known for $t > 2$.

(2) If $\ell > 1$, then $\binom{n-t}{w-t} = (q-1)^\ell$ and first we have to consider the solutions for this equation. We distinguish between three cases depending whether $w-t \in \{1, n-t-1\}$, $w-t \in \{2, 3, n-t-3, n-t-2\}$, or $3 < w-t < n-t-3$.

Case 2.1. If $w-t \in \{1, n-t-1\}$.
If $w-t = n-t-1$, then $w = n-1$ and one code is a generalized Steiner system $\text{GS}(t, n-1, n, q)$. By iteratively applying Lemma 3.4, we obtain a generalized Steiner system $\text{GS}(1, n-t, n-t+1, q)$. By Theorem 3.1 for such a code $n-t+1 \geq 1+(n-t-1)(q-1)$ and hence such a code does not exist.
If $w-t = 1$, then one code is a generalized Steiner system $\text{GS}(t, t+1, n, q)$ for which the minimum distance is 3. Hence, the related codes from the families [**F4**], [**F5**], and [**F6**] are only those considered in Cases 1.1., 1.2., and 1.3., respectively. Therefore, no such code will be found for $\ell > 1$.

Case 2.2. If $w-t \in \{2, n-t-2\}$.
When $w-t = 2$ or $w-t = n-t-2$, there are infinitely many such solutions which satisfy the recursion $a_m = 6a_{m-1} - a_{m-2}$ (where $q-1 = a_m$ and $\ell = 2$) with the initial conditions $a_1 = 1$ and $a_2 = 6$. Similar analysis to the previous cases shows that there is no diameter perfect code from two different families in this case.

Case 2.3. If $2 < w-t < n-t-2$.
In this case the equation has exactly one solution for $n-t = 50$ and $w-t = 3$ or $w-t = n-t-3$. In the region for this solution there is no code from two families.

Therefore, all those cases for which the anticodes $\mathcal{A}^{\text{s}}(n, w, t)$ and $\mathcal{A}^{\text{m}}(n, w, \delta)$ have the same size and different structure (except for the anticodes in Case 1.1. given in Example 9.4 which follows, they are not related to a diameter perfect codes from two different families. Moreover, they might not be of maximum size. Note also, that the analysis could have taken other parameters into account. For example, by the diameter of the anticodes we have that $\delta = 2(w - t)$.

In general, one can decide based on the size of a maximum size anticode if the given parameters are in the range of a generalized Steiner system, an MDS-CW code, or an MOA-CW code. Each such code is an optimal nonbinary constant-weight code that meets the value of $A_q(n, d, w)$. In some cases these codes coincide as illustrated again in the following example

Example 9.4. Let $\mathcal{C}$ be a linear 1-perfect code of length $q + 1$, dimension $q - 1$, and minimum Hamming distance 3, over $\mathbb{F}_q$. By its parameters, this code is also an MDS code. The codewords of weight three of $\mathcal{C}$ form a generalized Steiner system $\mathrm{GS}(2, 3, q + 1, q)$ and also a $(q + 1, 3, q)$ MDS-CW code. The related maximum size anticodes are $\mathcal{A}^{\mathrm{s}}(q+1, 3, 2)$ and $\mathcal{A}^{\mathrm{m}}(q + 1, 3, 2)$ which are of the same size $(q - 1)^2$, but different structures.

If $n = 5$, $w = 3$, and $q = 4$, then the two anticodes have nine anticodewords and the following structures:

$$\mathcal{A}^{\mathrm{s}}(5, 3, 2) = \{11100, 11010, 11001, 11200, 11020, 11002, 11300, 11030, 11003\},$$

$$\mathcal{A}^{\mathrm{m}}(5, 3, 2) = \{11100, 12100, 13100, 21100, 22100, 23100, 31100, 32100, 33100\}.$$

We conclude this subsection with some more intriguing problems which arise from our discussion.

Problem 9.9. Characterize all parameters for which $\mathcal{A}^{\mathrm{s}}(n, w, t)$ is a maximum size anticode. Such a proof can be done by using extremal combinatorics for such anticodes related to binary words with constant weight. Similarly, characterize all parameters for which $\mathcal{A}^{\mathrm{m}}(n, w, \delta)$ is a maximum size anticode.

Problem 9.10. Are there $(q + 1, 3, q)$ MDS-CW codes which are also generalized Steiner systems $\mathrm{GS}(2, 3, q + 1, q)$ beside those for prime power q?

Problem 9.11. Is there any $\mathrm{GS}(3, 4, q+2, q)$ for a prime power q? and for non-prime power q?

Problem 9.12. Define the anticodes $\mathcal{A}^{\mathrm{s}}(n, w, t)$ and $\mathcal{A}^{\mathrm{m}}(n, w, \delta)$ in terms of t-intersecting families. Find the maximum size of these t-intersecting families (in other words, the maximum size of these anticodes) for all parameters, including the cases where $d > w + 1$.

9.10 Notes

Section 9.1. The ternary perfect constant-weight codes were constructed independently using the same technique in [van Lint and Tolhuizen (1999)] and in [Svanström (1999a)]. The generalization for an alphabet of size $q+1$, where q is a power of 2, was presented in [Etzion and van Lint (2001)]. A perfect code with parameters corresponding to Lemma 9.2, however, could still exist. Using ad hoc arguments one can exclude the case of 4-ary alphabet, weight 4, and length 5 [Etzion and van Lint (2001)].

Constructions similar to the one used to construct nonlinear binary perfect codes were used in [Krotov (2001b)] to obtain $2^{2^{2^{n-1}}-2}$ different such codes. Other nonisomorphic 1-perfect ternary constant-weight codes were constructed in [Krotov (2008)]. In both [Krotov (2001b, 2008)] it was proved that a family 1-perfect constant-weight codes is equivalent to a specific family of perfect matchings of the n-dimensional cube, i.e., the graph whose vertices are the words of $\mathbb{F}_2^n$ and two vertices $x, y \in \mathbb{F}_2^n$ are connected by an edge if and only if $d(x,y) = 1$.

A completely different approach for the nonbinary perfect constant-weight codes problem is to generalize the Johnson distance in terms of subsets. This approach was used in [Schwartz (2004)]. The codewords were taken as subsets as follows. Let Q be an alphabet with $q+1$ symbols including the *zero* symbol. A word of length n and weight w over Q is represented by a w-subset from $[n] \times Q^-$, i.e., pairs of coordinates and values, under the restriction that no coordinate appears twice in the w-subset.

Using this representation, the distance between two words x_1 and x_2 is given by

$$d(x_1, x_2) = w - |x_1 \cap x_2| \ .$$

When Q is the binary alphabet, this is simply the Johnson distance on binary words of length n and weight w. When $w = n$ the distance degenerates to the Hamming distance and the graph is isomorphic to the Hamming graph with alphabet Q^-, similarly to Theorem 9.1. Therefore, w.l.o.g. we assume that $w < n$ and $|Q| > 2$.

Denote by $\mathrm{CW}(n, w, q+1, d)$ a constant-weight code of length n, weight w, over an alphabet Q of size $q+1$, and minimum distance d.

Lemma 9.17. *If $\mathcal{C}$ is a* $\mathrm{CW}(n, w, q+1, d)$*, then*

$$|\mathcal{C}| \binom{w}{w-d+1} \leq \binom{n}{w-d+1} q^{w-d+1} \ .$$

Proof. Take any projection of $\mathcal{C}$ onto $w-d+1$ coordinates and remove codewords containing a zero in any of these coordinates. Of the remaining codewords, there cannot be any two that agree on all $w-d+1$ coordinates, since then their distance is at most $d-1$. If we count the total number of such nonzero strings of length $w-d+1$, each original codeword contributes $\binom{w}{w-d+1}$ strings, while the total number of such strings is $\binom{n}{w-d+1}q^{w-d+1}$, and the claim in the lemma follows immediately. □

For $d = 3$, the size of a sphere with radius one is given by $1+(k-1)w+w(n-w)q$, so for a 1-perfect $\mathrm{CW}(n,w,q+1,3)$ to exist, the following must hold:

$$\frac{\binom{n}{w}q^w}{1+(q-1)w+w(n-w)q}\binom{w}{w-d+1} \leq \binom{n}{w-d+1}q^{w-d+1} ,$$

which implies that

$$(n-w+1)(n-w+2)q^2 \leq 2(1+(q-1)w+w(n-w)q .$$

Note that for fixed q and w, and n tending to infinity, the left side of the inequality grows as n^2 while the right side of the inequality grows as n, so there are no asymptotic 1-perfect constant-weight codes. On the other hand, if q and $n-w$ are fixed, and w tends to infinity, the right side grows as w while the left side is a constant and hence there are also no perfect constant-weight codes in this case. This does not exclude possible perfect codes for specific parameters or when neither n nor w tends to infinity. Similar arguments can be made for $d > 3$.

Finally, a perfect $\mathrm{CW}(6,5,3,3)$ is given by the following set of codewords:

$$\begin{array}{cccc} 011221, & 022112, & 102121, & 201212, \\ 120211, & 210122, & 212011, & 121022, \\ 221101, & 112202, & 111110, & 222220. \end{array}$$

This code is the same code as the 3-diameter perfect code in $\mathrm{J}_3(6,5)$.

Problem 9.13. Are there more perfect $\mathrm{CW}(n,w,q,d)$? It would be interesting to have even a few sporadic examples of such codes or to prove that they cannot exist.

Problem 9.14. Provide nonexistence theorems for perfect $CW(n,w,q,d)$ and exclude as many parameters as possible.

Problem 9.15. Is there a connection between this family of nonbinary constant-weight codes and diameter perfect codes as the perfect CW(6, 5, 3, 3) which is a 3-diameter perfect code in $J_3(6,5)$.

Section 9.2. The discussion on the families of nonbinary diameter perfect constant-weight codes is taken from [Etzion (2021)]. In this paper there is a comprehensive analysis of diameter perfect constant-weight codes. The proof of Lemma 9.7 was also presented in this paper.

Section 9.4. Ternary diameter perfect constant-weight codes in $J_3(n, n-1)$ were mainly considered in [Krotov (2008)] who proved that such a code with diameter 3 exist only for $n = 6$. The code of length 6 was first presented in [Svanström (1999b)] and later in [Östergård and Svanström (2002)]. Theorem 9.7 used for the nonexistence proof was proved by [Fon-Der-Flaass (2007)]. The proof of Lemma 9.9 was provided for this book by Denis Krotov.

Ternary diameter perfect constant-weight codes with diameter 4 in $J_3(n, n-1)$ were considered first in [Krotov (2008)]. It was proved in this paper that if there exists an APN (almost perfect nonlinear) permutation, then there exists such a diameter perfect code. A bijection $f : \mathbb{F}_2^n \to \mathbb{F}_2^n$ is a ***almost perfect nonlinear*** function (permutation) of order n if and only if the system of equations

$$\alpha = x + y,$$

$$\beta = f(x) + f(y),$$

has either no solution or two solutions for every pair $(\alpha, \beta) \neq (0^n, 0^n)$. The connections between these bijections and coding theory were considered first in [Carlet, Charpin, and Zinoviev (1998)]. It was proved in [Krotov (2008)] that if there exists an APN permutation of order n, then there exists a 4-diameter perfect code in $J_3(2^n, 2^n - 1)$. APN functions exist for all odd orders [Carlet, Charpin, and Zinoviev (1998)] and hence there exists a 4-diameter perfect code in $J_3(2^n, 2^n - 1)$ if n is odd. When n is even the situation is more complicated. It is trivial to see that there is no 4-diameter perfect code in $J_3(4,3)$. It was proved in [Krotov, Östergård, and Pottonen (2016)] that there is no 4-diameter perfect code in $J_3(16, 15)$. On the other hand, an APN permutation of order 6 was demonstrated in [Browning, Dillon, McQuistan, and Wolfe (2009)]. Therefore, there exists a 4-diameter perfect code in $J_3(64, 63)$. This leads to the following question.

Problem 9.16. For which values of even $n > 6$ there exists a 4-diameter perfect code in $J_3(2^n, 2^n - 1)$? and for which such values there is no such diameter perfect code?

Problem 9.17. Does there exists a D-diameter perfect code in $J_3(n, n-1)$, where $D \geq 5$?

Problem 9.18. Does there exists another D-diameter perfect code in $J_q(n, n-1)$, where $q > 3$ and $D \geq 3$?

Finally, the number of anticodewords in a maximum size anticode with diameter 2 or 3 was considered in [Krotov (2008)].

Section 9.6. MDS-CW codes were introduced in [Etzion (1997)] and the results in this section on these codes were taken mainly from this paper.

Theorem 9.16 can also be obtained from the results in [Blanchard (1995)], but the proof of Theorem 9.16 is much simpler than the proof in [Blanchard (1995)], and the bounds are much better than the ones that can be obtained from the proofs in [Blanchard (1995)].

A Mersenne prime is a prime of the form $2^n - 1$. It is known that for such a Mersenne prime, n must be a prime. Only 51 primes of this form are known as of 2021, but it is conjectured that infinitely many exist. A Fermat prime is a prime of the form $2^n + 1$. It is known that for a Fermat prime, $n = 2^k$. There are five known Fermat primes, 3, 5, 17, 257, and 65537 and it is conjectured that there are no more such primes. The online encyclopedia on integer sequences that is regularly updated is an excellent reference to view the up-to-date status of this sequences of primes. This sequences of primes are mentioned in most books on number theory. There are many papers on these two types of primes. An example for such a paper is [Robinson (1954)].

The partition problem of Frobenius was used to find upper bounds on QMDS(n, w). This partition problem is stated as follows. Given k relatively prime positive integers $a_1, a_2, \ldots, a_k$, what is the largest integer $M(a_1, a_2, \ldots, a_k)$ which does not have a representation as $\sum_{i=1}^{k} r_i a_i$, where each r_i, $1 \leq i \leq k$, is a nonnegative integer. The solution for the problem when $k = 2$ can be used to obtain an upper bound on QMDS(n, w). It was proved in [Sylvester (1884)] that $M(a_1, a_2) = a_1 a_2 - (a_1 + a_2)$.

Sections 9.7, 9.8. The results in these sections about the fifth family, [**F5**], and the sixth family, [**F6**], of nonbinary diameter perfect constant-weight codes were taken from [Etzion (2021)]. This fifth family, [**F5**], of codes is very interesting and especially when finding bounds on $q_0(w, n)$. The partition problem, to obtain such a bound, was extensively considered in [Brouwer, Shearer, Sloane, and Smith (1990)] and a proof of Theorem 9.19 can be obtained from the codes and related partitions in [Graham and Sloane (1980)]. Other codes with the same parameters as the ones constructed in Theorem 9.18 were also presented in [Chee, Dau, Ling, and Ling (2008); Chee and Ling (2007)], by using different techniques from combinatorial designs.

As for the sixth family, [**F6**], of code, we believe that there is still lot of ground for further research on this family and the results in [Etzion (2021)] given in this section are just the foundation for this family.

Section 9.9. The families of maximum anticodes for nonbinary constant-weight codes that were defined in the section form a demonstration of the strength of the code-anticode bound. Usually, such anticodes are found by combinatorial methods related to extremal combinatorics. These anticodes are associated with t-intersecting families of maximum size, something that usually requires a lot of nontrivial combinatorial work. The existence of diameter perfect codes implies that for the given parameters, associated anticodes are of maximum size and also form the maximum size of the associated intersecting family.

For example, it was proved in [Krotov, Östergård, and Pottonen (2016)] that for ternary codes if $n = 2^m \geq 8$, $w = n-1$, and the diameter is 4, then the maximum size anticode has size n^2 and such an anticode can be defined by the union of the set of ternary words with a unique *zero* (n anticodewords) and all the other symbols are *ones* with the set of ternary words with a unique *zero* and a unique *two* and all the other symbols are *ones* ($n(n-1)$ anticodewords). We currently have no information about anticodes with larger diameter.

Problem 9.19. What is the size and the structure of maximum size anticodes in $\mathrm{J}_3(n, n-1)$ whose diameter is at least 5?

Problem 9.20. What is the size and the structure of maximum size anticodes in $\mathrm{J}_q(n, n-1)$, where $q > 3$?

Finally, the solution for the equation $\binom{n}{2} = \ell^2$, with the recursive formular $a_m = 6a_{m-1} - a_{m-2}$ with the initial condition $a_1 = 1$ and $a_2 = 6$, and the extensive literature can be found in the Online Encyclopedia on Integer Sequences, sequence A001109. The solution for the equation $\binom{n}{r} = \ell^2$, where $3 \leq r \leq n-3$ was proved by Erdös [Le Lionnais (1983), p. 48].

Chapter 10

Codes Over Subspaces

So far all the perfect codes we considered use the Hamming distance (the Johnson distance is no exception since it is exactly half of the Hamming distance). We now turn our attention to codes over subspaces with a different distance measure. Some codes related to subspaces were considered in Section 7.2 in constructions of 1-perfect byte correcting codes. These subspaces and their codes will also be important in the current chapter, which considers codes over subspaces. The subspaces will be considered first in the third important scheme (after the Hamming and the Johnson schemes) based on a distance-regular graph, the Grassmann scheme. The Grassmann scheme $\mathrm{G}_q(n, k)$ consists of all the k-subspaces of a given n-dimensional space over $\mathbb{F}_q$. The metric in this scheme is called the Grassmann metric. This scheme began getting a lot of attention in the 21st century, due to the important application of codes from this scheme in network coding. In Section 10.1 the basic definitions for this scheme will be presented and it will be proved that there are no perfect codes in the Grassmann scheme.

The setting for subsets related to codes in the Hamming scheme and the Johnson scheme should be transferred to setting for subspaces. This transformation from subsets to subspaces is known as a q-analog, where the q-analog of subsets are subspaces, the q-analog of the size is the dimension, the q-analog of the binomial coefficients are the q-binomial coefficients (known also as Gaussian coefficients), etc. Diameter perfect codes in this scheme are q-Steiner systems (and also systems with the dual subspaces of the subspaces in a q-Steiner system) that are the q-analog of Steiner systems. Other diameter perfect codes do not exist in this scheme. These systems will be the topic of Section 10.2. One family of q-Steiner systems is formed by spreads which were mentioned in Chapter 7. An important class of spreads are the normal spreads discussed in Section 10.3. A connection between these spreads and Hamming codes is proved in this section

In Section 10.4 the discussion will be expanded to subspaces, over $\mathbb{F}_q^n$, with no restriction on the dimension. The set of all subspaces in $\mathbb{F}_q^n$ will be called the projective space and will be denoted by $\mathcal{P}_q(n)$. The Grassmann distance is generalized for this space to a distance called the subspace distance. It will be proved that also in $\mathcal{P}_q(n)$, there are no nontrivial perfect codes. Next, in Section 10.5 a related metric, the rank metric, applied on matrices will be considered. This metric is associated with the bilinear forms scheme. The matrices in this scheme are used to construct codes over subspaces. In addition, this metric is also important in the context of network coding. This is also the motivation for the topic of Section 10.6, where constant-dimension MDS codes, i.e., subspace-MDS codes, will be considered. These codes form a family of codes that lie in-between the linear MDS codes and the nonlinear orthogonal arrays.

10.1 No Perfect Codes in the Grassmann Scheme

The ***Grassmann scheme*** (and the Grassmann Graph) $\mathrm{G}_q(n,k)$ consists of all the k-subspaces of an n-space over $\mathbb{F}_q$. For two elements $X, Y \in \mathrm{G}_q(n,k)$, the ***Grassmann distance*** $d_G(X,Y)$, is defined by

$$d_G(X,Y) = k - \dim(X \cap Y) \ .$$

We are also interested in a distance measure between subspaces that are not of the same dimension. The ***projective space*** $\mathcal{P}_q(n)$ consists of all subspace of $\mathbb{F}_q^n$. This space is a q-analog of the Hamming space $\mathcal{H}_q(n)$. The distance used for the projective space is a generalization of the Grassmann distance. For two subspace X and Y in $\mathcal{P}_q(n)$, the ***subspace distance*** $d_S(X,Y)$, is defined by

$$d_S(X,Y) \triangleq \dim X + \dim Y - 2\dim(X \cap Y).$$

The subspace distance is the q-analog of the Hamming distance, but contrary to the Hamming space together with the Hamming distance that define an association scheme, the projective space together with the subspace distance do not define an association scheme. This is easy to verify since the associated graph is not regular. The distinction between the subspace distance and the Grassmann distance is the same as the distinction between the Hamming distance and the Johnson distance, as one can readily verify.

Lemma 10.1. *The Grassmann distance of two k-subspaces is exactly half of their subspace distance.*

Our next step is to prove that the subspace distance is a metric. The first lemma is well known.

Lemma 10.2. *For any two subspaces $A, B \in \mathcal{P}_q(n)$ we have*

$$\dim(A \cap B) = \dim A + \dim B - \dim(A + B) \ .$$

Proposition 10.1. *The function*

$$d_S(X, Y) = \dim X + \dim Y - 2\dim(X \cap Y)$$

is a metric for the projective space $\mathcal{P}_q(n)$.

Proof. The coincidence and the symmetry axioms are trivial. Hence, it is sufficient to prove the triangle inequality. Let $X, Y, Z \in \mathcal{P}_q(n)$. By Lemma 10.2, we have that

$$\dim((X \cap Y) + (Y \cap Z)) = \dim(X \cap Y) + \dim(Y \cap Z) - \dim(X \cap Y \cap Z)$$

and hence,

$$\dim(X \cap Y) + \dim(Y \cap Z)$$

$$= \dim((X \cap Y) + (Y \cap Z)) + \dim(X \cap Y \cap Z)$$

$$\leq \dim Y + \dim(X \cap Z).$$

This implies by the definition of the subspace distance that

$$d_S(X, Y) + d_S(Y, Z)$$

$$= \dim X + \dim Z + 2\dim Y - 2\dim(X \cap Y) - 2\dim(Y \cap Z)$$

$$\geq \dim X + \dim Z - 2\dim(X \cap Z) = d_S(X, Z),$$

which completes the proof. □

Corollary 10.1. *The Grassmann distance forms a metric in $\mathrm{G}_q(n, k)$.*

We note that the Grassmann graph $\mathrm{G}_q(n, k)$ together with the Grassmann distance also define an association scheme. Moreover, the Grassmann scheme $\mathrm{G}_q(n, k)$ is the q-analog of the Johnson scheme $\mathrm{J}(n, k)$. To prove that it is a scheme, the important property to verify is that each intersection number $p^{\ell}_{i,j}$ for the Grassmann metric does not depend on the two subspaces $X, Y \in \mathrm{G}_q(n, k)$ for which $d_G(X, Y) = \ell$. For such a proof, a q-analog for the binomial coefficient is required. To accomplish this, we will make use of the q-***binomial coefficients***, known also as the ***Gaussian coefficients***, defined by

$$\begin{bmatrix} n \\ k \end{bmatrix}_q \triangleq \frac{(q^n-1)(q^{n-1}-1)\cdots(q^{n-k+1}-1)}{(q^k-1)(q^{k-1}-1)\cdots(q-1)} .$$

Many of the well-known equalities for binomial coefficients have q-analog equalities with the q-binomial coefficients. For example

$$\begin{bmatrix} n \\ k \end{bmatrix}_q = \begin{bmatrix} n \\ n-k \end{bmatrix}_q ;$$

for every $0 < k < n$,

$$\begin{bmatrix} n \\ k \end{bmatrix}_q = q^k \begin{bmatrix} n-1 \\ k \end{bmatrix}_q + \begin{bmatrix} n-1 \\ k-1 \end{bmatrix}_q , \quad \begin{bmatrix} n \\ k \end{bmatrix}_q = \begin{bmatrix} n-1 \\ k \end{bmatrix}_q + q^{n-k} \begin{bmatrix} n-1 \\ k-1 \end{bmatrix}_q ;$$

for every positive odd integer n,

$$\sum_{j=0}^{n} (-1)^j \begin{bmatrix} n \\ j \end{bmatrix}_q = 0;$$

and for any two non-negative integer n and m,

$$\sum_{j=0}^{n} q^j \begin{bmatrix} m+j \\ m \end{bmatrix}_q = \begin{bmatrix} n+m+1 \\ n \end{bmatrix}_q .$$

Similarly to the Johnson distance, it is not trivial to compute the intersection number $p_{i,j}^{\ell}$, but it is easy to verify that this number does not depend on the two k-subspaces X and Y for which $d_G(X,Y) = \ell$. The computations are similar to other computations that are performed later on.

Theorem 10.1. *For each two positive integers n and k such that $0 \leq k \leq n$, we have that*

$$|\mathrm{G}_q(n,k)| = \begin{bmatrix} n \\ k \end{bmatrix}_q .$$

Proof. We form a k-subspace of $\mathbb{F}_q^n$ by choosing an appropriate base for the subspace. We first choose the nonzero vectors of the basis in some order and at the end we cancel this order. The first nonzero vector can be chosen in $q^n - 1$ distinct ways; the linear span of this vector has size q. For the second one we can choose any nonzero vectors not in the linear span of the first vector. Hence, there are $q^n - q$ distinct ways to choose the second vector. The first two vectors span a subspace with q^2 vectors and hence the third nonzero vector can be chosen in $q^n - q^2$ distinct ways. We continue iteratively, where the last (k-th) nonzero vector can be chosen in $q^n - q^{k-1}$ distinct ways. Therefore, the total number of distinct ways to choose a k-subspace, with an ordered base, in this way is $\prod_{i=0}^{k-1}(q^n - q^i)$.

Now, we compute the number of times that a given k-subspace X is counted in this enumeration. The first nonzero vector of X can be chosen in $q^k - 1$ distinct ways. The second one we can choose in $q^k - q$ distinct ways and the third one in $q^k - q^2$ distinct ways. We continue iteratively, where the last (k-th) nonzero vector can be chosen in $q^k - q^{k-1}$ distinct ways. Therefore, the total number of distinct k-subspaces of $\mathbb{F}_q^n$ is

$$\frac{\prod_{i=0}^{k-1}(q^n - q^i)}{\prod_{i=0}^{k-1}(q^k - q^i)} = \frac{(q^n - 1)(q^{n-1} - 1)\cdots(q^{n-k+1} - 1)}{(q^k - 1)(q^{k-1} - 1)\cdots(q - 1)} = \begin{bmatrix} n \\ k \end{bmatrix}_q$$

and hence $|\mathrm{G}_q(n,k)| = \begin{bmatrix} n \\ k \end{bmatrix}_q$. □

Clearly, Theorem 10.1 is the q-analog for the assertion that $|\mathrm{J}(n,k)| = \binom{n}{k}$.

Lemma 10.3. *If $\mathcal{V}$ is an n-space over $\mathbb{F}_q$ and X is an m-subspace of $\mathcal{V}$, then the number of i-subspaces of $\mathcal{V}$ that intersect X in an ℓ-subspace is*

$$q^{(i-\ell)(m-\ell)} \begin{bmatrix} n-m \\ i-\ell \end{bmatrix}_q \begin{bmatrix} m \\ \ell \end{bmatrix}_q .$$

Proof. Let Z be an ℓ-subspace of a subspace $X \in \mathrm{G}_q(n,m)$. We complete Z to an i-subspace by adding $i - \ell$ linearly independent vectors of length n, which are not contained in X, to obtain an i-subspace whose intersection with X is exactly Z. For each ℓ-subspace Z, the number of such i-subspaces is

$$\frac{(q^n - q^m)(q^n - q^{m+1})\cdots(q^n - q^{m+i-\ell-1})}{(q^i - q^\ell)(q^i - q^{\ell+1})\cdots(q^i - q^{i-1})} = \begin{bmatrix} n-m \\ i-\ell \end{bmatrix}_q q^{(m-\ell)(i-\ell)} .$$

The number of ℓ-subspaces of X is $\begin{bmatrix} m \\ \ell \end{bmatrix}_q$ and each can be chosen as Z and hence the total number of i-subspaces of $\mathcal{V}$ that intersect X in an ℓ-subspace is

$$q^{(i-\ell)(m-\ell)} \begin{bmatrix} n-m \\ i-\ell \end{bmatrix}_q \begin{bmatrix} m \\ \ell \end{bmatrix}_q .$$

□

In the rest of this chapter we omit the q from the Gaussian coefficient $\begin{bmatrix} n \\ k \end{bmatrix}_q$ and denote it by $\begin{bmatrix} n \\ k \end{bmatrix}$.

Lemma 10.4. *The size of a ball with radius e around a subspace of $G_q(n,k)$ is*

$$\sum_{i=0}^{e} q^{i^2} \begin{bmatrix} k \\ i \end{bmatrix} \begin{bmatrix} n-k \\ i \end{bmatrix}.$$

Proof. Let X be a subspace in $\mathrm{G}_q(n,k)$. Now, apply Lemma 10.3, where $m = k$ and $i = k$. A k-subspace Y is in the ball with radius e around X if its intersection with X is of dimension between k (X itself) and $k-e$. Hence, in Lemma 10.3 ℓ takes all the values from $k-e$ to k. This implies that these k-subspaces are exactly all the k-subspaces that are within distance 0 to e from X, i.e., they form the ball with radius e around X. Hence, the size of the ball is

$$\sum_{\ell=k-e}^{k} q^{(k-\ell)(k-\ell)} \begin{bmatrix} n-k \\ k-\ell \end{bmatrix} \begin{bmatrix} k \\ \ell \end{bmatrix} = \sum_{i=0}^{e} q^{i^2} \begin{bmatrix} k \\ i \end{bmatrix} \begin{bmatrix} n-k \\ i \end{bmatrix}.$$

□

Let X be a $(k-2)$-subspace of an n-space $\mathcal{V}$. Let $\mathbb{T}$ be the subset of all the subspaces from $\mathrm{G}_q(n,k)$ that intersect X in a subspace whose dimension is at least $k-e-1$.

Lemma 10.5. *The subset $\mathbb{T}$ is an anticode in $\mathrm{G}_q(n,k)$ whose diameter is $2e$ and whose size is*

$$\sum_{i=0}^{e-1} q^{i(i+2)} \begin{bmatrix} k-2 \\ i \end{bmatrix} \begin{bmatrix} n-k+2 \\ i+2 \end{bmatrix}.$$

Proof. If $Y, Z \in \mathbb{T} \subset \mathrm{G}_q(n,k)$ and $X \in \mathrm{G}_q(n,k-2)$, then by the triangle inequality we have that

$$d_S(Y,Z) \le d_S(Y,X) + d_S(X,Z) \tag{10.1}$$

and by the definition of the subspace distance and since $Y \in \mathbb{T}$, it follows that

$$d_S(Y,X) = \dim Y + \dim X - 2\dim(Y \cap X) \le k + k - 2 - 2(k-e-1) = 2e.$$

Similarly, $d_S(X,Z) \le 2e$, which implies by (10.1) that $d_S(Y,Z) \le 4e$, and by Lemma 10.1, by taking $m = k-2$ and $i = k$, we have that $d_G(Y,Z) \le 2e$ and hence $\mathbb{T}$ is an anticode in $\mathrm{G}_q(n,k)$ with diameter $2e$.

By Lemma 10.3, the number of k-subspaces that intersect the $(k-2)$-subspace X in an ℓ-subspace is

$$q^{(k-\ell)(k-2-\ell)} \begin{bmatrix} n-k+2 \\ k-\ell \end{bmatrix} \begin{bmatrix} k-2 \\ \ell \end{bmatrix}.$$

Since ℓ can take any value between $k-e-1$ and $k-2$, it follows that the size of $\mathbb{T}$ is

$$\sum_{\ell=k-e-1}^{k-2} q^{(k-\ell)(k-2-\ell)} \begin{bmatrix} n-k+2 \\ k-\ell \end{bmatrix} \begin{bmatrix} k-2 \\ \ell \end{bmatrix}.$$

By changing the variable of the sum from ℓ to i, where $i = k-2-\ell$, the sum over i is between 0 and $e-1$. Moreover, the required size claimed in each product of the summation in the lemma is obtained and the lemma is proved. □

Theorem 10.2. *There is no nontrivial e-perfect code in the Grassmann scheme* $\mathcal{G}_q(n,k)$.

Proof. Assume the contrary, that there exists a nontrivial e-perfect code $\mathbb{C}$ in $\mathcal{G}_q(n,k)$. Since $\mathbb{C}$ is a nontrivial code, this implies that the minimum Grassmann distance of the code is $2e+1$ and hence $\mathbb{C}$ has at least two codewords and $k \geq 2e+1$. Let Y be a subspace of $\mathcal{G}_q(n,k)$ and X be a $(k-2)$-subspace of Y. As before, let $\mathbb{T}$ be the subset of all the subspaces from $\mathcal{G}_q(n,k)$ that intersect X in a subspace whose dimension is at least $k-e-1$. Let $\mathbb{S}$ be the set of all subspaces in $\mathcal{G}_q(n,k)$ that intersect both X and Y in a subspace whose dimension is $k-e-1$. Let $\mathbb{P}$ be the set of all subspaces of $\mathcal{G}_q(n,k)$ that intersect Y in a $(k-e)$-subspace and intersect X in a $(k-e-2)$-subspace. Formally,

$$\mathbb{T} \triangleq \{Z \,:\, Z \in \mathcal{G}_q(n,k),\ \dim(Z \cap X) \geq k-e-1\},$$

$$\mathbb{S} \triangleq \{Z \,:\, Z \in \mathcal{G}_q(n,k),\ \dim(Z \cap X) = \dim(Z \cap Y) = k-e-1\},$$

and

$$\mathbb{P} \triangleq \{Z \,:\, Z \in \mathcal{G}_q(n,k),\ \dim(Z \cap X) = k-e-2,\ \dim(Z \cap Y) = k-e\}.$$

To prove the claim of the theorem our goal is to show that the set $\mathbb{T}$ is an anticode with diameter $2e$ in $\mathcal{G}_q(n,k)$ whose size is larger than the ball of radius e in $\mathcal{G}_q(n,k)$. This will form a contradiction to the code-anticode bound and imply that there is no nontrivial e-perfect code in the $\mathcal{G}_q(n,k)$.

The first step is to find the diameter of $\mathbb{T}$. The dimension of X is $k-2$ if Z_1 and Z_2 are two subspace which intersect X in a subspace whose dimension is at least $k-e-1$, then their intersection is of dimension at least $2(k-e-1)-(k-2)=k-2e$. This implies that if $Z_1, Z_2 \in \mathrm{G}_q(n,k)$, then $d_G(Z_1,Z_2) \leq k-(k-2e)=2e$. Therefore, The diameter of $\mathbb{T}$ is $2e$.

Assume that Z is a k-subspace in $\mathbb{T}$ and not in $\mathcal{B}_e(Y)$. Since $Z \in \mathbb{T}$, it follows that $\dim(Z \cap X) \geq k-e-1$ and if $\dim(Z \cap Y) > k-e-1$, this will imply that $Z \in \mathcal{B}_e(Y)$. Therefore, $\dim(Z \cap Y) \leq k-e-1$ and since $X \subset Y$, it follows that $\dim(Z \cap X) = \dim(Z \cap Y) = k-e-1$ and hence $Z \in \mathbb{S}$. Clearly, by the definition of X, Y, and $\mathbb{S}$, each element of $\mathbb{S}$ satisfies this property of Z.

Assume now that Z is a k-subspace in $\mathcal{B}_e(Y)$ and not in $\mathbb{T}$. Assume further that $\ell = \dim(Z \cap Y)$ and since $Z \in \mathcal{B}_e(Y)$, it follows that $\ell \geq k=e$. Since $X \subset Y$ and $\dim X = \dim Y - 2$, it follows that $\dim(Z \cap X) \geq \ell - 2$. Recall that Y is a k-subspace and Z is not in $\mathbb{T}$ and hence $\dim(Z \cap X) < k-e-1$. Together, this implies that $\dim(Z \cap X) = k-e-2$, $\ell = k-e$, and $Z \in \mathbb{P}$. Clearly, each element of $\mathbb{P}$ satisfies this property.

Taking all these assertions into account, it follows that $\mathbb{S} = \mathbb{T} \setminus \mathcal{B}_e(Y)$ and $\mathbb{P} = \mathcal{B}_e(Y) \setminus \mathbb{T}$. Therefore, $|\mathbb{S}| = |\mathbb{T}| - |\mathbb{T} \cap \mathcal{B}_e(Y)|$ and $|\mathbb{P}| = |\mathcal{B}_e(Y)| - |\mathcal{B}_e(Y) \cap \mathbb{T}|$. This implies that $|\mathbb{T}| - |\mathcal{B}_e(Y)| = |\mathbb{S}| - |\mathbb{P}|$.

Applying arguments as in Lemma 10.3, one can find that

$$|\mathbb{S}| = q^{(e+1)^2} \begin{bmatrix} k-2 \\ e-1 \end{bmatrix} \begin{bmatrix} n-k \\ e+1 \end{bmatrix}$$

and

$$|\mathbb{P}| = q^{e(e+2)} \begin{bmatrix} k-2 \\ e \end{bmatrix} \begin{bmatrix} n-k \\ e \end{bmatrix}.$$

Therefore, we have that

$$|\mathbb{T}| - |\mathcal{B}_e(Y)| = |\mathbb{S}| - |\mathbb{P}| = q^{e(e+2)} \left(q \begin{bmatrix} k-2 \\ e-1 \end{bmatrix} \begin{bmatrix} n-k \\ e+1 \end{bmatrix} - \begin{bmatrix} k-2 \\ e \end{bmatrix} \begin{bmatrix} n-k \\ e \end{bmatrix} \right).$$

Nevertheless, if $q \geq 2$, $k \geq 2e+1$, and $n \geq 2k$, this difference is always positive, i.e., the size of the anticode $\mathbb{T}$ with diameter $2e$ is larger than the ball with radius e, which contradicts the code-anticode bound if $\mathbb{C}$ is an e-perfect code. Thus, there is no nontrivial e-perfect code in $\mathrm{G}_q(n,k)$. □

Note that in the proof of Theorem 10.2, we used the fact that the code-anticode bound is true for the Grassmann scheme. This fact was already proved by Delsarte as was mentioned in Chapter 2. Our proof of the code-anticode bound in Chapter 2 does not hold for the Grassmann scheme.

We do, however provide another proof for the bound based on the local inequality lemma for $\mathrm{G}_q(n,k)$, which will be presented in Lemma 10.9.

The idea in the proof of Theorem 10.2 can also be used to prove Theorem 8.6, but this proof for the Johnson scheme can be simplified as was done in the proofs of Theorems 8.6 and 8.10.

10.2 q-Steiner Systems

In Theorem 10.2 we proved that there is no nontrivial perfect code in the Grassmann scheme. The Grassmann scheme $\mathrm{G}_q(n,k)$ is the q-analog of the Johnson scheme $\mathrm{J}(n,k)$. In the Johnson scheme we were not able to resolve the existence problem of perfect codes, but we were able to exhibit a family of diameter perfect codes, the Steiner systems. Moreover, when we expanded our interest to perfect constant-weight codes over a nonbinary alphabet, some perfect codes were found and six families of diameter perfect codes, were presented. One of these families consists of the generalized Steiner systems. We would like to generalize these results for the Grassmann scheme.

Definition 10.1. A ***q-Steiner system*** $\mathrm{S}_q(t,k,n)$ is a set $\mathbb{S}$ of subspaces from $\mathrm{G}_q(n,k)$ (called blocks), such that each subspace of $\mathrm{G}_q(n,t)$ is contained in exactly one subspace of $\mathbb{S}$.

Similarly to the proof on the size of a Steiner system $\mathrm{S}(t,k,n)$ and its minimum Hamming (or Johnson) distance, the following results are proved.

Lemma 10.6. *The number of blocks in a q-Steiner system $\mathrm{S}_q(t,k,n)$ is $\begin{bmatrix} n \\ t \end{bmatrix} / \begin{bmatrix} k \\ t \end{bmatrix}$.*

Proof. Each k-subspace contains $\begin{bmatrix} k \\ t \end{bmatrix}$ distinct t-subspaces. The number of t-subspaces in an n-space is $\begin{bmatrix} n \\ t \end{bmatrix}$. Since each t-subspace is contained in exactly one k-subspace of a q-Steiner system $\mathrm{S}_q(t,k,n)$, the claim of the lemma follows. □

Proposition 10.2. *The minimum Grassmann distance of the code defined by the k-subspaces of a q-Steiner system $\mathrm{S}_q(t,k,n)$ is $k-t+1$.*

Proof. Two k-subspaces X and Y of $\mathrm{S}_q(t,k,n)$ can intersect in a subspace whose dimension is at most $t-1$. Therefore,

$$d_G(X,Y) \geq k-(t-1) = k-t+1.$$

□

Lemma 10.7. *If $\mathbb{C}$ is a code in $\mathrm{G}_q(n,k)$ with minimum Grassmann distance $k-t+1$ and $\begin{bmatrix} n \\ t \end{bmatrix} / \begin{bmatrix} k \\ t \end{bmatrix}$ codewords, then $\mathbb{C}$ is a q-Steiner system $\mathrm{S}_q(t,k,n)$.*

Proof. By the same arguments as in the proof of Lemma 10.6 on can verify that a q-Steiner system $\mathrm{S}_q(t,k,n)$ is the largest code in $\mathrm{G}_q(n,k)$ with minimum Grassmann distance $k-t+1$. □

For the rest of this chapter, we will assume now for simplicity, but w.l.o.g., that the n-space taken for $\mathcal{P}_q(n)$ and for $\mathrm{G}_q(n,k)$ is $\mathbb{F}_q^n$.

Lemma 10.8. *If there exists a q-Steiner system $\mathrm{S}_q(t,k,n)$, $t>1$, then there exists a q-Steiner system $\mathrm{S}_q(t-1,k-1,n-1)$.*

Proof. Let $\mathbb{S}$ be a q-Steiner system $\mathrm{S}_q(t,k,n)$ in $\mathbb{F}_q^n$, X be a one-subspace of $\mathbb{F}_q^n$, and U be an $(n-1)$-subspace of $\mathbb{F}_q^n$ such that $X \oplus U = \mathbb{F}_q^n$. Define the following set

$$\mathbb{S}' \triangleq \{Y \cap U \ : \ Y \in \mathbb{S}, \ X \subset Y\} \ .$$

We claim that $\mathbb{S}'$ is a q-Steiner system $\mathrm{S}_q(t-1,k-1,n-1)$ in U. Let $Z \subseteq U$ be a $(t-1)$-subspace. Since $\mathbb{S}$ is a q-Steiner system $\mathrm{S}_q(t,k,n)$, it follows that for a t-subspace $Z^e \triangleq X \oplus Z$, there exists a unique k-subspace Y such that $Y \in \mathbb{S}$ and $Z^e \subset Y$. Therefore, by the definition of $\mathbb{S}'$ we have that $Z = Z^e \cap U \subset Y \cap U \in \mathbb{S}'$, i.e., each $(t-1)$-subspace is contained in a $(k-1)$-subspace of $\mathbb{S}'$.

If Z is contained in two distinct blocks of $\mathbb{S}'$, i.e., $Z \subset Y_1$, $Z \subset Y_2$, $Y_1, Y_2 \in \mathbb{S}'$, then using the same arguments we have that $Z^e = Z \oplus X$ is contained in $Y_1^e \triangleq X \oplus Y_1$ and $Y_2^e \triangleq X \oplus Y_2$, which are in two distinct blocks of $\mathbb{S}$, a contradiction. This completes the proof. □

Corollary 10.2. *A necessary condition for the existence of a q-Steiner system $\mathrm{S}_q(t,k,n)$ is that all the numbers*

$$\frac{\begin{bmatrix} n-i \\ t-i \end{bmatrix}}{\begin{bmatrix} k-i \\ t-i \end{bmatrix}}, \quad 0 \le i \le t-1,$$

are integers.

Lemma 10.9. *Let $\mathbb{C}_{\mathcal{D}}$ be a code in $\mathrm{G}_q(n,k)$, where the distances between the codewords in $\mathbb{C}_{\mathcal{D}}$ are taken from a subset $\mathcal{D}$. Let $\mathbb{A}$ be a subset of $\mathrm{G}_q(n,k)$ and let $\mathbb{C}'_{\mathcal{D}} \subseteq \mathbb{A}$ be the largest code in $\mathbb{A}$ with distances between codewords of $\mathbb{C}'_{\mathcal{D}}$ are from $\mathcal{D}$. Then*

$$\frac{|\mathbb{C}_{\mathcal{D}}|}{\begin{bmatrix} n \\ k \end{bmatrix}} \le \frac{|\mathbb{C}'_{\mathcal{D}}|}{|\mathbb{A}|} \ . \tag{10.2}$$

Proof. Consider the set of pairs

$$\mathcal{P} \triangleq \{(X, M) \ : \ X \in \mathbb{C}_{\mathcal{D}}, \ M \in \mathbb{F}_q^{n\times n} \text{ is nonsingular}, \ \langle G(X)\cdot M\rangle \in \mathbb{A} \ \},$$

where $G(X)$ is any $k \times n$ generator matrix of X. For a fixed $X \in \mathbb{C}_{\mathcal{D}}$ and a fixed $Y \in \mathbb{A}$, the number of nonsingular matrices, such that $Y = \langle G(X) \cdot M\rangle$, is computed as follows. Choose an arbitrary generator matrix $G(X)$ for X. The number of possible distinct generator matrices for Y equals the number of nonsingular $k \times k$ matrices over $\mathbb{F}_q$, which is equal to

$$\prod_{i=0}^{k-1}(q^k - q^i) = \prod_{i=0}^{k-1}(q^i(q^{k-i} - 1)) = q^{k(k-1)/2}\prod_{i=0}^{k-1}(q^{k-i} - 1) \ .$$

Given one such generator matrix $G(Y)$ for Y, the number of nonsingular $n \times n$ matrices, for M, such that $G(Y) = G(X) \cdot M$, is equal to the number of distinct ways to complete $G(X)$ to a nonsingular $n \times n$ matrix. This claim can be observed from the fact that for each two nonsingular $n \times n$ matrices $\hat{X}$ and $\hat{Y}$, there exists a unique nonsingular $n \times n$ matrix M such that $\hat{Y} = \hat{X}M$. Hence, this number equals

$$\prod_{i=k}^{n-1}(q^n - q^i) = \prod_{i=0}^{n-k-1}(q^n - q^{k+i}) = q^{k(n-k)}\prod_{i=0}^{n-k-1}(q^{n-k} - q^i) \ .$$

Therefore, the number of pairs in $\mathcal{P}$ equals

$$|\mathbb{C}_{\mathcal{D}}| \cdot |\mathbb{A}| \cdot \prod_{i=0}^{k-1}(q^k - q^i) \cdot q^{k(n-k)}\prod_{i=0}^{n-k-1}(q^{n-k} - q^i) \ .$$

Note that for each nonsingular $n \times n$ matrix M and two subspaces $X, X' \in \mathrm{G}_q(n, k)$, we have that $d(\langle G(X) \cdot M\rangle, \langle G(X') \cdot M\rangle) = d(X, X')$. This implies that a fixed nonsingular matrix can transfer the elements of $\mathbb{C}_{\mathcal{D}}$ into at most $|\mathbb{C}'_{\mathcal{D}}|$ elements of $\mathbb{A}$. Since there are $\prod_{i=0}^{n-1}(q^n - q^i)$ nonsingular $n \times n$ matrices, it follows that the number of pairs in $\mathcal{P}$ is at most $|\mathbb{C}'_{\mathcal{D}}| \prod_{i=0}^{n-1}(q^n - q^i)$. This implies that

$$|\mathbb{C}_{\mathcal{D}}| \cdot |\mathbb{A}| \cdot \prod_{i=0}^{k-1}(q^k - q^i) \cdot q^{k(n-k)}\prod_{i=0}^{n-k-1}(q^{n-k} - q^i) \leq |\mathbb{C}'_{\mathcal{D}}| \prod_{i=0}^{n-1}(q^n - q^i) \ ,$$

which proves the claim of the lemma. □

Let $\mathcal{A}_q(n, 2\delta, k)$ be the size of the largest code in $\mathrm{G}_q(n, k)$ with a minimum subspace distance 2δ (Grassmann distance δ).

Corollary 10.3. *Let $\mathbb{C}_\mathcal{D}$ be a code in $\mathrm{G}_q(n,k)$ with distances between the codewords of $\mathbb{C}_\mathcal{D}$ are taken from the range $[2\delta, n]$. Let $\mathbb{A}$ be a subset of $\mathrm{G}_q(n,k)$ and let $\mathbb{C}'_\mathcal{D} \subseteq \mathbb{A}$ be the largest code in $\mathbb{A}$ with distances taken from $[2\delta, n]$. Then*

$$\mathcal{A}_q(n, 2\delta, k) \leq \frac{\begin{bmatrix} n \\ k \end{bmatrix}_q |\mathbb{C}'_\mathcal{D}|}{|\mathbb{A}|} . \tag{10.3}$$

Corollary 10.4.

$$\mathcal{A}_q(n, 2\delta, k) \leq \left\lfloor \frac{q^n - 1}{q^k - 1} \mathcal{A}_q(n-1, 2\delta, k-1) \right\rfloor .$$

Proof. In (10.3) take $\mathbb{A}$ to be the set of all the subspaces in $\mathrm{G}_q(n,k)$ that contain a given one-subspace X of $\mathbb{F}_q^n$. The size of $\mathbb{A}$ in this case is

$$\frac{(q^n - q)(q^n - q^2) \cdots (q^n - q^{k-1})}{(q^k - q)(q^k - q^2) \cdots (q^k - q^{k-1})} = \begin{bmatrix} n-1 \\ k-1 \end{bmatrix} .$$

Let U be any $(n-1)$-subspace in $\mathbb{F}_q^n$ such that $X \oplus U = \mathbb{F}_q^n$ and let $\mathbb{C} \triangleq \{Y \cap U \,:\, Y \in \mathbb{C}'_\mathcal{D}\}$. Clearly, each subspace in $\mathbb{C}$ is a $(k-1)$-subspace of the $(n-1)$-subspace U. Moreover, $|\mathbb{C}| = |\mathbb{C}'_\mathcal{D}|$ and since $d_S(\mathbb{C}'_\mathcal{D}) = 2\delta$, it follows that $d_S(\mathbb{C}) = 2\delta$. Therefore, $|\mathbb{C}| \leq \mathcal{A}_q(n-1, 2\delta, k-1)$ (in fact, equality can be proved, but it is not required) and the claim follows. $\square$

Corollary 10.3 is the q-analog of Corollary 8.16 proved for the Johnson scheme. Similarly, Corollary 10.4 and its proof are the q-analog of Corollary 8.17 and its proof for the Johnson scheme. Similarly, we can have a q-analog for Corollary 8.18 as follows.

Corollary 10.5.

$$\mathcal{A}_q(n, 2\delta, k) \leq \left\lfloor \frac{q^n - 1}{q^{n-k} - 1} \mathcal{A}_q(n-1, 2\delta, k) \right\rfloor .$$

Given an n-space $\mathcal{V}$ and an integer k, $0 \leq k \leq n$, let $\begin{bmatrix} \mathcal{V} \\ k \end{bmatrix}$ denote the set of all k-subspace of an n-space $\mathcal{V}$. A t-intersecting family $\mathbb{F}$ in $\mathrm{G}_q(n,k)$ consists of k-subspace such that for each two distinct subspace $X, Y \in \mathbb{F}$, we have that $\dim(X \cap Y) \geq t$. This implies that $d_G(X,Y) \leq k-t$ and hence a t-intersecting family is an anticode with diameter $k-t$. As noted before, finding the largest size of a t-intersecting family is an important problem in extremal combinatorics. The problem has been well treated and a complete solution is known. This solution is presented in the following theorem.

Theorem 10.3. *Let $\mathbb{P} \subset \begin{bmatrix} \mathcal{V} \\ k \end{bmatrix}$ be a t-intersecting family. Then*

$$|\mathbb{P}| \leq \max\left\{ \begin{bmatrix} n-t \\ k-t \end{bmatrix}, \begin{bmatrix} 2k-t \\ k \end{bmatrix} \right\}.$$

- If $2k - t < n < 2k$, then

$$|\mathbb{P}| \leq \begin{bmatrix} 2k-t \\ k \end{bmatrix}_q,$$

and this bound is attained with equality in a unique way by taking any $(2k-t)$-subspace $X \in \begin{bmatrix} \mathcal{V} \\ 2k-t \end{bmatrix}$ and defining

$$\mathbb{P} \triangleq \{F \in \begin{bmatrix} \mathcal{V} \\ k \end{bmatrix} \ : \ \dim(F \cap X) \geq k\}.$$

- If $2k < n$, then

$$|\mathbb{P}| \leq \begin{bmatrix} n-t \\ k-t \end{bmatrix},$$

and this bound is attained with equality in a unique way by taking any t-subspace $X \in \begin{bmatrix} \mathcal{V} \\ t \end{bmatrix}$ and defining

$$\mathbb{P} \triangleq \{F \in \begin{bmatrix} \mathcal{V} \\ k \end{bmatrix} \ : \ \dim(F \cap X) \geq k\}.$$

- If $n = 2k$, then

$$|\mathbb{P}| \leq \begin{bmatrix} n-t \\ k-t \end{bmatrix} = \begin{bmatrix} 2k-t \\ k \end{bmatrix}.$$

In the Johnson scheme we have not ruled out possible e-perfect codes, while in its q-analog scheme, the Grassmann scheme, such e-perfect codes were ruled out. Now, we continue and consider diameter perfect codes. We have found some families of such codes in the Johnson scheme, but did not rule out other possible families. In the Grassmann scheme we are going to characterize all the possible diameter perfect codes.

Theorem 10.4. *When $n \geq 2k$, a code $\mathbb{C}$ is a $(k-t)$-diameter perfect code in the Grassmann scheme $\mathrm{G}_q(n,k)$, if and only if $\mathbb{C}$ is a q-Steiner system $\mathrm{S}_q(t,k,n)$.*

Proof. Assume first that $n \geq 2k$ and $\mathbb{C}$ is a $(k-t)$-diameter perfect code, which implies that $\mathbb{C}$ has minimum Grassmann distance $k-t+1$. By Theorem 10.3, when $2k \leq n$, the size of a maximum anticode with diameter $k-t$ in $\mathrm{G}_q(n,k)$ is $\begin{bmatrix} n-t \\ k-t \end{bmatrix}$ and hence by the code-anticode bound, the size of a code $\mathbb{C}$ with minimum distance $k-t+1$ in $\mathrm{G}_q(n,k)$ is at most

$$\frac{\begin{bmatrix} n \\ k \end{bmatrix}}{\begin{bmatrix} n-t \\ k-t \end{bmatrix}} = \frac{\begin{bmatrix} n \\ t \end{bmatrix}}{\begin{bmatrix} k \\ t \end{bmatrix}}.$$

Since by Lemma 10.6, the size of a q-Stiener system $\mathrm{S}_q(t,k,n)$ is $\begin{bmatrix} n \\ t \end{bmatrix} / \begin{bmatrix} k \\ t \end{bmatrix}$, by Proposition 10.2 its minimum Grassmann distance is $k-t+1$, and by Lemma 10.7 each code with these parameters is an $\mathrm{S}_q(t,k,n)$, it follows that $\mathbb{C}$ is a q-Steiner system $\mathrm{S}_q(t,k,n)$.

Assume now that $\mathbb{C}$ is $\mathrm{S}_q(t,k,n)$, i.e., $\mathbb{C}$ is a code in $\mathrm{G}_q(n,k)$ with minimum Grassmann distance $k-t+1$ and $\begin{bmatrix} n \\ t \end{bmatrix} / \begin{bmatrix} k \\ t \end{bmatrix}$ codewords. Since the maximum anticode with diameter $k-t$ in $\mathrm{G}_q(n,k)$ is $\begin{bmatrix} n-t \\ k-t \end{bmatrix}$ it follows by the code-anticode bound that $\mathbb{C}$ is a $(k-t)$-diameter perfect code. □

To examine diameter perfect codes in $\mathrm{G}_q(n,k)$, where $2k > n$, we have to introduce the concept of ***orthogonal complement*** $\mathbb{C}^\perp$ for a code $\mathbb{C} \in \mathcal{P}_q(n)$ defined by

$$\mathbb{C}^\perp \triangleq \{X^\perp \,:\, X \in \mathbb{C}\}\ .$$

The following lemma is an immediate result of this definition.

Lemma 10.10. *If $\mathbb{C}$ is a code in $\mathrm{G}_q(n,k)$, then $\mathbb{C}^\perp$ is a code in $\mathrm{G}_q(n,n-k)$.*

Lemma 10.11. *If $X, Y \in \mathcal{P}_q(n)$, then $X^\perp \cap Y^\perp = (X+Y)^\perp$.*

Proof. For $u, v \in \mathbb{F}_q^n$, let $u \cdot v$ denote the inner product of the vectors u and v.

If $u \in X^\perp \cap Y^\perp$, then $u \cdot x = \mathbf{0}$ and $u \cdot y = \mathbf{0}$ for every $x \in X$ and $y \in Y$. This implies that $u \cdot (\alpha x + \beta y) = \mathbf{0}$ for each $\alpha, \beta \in \mathbb{F}_q$ and hence $u \in (X+Y)^\perp$.

If $u \in (X+Y)^\perp$, then $u \in X^\perp$ and $u \in Y^\perp$ which implies that $u \in X^\perp \cap Y^\perp$. □

Lemma 10.12. *If $X, Y \in \mathcal{P}_q(n)$, then*

$$\dim(X^\perp \cap Y^\perp) = n - \dim X - \dim Y + \dim(X \cap Y).$$

Proof. By Lemma 10.11 we have that

$$\dim(X^\perp \cap Y^\perp) = \dim(X+Y)^\perp = n - \dim(X+Y),$$

and the claim follows now from Lemma 10.2. □

Corollary 10.6. *If $\mathbb{C}$ is a code in $\mathcal{P}_q(n)$, then $d_S(\mathbb{C}^\perp) = d_S(\mathbb{C})$.*

Proof. If $X, Y \in \mathcal{P}_q(n)$, then by the definition of the subspace distance and by Lemma 10.12 we have that

$$d_S(X^\perp, Y^\perp) = \dim X^\perp + \dim Y^\perp - 2\dim(X^\perp \cap Y^\perp)$$

$$= n - \dim X + n - \dim Y - 2(n - \dim X - \dim Y + \dim(X \cap Y))$$

$$= \dim X + \dim Y - 2\dim(X \cap Y) = d_S(X, Y).$$

This implies that if $\mathbb{C}$ is a code in $\mathcal{P}_q(n)$, then $d_S(\mathbb{C}^\perp) = d_S(\mathbb{C})$. □

The same arguments used in Corollary 10.6 implies the following result.

Corollary 10.7. *If $\mathbb{A}$ is an anticode in $\mathcal{P}_q(n)$, then the maximum distance of $\mathbb{A}$ equals the maximum distance of $\mathbb{A}^\perp$.*

By Theorem 10.4 and Corollaries 10.6 and 10.7, and since $\begin{bmatrix} n \\ k \end{bmatrix} = \begin{bmatrix} n \\ n-k \end{bmatrix}$, we have the following theorem.

Theorem 10.5. *When $n < 2k$, a code $\mathbb{C}$ is an $(n-k-t)$-diameter perfect code in the Grassmann scheme $\mathrm{G}_q(n, k)$, if and only if $\mathbb{C}^\perp$ is a q-Steiner system $\mathrm{S}_q(t, n-k, n)$. When $n < 2k$, the maximum size anticode with diameter $n-k-t$ has size $\begin{bmatrix} n-t \\ n-k-t \end{bmatrix}_q$.*

Proof. If $n < 2k$, then $2n - 2k < n$, i.e., $2(n-k) < n$ and the claim follows from Theorem 10.4. □

In contrast to Steiner systems, which have been studied extensively over the years and for which there are many constructions, bounds, and interesting properties, much less is known about q-Steiner systems. The main reason is that constructing q-Steiner systems is an extremely difficult task. Many efforts have been made, but few results have accrued. The efforts became more intensive at the beginning of the 21st century after the introduction of error-correction in random network coding. It appears that error-correction for random network coding is performed efficiently with codes whose codewords are subspaces.

There is one well-known family of q-Steiner systems, the q-Steiner systems $\mathrm{S}_q(1, k, n)$. This family exists, for any finite field $\mathbb{F}_q$, if and only if k divides n. These q-Steiner systems were already considered in Chapter 7 in the context of 1-perfect byte-correcting codes, where all the bytes are of the same size k. For the parity-check matrix of such a code, each byte of size k is represented by a k-subspace. The collection of the k-subspaces that form the parity-check marix for such a 1-perfect byte-correcting code in $\mathbb{F}_q^n$, alsp form a q-Steiner system $\mathrm{S}_q(1, k, n)$. Recall that these q-Steiner systems are called k-spreads, and it was mentioned in Section 7.2 that there are non-isomorphic such codes (systems). This also leads to the following observation.

Theorem 10.6. *A k-spread in $\mathbb{F}_q^n$ exists if and only if k divides n. The size of a k-spread in $\mathbb{F}_q^n$ is $(q^n - 1)/(q^k - 1)$.*

Except for the family of q-Steiner systems $\mathrm{S}_q(1, k, n)$, i.e., k-spreads, the only known q-Steiner systems have the parameters of a q-Steiner system $\mathrm{S}_2(2, 3, 13)$. Let $\alpha \in \mathbb{F}_{2^{13}}$ be a root of the primitive polynomial $x^{13} + x^{12} + x^{10} + x^9 + 1$. Let $\mathbb{S}$ be a system of 3-subspaces of $\mathbb{F}_{2^{13}}$ with the following two properties:

[P1] If $\{0, \alpha^{i_1}, \alpha^{i_2}, \alpha^{i_3}, \alpha^{i_4}, \alpha^{i_5}, \alpha^{i_6}, \alpha^{i_7}\}$ is a 3-subspace in $\mathbb{S}$, then $\{0, \alpha^{i_1+1}, \alpha^{i_2+1}, \alpha^{i_3+1}, \alpha^{i_4+1}, \alpha^{i_5+1}, \alpha^{i_6+1}, \alpha^{i_7+1}\}$ is also a 3-subspace in $\mathbb{S}$. This property is related to a ***cyclic mapping*** on the subspaces.

[P2] If $\{0, \alpha^{i_1}, \alpha^{i_2}, \alpha^{i_3}, \alpha^{i_4}, \alpha^{i_5}, \alpha^{i_6}, \alpha^{i_7}\}$ is a 3-subspace in $\mathbb{S}$, then $\{0, \alpha^{2i_1}, \alpha^{2i_2}, \alpha^{2i_3}, \alpha^{2i_4}, \alpha^{2i_5}, \alpha^{2i_6}, \alpha^{2i_7}\}$ is also a 3-subspace in $\mathbb{S}$. This property is related to a ***Frobenius mapping*** on the subspaces.

Properties [P1] and [P2] are related to the application of the ***Singer subgroup*** on the set of 3-subspaces. We start with fifteen 3-subspaces of $\mathbb{F}_{2^{13}}$ associated with the following sets (the seven elements in each set represent the powers of α of the elements in the 3-subspaces). Each set contains eight elements of $\mathbb{F}_{2^{13}}$, which together form a 3-subspace.

$$\begin{array}{ll}
\{0,1,1249,5040,7258,7978,8105\}, & \{0,7,1857,6681,7259,7381,7908\},\\
\{0,9,1144,1945,6771,7714,8102\}, & \{0,11,209,1941,2926,3565,6579\},\\
\{0,12,2181,2519,3696,6673,6965\}, & \{0,13,4821,5178,7823,8052,8110\},\\
\{0,17,291,1199,5132,6266,8057\}, & \{0,20,1075,3939,3996,4776,7313\},\\
\{0,21,2900,4226,4915,6087,8008\}, & \{0,27,1190,3572,4989,5199,6710\},\\
\{0,30,141,682,2024,6256,6406\}, & \{0,31,814,1161,1243,4434,6254\},\\
\{0,37,258,2093,4703,5396,6469\}, & \{0,115,949,1272,1580,4539,4873\},\\
\multicolumn{2}{c}{\{0,119,490,5941,6670,6812,7312\}.}
\end{array}$$

On these fifteen 3-subspace we apply the Singer subgroup, i.e., apply the mappings induced by [P1] and [P2].

This construction leads to the following conjecture.

Conjecture 10.1. *If $p \equiv 1 \pmod 6$, $p > 7$, is a prime, then there exists a q-Steiner system $S_2(2, 3, p)$ whose subspaces satisfy properties* [P1] *and* [P2].

For $p = 7$ such a construction does not work. Three representative are required to construct a q-Steiner system $S_2(2,3,7)$, but only two such representatives exist. Nevertheless, the number of potential representative is relatively small compared to large p. It is highly probable that as p increased more nonequivalent systems exist, but unfortunately as p increased the search for such systems is more complicated. The Steiner system $S(2,3,7)$ is known as the ***Fano plane*** (see Fig. 3.2). It is embedded in the binary Hamming code of length 7. Does there exist a q-Fano plane? A lot of research was done in this direction, especially for $q = 2$, and most people believe that even if it exists, it will not be found in the near future or even the unseen future.

It is worth to mention that each codeword of weight three in the Hamming code $\mathcal{H}(r)$ is associated with a two-subspace X of $G_2(r,2)$, which contains the corresponding three nonzero column vectors of X in the parity-check matrix.

So far we have realized that there are no nontrivial perfect codes in the Grassmann scheme and the only diameter perfect codes are q-Steiner systems and their orthogonal complements, for which the only known ones are the spreads (which many will claim are trivial q-Steiner systems) and q-Steiner systems with the parameters of the q-Steiner system $S_2(2,3,13)$. Although we conjecture that there are infinitely many sets of parameters for q-Steiner systems, we just barely believe that they will be found, with a possible exception of a small number of such systems, if any. This motivates us to search for q-perfect sets (see the definition in Section 2.4) and, accordingly, we have the following three research problems.

Problem 10.1. Prove or disprove the existence of a q-Fano plane $S_q(2,3,7)$, where q is a power of some prime. Such a result for any value of q will be a major breakthrough.

Problem 10.2. Are there any nontrivial perfect sets that are not q-Steiner systems in the Grassmann scheme?

Problem 10.3. Develop a theory for the existence and nonexistence of perfect sets in the Grassmann scheme.

10.3 Normal Spreads

Recall that a ***k-spread*** in $\mathbb{F}_q^n$ is a partition of the elements of $\mathbb{F}_q^n$ into k-subspaces (where the *zero* element of $\mathbb{F}_q^n$ and the k-subspaces is ignored).

A k-spread $\mathbb{S}$ is called a ***normal spread*** (also a ***geometric spread***) if every subspace spanned by k-subspaces of $\mathbb{S}$ is partitioned by elements of $\mathbb{S}$. In other words, $\mathbb{S}$ is a normal spread if it satisfies the following requirements:

(1) $\mathbb{S}$ is a k-spread.
(2) If $Y_1, Y_2, \ldots, Y_m$ are k-subspaces in $\mathbb{S}$ and Z is the subspace spanned by $Y_1, Y_2, \ldots, Y_m$, then there exists a set $\{X_1, X_2, \ldots, X_\ell\}$ of k-subspaces in $\mathbb{S}$ such that $Z = \bigcup_{i=1}^{\ell} X_i$.

The next lemma is a trivial observation.

Lemma 10.13. *Each k-spread in $\mathbb{F}_q^{2k}$ is a normal spread.*

Another simple lemma considers the dimension of any subspace spanned by any set of elements of a normal spread.

Lemma 10.14. *If $\mathbb{S}$ is a normal k-spread in $\mathbb{F}_q^{rk}$, then the dimension of any subspace spanned by any subset of k-subspaces of $\mathbb{S}$ is a multiple of k.*

Proof. Assume the contrary and let $\mathbb{B} = \{Y_1, \ldots, Y_{m-1}, Y_m\}$ be the smallest subset of k-subspaces from $\mathbb{S}$ which span a subspace whose dimension is not a multiple of k. Hence, the k-subspaces $Y_1, \ldots, Y_{m-1}$ span a subspace X whose dimension is $(m-1)k$. Clearly, Y_m is not contained in X. Since $\mathbb{S}$ is a normal spread, it follows that X is partitioned by elements of $\mathbb{S}$ which do not contain Y_m and hence $Y_m \cap X = \{\mathbf{0}\}$. Therefore, X and Y_m span a subspace of dimension mk, a contradiction and the claim of the lemma follows. □

Theorem 10.7. *If $\mathbb{S}$ is a k-spread in $\mathbb{F}_q^{rk}$ in which each $(2k)$-subspace spanned by two k-subspaces of $\mathbb{S}$ is partitioned by elements of $\mathbb{S}$, then $\mathbb{S}$ is a normal spread.*

Proof. The proof is by induction, where the claim is that for $m \leq r$, any (mk)-subspace spanned by elements of $\mathbb{S}$ is partitioned by elements of $\mathbb{S}$. The basis for the induction is $m = 2$, which is a property of $\mathbb{S}$ as given in the theorem. Assume that the claim is true for some $2 \leq m < r$. Let X be an $((m+1)k)$-subspace, where $m+1 \leq r$, spanned by the $m+1$ elements of $\mathbb{S}$, say, the $m+1$ elements in the set $\mathbb{B} = \{Y_1, Y_2, \ldots, Y_{m+1}\}$. By the induction hypothesis, each (mk)-subspace spanned by m of the $m+1$ subspaces in $\mathbb{B}$ is partitioned by elements of $\mathbb{S}$. Let $\mathbb{S}_1$ be the set of all subspaces that partition these (mk)-subspaces. Let Z be any k-subspace of the (mk)-subspace $\mathbb{S}$ spanned by $\{Y_1, Y_2, \ldots, Y_m\}$, which is not spanned by any proper subset of

$\{Y_1, Y_2, \ldots, Y_m\}$. By the basis of the induction, the $(2k)$-subspace spanned by Z and Y_{m+1} is partitioned by elements of $\mathbb{S}$. Add these subspaces of $\mathbb{S}$ to $\mathbb{S}_1$ and do the same for each such k-subspace Z. It can be verified easily that $\mathbb{S}_1$ is a partition of the $((m+1)k)$-subspace X and this completes the proof. □

We continue to show a connection between normal spreads in $\mathbb{F}_q^n$ and 1-perfect codes over $\mathbb{F}_q$ in the Hamming scheme.

Theorem 10.8. *Let $\mathbb{S}$ be a normal k-spread in $\mathbb{F}_q^{rk}$ for some $r \geq 3$. Let $\mathbb{T}$ be the set of all $(2k)$-subspaces spanned by all the pairs of k-subspaces of $\mathbb{S}$, where each $(2k)$-subspace X is represented by the $\frac{q^{2k}-1}{q^k-1}$ k-subspaces of $\mathbb{S}$ contained in X. Then, $\mathbb{T}$ form a (Q, B) Steiner system $\mathrm{S}(2, q^k+1, \frac{q^{rk}-1}{q^k-1})$, where $Q = \mathbb{S}$, $B = \mathbb{T}$, and a block X of B contains the points (k-subspaces of $\mathbb{S}$), which are contained in X.*

Proof. First, consider the parameters of the system. The total number of k-subspaces in $\mathbb{S}$ is $\frac{q^{rk}-1}{q^k-1}$, which is the number of points in $\mathbb{T}$. Two disjoint k-subspaces in $\mathbb{S}$ span a $(2k)$-subspace that consists of $\frac{q^{2k}-1}{q^k-1} = q^k + 1$ distinct k-subspaces and hence the size of a block is $q^k + 1$.

Let $\{X, Y\}$ be a pair of two disjoint k-subspaces of $\mathbb{S}$. X and Y span a unique $(2k)$-subspace and hence $\mathbb{T}$ yields a Steiner system $\mathrm{S}(2, q^k+1, \frac{q^{rk}-1}{q^k-1})$. □

Let

$$c(x) = x^k - \sum_{i=1}^{k} c_i x^{k-i}, \quad c_i \mathbb{F}_q \tag{10.4}$$

be a primitive polynomial over $\mathbb{F}_q$. The $k \times k$ matrix

$$C = \begin{bmatrix} 0 & 0 \cdots 0 & c_k \\ 1 & 0 \cdots 0 & c_{k-1} \\ 0 & 1 \cdots 0 & c_{k-2} \\ \vdots & \vdots \ddots \vdots & \vdots \\ 0 & 0 \cdots 1 & c_1 \end{bmatrix}$$

is called the ***companion matrix*** of the polynomial $c(x)$ (or the companion matrix of β, where β is any root of $c(x)$).

If β is a root of $c(x)$, then by (10.4) we have

$$\beta^k = \sum_{i=1}^{k} c_i \beta^{k-i} = \sum_{i=0}^{k-1} c_{k-i} \beta^i . \tag{10.5}$$

If an element β^m has the vector representation

$$(\beta^m) = \begin{pmatrix} b_0 \\ b_1 \\ \vdots \\ b_{k-1} \end{pmatrix},$$

then

$$\beta^m = \sum_{i=0}^{k-1} b_i \beta^i .$$

By plugging (10.5) this implies that

$$\beta^{m+1} = \sum_{i=0}^{k-1} b_i \beta^{i+1} = \sum_{i=1}^{k-1} b_{i-1} \beta^i + b_{k-1} \beta^k$$

$$= \sum_{i=1}^{k-1} b_{i-1} \beta^i + b_{k-1} \sum_{i=0}^{k-1} c_{k-i} \beta^i = b_{k-1} c_k + \sum_{i=1}^{k-1} (b_{i-1} + b_{k-1} c_{k-i}) \beta^i ,$$

and hence $(\beta^{m+1}) = C \cdot (\beta^m)$. Therefore,

$$(\beta^m) = C^m \cdot (\beta^0) \Rightarrow [(\beta^m)(\beta^{m+1}) \ \cdots \ (\beta^{m+k-1})] = C^m ,$$

which implies that

$$\beta^i + \beta^j = \beta^\ell \Rightarrow \beta^{i+m} + \beta^{j+m} = \beta^{\ell+m} \Rightarrow C^i + C^j = C^\ell .$$

Thus, we have the following theorem.

Theorem 10.9. *There is an isomorphism between the finite field $\mathbb{F}_{q^k}$ and the $q^k - 1$ consecutive powers of the companion matrix C together with the $k \times k$ all-zero matrix.*

Corollary 10.8. *Let $c(x)$ be a primitive polynomial of degree k over $\mathbb{F}_q$, β be its root, and C the associated $k \times k$ companion matrix. Let γ be any nonzero element of $\mathbb{F}_{q^k}$ and let Γ be its representation as a column vector of length k. Then, for each i, $0 \le i \le q^k - 2$, we have that the vector $C^i \Gamma$ is the representation of the element $\beta^i \gamma$ in $\mathbb{F}_{q^k}$ as a k-ary vector over $\mathbb{F}_q$.*

Recall that the parity-check matrix H of the Hamming code of length $\frac{q^{rk}-1}{q^k-1}$ over $\mathbb{F}_{q^k}$ consists of pairwise linearly independent nonzero column vectors of length r with elements of $\mathbb{F}_{q^k}$, i.e., the points of $\mathrm{PG}(r-1, q^k)$. Now, let H' be the $(rk) \times \left(\frac{q^{rk}-1}{q^k-1}k\right)$ matrix formed from H by replacing each element β^i, $0 \leq i \leq q^r - 2$, of $\mathbb{F}_{q^k}$ in H by C^i in H' and replacing any *zero* in H by the $k \times k$ all-zero matrix in H'.

Theorem 10.10. *Each k consecutive columns of H' originating from the same column of H is a basis of a k-subspace. The set of these $\frac{q^{rk}-1}{q^k-1}$ subspaces is a normal k-spread.*

Proof. Clearly, $C^i C^{q^k-1-i} = C^{k-1} = C^0 = I_k$ and hence each C^i is a nonsingular matrix, which implies that each k consecutive columns of H' originating from a column of H are linearly independent and hence they form a base of a k-subspace.

Consider now the isomorphism, implied by Theorem 10.9, between the nonzero elements of $\mathbb{F}_{q^k}$ and the q^k-1 consecutive powers of the companion matrix C. This isomorphism implies that each block of the Steiner system $\mathrm{S}(2, q+1, \frac{q^{rk}-1}{q^k-1})$ defined in Theorem 10.8 is transferred into a $(2k)$-subspace spanned by any two k-subspaces of H'. The claim is now implied by Theorem 10.7. □

10.4 Nonexistence of Perfect Codes in the Projective Space

We have proved in Theorem 10.2 that there are no nontrivial perfect codes in $\mathrm{G}_q(n,k)$. Are there any e-perfect codes in $\mathcal{P}_q(n)$? Except for the usual trivial perfect codes, there exists another trivial perfect code. When n is odd, i.e., $n = 2e+1$, there exists an e-perfect code that contains exactly two disjoint codewords, the null space $\{\mathbf{0}\}$ and $\mathcal{P}_q(n)$. This is the q-analog of the binary e-perfect code of length $n = 2e+1$ with two codewords, the all-zero word and the all-one word. Finally, we note that the graph of $\mathcal{P}_q(n)$ is not a regular graph and hence the subspace distance is not a regular metric since there are balls with the same radius and different sizes.

For the proof of the nonexistence of perfect codes in $\mathcal{P}_q(n)$, we will first need the following lemma.

Lemma 10.15.

$$\mathcal{A}_q(n, 2k, k) \leq \left\lfloor \frac{q^n - 1}{q^k - 1} \right\rfloor - 1 \qquad \textit{if} \;\; n \not\equiv 0 \pmod{k} .$$

Proof. Divide k into n to write $n = mk + r$, where the remainder r is nonzero by assumption and $r < k$. It is easy to verify that

$$q^n - 1 = q^r(q^{(m-1)k} + q^{(m-2)k} + \cdots q^k + 1)(q^k - 1) + q^r - 1\,. \qquad (10.6)$$

Now assume to the contrary that there exists a code $\mathbb{C}$ in $\mathrm{G}_q(n,k)$ with $M = \left\lfloor \frac{q^n-1}{q^k-1} \right\rfloor$ codewords. Further, let $Z_1, Z_2, \ldots, Z_M$ denote the codewords of $\mathbb{C}$, and observe that $Z_i \cap Z_j = \{\mathbf{0}\}$ for all $i \neq j$. Hence, we can partition $\mathbb{F}_q^n \setminus \{\mathbf{0}\}$ into $M+1$ disjoint sets as follows:

$$\mathbb{F}_q^n \setminus \{\mathbf{0}\} = Z_1^- \cup Z_2^- \cup \cdots Z_M^- \cup X\,, \qquad (10.7)$$

where X denotes the set of all vectors in $\mathbb{F}_q^n$ that are not contained in any codeword of $\mathbb{C}$. Thus,

$$|X| = q^n - 1 - M(q^k - 1) = q^r - 1\,.$$

in view of (10.6) and (10.7). Given a fixed nonzero vector $u \in \mathbb{F}_q^n$ and a set $\mathcal{S} \subseteq \mathbb{F}_q^n$, let $\eta_u(\mathcal{S})$ denote the number of vectors in $\mathcal{S}$ that are not orthogonal to u; that is,

$$\eta_u(\mathcal{S}) \triangleq |\{x \in \mathcal{S} \,:\, \langle x, u\rangle \neq 0\}|,$$

where the inner product is over $\mathbb{F}_q$. Note that $\eta_u(Z_i^-) = \eta_u(Z_i)$ is either 0 or $(q-1)q^{k-1}$ for all i, since Z_i is a vector space of dimension k. This claim can be proved by induction. This also implies that $\eta_u(\mathbb{F}_q^n) = (q-1)q^{n-1}$. Hence,

$$\eta_u(X) = \eta_u(\mathbb{F}_q^n \setminus \{\mathbf{0}\}) - \sum_{\mathbf{i=1}}^{\mathbf{M}} \eta_{\mathbf{u}}(\mathbf{Z_i^-})$$

is divisible by q^{k-1}. But $|X| = q^r - 1 < q^{k-1}$, which implies that $\eta_u(X) = 0$. Since this is true for all nonzero $u \in \mathbb{F}_q^n$, the set X cannot contain any nonzero vectors, a contradiction.

Thus, $M \leq \left\lfloor \frac{q^n-1}{q^k-1} \right\rfloor - 1$. □

Recall that for an e-perfect code, we say that X covers Y if $d(X,Y) \leq e$.

Theorem 10.11. *For all q and n, there are no nontrivial perfect codes in the projective space $\mathcal{P}_q(n)$.*

Proof. Let us assume to the contrary that $\mathbb{C}$ is an e-perfect code in $\mathcal{P}_q(n)$. Let $d = 2e+1$, and define $\mathbb{C}_k \triangleq \mathbb{C} \cap \mathrm{G}_q(n,k)$ for all $k = 0, 1, \ldots, n$. We distinguish between two cases depending on whether $\{\mathbf{0}\}$ is a codeword in $\mathbb{C}$ or $\{\mathbf{0}\} \notin \mathbb{C}$.

Case 1. $\{\mathbf{0}\} \in \mathbb{C}$.

Clearly, $\mathbb{C}_1 = \mathbb{C}_2 = \cdots = C_{2e} = \varnothing$, and all the subspaces in $\mathrm{G}_q(n, e+1)$ must be covered by the codewords of $\mathbb{C}_d$. This implies that $\mathbb{C}_d$ is a q-Steiner system $\mathrm{S}_q(e+1, 2e+1, n)$ and, hence by Lemma 10.6, we have that $|\mathbb{C}_d| = \begin{bmatrix} n \\ e+1 \end{bmatrix} / \begin{bmatrix} 2e+1 \\ e+1 \end{bmatrix}$. Each subspace of $\mathbb{C}_d$ covers $\begin{bmatrix} 2e+1 \\ e+2 \end{bmatrix}$ subspaces of $\mathrm{G}_q(n, e+2)$. This leaves $\begin{bmatrix} n \\ e+2 \end{bmatrix} - |\mathbb{C}_d| \begin{bmatrix} 2e+1 \\ e+2 \end{bmatrix}$ uncovered subspaces of $\mathrm{G}_q(n, e+2)$, and each one must be covered by a codeword of $\mathbb{C}_{d+1}$. Furthermore, each codeword of $\mathbb{C}_{d+1}$ covers exactly $\begin{bmatrix} 2e+2 \\ e+2 \end{bmatrix}$ subspaces of $\mathrm{G}_q(n, e+2)$. Putting all this together implies that

$$|\mathbb{C}_{d+1}| = \left(\begin{bmatrix} n \\ e+2 \end{bmatrix} - \begin{bmatrix} 2e+1 \\ e+2 \end{bmatrix} \cdot \begin{bmatrix} n \\ e+1 \end{bmatrix} \Big/ \begin{bmatrix} 2e+1 \\ e+1 \end{bmatrix} \right) \Big/ \begin{bmatrix} 2e+2 \\ e+2 \end{bmatrix}$$

$$= \frac{(q^n - 1)(q^{n-1} - 1) \cdots (q^{n-e} - 1)}{(q^{d+1} - 1)(q^d - 1) \cdots (q^{e+1} - 1)} (q^{n-e-1} - q^e) \, .$$

Observe that $\mathcal{A}_q(n, 2e+2, 2e+2) = \mathcal{A}_q(n, d+1, d+1) \geq |\mathbb{C}_{d+1}|$. Starting with this, and applying Corollary 10.4 iteratively $e+1$ times, we obtain

$$\mathcal{A}_q(m, 2(e+1), e+1) \geq \frac{q^m - q^{k-1}}{q^k - 1} \, , \tag{10.8}$$

where $m = n - (e+1)$ and $k = e+1$. Moreover, the fact that $\mathbb{C}_d$ is a q-Steiner system $\mathrm{S}_q(e+1, 2e+1, n)$ implies, by Lemma 10.8, that there exists an $\mathrm{S}_q(1, e+1, n-e)$ and hence $k = e+1$ divides $n - e = m+1$. This further implies that

$$\frac{q^m - q^{k-1}}{q^k - 1} = q^{m-k} + q^{m-2k} + q^{m-2k} + \cdots + q^{2k-1} + q^{k-1} = \left\lfloor \frac{q^m - 1}{q^k - 1} \right\rfloor .$$

Also, since $k = e+1$ divides $m+1$, it cannot divide m. This establishes a contradiction between (10.8) and Lemma 10.15.

Case 2. $\{\mathbf{0}\} \notin \mathbb{C}$.

Our proof for this case is based upon constructing a certain partition of $\mathbb{F}_q^n$, and then applying a counting argument to this partition to arrive at a contradiction. For the counting argument, let us introduce a function ξ from subsets of $\mathbb{F}_q^n$ to the natural numbers, defined as follows: given a set $\mathcal{S} \subseteq \mathbb{F}_q^n$, let $\xi(\mathcal{S})$ denote the number of vectors $(x_1, x_2, \ldots, x_n)$ in $\mathcal{S}$ such that $x_1 = 1$. Note that if $\mathcal{S}$ is a vector space of dimension ℓ and $\xi(\mathcal{S}) \neq 0$, then each symbol occurs the same amount of times in the first coordinate and hence $\xi(\mathcal{S}) = q^{\ell-1}$. Now let $X \in \mathbb{C}$ be a codeword of the smallest dimension k among all the codewords in $\mathbb{C}$. Since $X \neq \{\mathbf{0}\}$, we can assume w.l.o.g. that $\xi(X) \neq 0$ (otherwise, permute the coordinates of the ambient space $\mathbb{F}_q^n$ or the code $\mathbb{C}$ so that X is not entirely zero on the first coordinate). The partition of $\mathbb{F}_q^n$ is constructed as follows. Since X must cover the null-space $\{\mathbf{0}\}$, it follows that $k \leq e$. Find a subspace Z in $\mathcal{P}_q(n)$ that satisfies the following conditions:

$$\dim Z = e - k, \qquad X \cap Z = \{\mathbf{0}\}, \qquad \xi(Z) = 0. \tag{10.9}$$

It is easy to verify that such a subspace Z always exists. Next, define $W = X \oplus Z$. In view of (10.9), we have that $\dim W = e$, and since $\xi(X) \neq 0$, it follows that $\xi(W) \neq 0$, i.e., $\xi(W) = q^{e-1}$. Finally, define a sub-code $\mathbb{C}'$ of $\mathbb{C}$ as follows:

$$\mathbb{C}' \triangleq \{Y \in \mathbb{C} \,:\, Z \subset Y \ \text{ and } \ \dim Y = 2e + 1 - k\}.$$

Suppose that $\mathbb{C}'$ contains M codewords $Y_1, Y_2, \ldots, Y_M$. For all $i = 1, 2, \ldots, M$, let $Y_i^{\times} = Y_i \setminus Z$. We claim that

$$\{Y_1^{\times}, Y_2^{\times}, \ldots, Y_M^{\times}, W\} \,. \tag{10.10}$$

is a partition of $\mathbb{F}_q^n$. Assuming that (10.10) is, indeed, a partition of $\mathbb{F}_q^n$, we easily arrive at a contradiction. Since $\dim Y_i = d - k$ and $\xi(Z) = 0$, we have that $\xi(Y_i^{\times}) = \xi(Y_i)$ is either 0 or $q^{d-k-1} = q^{2e-k}$ for all i. Also, $\xi(\mathbb{F}_q^n) = q^{n-1}$ and, therefore,

$$\xi(W) = \xi(\mathbb{F}_q^n) - \sum_{i=1}^{M} \xi(Y_i^{\times})$$

must be divisible by q^{2e-k}. This is a contradiction, since we have already shown that $\xi(W) = q^{e-1}$, but $e - 1 < 2e - k$ for all $k \leq e$. To complete the proof, it remains to establish that (10.10) is indeed a partition.

Claim 10.1. Let u be a vector of $\mathbb{F}_q^n$ that lies outside of W. Then there exists a $Y_i \in \mathbb{C}'$ such that $u \in Y_i$.

Proof. If $U = Z \oplus \{\mathbf{0}, u\}$, then U is a subspace of dimension $e - k + 1$ that must be covered by some codeword of $\mathbb{C}$. This codeword is not X since $U \cap X = \{\mathbf{0}\}$, and hence

$$d_S(U, X) = \dim U + \dim X = (e - k + 1) + k = e + 1 \,.$$

Let $Y \in \mathbb{C}$ be the codeword that covers U, i.e., $d_S(U, Y) \leq e$. Since $X \in \mathbb{C}$, $\dim X = k$, and $d_S(\mathbb{C}) = d$, it follows that $d_S(X, Y) \geq d$ which implies that $\dim Y \geq d - k$. From the fact that Y covers U, we obtain that

$$d_S(U, Y) = \dim U + \dim Y - 2\dim(U \cap Y) \leq e \,.$$

This implies that $e-k+1+\dim Y-2\dim(U\cap Y)\le e$ and hence

$$d-k-2\dim(U\cap Y)\le \dim Y-2\dim(U\cap Y)\le k-1\ .$$

Therefore, $d-2k+1\le 2\dim(U\cap Y)$, i.e., $e+1-k\le \dim(U\cap Y)$ and since $\dim U=e+1-k$, it follows that $\dim Y=2e-k$ and

$$\dim(U\cap Y)=\dim U=e-k+1\ . \tag{10.11}$$

(10.11), the definition of U, and $\dim Y=2e+1-k$, however, imply that $Z\subset U\subset Y$ and, therefore, $Y\in\mathbb{C}'$. Finally, $U\subset Y$ also implies that $u\in Y$, which completes the proof of the claim. ■

If u lies outside of $W=X\oplus Z$ and $u\in Y_i$, then clearly u must belong to $Y_i^\times=Y_i\setminus Z$. Hence, Claim 10.1 shows that the set union $Y_1^\times\cup Y_2^\times\cup\cdots\cup Y_M^\times\cup W$ indeed contains all of $\mathbb{F}_q^n$.

Claim 10.2. The sets $Y_1^\times,Y_2^\times,\ldots,Y_M^\times$ and W are pairwise disjoint.

Proof. Given any two codewords Y_i and Y_j in $\mathbb{C}'$, we have that

$$d_S(Y_i,Y_j)=2(d-k)-2\dim(Y_i\cap Y_j)=2(d-k)-2(e-k)=2(e+1)>d.$$

This implies that $\dim(Y_i\cap Y_j)\le e-k=\dim Z$ and, therefore, $Y_i\cap Y_j=Z$. Consequently, the sets $Y_1^\times,Y_2^\times,\ldots,Y_M^\times$ are disjoint.

Now assume to the contrary that there exists a nonzero vector y in the intersection $Y_i^\times\cap W$ for some i. Then $y\in Y_i$, and $y=x+z$ for some nonzero $x\in X$ and some $z\in Z$. Y_i, however, is a vector space that contains Z as a subspace. Therefore, Y_i also contains the vector $y-z=x$, and hence $\dim(X\cap Y_i)\ge 1$. This clearly contradicts the minimum distance of $\mathbb{C}$, since then $d_S(X,Y_i)=k+(d-k)-2\dim(X\cap Y_i)\le d-2$. ■

Claims 10.1 and 10.2 complete the proof that (10.10) is, indeed, a partition of $\mathbb{F}_q^n$. This, in turn, completes the proof of the theorem. □

10.5 Rank-Metric Codes

Rank-metric codes are highly related to subspaces (as a concept and in their applications in random network coding) and hence these codes are considered with subspaces.

For two $k\times m$ matrices A and B over $\mathbb{F}_q$, the ***rank distance*** is defined by

$$d_R(A,B)\triangleq \operatorname{rank}(A-B)\ .$$

A $[k\times m,\varrho,\delta]$ ***rank-metric code*** $\mathcal{C}$ is a linear code, whose codewords are $k\times m$ matrices over $\mathbb{F}_q$; they form a linear subspace with dimension ϱ

of $\mathbb{F}_q^{k\times m}$, and for each two distinct codewords A and B, we have that $d_R(A,B) \geq \delta$. Similarly, we define a $(k \times m, M, \delta)_q$ rank-metric code to be a code of size M whose codewords are $k \times m$ matrices over $\mathbb{F}_q$ and the rank distance between any two distinct matrices is at least δ. The related graph $\mathrm{B}_q(k,m)$, $k \leq m$, called the ***bilinear forms graph***, is the graph whose vertices are all the $q^{k\cdot m}$ distinct $k \times m$ matrices over $\mathbb{F}_q$ and two vertices are joined by an edge if and only if the difference between their related matrices is a matrix whose rank is one. The following theorem is not difficult to prove.

Theorem 10.12. *The bilinear forms graph $B_q(k,m)$ defines an association scheme.*

Theorem 10.13. *If $\mathcal{C}$ is a $(k \times m, M, \delta)_q$ rank-metric code, then $M \leq q^{\varrho}$, where*

$$\varrho \leq min\{k(m-\delta+1), m(k-\delta+1)\} . \tag{10.12}$$

Proof. Let $\mathcal{C}$ be a $(k \times m, M, \delta)_q$ rank-metric code and w.l.o.g. assume that $k \leq m$, which implies that $m(k-\delta+1) \leq k(m-\delta+1)$. Consider the first $k-\delta+1$ rows of the M codewords of $\mathcal{C}$. If the entries in these rows of two such matrices A and B, where $A, B \in \mathcal{C}$, are equal, then the matrix $A-B$ has $k-\delta+1$ rows with *zeroes* and at most $\delta-1$ nonzero rows. This implies that $\text{rank}(A-B) \leq \delta-1$, a contradiction. Thus, in each two such matrices, the $m(k-\delta+1)$ entries in the first $k-\delta+1$ rows of the M matrices are different, i.e., $M \leq q^{m(k-\delta+1)}$ and the proof is completed. □

Corollary 10.9. *If $\mathcal{C}$ is a $[k \times m, \varrho, \delta]$ rank-metric code, then*

$$\varrho \leq min\{k(m-\delta+1), m(k-\delta+1)\} . \tag{10.13}$$

The bound of Theorem 10.13 (for nonlinear codes) and the bound of Corollary 10.9 (for linear codes), is called the Singleton bound for the rank distance. This bound is attained with equality for all feasible parameters. The codes that meet this bound are called ***maximum rank-distance codes*** (or ***MRD codes*** in short). A generator matrix for a $[k \times (n-k), (n-k)(k-\delta+1), \delta]$ MRD code $\mathcal{C}$, $n-k \geq k$, can be represented as follows.

$$G = \begin{pmatrix} g_1 & g_2 & \cdots & g_k \\ g_1^q & g_2^q & \cdots & g_k^q \\ \vdots & \vdots & \cdots & \vdots \\ g_1^{q^{k-\delta}} & g_2^{q^{k-\delta}} & \cdots & g_k^{q^{k-\delta}} \end{pmatrix},$$

where the g_i's, $1 \leq i \leq k$, $g_i \in \mathbb{F}_{q^{n-k}}$, are linearly independent over $\mathbb{F}_q$ and hence a row in G is a $k \times (n-k)$ matrix. If the last i rows, $1 \leq i \leq k-\delta$ are removed from G, then the outcome is a generator matrix for a $[k \times (n-k), (n-k)(k-\delta+1-i), \delta+i]$ MRD sub-code of $\mathcal{C}$.

Theorem 10.14. *The maximum size anticode whose codewords are $k \times m$, $k \leq m$, matrices, having maximum rank-distance $\delta - 1$, has $q^{m(\delta-1)}$ codewords. A $[k \times m, m(k-\delta+1), \delta]$ MRD code $\mathcal{C}$ is a $(\delta-1)$-diameter perfect code.*

Proof. Let $\mathcal{A}$ be the set of all $k \times m$ matrices, with nonzero entries only in the last $\delta - 1$ rows. There are $q^{m(\delta-1)}$ such distinct matrices and the rank distance between any two such matrices is at most $\delta - 1$. Therefore, a $[k \times m, m(k-\delta+1), \delta]$ MRD code $\mathcal{C}$ and the anticode $\mathcal{A}$ meet the code-anticode bound, i.e.,

$$|\mathcal{C}| \cdot |\mathcal{A}| = q^{m(k-\delta+1)} q^{m(\delta-1)} = q^{m \cdot k} .$$

Thus, $\mathcal{A}$ is a maximum size anticode in $\mathrm{B}_q(k,m)$, $k \leq m$, with diameter $\delta - 1$ and size $q^{m(\delta-1)}$ and the $[k \times m, m(k-\delta+1), \delta]$ MRD code $\mathcal{C}$ is a $(\delta-1)$-diameter perfect code. □

Corollary 10.10. *The set of anticodewords in an anticode with diameter δ, $\delta \geq 2$, defined for the bilinear forms scheme, contains the set of anticodewords in an anticode with diameter $\delta - 1$, defined for the bilinear forms scheme,*

In view of Corollary 10.10 we consider the anticodes defined in Theorem 10.14 as the maximum size anticodes considered in the rest of this section.

The code-anticode bound of Corollary 2.15 holds for the rank-metric since the conditions of Lemma 2.14 are satisfied for this metric. The code-anticode bound and the fact that for each set of parameters there exists an MRD code which meets the bound in (10.12) can also be used to prove that there are no nontrivial perfect codes in the bilinear forms graph. This claim will be proved in the rest of this section. The idea is to consider the last sphere in a ball and the "almost" sphere in a maximum size anticode, where an "almost" sphere is the set of elements that are contained in a maximum size anticode with diameter δ, but are not contained in its sub-anticode of maximum size whose diameter is $\delta - 2$.

Lemma 10.16. *The size of a sphere with radius t of in $B_q(k,m)$ is*

$$\prod_{j=0}^{t-1} \frac{(q^k - q^j)(q^m - q^j)}{q^t - q^j}.$$

Proof. The size of a sphere with radius t does not depend on its center and hence it equals the number of $k \times m$ matrices of rank t over $\mathbb{F}_q$. Let $F(k,t)$ be the number of such matrices. We claim that

$$F(k,t) = q^t F(k-1,t) + (q^m - q^{t-1})F(k-1,t-1) \ , \tag{10.14}$$

where $F(k,0) = 1$ for $k \geq 1$, and $F(k,k) = \prod_{i=0}^{k-1}(q^m - q^i)$.

Given a $(k-1) \times m$ matrix M of rank t, a $k \times m$ matrix of rank t is obtained by adding a row of length m, which is formed by one of the q^t distinct linear combinations of the $k-1$ rows of M. Hence, the number of $k \times m$ matrices of rank t whose first $k-1$ rows form a matrix of rank t is $q^t F(k-1,t)$.

Given a $(k-1) \times m$ matrix M of rank $t-1$, a $k \times m$ matrix of rank t is obtained by adding a row of length m that is not in the linear span of the first $k-1$ rows of M. There are q^m possible rows of length m and q^{t-1} linear combinations of these $k-1$ rows of M since their rank is $t-1$. Hence, the number of $k \times m$ matrices of rank t whose first $k-1$ rows form a matrix of rank $t-1$ is $(q^m - q^{t-1})F(k-1,t-1)$.

This implies the recursion in (10.14).

As for the initial conditions in (10.14), $F(k,0)$ is readily verified, while $F(k,k)$ is the number of $k \times m$ matrices of full-rank k.

To complete the proof, only a simple computation using induction is needed to verify that $F(k,t) = \prod_{j=0}^{t-1} \frac{(q^k - q^j)(q^m - q^j)}{q^t - q^j}$ is the solution for this recursion with the initial conditions. □

Using Theorem 10.14 we can compute the size of an "almost" sphere in a maximum size anticode. This size is computed in the following lemma.

Lemma 10.17. *The difference in the size of a maximum size anticode with diameter $2e$ and the size of a maximum size anticode with diameter $2e-2$ in $B_q(k,m)$ is*

$$q^{2me} - q^{m(2e-2)} = q^{m(2e-2)}(q^{2m} - 1) \ .$$

Proof. By Theorem 10.14, a maximum size anticode with diameter $2e$ in $\mathrm{B}_q(k,m)$ has $q^{m \cdot 2e}$ codewords. By Theorem 10.14, a maximum size anticode with diameter $2e-2$ in $\mathrm{B}_q(k,m)$ has $q^{m\cdot(2e-2)}$ codewords. This implies the claim in the lemma. □

Lemma 10.18. *If $m \geq k \geq e \geq 1$, then*

$$q^{m(2e-2)}(q^{2m}-1) > \prod_{j=0}^{e-1} \frac{(q^k-q^j)(q^m-q^j)}{q^e-q^j}\ .$$

Proof. The proof is by induction on e. Let

$$DA(e) \triangleq q^{2me} - q^{m(2e-2)} = q^{m(2e-2)}(q^{2m}-1)$$

and

$$DB(e) \triangleq \prod_{j=0}^{e-1} \frac{(q^k-q^j)(q^m-q^j)}{q^e-q^j}.$$

When $e=1$, $DA(1) = q^{2m}-1$ and $DB(1) = \frac{(q^k-1)(q^m-1)}{q-1}$ and since $m \geq k$, it follows that $DA(1) > DB(1)$ and the basis of the induction is proved. Assume now that $DA(e) > DB(e)$, where $k > e \geq 1$. It is easy to verify that

$$DA(e+1) = DA(e)q^{2m}$$

and

$$DB(e{+}1) = DB(e)(q^k{-}q^e)(q^m{-}q^e)\frac{\prod_{j=0}^{e-1}(q^e-q^j)}{\prod_{j=0}^{e}(q^{e+1}-q^j)} < DB(e)(q^k{-}q^e)(q^m{-}q^e).$$

Since $q^{2m} > (q^m - q^e)(q^k - q^e)$ and by the induction hypothesis $DA(e) > DB(e)$, it follows that $DA(e+1) > DB(e+1)$ and the proof of the claim is completed. □

Theorem 10.15. *There are no nontrivial perfect codes in the bilinear forms scheme $B_q(k,m)$.*

Proof. Clearly, by (10.13) if any perfect code exists in $B_q(k,m)$, then it should be an MRD code. Assume $\mathcal{C}$ is a $[k \times m, \varrho, \delta]$ e-perfect rank-metric code. Since $\mathcal{C}$ is a perfect code, it follows that $\delta = 2e+1$. By Theorem 2.11 (and also by Theorem 10.14), $\mathcal{C}$ is also a $(2e)$-diameter perfect code. This implies that the size of ball with radius e, $\mathcal{B}_e(k \times m)$, is equal to the size of the largest anticode with diameter $2e$, $q^{m(\delta-1)}$, i.e.,

$$|\mathcal{B}_e(k \times m)| = q^{m(\delta-1)}. \tag{10.15}$$

Consider now a $[k \times m, \varrho, \delta - 2]$ code $\mathcal{C}'$ of maximum size. The code $\mathcal{C}'$ is also an MRD code, which is a $(2e-2)$-diameter perfect code whose anticode has size is $q^{m(2e-2)}$ and, therefore, by the code-anticode bound, $|\mathcal{C}'| \cdot q^{m(\delta-3)} = q^{km}$. For the related ball, $\mathcal{B}_{e-1}(k \times m)$, we have that $|\mathcal{C}'| \cdot |\mathcal{B}_{e-1}(k \times m)| \leq q^{km}$. This implies that

$$|\mathcal{B}_{e-1}(k \times m)| \leq \frac{q^{km}}{|\mathcal{C}'|} = q^{m(\delta-3)} ,$$

which by (10.15) implies that

$$|\mathcal{B}_e(k \times m)| - |\mathcal{B}_{e-1}(k \times m)| \geq q^{m(\delta-1)} - q^{m(\delta-3)} . \tag{10.16}$$

By Lemma 10.16 we have that

$$|\mathcal{B}_e(k \times m)| - |\mathcal{B}_{e-1}(k \times m)| = \prod_{j=0}^{e-1} \frac{(q^k - q^j)(q^m - q^j)}{q^e - q^j} .$$

By Lemma 10.18 we have that $DB(e) < DA(e)$ and hence

$$|\mathcal{B}_e(k \times m)| - |\mathcal{B}_{e-1}(k \times m)| < q^{m(\delta-1)} - q^{m(\delta-3)} ,$$

which contradicts (10.16).

Thus, there are no nontrivial perfect codes in the bilinear forms scheme $\mathrm{B}_q(k, m)$. □

10.6 Constant-Dimension MDS Codes

In the Hamming scheme, we have proved that orthogonal arrays (or MDS codes) form diameter perfect codes. In $\mathrm{J}_q(n, w)$, the MDS-CW codes are also diameter perfect codes. Are there some similar subspace-MDS codes (constant-dimension MDS codes)? The answer is that there do exist such similar subspace-MDS codes. Moreover, these codes can be generated from MDS codes and they also yield a family of orthogonal arrays (Lemmas 10.19 and 10.20, respectively). In other words, they form a family between the linear MDS codes and the nonlinear orthogonal arrays.

An $(n, t, k)_q$ ***subspace-MDS code*** $\mathbb{C}$ is a set of n subspaces of $\mathrm{G}_q(kt, t)$ such that each k subspaces of $\mathbb{C}$ span $\mathbb{F}_q^{kt}$. Before elaborating why this family of code lies between the nonlinear orthogonal arrays and the linear MDS codes, we consider a slightly larger family of codes and a bound on the size of such a family.

Definition 10.2. Let t, k, α be three positive integers, where $\alpha \leq k$ and $t \geq 1$. A $(t; k, \alpha)_q$-***independent configuration (IC)*** is a set $\mathbb{C} = \{U_1, \ldots, U_n\} \subseteq \mathrm{G}_q(kt, t)$, such that for all $1 \leq i_1 < i_2 < \cdots < i_\alpha \leq n$,

$$\dim(U_{i_1} + U_{i_2} + \cdots + U_{i_\alpha}) = \alpha t .$$

We say that $|\mathbb{C}| = n$ is the ***size*** of the IC.

Theorem 10.16. *If $\mathbb{C}$ is a $(t;k,\alpha)_q$-IC, where $\alpha \geq 2$, then*

$$|\mathbb{C}| \leq \frac{q^{(k-\alpha+2)t}-1}{q^t-1}+\alpha-2.$$

Proof. If $\alpha = 2$, the claim is immediate by considering the size of a t-spread in $\mathbb{F}_q^{kt}$ (see Theorem 10.6).

Assume now that $\alpha > 2$ and let $\mathbb{C} \triangleq \{U_1, U_2, \dots, U_n\}$ be a $(t;k,\alpha)_q$-IC. Define

$$W_1 \triangleq U_1 + U_2 + \cdots + U_{\alpha-2},$$

where $\dim W_1 = (\alpha-2)t$. By the definition of an IC, we have that there exists a $((k-\alpha+2)t)$-subspace W_2 of $\mathbb{F}_q^{kt}$ such that $\mathbb{F}_q^{kt} = W_1 + W_2$. It follows that any vector $u \in U_j$, $\alpha - 1 \leq j \leq n$, may be written uniquely as $u = u_1 + u_2$, where $u_1 \in W_1$ and $u_2 \in W_2$. We now define

$$U_j' \triangleq \{u_2 \ : \ u_1 + u_2 \in U_j,\ u_1 \in W_1,\ u_2 \in W_2\},$$

for all $\alpha - 1 \leq j \leq n$. It is easily verified that $\dim U_j' = t$.

Furthermore, for any $\alpha - 1 \leq j_1 < j_2 \leq n$,

$$\dim(W_1 + U_{j_1}' + U_{j_2}') = \alpha t \Rightarrow \dim(U_{j_1}' + U_{j_2}') = 2t.$$

Thus, the set $\{U_i' \ : \ \alpha - 1 \leq i \leq n\}$ contains $|\mathbb{C}| - \alpha + 2$ pairwise disjoint t-subspaces of W_2. The number of such subspaces is upper bounded by the size of a t-spread (see Theorem 10.6), and thus,

$$|\mathbb{C}| - \alpha + 2 \leq \frac{\begin{bmatrix}(k-\alpha+2)t\\1\end{bmatrix}}{\begin{bmatrix}t\\1\end{bmatrix}}.$$

□

When $t = 1$, bounding the size of $(1;k,k)_q$-IC is equivalent to finding the longest MDS codes, and hence it is related to the MDS conjecture. Thus, Theorem 10.16 forms a generalization of an upper bound on the length of an MDS code (see Theorem 3.5). When $\alpha = k$, a $(t;k,k)_q$-IC of size n is an $(n,t,k)_q$ subspace-MDS code. To end this section we prove the connections between subspace-MDS codes and orthogonal arrays on one side and MDS codes on the other side.

Lemma 10.19. *If there exists an $[n,k,d]_{q^t}$ MDS code, then there exists an $(n,t,k)_q$ subspace-MDS code.*

Proof. Let G be the $k \times n$ generator matrix of an $[n,k,d]_{q^t}$ MDS code, let α be a primitive element of $\mathbb{F}_{q^t}$, and let C be the related $t \times t$ companion matrix. From the i-th column vector v of length k in G we form a $(kt) \times t$ matrix, $\hat{G}_i$, by replacing each element α^j in v by the matrix C^j; and each *zero* in the column vector v of G is replaced by the $t \times t$ all-zero matrix. Let $G_i \triangleq \hat{G}_i^{\text{tr}}$ be a $t \times (kt)$ matrix. Since each power of the companion matrix is of full-rank, it follows that G_i is a generator matrix for a t-subspace of $\mathbb{F}^{kt}$. Since G is a $k \times n$ matrix, it follows that from the columns of G, the matrices $G_1, G_2, \ldots, G_n$ form n subspaces of dimension t. Hence, to complete the proof it is sufficient to prove that the $(kt) \times (kt)$ matrix formed by concatenations of any k distinct matrices from the set $\{\hat{G}_i \; : \; 1 \leq i \leq n\}$ is of full rank. Let A be a $k \times k$ matrix formed by any projection on k columns of G. Since G is a generator matrix of an $[n,k,d]_{q^t}$ MDS code, it follows that A is of full-rank. Let $\hat{A}$ be the $(kt) \times (kt)$ matrix formed from A by replacing each element α^j in A by the matrix C^j; and each *zero* in the matrix A of G is replaced by the $t \times t$ all-zero matrix. Assume that $u = (u_1, u_2, \ldots, u_k) \in \mathbb{F}_q^{kt}$ is a nonzero vector such that $\hat{A} \cdot u^{\text{tr}} = \mathbf{0}$, where u_j is a word of length t for each $1 \leq j \leq k$. By Corollary 10.8, if $(\alpha_1, \alpha_2, \ldots, \alpha_k) \in \mathbb{F}_{q^t}^k$, where α_j is the element in $\mathbb{F}_{q^t}$ that is represented by the vector u_j of length t, then $A \cdot (\alpha_1, \alpha_2, \ldots, \alpha_k)^{\text{tr}} = \mathbf{0}$. This is a contradiction of the fact that G is the $k \times n$ generator matrix of an $[n,k,d]_{q^t}$ MDS code, where any k column vectors are linearly independent. Therefore, the set $\{\langle G_i \rangle\}_{i=1}^n$ is an $(n,t,k)_q$ subspace-MDS code. □

Lemma 10.20. *If there exists an $(n,t,k)_q$ subspace-MDS code, then there exists an $OA(k,n,q^t)$.*

Proof. Let $\mathbb{C}$ be an $(n,t,k)_q$ subspace-MDS code whose n subspaces are represented by $t \times (kt)$ generator matrices. For a given such matrix G_i, $1 \leq i \leq n$, and a word $x = (x_1, x_2, \ldots, x_k) \in \mathbb{F}_{q^t}^k$, where x_j is a sub-word of length t over $\mathbb{F}_q$, we form the word $u_i(x) = G_i \cdot x^{\text{tr}}$. For each such word $u_i(x)$, of length t over $\mathbb{F}_q$, let $\gamma_i(x)$ be the element of $\mathbb{F}_{q^t}$ whose q-ary representation is $u_i(x)$. We form a row vector $(\gamma_1(x), \gamma_2(x), \;\cdots\;, \gamma_n(x))$ in an array M. We claim that all these row vectors of M form an $\text{OA}(k,n,q^t)$. There are q^{tk} distinct words that can be taken as x and each will generate one row in M and hence M has the required number of rows and columns of the orthogonal array, i.e., M is a $q^{tk} \times n$ array.

Therefore, to complete the proof it is sufficient to show that the projection of each k columns of M contains each word of length k over $\mathbb{F}_{q^t}$ exactly once. Assume to the contrary, that some word of length k over $\mathbb{F}_{q^t}$ appears at least twice in some projection. W.l.o.g. we can assume that this is the projection on the first k columns of M. Let G be the $(kt) \times (kt)$ matrix formed from the rows of the k generator matrices $G_1, G_2, \ldots, G_k$. Since the projection of the first columns of M form two equal k-tuples this implies that there exist two distinct words $x, y \in \mathbb{F}_{q^t}^k$ such that $G \cdot x^{\text{tr}} = G \cdot y^{\text{tr}}$, i.e., $G \cdot (y^{\text{tr}} - x^{\text{tr}}) = \mathbf{0}$. Since $y \neq x$, this implies that G is not a matrix of full rank and hence the related first k subspaces do not span $\mathbb{F}^{kt}$. Therefore, $\mathbb{C}$ is not an $(n, t, k)_q$ subspace-MDS code, a contradiction.

Thus, M is an orthogonal array $\text{OA}(k, n, q^t)$ and the proof is completed. □

The following intriguing research problems seem to be very difficult.

Problem 10.4. Are there subspace-MDS codes with parameters that cannot be obtained from MDS codes?

Problem 10.5. Are there orthogonal arrays over an alphabet which is a power of a prime that cannot be obtained from subspace-MDS codes?

Problem 10.6. Can a subspace-MDS code be described as a diameter perfect code without its translation into an orthogonal array?

10.7 Notes

The projective space and the Grassmann scheme are of particular interest in coding theory. Representation of subspaces and codes defined by subspaces can be a key to obtain results in coding theory and in particular on codes in the projective space and the Grassmann scheme. In particular, one can observe that the $\begin{bmatrix} n \\ k \end{bmatrix}_q$ subspaces of $\text{G}_q(n, k)$ are exactly all the linear codes of dimension k over $\mathbb{F}_q$. Hence, better understanding of the projective space can influence on the knowledge about linear codes. For example, encoding and decoding $\text{G}_q(n, k)$ is equivalent for enumerating and ordering all $[n, k]_q$ codes. This problem was considered first in [Silberstein and Etzion (2011)], where several representation of k-subspaces are presented and as a result a few encoding and decoding algorithms for $\text{G}_q(n, k)$ are presented.

Section 10.1. The nonexistence proof for perfect codes in the Grassmann scheme is due to [Martin and Zhu (1995)]. This proof is a shorter proof than the one given earlier by [Chihara (1987)]. In [Ahlswede, Aydinian, and Khachatrian (2001)] it was asserted that the most interesting association schemes in coding theory are the Hamming scheme, the Johnson scheme, and the Grassmann scheme. The projective space was defined first in the connection of error-correcting codes for random network coding [Koetter and Kschischang (2008)]. Related computations on the intersection between subspaces and enumerations for intersection numbers were done in [Etzion and Vardy (2011)]. This paper was the first that analyzed codes in the projective space after they were found applicable to error-correction for random network coding [Koetter and Kschischang (2008)]

Section 10.2. The classic theory for *q-analogs* of mathematical objects and functions has its beginnings in the early work of Leonard Euler [Euler (1750 - 51)]; see also [Koelink and van Assche (2009)]. It is, therefore, natural to ask which combinatorial structures can be generalized from sets to vector spaces over $\mathbb{F}_q$. For t-designs and Steiner systems, this question was first studied in [Cameron (1974a,b)] and [Delsarte (1976)]. While there has been a lot of progress in constructing q-analogs of block designs (see [Etzion and Storme (2016)] and references therein), there has been almost no progress in constructing nontrivial q-Steiner systems (recall that spreads are considered to be trivial q-Steiner systems). The only major breakthrough was done by [Braun, Etzion, Östergård, Vardy and Wassermann (2016)], where a construction of the q-Steiner system $\mathrm{S}_2(2,3,13)$ was given. Many such non-isomorphic systems were found by sophisticated computer search. Although q-analog of designs are more difficult to find than designs over sets, there has been a lot of progress in this direction from the beginning of the 21st century. Apart from the work by [Braun, Etzion, Östergård, Vardy and Wassermann (2016)], there has been hardly any progress, on q-Steiner systems, since the problem was tackled in [Thomas (1996)]. In [Etzion and Hooker (2018)], a construction of codes, which are similar to a punctured q-Fano plane $\mathrm{S}_q(2,3,7)$ for $q=2$, was presented. Their code has the same parameters and properties as the ones expected from the punctured q-Fano plane. Another interesting result was presented in [Kiermaier and Laue (2015)], where the derived and the residual designs for q-analog of designs are considered. The q-analog of designs are also called ***subspace designs***.

It was observed in [Ahlswede, Aydinian, and Khachatrian (2001)] that q-Steiner systems are diameter perfect codes in the Grassmann scheme. The size of maximum size anticodes in the Grassamnn scheme was found in [Frankl and Wilson (1986)] in the connection of finding the maximum size of t-intersecting families and a generalization of the Erdös-Ko-Rado theorem for vector spaces. Their results were used by [Schwartz and Etzion

(2002)] to consider q-Steiner systems and tilings of the Grassmann space with these maximum size anticodes in the Grassmann scheme.

Finally, normal spreads were characterized in [Beutelspacher and Ueberberg (1991)] where the following theorem is proved.

Theorem 10.17. *There exists a unique normal k-spread in $\mathbb{F}_q^{rk}$ for each $r \geq 3$.*

Clearly, Theorem 10.17 does not hold for $r = 2$ as it was proved in Lemma 10.13 that all k-spreads in $\mathbb{F}_q^{2k}$ are normal. Normal spreads were further considered in [Lunardon (1999)].

Various applications of the companion matrix and its powers to network coding and subspaces were suggested in [Etzion and Wachter-Zeh (2018)]. The connections between Hamming codes, normal spreads, companion matrices, and Steiner systems were presented by the author of this book in an invited talk presented at Combinatorics 2016, held in Maratea, Italy.

Section 10.4. The nonexistence proof for perfect codes in $\mathcal{P}_q(n)$, with the subspace distance, was presented in [Etzion and Vardy (2011)].

Section 10.5. The Singleton bound and the constructions of rank-metric codes, which meet this bound, were found using different approaches by [Delsarte (1978); Gabidulin (1985); Roth (1991)]. This bound was generalized in [Etzion, Gorla, Ravagnani and Wachter-Zeh (2016)] to Ferrers-diagram rank-metric codes. They also extended the work on anticodes and maximum size rank-metric codes (which are diameter perfect codes) to Ferrers-diagram rank-metric codes, which were defined in [Etzion and Silberstein (2009)] in the context of codes for random network coding.

The recursive solution in the proof of Lemma 10.16 was given by [Landsberg (1893)]. Previous proofs for the nonexistence of perfect rank-metric codes are slightly more complicated than the one given in Theorem 10.15. Such proofs were given, for example, in [Chihara (1987); Martin and Zhu (1995); Loidreau (2014)].

There are many strong connections between codes in the bilinear forms scheme $\mathrm{B}_q(k, m)$ and the Grassmann scheme $\mathrm{G}_q(m+k, k)$. Given a $k \times m$ matrix M over $\mathbb{F}_q$, we form a k-subspace in $\mathrm{G}_q(m+k, k)$ whose $k \times (m+k)$ generator matrix $[I_k \; M]$, is called the ***lifting*** of M.

For a $[k \times m, \varrho, \delta]$ rank-metric code $\mathcal{C}$, the code

$$\mathbb{C} \triangleq \{\langle [I_k \; M] \rangle \; : \; M \in \mathcal{C}\}$$

is called the ***lifted code*** of $\mathcal{C}$. This code $\mathbb{C}$ in $\mathrm{G}_q(m+k, k)$ has minimum subspace distance 2δ, i.e., minimum Grassmann distance δ. If $\mathcal{C}$ is an MRD code, the corresponding lifted code is called the ***lifted MRD code*** $\mathbb{C}^{\text{MRD}}$ of $\mathcal{C}$. This construction was first observed in [Silva, Kschischang, and Koetter (2008)]. A generalization with Ferrers-diagram rank-metric codes was done first in [Etzion and Silberstein (2009)] and later explored in [Etzion and Silberstein (2013)]. The paper [Etzion and Silberstein (2013)] also tie together the code $\mathbb{C}^{\text{MRD}}$ with combinatorial designs and in particular q-analog of designs.

Section 10.6. MDS codes over subspaces were considered first as array codes by various authors (see [Blaum, Bruck, and Vardy (1996); Raviv, Silberstein, and Etzion (2017); Silberstein, Etzion, and Schwartz (2019)] and references therein), where the representation was not always given using subspaces (the representation usually used arrays). It was considered later for network coding solutions with subspaces in [Etzion and Wachter-Zeh (2018)] and in [Etzion and Zhang (2019)]. The MDS bound in Theorem 10.16 is due to [Cai, Chrisnata, Etzion, Schwartz, and Wachter-Zeh (2020)]. Various metrics for network coding with matrices and subspaces were considered by [Silva and Kschischang (2009)].

Chapter 11

The Lee and the Manhattan Metrics

The Lee metric and the Manhattan metric are two important metrics from both theoretical and practical points of view. In this chapter codes in these metrics are considered. Section 11.1 is devoted to the definitions of the two metrics, their codes, and anticodes. Section 11.2 introduces the concept of lattice tiling, which replaces the concept of linear perfect codes when we consider linear codes in $\mathbb{Z}^n$. This concept can also be used for linear codes in $\mathbb{Z}_m^n$. It will be used in this chapter and in Chapter 12. Section 11.3 is devoted to three constructions of perfect codes that form two infinite sets of parameters. Section 11.4 deals with constructions and properties of diameter perfect codes. All codes which are constructed in this section have periodicity properties. In Section 11.5, nonperiodic codes are constructed and some constructions for a large set of inequivalent perfect codes and inequivalent diameter perfect codes are presented. Finally, in Section 11.6 we present one of the known techniques to exclude the existence of perfect codes in the Lee metric. This existence problem of perfect codes in the Lee metric and perfect codes in the Manhattan metric is the main open problems in this research area.

11.1 The Lee and the Manhattan Distances

The Lee metric was introduced for transmission of signals taken from $\mathbb{F}_p$, where p is a prime, over some certain noisy channels. It was later generalized for $\mathbb{Z}_m$, where $m > 1$ is any positive integer. Our assumption throughout this chapter, when e-perfect codes are considered, is that $m \geq 2e + 1$.

The ***Lee distance*** $d_L(x, y)$ between two words $x = (x_1, x_2, \ldots, x_n)$ and

$y = (y_1, y_2, \ldots, y_n)$ in $\mathbb{Z}_m^n$ is given by

$$d_L(x, y) \triangleq \sum_{i=1}^{n} \min\{x_i - y_i \pmod{m}, y_i - x_i \pmod{m}\} ,$$

where the outcome from the computation modulo m is taken as the residue modulo m between 0 and $m-1$. The ***Manhattan distance*** (known also as the L_1 distance, the rectilinear distance, and the taxicab distance), is a related distance defined over $\mathbb{Z}^n$. This is the first time, in this book, where the space being considered is not a finite set. For two words $x = (x_1, x_2, \ldots, x_n)$ and $y = (y_1, y_2, \ldots, y_n)$ in $\mathbb{Z}^n$, the Manhattan distance between x and y is defined by

$$d_M(x, y) \triangleq \sum_{i=1}^{n} |x_i - y_i| .$$

It is easy to verify, that the triangle inequality holds for the Lee distance and also for the Manhattan distance, and hence these distances define metrics. Neither of these two metrics defines an association scheme. For the Manhatten metric, this is trivial as it is defined on an infinite space. The Lee metric is also not a scheme. This can be verified by considering a simple example for the possible intersection numbers. Consider the following three words $x = (0, 0, 0, \ldots, 0)$, $y = (2, 0, 0, \ldots, 0)$, and $z = (1, 1, 0, \ldots, 0)$ of length n over $\mathbb{Z}_m$, $m \geq 5$. Clearly, $d_L(x, y) = d_L(x, z) = 2$. There exists exactly one word u for which $d_L(x, u) = d_L(y, u) = 1$ (this word is $u = (1, 0, 0, \ldots, 0)$), while the number of words for which $d_L(x, u) = d_L(z, u) = 1$ is two, where $u \in \{(1, 0, 0, \ldots, 0), (0, 1, 0, \ldots, 0)\}$. This implies that the Lee metric does not define an association scheme for $m \geq 5$. The same is true for $m = 4$ and a proof for this is left as an exercise to the reader. What about $m < 4$? When $m = 2$ and when $m = 3$, the Lee metric coincides with the Hamming metric and hence in both cases the Lee metric defines a scheme. As was mentioned before, it will be assumed that $m \geq 2e+1$ and hence $m \leq 3$ does not introduce interesting new results. The reason for choosing $m \geq 2e+1$ is that, geometrically, the balls with radius e will be the same in $\mathbb{Z}^n$ and $\mathbb{Z}_m^n$, with the only difference that in $\mathbb{Z}_m^n$ there might be a wrap around.

An ***n-dimensional Lee ball*** (also called a ***Lee sphere***) with radius e, centered at the point $z = (z_1, z_2, \ldots, z_n) \in \mathbb{Z}^n$, is the shape defined by

$$\mathcal{B}_e(z) \triangleq \{(x_1, x_2, \ldots, x_n) \ : \ (x_1, x_2, \ldots, x_n) \in \mathbb{Z}^n, \ \Sigma_{i=1}^{n} |x_i - z_i| \leq e\} .$$

In other words, $\mathcal{B}_e(z)$ consists of all points in $\mathbb{Z}^n$ whose Manhattan distance from the given point $(z_1, z_2, \ldots, z_n)$ is at most e. If we are in the Lee metric, i.e., $\mathcal{B}_e(z)$ ($\mathcal{B}_e(n)$) is in $\mathbb{Z}_m^n$, then the points are taken modulo m. The shapes are the same in both metrics since $m \geq 2e+1$, but the presentation in $\mathbb{Z}^n$ makes it simpler to understand the shape (the e-ball). Moreover, we can associate each point $x \in \mathbb{Z}^n$ with an n-dimensional unit cube whose center is in x. This will turn our coding problems into geometrical problems. We further emphasize that geometrically the ball with radius e in the Lee metric and the Manhattan metric are equal. The only difference is that in the Lee metric it can be wrapped around depending on the alphabet $\mathbb{Z}_m$ in the Lee metric. A 2-dimensional and a 3-dimensional Lee balls with radius one are depicted in Fig. 11.1, where the center of the balls has no specific point in $\mathbb{Z}^n$.

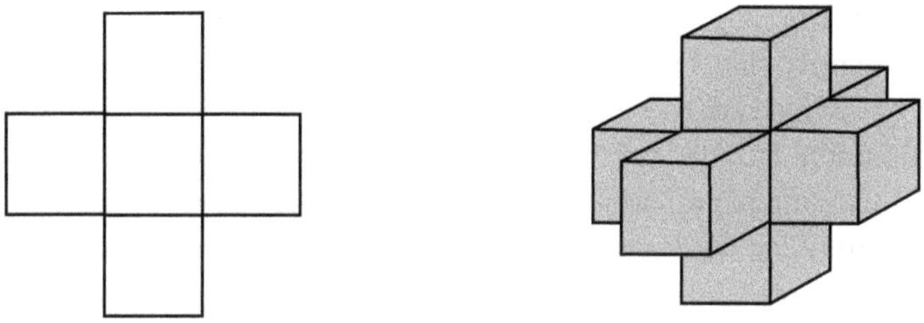

Fig. 11.1 A 2-dimensional and a 3-dimensional Lee balls with radius one.

Theorem 11.1.

$$|\mathcal{B}_e(n)| = \sum_{i=0}^{\min\{n,e\}} 2^i \binom{n}{i}\binom{e}{i} . \tag{11.1}$$

Proof. We consider the coordinates of each codeword as "boxes" and there are at most e "elements" to be distributed to all the boxes. For each $i \leq e$, we consider the number of different distributions of elements to exactly i boxes. There are $\binom{n}{i}$ distinct ways to choose i boxes, where $i \leq n$. Each such box has at least one element and hence we have to distribute all the other $e-i$ elements (or some of them) to these i boxes. This is a simple partition problem of $e-i$ identical elements into $i+1$ distinct boxes, with an unlimited numbers of elements in a box. For this purpose, there are $\binom{e-i+i+1-1}{e-i} = \binom{e}{i}$ different distributions. The absolute value in each coordinate is determined by the number of elements distributed to the associated boxes. Each such value can be positive or negative and hence there are 2^i distinct ways to determine the signs. Multiplying these three factors yields the formula. □

Corollary 11.1. *For* $n \geq 1$, $|\mathcal{B}_1(n)| = 2n+1$.

Corollary 11.2. *For* $e \geq 1$, $|\mathcal{B}_e(2)| = 2e^2+2e+1$.

An $(n,d,m)_L$ ***Lee code*** is a code of length n, minimum Lee distance d, over $\mathbb{Z}_m$. The following theorem can easily be verified from the sphere-packing bound.

Theorem 11.2. *A Lee code $\mathcal{C}$ of length n over $\mathbb{Z}_m$ has minimum Lee distance 3 and size $\frac{m^n}{1+2n}$ if and only if $\mathcal{C}$ is a 1-perfect code.*

What is the size of the largest anticode with diameter D in $\mathbb{Z}^n$, $n \geq 2$? It is easy to verify that the ball $\mathcal{B}_e(n)$ is an anticode with diameter $D = 2e$ and it will be defined as the anticode $\mathcal{A}_{2e}(n)$. For odd D, we define anticodes with diameter $D = 2e+1$, $\mathcal{A}_{2e+1}(n)$ as follows. For $e = 0$, let $\mathcal{A}_1(n)$ be a shape consisting of two adjacent points of $\mathbb{Z}^n$, where two points are adjacent if the Manhattan distance between them is one. These two points are the ***core*** of the anticode. The anticode $\mathcal{A}_{2e+1}(n)$ is defined as all the points in $\mathbb{Z}^n$ whose distance from the core $\mathcal{A}_1(n)$ is at most e. A 2-dimensional and a 3-dimensional anticodes with diameter three are depicted in Fig. 11.2.

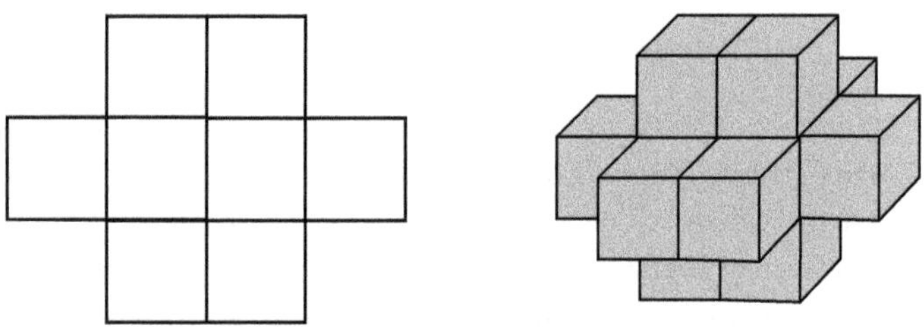

Fig. 11.2 A 2-dimensional and a 3-dimensional anticodes with diameter three.

Theorem 11.3.

$$|\mathcal{A}_{2e+1}(n)| = \sum_{i=0}^{\min\{n-1,e\}} 2^{i+1}\binom{n-1}{i}\binom{e+1}{i+1}. \qquad (11.2)$$

Proof. Assume that the two adjacent points are $(0,\dots,0,0)$ and $(0,\dots,0,1)$. We compute the number of different ways to obtain elements of $\mathcal{A}_{2e+1}(n)$. Each element has n coordinates. The last coordinate is initially 0.5 and should be changed (either increased or decreased) by at least 0.5. There are two distinct ways for the direction change (plus or minus) of this coordinate. Except for the last coordinate, another i coordinates are changed, but not more than $n-1$ coordinates and not more than e coordinates by the definition of $\mathcal{A}_{2e+1}(n)$. For these i coordinates, there are 2^i ways to determine if the change is positive or negative. Now we have to distribute up to e identical elements into $i+2$ boxes (note that one box represents the last coordinate and one box represents the elements that are not contained in the other $i+1$ boxes), where exactly i specific boxes are not empty. This can be done in $\binom{e-i+i+2-1}{e-i} = \binom{e+1}{i+1}$ distinct ways. □

Corollary 11.3. *For* $n \geq 1$, $|\mathcal{A}_3(n)| = 4n$.

Corollary 11.4. *For* $2e+1 \geq 1$, $|\mathcal{A}_{2e+1}(2)| = 2(e+1)^2$.

The following fact is known about the maximum size anticode for each diameter.

Theorem 11.4. *The anticode* $\mathcal{A}_{2e+1}(n)$ *is a maximum size anticode with diameter* $2e+1$ *and the ball* $\mathcal{B}_e(n)$ *is a maximum size anticode with diameter* $2e$.

Finally, the conditions of Corollary 2.15 are clearly satisfied for the Lee metric, where addition is the binary operation, and hence the code-anticode bound holds for the Lee metric.

Corollary 11.5. *If a code* $\mathcal{C}$ *of length* n *over* $\mathbb{Z}_m$ *has minimum Lee distance 4 and size* $\frac{m^n}{4n}$, *then* $\mathcal{C}$ *is a 3-diameter perfect code.*

11.2 Lattice Tiling

Lattices are a very important concept for packing and tiling of the n-dimensional Euclidian space with a given n-dimensional shape. They are most useful in describing linear codes for the Lee metric and for the Manhatten metric. Lattices replace the concept of linear codes for these metrics.

A ***lattice*** Λ is an additive subgroup of the real n-space $\mathbb{R}^n$, defined by

$$\Lambda \triangleq \{a_1 v_1 + a_2 v_2 + \cdots + a_n v_n \;:\; a_1, a_2, \dots, a_n \in \mathbb{Z}\}$$

where $\{v_1, v_2, \ldots, v_n\}$ is a set of linearly independent vectors in $\mathbb{R}^n$, i.e., the lattice has ***rank*** n. The set of vectors $\{v_1, v_2, \ldots, v_n\}$ is called ***the base*** for Λ, and the $n \times n$ matrix

$$\mathbf{G} \triangleq \begin{bmatrix} v_1 \\ v_2 \\ \vdots \\ v_n \end{bmatrix} = \begin{bmatrix} v_{11} & v_{12} & \cdots & v_{1n} \\ v_{21} & v_{22} & \cdots & v_{2n} \\ \vdots & \vdots & \ddots & \vdots \\ v_{n1} & v_{n2} & \cdots & v_{nn} \end{bmatrix}$$

having these vectors as its rows is said to be the ***generator matrix*** for Λ and it is also denoted by $G(\Lambda)$. The lattice Λ will be denoted also by $\Lambda(\mathbf{G})$.

The ***volume*** of a lattice Λ, denoted by $V(\Lambda)$, is inversely proportional to the number of lattice points per a unit volume. More precisely, $V(\Lambda)$ may be defined as the volume of the ***fundamental parallelogram*** (fundamental region) $\Pi(\Lambda)$, which is given by

$$\Pi(\Lambda) \triangleq \{\xi_1 v_1 + \xi_2 v_2 + \cdots + \xi_n v_n \ : \ 0 \leq \xi_i < 1, \ 1 \leq i \leq n\} .$$

There is a simple expression for the volume of Λ, namely, $V(\Lambda) = |\det \mathbf{G}|$.

A set of points $\mathcal{T}$ in $\mathbb{Z}^n$ (or $\mathbb{Z}_m^n$) is a ***tiling*** for a shape $\mathcal{S}$, $\mathcal{S} \in \mathbb{Z}^n$, if $\mathcal{T}+\mathcal{S}$ is equal to $\mathbb{Z}^n$ ($\mathbb{Z}_m^n$, respectively) and each point of $\mathbb{Z}^n$ ($\mathbb{Z}_m^n$, respectively), can be represented in a unique way as $x + y$, where $x \in \mathcal{T}$ and $y \in \mathcal{S}$. Recall that with this property, the pair $(\mathcal{T}, \mathcal{S})$ is a tiling. This definition can be generalized to non-discrete shapes in $\mathbb{R}^n$, but these tilings will be considered in our discussion only briefly in Section 12.5. Moreover, these shapes in $\mathbb{Z}^n$ and also in $\mathbb{Z}_m^n$ can be described by a collection of unit cubes in the n-dimensional Euclidian space, whose centers are in the integer points of the shape. This will form a one-to-one correspondence between tilings in $\mathbb{Z}^n$ (or $\mathbb{Z}_m^n$) and some sets of tilings in the n-dimensional Euclidian space. Nevertheless, it should be noted that these shapes can tile the n-dimensional Euclidian space in other ways (where only some points in the tilings are integer points). These tilings will not be discussed in our context. The only lattices in $\mathbb{R}^n$ that will be discussed in Section 12.5 are those of shapes that cannot be described as shapes in $\mathbb{Z}^n$.

A lattice Λ is a ***lattice tiling*** for a shape $\mathcal{S}$ if the points of Λ form a tiling for $\mathcal{S}$. A lattice tiling Λ is an ***integer lattice tiling*** for $\mathcal{S}$ if all entries of $G(\Lambda)$ are integers. The following lemma is well known and can be verified from the given definitions.

Lemma 11.1. *If Λ defines a lattice tiling (over $\mathbb{R}^n$ or $\mathbb{Z}^n$) with the shape $\mathcal{S}$, then $V(\Lambda) = |\mathcal{S}|$, where $|\mathcal{S}|$ denote the volume of the shape $\mathcal{S}$.*

Related to a tiling of a shape $\mathcal{S}$ is a packing of a shape $\mathcal{S}$, where no point in the space is covered more than once by a translate of the shape $\mathcal{S}$. Lemma 11.1 is relaxed in the following way.

Lemma 11.2. *A necessary condition for a lattice* Λ *(over* $\mathbb{R}^n$ *or* $\mathbb{Z}^n$*) to define a lattice packing with a shape* $\mathcal{S}$ *is that* $V(\Lambda) \geq |\mathcal{S}|$. *A sufficient condition for a lattice packing* Λ*, for the shape* $\mathcal{S}$*, to define a lattice tiling of the shape* $\mathcal{S}$ *is that* $V(\Lambda) = |\mathcal{S}|$.

A lattice $\Lambda \subseteq \mathbb{Z}^n$ has ***period*** $(t_1, \ldots, t_n) \in \mathbb{Z}^n$ if whenever $v \in \Lambda$, also $v \pm t_i \mathbf{e}_i \in \Lambda$ for each i, $1 \leq i \leq n$. Lattices are always periodic, and the *minimum period* in the i-th direction, t_i, is the smallest positive integer for which $t_i \mathbf{e}_i \in \Lambda$. We also say that the period of the lattice is t, where t equals to the least common multiplier of $t_1, t_2, \ldots, t_n$. Note that this definition implies that if a lattice has period t, then it also has period $r \cdot t$ for each positive integer r.

11.3 Constructions of Perfect Codes

This section is devoted to constructions of perfect codes in the Lee metric. As usual, the codes that will be generated and the codes that will be considered contain the all-zero codeword. We note that an e-perfect code in the Lee metric over $\mathbb{Z}_m^n$ induces an e-perfect code in the Manhattan metric over $\mathbb{Z}^n$. This e-perfect code in $\mathbb{Z}^n$ has periodicity m. Similarly, an e-perfect code in the Manhattan metric over $\mathbb{Z}^n$, which has periodicity m in all dimensions, induces an e-perfect code in the Lee metric over $\mathbb{Z}_m^n$.

Theorem 11.5. *The set*

$$\mathcal{C} \triangleq \left\{ (c_1, c_2, \ldots, c_n) \; : \; \sum_{i=1}^{n} i \cdot c_i \equiv 0 \pmod{2n+1} \right\}$$

is a 1-perfect code over $\mathbb{Z}_{2n+1}$.

Proof. By Corollary 11.1, the size of a ball with radius one is $m = 2n + 1$. The number of solutions to the congruence in the definition of $\mathcal{C}$ is m^{n-1}, since for any choice of $c_2, c_3, \ldots, c_n$, there is a unique value of c_1 modulo $2n + 1$ for which the congruence equals *zero*. Hence, by the sphere-packing bound, to prove that the code is a 1-perfect code, it suffices to show that the minimum distance of $\mathcal{C}$ is three or that the covering radius of $\mathcal{C}$ is one. Therefore, to complete the proof, it is sufficient to prove that for each word $b = (b_1, b_2, \ldots, b_n)$, there is one codeword within distance one of b. Let

$$\sum_{i=1}^{n} i \cdot b_i \equiv k \pmod{2n+1} .$$

If $k = 0$, then b is a codeword. Otherwise, let $c = (c_1, c_2, \ldots, c_n)$, where $c_i = b_i$ for $i \neq k$ and $c_k = b_k - 1$ (taken modulo $m = 2n+1$ in $\mathbb{Z}_m$). Clearly,

$$\sum_{i=1}^{n} i \cdot c_i \equiv \sum_{i=1}^{n} i \cdot b_i - k \equiv 0 \pmod{2n+1},$$

and hence c is a codeword for which $d_L(c, b) = 1$ and the proof is completed. □

Theorem 11.6. *The set*

$$\mathcal{C} \triangleq \{(\alpha, (2e+1)\alpha) \ : \ 0 \leq \alpha \leq 2e^2 + 2e\}$$

is an e-perfect code over $\mathbb{Z}_m^2$*, where* $m = 2e^2 + 2e + 1$.

Proof. By Corollary 11.2, the size of a ball with radius e in $\mathbb{Z}_m^2$ is $|\mathcal{B}_e(2)| = 2e^2 + 2e + 1$. The size of the code $\mathcal{C}$ is $2e^2 + 2e + 1$ since there are $2e^2 + 2e + 1$ distinct choices for α in the definition of $\mathcal{C}$. Hence,

$$|\mathcal{C}| \cdot |\mathcal{B}_e(2)| = (2e^2 + 2e + 1)^2 = \left|\mathbb{Z}_m^2\right| .$$

Thus, by the sphere-packing bound, to prove that the code is an e-perfect, it is sufficient to show that the minimum distance of $\mathcal{C}$ is $2e + 1$. Since $\mathcal{C}$ is linear, it follows that it is sufficient to show that the Lee weight of each codeword is at least $2e + 1$. Let $(\alpha, (2e + 1)\alpha)$ be a codeword. If $2e < \alpha < 2e^2 + 1$, then the claim is trivial and hence assume that $0 < \alpha < 2e + 1$. If $0 < \alpha < e$, then $2e + 1 \leq (2e + 1)\alpha < 2e^2 - e$ and hence $\mathrm{wt}((\alpha, (2e + 1)\alpha)) > 2e + 1$. If $\alpha = e$, then $(\alpha, (2e + 1)\alpha) = (e, -(e + 1))$ and hence $\mathrm{wt}((\alpha, (2e + 1)\alpha)) = 2e+1$. If $e < \alpha \leq 2e$, then note that we can write $(\alpha, (2e + 1)\alpha)$ as $(-(2e + 1)\beta, \beta)$ (by assuming that $\alpha = -(2e + 1)\beta$, which implies that $\beta = (2e+1)\alpha$, where the computation is performed modulo $2e^2 + 2e + 1$), and the proof is similar. Finally, if $2e^2 < \alpha < 2e^2 + 2e + 1$, then the proof is symmetric to the case when $0 < \alpha < 2e + 1$. □

Clearly, $(1, 2e+1)$ is a codeword in the code $\mathcal{C}$ defined in Theorem 11.6. Since $m = 2e^2+2e+1$ is a period of $\mathcal{C}$ by the definition of the code, it follows that $(0, 2e^2 + 2e + 1)$ is also a codeword of $\mathcal{C}$, where $(0, 2e^2 + 2e + 1) = (0, 0)$ over $\mathbb{Z}_m^2$. Since these two codewords of $\mathcal{C}$ are linearly independent and clearly, by its definition, the code is linear, it follows that the lattice of the code $\mathcal{C}$, defined in Theorem 11.6, has the generator matrix

$$\begin{bmatrix} 1 & 2e+1 \\ 0 & 2e^2+2e+1 \end{bmatrix}.$$

This lattice induces a tiling of the Lee ball with radius e, $\mathcal{B}_e(2)$, whose size by Corollary 11.2 is $2e^2+2e+1$.

At this point we will use an adaptation of the general product construction to construct 1-perfect error-correcting Lee codes.

Let $\mathcal{C}^1$ be a 1-perfect Lee code of length $n = \frac{q^r-1}{2}$ (i.e., $q^r = 2n+1$), q odd, over an alphabet with $\nu(2n+1)$ symbols (i.e., $\mathbb{Z}_{\nu(2n+1)}$), where $\nu \geq 1$, which has a total of q^r translates (the size of the Lee ball with radius one, which is $\mathcal{B}_1(n)$ in this case), including $\mathcal{C}^1$ itself. Let $\pi_t = (\pi_t(1) = 1, \pi_t(2), \ldots, \pi_t(2n+1))$, $1 \leq t \leq \ell$, be a permutation of $\{1, 2, \ldots, q^r\}$. Consider ℓ permutations, where the first permutation is defined to be the identity permutation. Let $\mathcal{C}^2$ be a 1-perfect code of length $\ell = \frac{q^{rs}-1}{q^r-1}$, in the Hamming scheme, over an alphabet with q^r symbols. Let $\mathcal{C}_i^1$, $1 \leq i \leq q^r$, be the i-th translate of $\mathcal{C}^1$, where $\mathcal{C}_1^1 = \mathcal{C}^1$. We construct the following code $\hat{\mathcal{C}}$

$$\hat{\mathcal{C}} \triangleq \{(x_{i_1}, \ldots, x_{i_\ell}) : x_{i_t} \in \mathcal{C}^1_{\pi_t(i_t)},\ (i_1, \ldots, i_\ell) \in \mathcal{C}^2\}.$$

Theorem 11.7. *The code $\hat{\mathcal{C}}$ is a 1-perfect error-correcting Lee code of length $\frac{q^{rs}-1}{2}$ over an alphabet of size $\nu(2n+1)$.*

Proof. Clearly, the length of the codewords from $\hat{\mathcal{C}}$ is $\frac{q^r-1}{2}\frac{q^{rs}-1}{q^r-1} = \frac{q^{rs}-1}{2}$. Hence, the size of a ball with radius one is q^{rs}. The proof that the code $\hat{\mathcal{C}}$ has minimum Lee distance three is identical to the related proof in Theorem 5.4.

The perfect code $\mathcal{C}_2$ is in the Hamming scheme and hence its size is $\frac{q^{r\ell}}{1+(q^r-1)\ell} = q^{r\ell-rs}$. The size of $\mathcal{C}^1$ is $\nu^n(2n+1)^{n-1}$, where $2n+1 = q^r$. Clearly,

$$\left|\hat{\mathcal{C}}\right| = \left|\mathcal{C}^2\right| \cdot \left|\mathcal{C}^1\right|^\ell = q^{r\ell-rs}\nu^{n\ell}(2n+1)^{(n-1)\ell}$$

$$= q^{r\ell-rs}\nu^{n\ell}q^{r(n-1)\ell} = q^{rn\ell-rs}\nu^{n\ell} = \frac{\nu^{n\ell}(2n+1)^{n\ell}}{q^{rs}}.$$

This implies, by Theorem 11.2, that $\hat{\mathcal{C}}$ is a 1-perfect Lee code of length $\frac{q^{rs}-1}{2}$ over an alphabet of size $\nu(2n+1)$. □

Finally, it worth mentioning that the construction of $\hat{\mathcal{C}}$ can be applied in various ways, since there are $(2n)!^{\ell-1}$ different ways to choose the ℓ permutations, to obtain many nonisomorphic 1-perfect Lee codes.

11.4 Diameter Perfect Codes

There are only two known families of perfect Lee codes, having two sets of parameters (presented in Theorems 11.5 and 11.6), and it is also easy to verify that the related codes in dimension two are unique. The situation for diameter perfect codes is similar, as only two families of diameter perfect codes are known, but the codes in dimension two are not unique. Recall again that the code-anticode bound holds for the Lee metric. The first theorem is a consequence of the code-anticode bound.

Theorem 11.8. *Let Λ be a lattice that forms a code $\mathcal{C} \subset \mathbb{Z}_m^n$ with minimum Lee distance d. Then the size of any anticode of length n with maximum distance $d-1$ is at most $V(\Lambda)$.*

Recall that any lattice in $\mathbb{Z}^n$, obtained from a lattice Λ, is reduced to a Lee code of length n over $\mathbb{Z}_m$, where $m = V(\Lambda)$, and similarly a Lee code of length n over $\mathbb{Z}_m$ can be expanded to a code in $\mathbb{Z}^n$.

Corollary 11.6. *Let Λ be a lattice that forms a code $\mathcal{C} \subset \mathbb{Z}^n$ with minimum Manhattan distance d. Then the size of any anticode of length n with maximum distance $d-1$ is at most $V(\Lambda)$.*

Since the definition of a diameter perfect code is based on the code-anticode bound, it is required to have a different definition for the Manhattan distance. The definition is straightforward if the code is defined by a lattice, since in this case the code can be reduced to a code over a finite space with the Lee distance. When the code is not based on a lattice, then we can use a definition based on the density of the code. This will not be necessary in the discussion that follows. We will use a simpler definition based on the straightforward observation that the code and translates of the anticode based on the code form a tiling of $\mathbb{Z}^n$. In other words, the code $\mathcal{C}$ is a D-diameter perfect code in $\mathbb{Z}^n$ if its minimum Manhattan distance is $D+1$ and if it induces a tiling of $\mathbb{Z}^n$ with $\mathcal{A}_D(n)$.

By Corollary 11.3, there exists a maximum size anticode of length n, with diameter three over $\mathbb{Z}^n$ in the Manhattan metric, whose size is $4n$. Therefore, by Corollary 11.6, a related diameter perfect code with minimum distance four can be formed from a lattice whose volume is $4n$ in which the minimum Manhattan distance is four. Consider the lattice $\Lambda(G_n)$ defined by the following generator matrix

$$G_n = \begin{bmatrix} A_n & B_n \\ C_n & D_n \end{bmatrix},$$

where $A_n = I_{n-1}$, B_n is the $(n-1) \times 1$ for which $B_n^{\text{tr}} = [3\ 5\ \cdots\ 2n-1]$, and C_n is an $1 \times (n-1)$ all-zero matrix, and $D_n = [4n]$.

Example 11.1. For $n = 6$, G_6 is the following generator matrix

$$G_6 = \begin{bmatrix} 1\,0\,0\,0\,0 & 3 \\ 0\,1\,0\,0\,0 & 5 \\ 0\,0\,1\,0\,0 & 7 \\ 0\,0\,0\,1\,0 & 9 \\ 0\,0\,0\,0\,1 & 11 \\ 0\,0\,0\,0\,0 & 24 \end{bmatrix}.$$

Note that the lattice is reduced to a code with Lee distance 4 over $\mathbb{Z}_{24}^6$. The period of the lattice is 24, but considering each coordinate separately, its period is $(8, 24, 24, 8, 24, 24)$.

The volume of $\Lambda(G_n)$ is $4n$ and it is easy to verify that the minimum Manhattan distance of the code defined by $\Lambda(G_n)$ is four. Moreover, $|\mathcal{A}_3(n)| = 4n$ and it is readily verified that $\Lambda(G_n)$ is reduced to a Lee code over $\mathbb{Z}_{4n}$. Hence, we have the following theorem.

Theorem 11.9. *The code defined by the lattice $\Lambda(G_n)$ is a 3-diameter perfect code in the Manhattan metric. The code can be reduced to a Lee code $\mathcal{C}$ over $\mathbb{Z}_{4n}^n$, which is a 3-diameter perfect code.*

The 3-diameter perfect code defined by $\Lambda(G_n)$ is not unique. There are other 3-diameter perfect codes of length n in the Lee metric. The main goal is to construct many such codes over alphabets with the smallest possible size, i.e., with the smallest period.

For $n = 2$, there are $(2e+1)$-diameter perfect codes for each $e \geq 1$. By Theorem 11.4 and Corollary 11.4, the size of a maximum size anticode with diameter $2e+2$ over $\mathbb{Z}^2$ is $2(e+1)^2$. For each i, $0 \leq i \leq e$, the following generator matrix forms a lattice that generates a diameter perfect code whose minimum distance is $2e+2$.

$$G_i^e = \begin{bmatrix} e+1+i & e+1-i \\ i & 2(e+1)-i \end{bmatrix}. \tag{11.3}$$

There are nonequivalent two-dimensional $(2e+1)$-diameter perfect codes in this case. For example, $\Lambda(G_0^e)$ is not equivalent to $\Lambda(G_i^e)$ for any $1 \leq i \leq e$.

This can easily be verified from the fact that $\Lambda(G_0^e)$ contains the points $(0,0)$, $(2e+2,0)$, while the other lattices do not contain two points with a similar distance horizontally and vertically.

Cosets and translates of a code $\mathcal{C}$ with the Lee distance or Manhattan distance are defined in the same way that they are defined in the Hamming scheme. A translate of a code $\mathcal{C}$ is said to be an even translate if all its words have even Lee (Manhattan, respectively) weight. A translate is said to be an odd translate if all its words have odd Lee (Manhattan, respectively) weight.

Similarly to the cosets and translates of extended 1-perfect error-correcting codes in the Hamming scheme, there are even translates and odd translates for the two families of diameter perfect codes in the Lee metric. This is proved in the following results. The following lemma can be readily verified.

Lemma 11.3. *A $(d-1)$-diameter perfect code with minimum Lee distance d over $\mathbb{Z}_m^n$ can be expanded to a $(d-1)$-diameter perfect code with minimum Manhattan distance d over $\mathbb{Z}^n$.*

For the next lemma, we have to show that the code-anticode bound, in the Lee metric and the Manhattan metric, induces a tiling.

Lemma 11.4. *A $(d-1)$-diameter perfect code, in $\mathbb{Z}_m^n$, with Lee distance d (or in $\mathbb{Z}^n$ with Manhattan distance d, respectively), induces a tiling of $\mathbb{Z}_m^n$ ($\mathbb{Z}^n$, respectively) with the related anticode whose diameter is $d-1$.*

Proof. Assume the contrary, that $\mathcal{C}$ is a $(d-1)$-diameter perfect code in $\mathbb{Z}^n$ and $\mathcal{A}$ is its related anticode with diameter $d-1$ and $\mathcal{C}$ does not induce a tiling of $\mathbb{Z}^n$ with $\mathcal{A}$. This implies that there are two codewords of $\mathcal{C}$, c_1 and c_2 such that $(c_1+\mathcal{A}) \cap (c_2+\mathcal{A}) \neq \varnothing$. Assume that $x \in (c_1+\mathcal{A}) \cap (c_2+\mathcal{A})$. Since $x \in c_1+\mathcal{A}$, it follows that $x-c_1 \in \mathcal{A}$, and hence $c_2+x-c_1 \in c_2+\mathcal{A}$. Clearly, $c_1+x-c_1 = x \in c_2+\mathcal{A}$ and hence $d_M(c_2+x-c_1,x) \leq d-1$. But, $d_M(c_2+x-c_1,x) = d_M(c_2,c_1) \geq d$, a contradiction. Therefore, $\mathcal{C}$ induces a tiling of $\mathbb{Z}^n$ with $\mathcal{A}$. If $\mathcal{C}$ is a $(d-1)$-diameter perfect code in $\mathbb{Z}_m^n$, it can be extended periodically to a $(d-1)$-diameter perfect code in $\mathbb{Z}^n$ and the same proof follows. □

Lemma 11.5. *If $\mathcal{C}$ is a $(d-1)$-diameter perfect code in the Manhattan metric, where d is even, then all the codewords of $\mathcal{C}$ have even Manhattan weight.*

Proof. Let $\mathcal{C}$ be a $(2e+1)$-diameter perfect code in the Manhattan metric. Assume the contrary, that there exists a codeword with odd Manhattan weight. Since $\mathcal{C}$ is a $(2e+1)$-diameter perfect code, it follows by Lemma 11.4 that there exists a tiling of $\mathbb{Z}^n$ with the anticode $\mathcal{A}_{2e+1}(n)$. W.l.o.g. we can assume that the two adjacent points in $\mathcal{A}_{2e+1}(n)$ (the core of each $\mathcal{A}_{2e+1}(n)$ in this tiling) differ in the last coordinate, one of these points $(z_1, \ldots, z_{n-1}, z_n)$ is a codeword in $\mathcal{C}$ and the second point is $(z_1, \ldots, z_{n-1}, z_n + 1)$. Since there exists a codeword with odd Manhattan weight, it follows that for this tiling, there exist two points $x = (x_1, \ldots, x_{n-1}, x_n) \in \mathbb{Z}^n$ and $y = (y_1, \ldots, y_{n-1}, y_n) \in \mathbb{Z}^n$, such that $d_M(x, y) = 1$, x is in an anticode containing a codeword $\alpha \in \mathcal{C}$, and y is in an anticode containing a codeword $\beta \in \mathcal{C}$; furthermore, α has even Manhattan weight and β has odd Manhattan weight. Clearly, $d_M(x, \alpha), d_M(y, \beta) \in \{e, e+1\}$, $d_M(\alpha, \beta)$ is odd and at least $2e+2$. This implies that $d_M(x, \alpha) = d_M(y, \beta) = e+1$ and $d_M(\alpha, \beta) = 2e+3$. (recall that $\alpha + (0, \ldots, 0, 1)$ is an element in $\alpha + \mathcal{A}_{2e+1}(n)$ and $\beta + (0, \ldots, 0, 1)$ is an element in $\beta + \mathcal{A}_{2e+1}(n)$.) Note that $d_M(x, \alpha) = d_M(y, \beta) = e+1$, the facts that in the tiling, the point x is contained in the anticode containing α and the point y is contained in the anticode containing β, implies that x_n is greater than the last entry of α and y_n is greater than the last entry of β. Therefore, $d_M((x_1, \ldots, x_{n-1}, x_n - 1), \alpha) = d_M((y_1, \ldots, y_{n-1}, y_n - 1), \beta) = e$ and since $d_M(x, y) = 1$, it follows that $d_M(\alpha, \beta) \leq 2e+1$, a contradiction.

Thus, if $\mathcal{C}$ is a $(d-1)$-diameter perfect code in the Manhattan metric, where d is even, then all the codewords of $\mathcal{C}$ have even Manhattan weight. □

Lemma 11.6. *If $\mathcal{C}$ is a $(d-1)$-diameter perfect code, over $\mathbb{Z}_m^n$, with minimum Lee distance d, where d is even, then m is even.*

Proof. By Theorem 11.3, it can easily be verified that the size of the anticode with maximum distance $d-1$ is an even integer. This implies by the code-anticode bound that the size of the space, m^n, is even. Thus, m is even. □

Corollary 11.7. *If $\mathcal{C}$ is a $(d-1)$-diameter perfect code in the Lee metric where d is even, then all the codewords of $\mathcal{C}$ have even Lee weight.*

Theorem 11.10. *Each translate of a D-diameter perfect code in the Lee metric, where D is odd, is either an even translate or an odd translate. The number of even translates is equal the number of odd translates.*

Proof. Let $\mathcal{C}$ be a diameter perfect code over $\mathbb{Z}_m^n$ in the Lee metric. By Corollary 11.7, each translate of $\mathcal{C}$ is either an even translate or an odd translate. By Lemma 11.6 the period of the code m is even which implies that the number of points of even weight is equal to the number of points of odd weight. Therefore, the number of even translates of $\mathcal{C}$ is equal to the number of odd translates of $\mathcal{C}$. $\square$

We note one important difference between binary perfect codes and binary diameter perfect codes in the Hamming scheme and perfect codes and diameter perfect codes in the Lee metric. Binary 1-perfect error-correcting codes in the Hamming scheme can be extended to binary diameter perfect codes (and vice versa via puncturing, respectively) by adding a parity bit (removing a coordinate, respectively) and thus increasing (decreasing, respectively) the distance by one. There is no extension and puncturing with similar properties in the Lee metric, i.e., an extended code with increased distance of one can be defined, but puncturing can decrease the distance by more than one. Moreover, these operations do not induce a connection between perfect codes and diameter perfect codes.

Finally, there is one more known diameter perfect code in the Lee and Manhattan metrics. It is formed by the celebrated Minkowski lattice with the generator matrix

$$G = \begin{bmatrix} 1 & -2 & 3 \\ -2 & 3 & 1 \\ 3 & 1 & -2 \end{bmatrix}.$$

It is readily verified that the minimum Manhattan distance of the code $\mathcal{C}$ implied by $\Lambda(G)$ is 6. Simple calculation implies that $|\det G| = 38$ and $|\mathcal{A}_5(3)| = 38$ and hence $\mathcal{C}$ is a 5-diameter perfect code over $\mathbb{Z}_{38}^3$.

Let $\mathcal{C}^1$ and $\mathcal{C}^2$ be two $(n, 4, m)_L$ 3-diameter perfect codes. Each code has $4n$ translates of which $2n$ are even translates. Let $\mathcal{C}_1^\ell = \mathcal{C}^\ell, \mathcal{C}_2^\ell, \ldots, \mathcal{C}_{2n}^\ell$, $\ell = 1, 2$, be these $2n$ even translates, for each code. Let $\pi = (\pi(1) = 1, \pi(2), \ldots, \pi(2n))$ be a permutation of $\{1, 2, \ldots, 2n\}$. The following theorem is based on the direct product construction.

Theorem 11.11. *The code* $\mathcal{C}^\times$ *defined by*

$$\mathcal{C}^\times \triangleq \{(x, y) \; : \; x \in \mathcal{C}_i^1, \; y \in \mathcal{C}_{\pi(i)}^2, \; 1 \leq i \leq 2n\}$$

is a 3-diameter perfect Lee code of length $2n$, *over* $\mathbb{Z}_m$.

Proof. By Corollary 11.5, the size of the code $\mathcal{C}^\ell$, $\ell = 1, 2$, is $\frac{m^n}{4n}$ and hence the size of $\mathcal{C}^\times$ is $2n \frac{m^{2n}}{16n^2} = \frac{m^{2n}}{8n}$. We claim that $d_L(\mathcal{C}^\times) = 4$. Let (x_1, y_1)

and (x_2, y_2) be two distinct codewords in $\mathcal{C}^\times$ such that $x_1 \in \mathcal{C}_i^1$ and $x_2 \in \mathcal{C}_j^1$. We distinguish between two cases depending on whether $i = j$ or $i \neq j$.

Case 1. $i \neq j$.

This implies that $\mathcal{C}_i^1 \neq \mathcal{C}_j^1$ and $\mathcal{C}_{\pi(i)}^2 \neq \mathcal{C}_{\pi(j)}^2$. Hence $d_L(x_1, x_2) \geq 2$, $d_L(y_1, y_2) \geq 2$, which implies that $d_L((x_1, y_1), (x_2, y_2)) \geq 4$.

Case 2. $i = j$.

This implies that $x_1, x_2 \in \mathcal{C}_1^i$ and $y_1, y_2 \in \mathcal{C}_{\pi(i)}^2$. Since $x_1 \neq x_2$ or $y_1 \neq y_2$, we have that $d_L(x_1, x_2) \geq 4$ or $d_L(y_1, y_2) \geq 4$, respectively. Hence, $d_L((x_1, y_1), (x_2, y_2)) \geq 4$.

Thus, $d_L(\mathcal{C}^\times) \geq 4$.

The code $\mathcal{C}^\times$ is defined over $\mathbb{Z}_m^{2n}$ and hence the size of its space is m^{2n}. Since, by Corollary 11.3, the size of the related maximum size anticode with diameter three is $8n$ and the minimum Lee distance of $\mathcal{C}^\times$ is four, it follows, by Corollary 11.5, that $\mathcal{C}^\times$ is a 3-diameter perfect Lee code over $\mathbb{Z}_m^{2n}$. □

Combining Theorems 11.9 and 11.11 implies the following consequence.

Corollary 11.8. *For each two integers $n \geq 4$ and $r \geq 0$, there exists a $(2^r n, 4, 4n)_L$ 3-diameter perfect code.*

Note, that in Corollary 11.8 we can use n equal to a prime p and the codes will have the same minimum periodicity $4p$ in all dimensions.

The lattice of the following generator matrix

$$\begin{bmatrix} 2 & 2 \\ 0 & 4 \end{bmatrix},$$

forms a $(2, 4, 4)_L$ 3-diameter perfect code. By applying Theorem 11.11 iteratively, we obtain the following theorem.

Theorem 11.12. *For each $n = 2^r$, $r \geq 1$, there exists an $(n, 4, 4)_L$ 3-diameter perfect code.*

Theorem 11.11 can be modified and applied on codes in $\mathbb{Z}^n$ with the Manhattan metric.

Theorem 11.13. *Let $\mathcal{C}^1$ and $\mathcal{C}^2$ be a 3-diameter perfect codes of length n in the Manhattan metric. Each code has $4n$ translates of which $2n$ are even translates. Let $\mathcal{C}_1^i = \mathcal{C}^i, \mathcal{C}_2^i, \ldots, \mathcal{C}_{2n}^i$, $i = 1, 2$, be these $2n$ even translates. Let $\pi = (\pi(1) = 1, \pi(2), \ldots, \pi(2n))$ be a permutation of $\{1, 2, \ldots, 2n\}$. The code $\mathcal{C}^\times$ defined by*

$$\mathcal{C}^\times \triangleq \{(x, y) \; : \; x \in \mathcal{C}_i^1, \; y \in \mathcal{C}_{\pi(i)}^2, \; 1 \leq i \leq 2n\} \tag{11.4}$$

is a 3-diameter perfect Manhattan code of length $2n$, over $\mathbb{Z}$.

11.5 Nonperiodic Codes and Enumeration of Codes

In this section we consider two different problems with a related solution. The first one is the number of nonequivalent perfect codes in the Lee metric, which was considered in Section 5.7 for the Hamming scheme. For the Hamming scheme, it was proved that for any given $0 < \epsilon < \frac{1}{q}$, the number of nonequivalent 1-perfect codes of length n over $\mathbb{F}_q$ is at least $q^{q^{cn}}$, where $c = \frac{1}{q} - \epsilon$. The second problem is whether there exist nonperiodic perfect codes and nonperiodic diameter perfect codes in $\mathbb{Z}^n$. The question on nonperiodic codes is asked only for the Manhattan metric. The same question for the Lee metric is not relevant since we can always reduce the alphabet size in the Lee metric to the period of the code. The two questions are somehow related as will be demonstrated in this section. We will prove that there exists many different nonlinear perfect codes in the Lee metric.

Clearly, the product construction presented in Theorem 11.11 yields many 3-diameter perfect codes by using different permutations. We will not go into the exact computations on the number of nonequivalent perfect and diameter perfect codes. This computation is left as an exercise for the reader. We will just count the number of different perfect codes. The two product constructions, introduced in Section 11.3 and in Section 11.4, can be used to provide lower bounds on the number of perfect codes and diameter perfect codes in the Lee metric. Given a 3-diameter perfect code of length n, $n \geq 4$, the size of the related anticode is $4n$, and hence this code has $2n$ even translates. Therefore, there are $(2n-1)!$ different perfect codes of length $2n$, which can be generated by the construction of Theorem 11.11. By applying iteratively the construction of Theorem 11.11, we obtain

$$\prod_{i=1}^{\ell} (2^i n - 1)!^{2^{\ell-i}}$$

different 3-diameter perfect codes of length $2^\ell n$ over $\mathbb{Z}_{4n}$.

Similarly, a bound on the number of perfect Lee codes can be obtained from the general product construction and its proof in Theorem 11.7 presented in Section 11.3. In this computation, for the number of perfect Lee codes, we can also take into account the bounds on the number of perfect codes in the Hamming scheme, which are used in the construction.

Before we continue our discussion of nonperiodic perfect codes and nonperiodic diameter perfect codes in $\mathbb{Z}^n$, we consider a question related to both problems discussed in this section. Can we also apply the switching method, that was used several times in Chapter 5 for the Hamming scheme, to the Lee metric and the Manhattan metric?

Let $\mathcal{C}_1$ and $\mathcal{C}_2$ be two distinct sub-codes of a subspace $\mathcal{U} \subseteq \mathbb{Z}^n$, such that any elements in $\mathcal{U}$ is within Manhattan distance one from a unique codeword of $\mathcal{C}_1$ and a unique codeword of $\mathcal{C}_2$. Also, each element which is within distance one from $\mathcal{C}_1$ (also $\mathcal{C}_2$) is contained in $\mathcal{U}$. If we consider anticodes instead of balls, then the elements of $\mathcal{C}_1$ and $\mathcal{C}_2$ are the balanced points (taken instead of the centers of the balls) of the anticodes that form a tiling of $\mathcal{U}$. If $\mathcal{C}_1$ is contained in a 1-perfect Manhattan code (or a 3-diameter perfect code in the Manhattan metric, respectively) $\mathcal{C}$, then it can be replaced by $\mathcal{C}_2$ to obtain a new different 1-perfect Manhattan code $\mathcal{C}'$ (or a 3-diameter perfect code in the Manhattan metric, respectively). This switching method technique was used in Section 5.7 to form a large number of nonequivalent perfect codes in the Hamming scheme. The same technique can be considered for the Lee metric as follows.

The previous two product constructions can provide such a subspace $\mathcal{U}$ and codes $\mathcal{C}_1$ and $\mathcal{C}_2$. Assume that we take two permutations that differ in one transposition (a transposition exchanges two adjacent elements in the permutation) to apply the construction in Theorem 11.11. It is easily verified that the intersection of the two generated codes is relatively large (exactly $\frac{\eta-2}{\eta}$ of the code size, where $\eta = 2n$). The union of the two direct products related to this transposition in the two generated codes, in Theorem 11.11, for the construction of the 3-diameter perfect code are used to tile the same subset of $\mathbb{Z}_m^{2n}$. Similarly, we can define a larger set of disjoint transpositions to obtain a large set of different codes, which are also nonlinear codes in the Lee metric and the Manhattan metric. Different permutations can be also used in Theorem 11.7 to obtain nonlinear codes and also some linear codes for which we can compute the exact intersection between the perfect codes. Also, in this case, the obtained codes are nonlinear 1-perfect Lee codes for which we can find two codes $\mathcal{C}_1$ and $\mathcal{C}_2$ that perfectly cover the same finite subset of $\mathbb{Z}_m^n$. Unfortunately, such a finite space $\mathcal{U} \subset \mathbb{Z}^n$, and codes $\mathcal{C}_1$ and $\mathcal{C}_2$ cannot exist for the Manhattan metric as will be proved in the following theorem.

Recall that a subset $\mathcal{U}$ is ***perfectly e-covered*** by a code $\mathcal{C}$ if

(1) for each element $u \in \mathcal{U}$, there is a unique element $c \in \mathcal{C}$ such that $d(u, c) \leq e$;
(2) all the words within radius e from $\mathcal{C}$ are contained in $\mathcal{U}$.

When the Manhattan distance is applied, we will also have the natural requirement that the number of words covered by each codeword is the size of the related Lee ball.

Theorem 11.14. *There is no finite subset of $\mathbb{Z}^n$ that can be perfectly e-covered by two different codes, using the Manhattan metric.*

Proof. Assume the contrary, that there exists a subset S of $\mathbb{Z}^n$, for which there exist two codes $\mathcal{C}_1$ and $\mathcal{C}_2$ that perfectly e-cover S and assume further that S is the smallest subset with this property. Let k be the largest integer for which there exists a codeword in $\mathcal{C}_1$ or $\mathcal{C}_2$ with the value k in one of the n coordinates. W.l.o.g. we can assume that such a codeword is $(k, x_2, \ldots, x_n) \in \mathcal{C}_1$. This codeword covers the word $(k+e, x_2, \ldots, x_n) \in S$. Since k can be the largest integer in a codeword of $\mathcal{C}_2$, it follows that the only codeword of $\mathcal{C}_2$ that can cover $(k+e, x_2, \ldots, x_n) \in S$ is $(k, x_2, \ldots, x_n)$. Thus, $\mathcal{C}_1 \setminus \{(k, x_2, \ldots, x_n)\}$ and $\mathcal{C}_2 \setminus \{(k, x_2, \ldots, x_n)\}$ perfectly e-cover a subset $S' \subset S$, a contradiction to the assumption that S is such a subset with the smallest size. Thus, there is no finite subset in $\mathbb{Z}^n$ that can be perfectly e-covered by two different codes. □

For each $n = 2^\ell$, $\ell \geq 1$, there exists a nonperiodic 3-diameter perfect code over $\mathbb{Z}^n$. If $n = 2$, then nonperiodic $(2e+1)$-diameter perfect code exists for each $e \geq 1$. For this purpose we can interleave the $e+1$ lattices defined in (11.3) with $i = 0$ and that were used to construct diameter perfect codes with minimum Manhattan distance $2e+2$. This interleaving is defined as follows. For a given $e \geq 1$, let $S = \{s_i\}_{i=-\infty}^{\infty}$ be an infinite sequence, where $s_0 = 0$ and $s_i \in \mathbb{Z}_{e+1}$ for $i \neq 0$. Given the sequence S, we construct the following set $\mathcal{T}$ of points:

$$\mathcal{T} \triangleq \{(2(e+1)i + (e+1)j + s_i, (e+1)j + s_i) : s_i \in S,\ i, j \in \mathbb{Z}\}.$$

The sequence S will be called *nonperiodic* if there is no nonzero integer ρ and an integer τ such that $s_i = s_{i+\tau} + \rho \pmod{e+1}$ for each $i \in \mathbb{Z}$.

Theorem 11.15. *If the points of the set $\mathcal{T}$ are taken as balanced points for the translates of the anticode $\mathcal{A}_{2e+1}(2)$, then a tiling is obtained. The tiling is nonperiodic if the sequence S is nonperiodic.*

Proof. First note that the set of points $\{((e+1)j, (e+1)j) : j \in \mathbb{Z}\} \subset \mathcal{T}$ is taken as balanced points for the translates of the anticode $\mathcal{A}_{2e+1}(2)$. These translates are nonintersecting since their mutual distance is at least $2e+2$ and the diameter of the anticode is $2e+1$. Moreover, they form a connected diagonal strip of translates of the anticode $\mathcal{A}_{2e+1}(2)$. The same is true for the set of balanced points

$\{(2(e+1)i+(e+1)j+s_i,(e+1)j+s_i) \; : \; j \in \mathbb{Z}\} \subset \mathcal{T}$ for any given fixed $i \in \mathbb{Z}$. The set of points $\{(2(e+1)i+(e+1)j,(e+1)j) \; : \; i \in \mathbb{Z}, \; j \in \mathbb{Z}\}$ is exactly the set of points of the lattice formed by the generator matrix

$$\begin{bmatrix} e+1 & e+1 \\ 2(e+1) & 0 \end{bmatrix},$$

which forms a $(2e+1)$-diameter perfect code whose minimum distance is $2(e+1)$. Finally, replacing the set of balanced points $\{(2(e+1)i+(e+1)j,(e+1)j) \; : \; j \in \mathbb{Z}\}$ in this tiling by the set of points $\{(2(e+1)i+(e+1)j+s_i,(e+1)j+s_i) \; : \; j \in \mathbb{Z}\}$ is just a shift of length s_i of a diagonal strip, 45 degrees in the diagonal direction. This does not affect the fact that the set of these points forms a tiling with the anticodes $\mathcal{A}_{2e+1}(2)$.

Finally, it is readily verified that if the sequence S is periodic, then also the tiling generated from $\mathcal{T}$ is periodic. Similarly, iff the sequence S is nonperiodic, then also the tiling generated from $\mathcal{T}$ is nonperiodic. □

The construction presented in Theorem 11.15 for nonperiodic two-dimensional diameter perfect codes yields an uncountable number of diameter perfect codes. The number of nonequivalent $(2e+1)$-diameter perfect codes formed in this way is equal to the number of infinite nonperiodic sequences of the form $\{s_i\}_{i=-\infty}^{\infty}$ over the alphabet $\mathbb{Z}_{e+1}$. Of course, one has to consider shifts of the sequences that can coincide, but this does not affect the asymptotic number of nonequivalent codes.

Finally, it is easy to verify the following theorem.

Theorem 11.16. *Let $\mathcal{C}_1$ be a nonperiodic 3-diameter perfect code of length n in the Manhattan metric, and let $\mathcal{C}_2$ be a 3-diameter perfect code of length n in the Manhattan metric. The code $\mathcal{C}^{\times}$ defined in (11.4) is a nonperiodic 3-diameter perfect code of length $2n$ in the Manhattan metric.*

11.6 The Nonexistence of Perfect Codes

Do any other parameters of perfect codes in the Lee metric or in the Manhattan metric exist? A well-known conjecture from 1970, known as the Golomb-Welch conjecture is that there are no more such parameters. The conjecture is both for the Lee metric and the Manhattan metric. We note that a proof of the conjecture for the Manhattan metric implies that it also true for the Lee metric. Many attempts have been made to prove this conjecture using various techniques, some of them are ad hoc techniques and some are more general. We restrict ourselves to a general technique used for the Lee metric in any dimension, but we apply it only to the 3-dimensional space.

The basic idea is that for a perfect Lee code over $\mathbb{Z}_m^3$, the tiling with cubistic octahedra must have the property that the total number of unit cubes meeting in one point is always 8. The number of unit cubes belonging to the same cubistic octahedron (of a Lee ball) meet in one point can have the values 1, 4, and 7.

This number is called the ***type*** of that vertex. A vertex of type 4, which is adjacent (by a connecting edge on the 3-dimensional grid) to another vertex of type 4, is denoted as type 4* (the set of vertices of type 4 contains the set of vertices of type 4*). If we try to combine types, we find that there are only a few possible combinations of types meeting in a point, as follows. Two possible combinations of vertices from only two Lee balls, $[7, 1]$ and $[4, 4]$; one combination with five Lee balls $[4, 1, 1, 1, 1]$ and one combination with eight Lee balls $[1, 1, 1, 1, 1, 1, 1, 1]$. Accordingly, we make the following observations:

- Every vertex of type 7 requires a vertex of type 1 for combination up to 8.
- No two adjacent vertices of type 4 (which are, of course, of type 4*) can be completed by a vertex of type 4, i.e., at least half of the vertices of type 4* require four vertices of type 1 for a combination up to 8.

Figure 11.3 demonstrates the types of vertices in on octant of the Lee ball with radius 5.

Combining these two properties, we see that if a periodic tiling of $\mathbb{Z}_m^3$ with cubistic octahedra exists, then in every period box we have

$$t_1 - t_7 - \frac{1}{2}t_{4^*} \cdot 4 \geq 0,$$

where t_i stands for the number of vertices of type i in that period box. This is equivalent to the assertion

$$g_1 - g_7 - 2g_{4^*} \geq 0,$$

where g_i stands for the number of vertices of type i in one octant of a cubistic octahedron. These numbers are easily found (note the vertices of type 7 are exactly below vertices of type 4, which are exactly below the vertices of type 1. Vertices of type 4* are exactly on the boundaries of this octant Lee ball.):

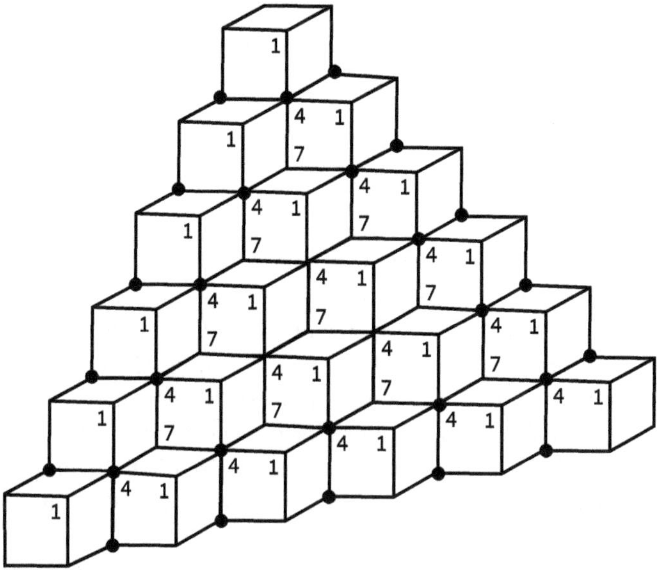

Fig. 11.3 Type of vertices of one octant with radius 5, where vertices of type 4^* are marked by •.

e	g_1	g_4	g_7	g_{4^*}
0	1	0	0	0
1	3	1	0	1
2	6	3	1	3
3	10	6	3	6
4	15	10	6	9
⋮	⋮	⋮	⋮	⋮
e	$\binom{e+2}{2}$	$\binom{e+1}{2}$	$\binom{e}{2}$	$3(e-1)$ $e\geq 2$

This implies that a necessary condition for the existence of an e-perfect Lee code over a large alphabet with parameters $(n, e) = (3, e)$, $e \geq 2$, is that

$$\binom{e+2}{2} - \binom{e}{2} - 6(e-1) \geq 0,$$

which implies that

$$-4e + 7 \geq 0.$$

This prove that there are no e-perfect Lee codes over $\mathbb{Z}_m$, for $(n, e) = (3, e)$, where $e \geq 2$.

11.7 Notes

Section 11.1. The Lee metric was introduced in [Ulrich (1957)] and [Lee (1958)] for transmission of signals taken from $\mathbb{F}_p$, p prime, over some noisy channels. It was generalized for $\mathbb{Z}_m$ in [Golomb and Welch (1970)]. Many applications were found for this metric in coding theory, e.g., [Roth and Siegel (1994); Blaum, Bruck and Vardy (1998); Etzion and Vardy (2002); Etzion and Yaakobi (2009); Etzion, Vardy, and Yaakobi (2013)], to name a few such applications. Theorem 11.4 on the size of the maximum anticode in the Manhattan metric was proved in [Ahlswede and Blinovsky (2008), pp. 30–41].

Section 11.2. Applications of lattice tilings with Lee spheres and maximum Lee anticodes were presented, for example, in [Blaum, Bruck and Vardy (1998); Etzion and Vardy (2002); Etzion (2011)]. These papers also include a short introduction to lattice tiling. A comprehensive book on lattices is the one by [Conway and Sloane (1988)]. It also includes tilings in $\mathbb{R}^n$ (which are not tilings in $\mathbb{Z}^n$), which are not considered in our exposition. Tilings in $\mathbb{R}^n$ were also considered in the book by [Stein and Szabó (1994)]. These tilings in $\mathbb{R}^n$ are also for shapes in $\mathbb{Z}^n$ and many interesting results are presented in the book.

Section 11.3. The two basic constructions for perfect Lee codes were presented in [Golomb and Welch (1970)]. The product constructions for Lee codes were presented in [Etzion (2011)].

Section 11.4. Diameter perfect codes in the Lee metric were considered in [Etzion (2011)]. The famous Minkowski lattice was presented in [Minkowski (1904)]. Codes with the same parameters as in Theorem 11.12 were also generated by Krotov [Krotov (2001a)]. A complete characterization of alphabet size for which there exist 3-diameter perfect codes was given in [Horak and AlBdaiwi (2012b)].

Section 11.5. The presentation of this section is also taken from [Etzion (2011)]. It was proved in [Horak and AlBdaiwi (2012b)] that there are uncountable distinct 3-diameter perfect codes for each dimension $n \geq 3$.

Section 11.6. The conjecture on the nonexistence of nontrivial perfect codes in the Manhattan metric was proposed by [Golomb and Welch (1970)]. They conjectured that no perfect code exists in the Lee metric (and the Manhattan metric), except for trivial ones, perfect codes with radius one, and two-dimensional perfect codes. They proved that for each $n \geq 3$, there exists an e_n, such that for each $e > e_n$, there is no e-perfect code in $\mathbb{Z}^n$.

Clearly, the nonexistence of e-perfect codes in $\mathbb{Z}^n$ implies the nonexistence of such codes in $\mathbb{Z}_m^n$ for each $m \geq 2e+1$. On the other hand, the nonexistence of a perfect codes in $\mathbb{Z}_m^n$ for each $m \geq 2e+1$ does not imply their nonexistence in $\mathbb{Z}^n$. The following conjecture was made in [Lagarias and Wang (1996)].

Conjecture 11.1. *If the shape $\mathcal{S}$ tiles $\mathbb{Z}^n$ by translations, then $\mathcal{S}$ admits a fully periodic tiling, i.e., $\mathcal{S}$ tiles $\mathbb{Z}_m^n$ for sufficiently large m.*

If Conjecture 11.1 is true, then the conjecture of Golomb and Welch for the nonexistence of perfect codes in the Lee metric and the Manhattan metric are equivalent. Conjecture 11.1 was proved in [Szegedy (1998)] for the case when $\mathcal{S}$ has a size of a prime. It was further proved in [Bhattacharya (2020)] for $\mathbb{Z}^2$. Nevertheless, in general, this interesting conjecture is far from resolved.

There are not many more results for the nonexistence of perfect codes in $\mathbb{Z}^n$. The nonexistence of such codes in $\mathbb{Z}^3$ was proved in [Gravier, Mollard, and Payan (1998, 2001)] using a technique that is based on the "picture" of the balls. In $\mathbb{Z}^4$ the nonexistence of such perfect codes was proved in [Špacapan (2007)] using a computer based exhaustive search. A completely different algebraic approach was used in [Horak (2009b)] to prove the nonexistence of perfect codes in $\mathbb{Z}^n$ for $3 \leq n \leq 5$. In fact, the proof in these papers also ruled out possible tiling with balls of different sizes. For $n = 6$, the nonexistence of 2-perfect codes in $\mathbb{Z}^6$ was proved in [Horak (2009a)]. The last result in this direction was done in [Kim (2017)] where it is proved that if the size $2n^2+2n+1$ of the ball with radius two is a prime and a certain number-theoretic condition is satisfied, then a 2-perfect code does not exist. This condition is not restrictive as, e.g., there are 12706 numbers for which $n \leq 10^5$, where $2n^2+2n+1$ is a prime. Only four values of n, out of these 12706 values, do not satisfy the given condition. It is not known, however, if there are infinity many values of n for which $2n^2+2n+1$ is a prime.

Even a restriction for perfect codes that are based on lattice tiling does not make the existence problem much easier. Nonexistence proofs and new techniques for such 2-perfect codes in $\mathbb{Z}^n$ for some values of n were presented first in [Horak and Grošek (2014)]. Other results in [Zhang and Ge (2017)] were improved later in [Leung and Zhou (2020)].

The proof for the nonexistence of e-perfect codes in $\mathbb{Z}_m^3$ presented in this section was done by [Post (1975)]. He has generalized his method and proved the following theorem.

Theorem 11.17. *There are no e-perfect codes in the Lee metric over* $\mathbb{Z}_m^n$, *where* $m \geq 2e+1$, *for*

(1) $3 \leq n \leq 5$, *where* $e \geq n-1$;
(2) $n \geq 6$, *where* $e \geq \frac{2n\sqrt{2}-3\sqrt{2}-2}{4}$.

These results of [Post (1975)] were asymptotically improved by [Lepistö (1981); Astola (1982a)]. A survey on the results related to this conjecture in the following fifty years was given in [Horak and Kim (2018)]. The survey includes all the various techniques that were used to exclude parameters where perfect codes can exist. This survey appeared in a special journal issue in memory of Solomon W. Golomb. They summarized the known results and their improved bounds in the following theorem.

Theorem 11.18. *There is no e-perfect code in* $\mathbb{Z}^n$ *the Manhattan metric for*

(1) $3 \leq n \leq 74$, *where* $2 \leq e$ *and* $\frac{\sqrt{2}}{2}n - \frac{3}{4}\sqrt{2} - \frac{1}{2} \leq e$.
(2) $75 \leq n \leq 405$, *where* $\max\{18, \sqrt{2n+40}\} \leq e \leq \frac{n-21}{3}$ *or* $\frac{\sqrt{2}}{2}n - \frac{3}{4}\sqrt{2} - \frac{1}{2} \leq e$.
(3) $406 \leq n \leq 876$, *where* $\sqrt{2n+40} \leq e \leq \frac{n-21}{3}$ *or* $285 \leq e$.
(4) $876 \leq n$, *where* $\sqrt{2n+40} \leq e$.

All the results mentioned above are for large alphabet sizes, i.e., e-perfect codes in $\mathbb{Z}_m^n$, where $m \geq 2e+1$. Smaller alphabet sizes were hardly considered. One such work in which parameters for e-perfect codes were excluded for small m was presented in [Astola (1982b)].

Chapter 12

Tiling with a Cluster of Unit Cubes

Tiling is a concept that is very closely related to perfect codes. The definition of a tiling in a finite space was given in Section 2.4. The definition was generalized for $\mathbb{Z}^n$ in Section 11.2, where the concept of lattices was also briefly introduced. A ball with radius e will be considered as a shape $\mathcal{S}$, which is a contiguous cluster of n-dimensional unit cubes. A tiling will be a tile of the n-dimensional Euclidian space with translates of $\mathcal{S}$. Tilings with the same shape will also be considered in $\mathbb{Z}_m^n$. Note again that a tiling in $\mathbb{Z}_m^n$ implies a tiling in $\mathbb{Z}^n$, but the reverse is not necessarily true. The Lee sphere $\mathcal{B}_e(n)$ and the Lee anticode $\mathcal{A}_{2e+1}(n)$ introduced in Chapter 11 are two types of such shapes and the related perfect codes in $\mathbb{Z}_m^n$ and $\mathbb{Z}^n$ form the related tilings. There are other shapes, which are not balls or anticodes, which are interesting in coding theory. These shapes will be called error spheres. Each ***error sphere*** represents some type of errors in some space.

There are a few types of tilings. We can distinguish between lattice tiling and non-lattice tilings. Another distinction is between integer tilings and non-integer tilings. For shapes which are contiguous clusters of n-dimensional unit cubes, only integer tilings, where the balanced points are on integer points, will be considered in this chapter. Another technique for describing a tiling is group splitting. This technique is equivalent to lattice tiling in $\mathbb{Z}^n$, but sometimes it can be described more efficiently. This technique is discussed in Section 12.1. The two most popular shapes that were considered for both tiling and coding are the cross and the semi-cross, which are considered in Section 12.2. Modern applications in flash memories in which such tilings were considered are discussed in Section 12.3. The first shape with application to flash memory is the quasi-cross, which is a generalization of the cross and the semi-cross. Tilings with this shape are considered in Section 12.4. A completely different shape is the notched cube whose tiling is discussed in Section 12.5.

12.1 Group Splitting

This section is devoted to two important concepts in tiling of various shapes, group splitting and generalized splitting.

A ***group splitting*** of a group G is a pair of sets, $M \in \mathbb{Z}^-$, called the ***multiplier set***, and $\mathcal{S} = \{s_1, s_2, \ldots, s_n\} \subseteq G$, called the ***splitter set***, such that the elements of the form $m \cdot s$, where

$$m \cdot s \triangleq \overbrace{s + s + \cdots + s}^{m \text{ summands}}, \tag{12.1}$$

$m \in M$, $s \in \mathcal{S}$, are all distinct nonzero elements and contain all the nonzero elements in G. In this case we say that the multiplier set M ***splits*** the group G.

Let G be an Abelian group and let $\beta = \beta_1, \beta_2, \ldots, \beta_n$ be a sequence with n elements of G. For every $X = (x_1, x_2, \ldots, x_n) \in \mathbb{Z}^n$, we define

$$X \cdot \beta = \sum_{i=1}^{n} x_i \cdot \beta_i,$$

as the inner product of X by β, where addition is performed in G and the multiplication is defined in (12.1).

A set $\mathcal{S} \subset \mathbb{Z}^n$ (which can be viewed as a discrete shape $\mathcal{S}$ in $\mathbb{Z}^n$) ***splits*** an Abelian group G with a ***splitting sequence*** $\beta = \beta_1, \beta_2, \ldots, \beta_n$, where $\beta_i \in G$, $1 \leq i \leq n$, if the set $\{\mathcal{E} \cdot \beta \ : \ \mathcal{E} \in \mathcal{S}\}$ contains $|\mathcal{S}|$ distinct elements from G. This operation is called ***generalized splitting***. Clearly, group splitting is a special case of generalized splitting, where the splitter set of the group splitting is the splitting sequence of the generalized splitting. Later on we will use the term "split" for both group splitting and generalized splitting. It will be understood from the context to which one we are referring.

In the rest of this section we examine the connection between generalized splitting and lattice tiling and prove that these concepts are equivalent when our shape $\mathcal{S}$ is a discrete shape. We note that a discrete shape consists of a collection of n-dimensional unit cubes that are connected by $(n-1)$-dimensional unit cubes. In fact, lattice tiling is also equivalent to group splitting and hence the two concepts of group splitting and generalized splitting are equivalent.

Lemma 12.1. *If Λ is a lattice packing of $\mathbb{Z}^n$ with a shape $\mathcal{S} \subset \mathbb{Z}^n$, then there exists an Abelian group G of order $V(\Lambda)$, such that $\mathcal{S}$ splits G.*

Proof. Let $G = \mathbb{Z}^n/\Lambda$ and let $\phi : \mathbb{Z}^n \to G$ be the group homomorphism that maps each element $X \in \mathbb{Z}^n$ to the coset $X + \Lambda$. Clearly, $|G| = V(\Lambda)$.

Let $\beta = \beta_1, \beta_2, \ldots, \beta_n$, be a sequence defined by $\beta_i = \phi(\mathbf{e}_i)$ for each i, $1 \leq i \leq n$. Clearly, for each $X \in \mathbb{Z}^n$ we have $\phi(X) = X \cdot \beta$.

Now assume that there exist two distinct elements $\mathcal{E}_1, \mathcal{E}_2 \in \mathcal{S}$, such that

$$\phi(\mathcal{E}_1) = \mathcal{E}_1 \cdot \beta = \mathcal{E}_2 \cdot \beta = \phi(\mathcal{E}_2) \ .$$

This implies that

$$\phi(\mathcal{E}_1 - \mathcal{E}_2) = (\mathcal{E}_1 - \mathcal{E}_2) \cdot \beta = \mathcal{E}_1 \cdot \beta - \mathcal{E}_2 \cdot \beta = 0 \ .$$

Since $\phi(X) = 0$ if and only if $X \in \Lambda$, it follows that there exists $X \in \Lambda$, where $X \neq \mathbf{0}$, such that

$$\mathcal{E}_1 = \mathcal{E}_2 + X \ .$$

Therefore, $\mathcal{S} \cap (X + \mathcal{S}) \neq \varnothing$, which contradicts the fact that Λ is a lattice packing of $\mathbb{Z}^n$ with the shape $\mathcal{S}$.

Thus, $\mathcal{S}$ splits G with the splitting sequence β. □

Lemma 12.2. *Let G be an Abelian group and let $\mathcal{S}$ be a shape in $\mathbb{Z}^n$. If $\mathcal{S}$ splits G with a splitting sequence β, then there exists a lattice packing Λ of $\mathbb{Z}^n$ with the shape $\mathcal{S}$, for which $V(\Lambda) \leq |G|$.*

Proof. Consider the group homomorphism $\phi : \mathbb{Z}^n \to G$ defined by

$$\phi(X) = X \cdot \beta.$$

Clearly, $\Lambda = \ker(\phi)$ is a lattice and its volume is $V(\Lambda) = |\phi(\mathbb{Z}^n)| \leq |G|$.

To complete the proof we have to show that Λ is a packing of $\mathbb{Z}^n$ with the shape $\mathcal{S}$. Assume the contrary, that there exists $X \in \Lambda$ such that $\mathcal{S} \cap (X + \mathcal{S}) \neq \varnothing$. Hence, there exist two distinct elements $\mathcal{E}_1, \mathcal{E}_2 \in \mathcal{S}$ such that $\mathcal{E}_1 = \mathcal{E}_2 + X$ and, therefore,

$$\phi(\mathcal{E}_1) = \phi(\mathcal{E}_2 + X) = \phi(\mathcal{E}_2) + \phi(X) = \phi(\mathcal{E}_2).$$

Accordingly, $\mathcal{E}_1 \cdot \beta = \mathcal{E}_2 \cdot \beta$, which contradicts the fact that $\mathcal{S}$ splits G with the splitting sequence β.

Thus, Λ is a lattice packing with the shape $\mathcal{S}$. □

Corollary 12.1. *A lattice tiling of $\mathbb{Z}^n$ with the shape $\mathcal{S} \subseteq \mathbb{Z}^n$ exists if and only if there exists an Abelian group G of order $|\mathcal{S}|$ such that $\mathcal{S}$ splits G.*

If our shape $\mathcal{S}$ is not discrete, i.e., cannot be represented as a shape in $\mathbb{Z}^n$, then its tiling might be represented with a lattice, but cannot be represented with a splitting sequence. If, however, our shape $\mathcal{S}$ is in $\mathbb{Z}^n$, then we can use both methods as they were proved to be equivalent. In fact, both methods are complementary. If we consider the matrix $H = [\beta_1 \ \beta_2 \ \cdots \ \beta_n]$, then the vector $X = (x_1, x_2, \ldots, x_n) \in \mathbb{Z}^n$ is contained in the related lattice if and only if $H \cdot X^{\text{tr}} = 0$. Therefore, H has some similarity to a parity-check matrix. The representation of a lattice with its generator matrix seems to be more practical. Sometimes, however, it is not easy to construct one. Moreover, the splitting sequence has, in many cases, a nice structure and from its structure, the general structure of the lattice can be found.

12.2 Crosses and Semi-Crosses

A (k, n)***-semi-cross*** (or a ***corner***) is an n-dimensional shape whose center unit can be considered w.l.o.g. as $\mathbf{0}$ and it has n arms of length k, an arm in each positive direction for each dimension. The i-th arm, $1 \leq i \leq n$, contains the k points $j \cdot \mathbf{e}_i$, $1 \leq j \leq k$.

A (k, n)***-cross*** is an n-dimensional shape whose center unit can be considered w.l.o.g. as $\mathbf{0}$ and it has $2n$ arms of length k, an arm in each direction for each dimension. The two arms in the i-th dimension, $1 \leq i \leq n$, contain the $2k$ points $j \cdot \mathbf{e}_i$, $j \in [-k, k]^- \triangleq [-k, k] \setminus \{0\}$.

An example of a $(2, 3)$-cross and a $(2, 3)$-semi-cross is presented in Fig. 12.1.

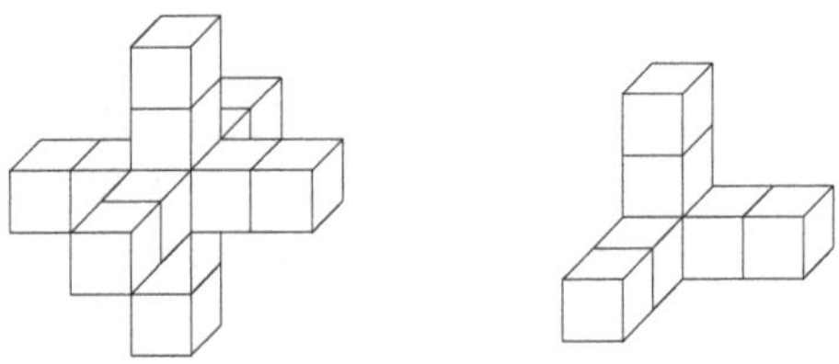

Fig. 12.1 A $(2, 3)$-cross on the left and a $(2, 3)$-semi-cross on the right.

We will be interested to know when there exists a tiling of $\mathbb{Z}^n$ with the cross? when there exist a tiling with the semi-cross? and when such tilings do not exist? These questions are related to 1-perfect codes in the Lee metric since the n-dimensional Lee sphere of radius one is a $(1, n)$-cross.

Theorem 12.1. *If* $n \geq 2$ *and* $k \geq 2n - 1$*, then there is no tiling of* $\mathbb{Z}^n$ *with the* (k, n)*-cross.*

Proof. Assume that $n \geq 2$ and that the (k,n)-cross tiles $\mathbb{Z}^n$ for some k. Consider any translate in $\mathbb{Z}^n$ of the $(k+1)^2$ points in the two-dimensional plane

$$\{(i,j,0,\ldots,0) \ : \ 0 \leq i,j \leq k\}.$$

Two (k,n)-crosses whose centers lie in any translate of such a plane must overlap in the two dimensions of the plane. Hence, each such translate contains at most one center of a cross from the tiling. It is readily verified that there exists a tiling of $\mathbb{Z}^n$ by translates of such a plane. Since each such translate contains at most one center of a the (k,n)-cross, it follows that the density of centers of the crosses, which cannot be larger, is at most $\frac{1}{(k+1)^2}$. The density of the centers of the (k,n)-crosses in a tiling, however, is $\frac{1}{2kn+1}$. Therefore,

$$\frac{1}{2kn+1} \leq \frac{1}{(k+1)^2}$$

which implies that $k \leq 2n-2$, and the theorem is proved. □

Let $S(k) \triangleq \{1,2,\ldots,k\}$ and $F(k) \triangleq \{\pm 1, \pm 2, \ldots, \pm k\}$. The following two theorems are implied by Corollary 12.1.

Theorem 12.2. *There is a lattice tiling of $\mathbb{Z}^n$ by (k,n)-semi-crosses if and only if $S(k)$ splits an Abelian group of order $kn+1$.*

Theorem 12.3. *There is a lattice tiling of $\mathbb{Z}^n$ by (k,n)-crosses if and only if $F(k)$ splits an Abelian group of order $2kn+1$.*

The following lemma is an immediate consequence.

Lemma 12.3. *If the (k,n)-cross tiles $\mathbb{Z}^n$, then the $(k,2n)$-semi-cross tiles $\mathbb{Z}^{2n}$.*

Theorem 12.4. *If $n \geq 2$ and $k > n-1$, then there is no lattice tiling of $\mathbb{Z}^n$ with the (k,n)-cross.*

Proof. The claim is trivial for $n=2$ and hence assume that $n \geq 3$. Assume to the contrary, that $F(k)$ splits the finite Abelian group G with a splitter set $\{s_1, s_2, \ldots, s_n\}$.

We first show that for each integer i, $2 \leq i \leq n$, there exist integers x_i, y_i, such that $k+1 \leq x_i \leq 2n-1$, $|y_i| \leq k$ and $x_i s_1 + y_i s_i = 0$.

For simplicity, let $i=2$ and consider the $2n(k+1)$ elements $a_1 s_1 + a_2 s_2$, $0 \leq a_1 \leq 2n-1$, $0 \leq a_2 \leq k$. Since $2kn+1 < 2n(k+1)$, i.e., the number of elements is greater than the order of G, it follows that there exist two distinct pairs (b_1, b_2) and (c_1, c_2), $0 \leq b_1, c_1 \leq 2n-1$, $0 \leq b_2, c_2 \leq k$ such that $b_1 s_1 + b_2 s_2 = c_1 s_1 + c_2 s_2$. W.l.o.g. assume that $b_1 \geq c_1$ and let $d_1 = b_1 - c_1$ and $d_2 = b_2 - c_2$. Clearly, $d_1 s_1 + d_2 s_2 = 0$, where $(d_1, d_2) \neq (0,0)$, $0 \leq d_1 \leq 2n-1$, and $|d_2| \leq k$. If $0 \leq d_1 \leq k$, then $d_1 s_1 = -d_2 s_2$, which contradicts the fact that s_1 and s_2 are part of a splitter set.

Therefore, for each integer i, $2 \leq i \leq n$, there exists a pair of integers (x_i, y_i) such that $k+1 \leq x_i \leq 2n-1$, $|y_i| \leq k$, and $x_i s_1 + y_i s_i = 0$.

Assume that there are distinct integers i and j such that $x_i = x_j$. Since $x_i s_1 + y_i s_i = 0$ and $x_j s_1 + y_j s_j = 0$, it follows that $y_i s_i = y_j s_j$. This contradicts the condition of the group splitting, unless $y_i = y_j = 0$. This implies that $x_i s_1 = 0$.

Note that the $2k+1$ elements

$$-k \cdot s_1, \ldots, -s_1, 0, s_1, \ldots, k \cdot s_1$$

are distinct since s_1 is an element of a splitter set. Hence, the order of s_1 in G is at least $2k+1$. Since $x_i s_1 = 0$, it follows that the order of s_1 divides x_i and hence $x_i \geq 2k+1$. But, $x_i \leq 2n-1$ and, therefore, $2k+1 \leq 2n-1$, i.e., $k \leq n-1$

If all the x_i's are distinct, then

$$n - 1 \leq 2n - 1 - (k+1) + 1,$$

since all the x_i's lie in the interval $[k+1, 2n-1]$ and hence $k \leq n$. Consider the $(2n-1)(k+1)$ elements of the form $b_1 s_1 + b_2 s_i$, where $0 \leq b_1 \leq 2n-2$, $0 \leq b_2 \leq k$. Now,

$$(2n-1)(k+1) = 2kn + 2n - k - 1 \geq 2kn + n - 1 \geq 2kn + 2 > 2kn + 1 \ .$$

With the same arguments, we can conclude that for each integer i, $2 \leq i \leq n$, there are integers x_i and y_i such that $k+1 \leq x_i \leq 2n-2$, $|y_i| \leq k$, and $x_i s_1 + y_i s_i = 0$.

If all $n-1$ of the x_i's are distinct, then from the fact that they are in the interval $[k+1, 2n-2]$, we have that $n-1 \leq 2n-2-(k+1)+1$, which implies that $k \leq n-1$.

If not all the x_i's are distinct, then we argue as before, starting with $y_i s_i = y_j s_j$. This time we conclude that $2k+1 \leq 2n-2$, and hence $k \leq n-2$, which concludes the proof. □

Theorem 12.4, will be extended later in Theorem 12.12 which uses a similar, but not identical, technique. It will imply another proof for Theorem 12.4.

We would like to know when $S(k)$ splits a group G of order $nk+1$ and when $F(k)$ splits a group G of order $2nk+1$. The motivation is to find when we can tile $\mathbb{Z}^n$ with a cross or a semi-cross. This motivates the rest of this section.

Lemma 12.4. *Let n and k be integers, where $n \geq 3$ and $k \geq n-1$. Assume that $S(k)$ splits an Abelian group G of order $nk+1$. If s and s' are two elements of the splitter set, then one of the following two conditions holds*

(c.1) *There are integers x and y, $1 \leq x \leq n-2$, $1 \leq y \leq k$, such that $xs + ys' = 0$.*

(c.2) *$s' = (1-n)s$ and G is a cyclic group with a generator s.*

Proof. Define a mapping $f : \mathbb{Z} \times \mathbb{Z} \to G$ by $f(i,j) = is + js'$ and let $A \triangleq \{(i,j) \,:\, 0 \leq i \leq n-2,\ 0 \leq j \leq k\}$.

If the mapping $f : A \to G$ is not one-to-one, then there exist integers x and y that satisfy (c.1). If it is one-to-one, note that it is also one-to-one on $B = A \cup \{(n-1,0), (n,0), \ldots, (k,0)\}$. Now, $|B| = (n-1)(k+1) + k - (n-2) = nk+1$. Thus, f, restricted to B, is one-to-one from B onto G. Hence, there exists a tiling of $\mathbb{Z}^2$ by translates of B using the points in the kernel of f as the points of the tiling for the balanced points of the translated copies of B. The $nk+1$ points in B form an L-shaped set, as demonstrated in Fig. 12.2.

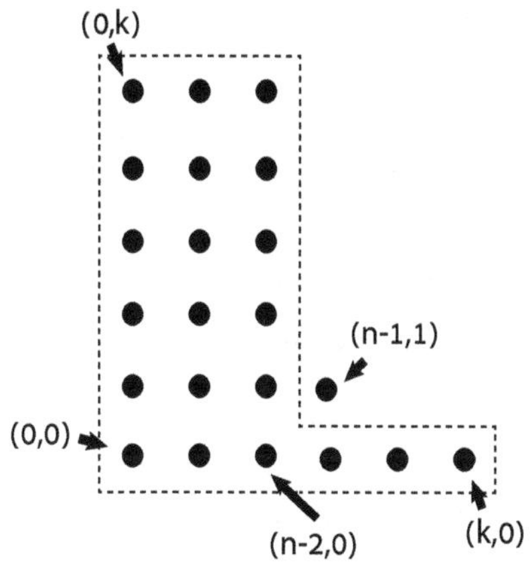

Fig. 12.2 The L-shaped set.

Since translates of B tile $\mathbb{Z}^2$, it follows by analyzing, for example, Fig. 12.2, that the point $(n-1,1)$ is a point in the lattice which tiles the translates of B. In other words, $(n-1,1)$ lies in the kernel of f. This means that $(n-1)s+s'=0$.

Since s and s' generate G and $s'=-(n-1)s$, it follows that s generates G, and, therefore, G is a cyclic group. □

Theorem 12.5. *Let n and k be integers, such that $n \geq 3$. If $S(k)$ splits a group G of order $nk+1$, then $k \leq n-2$.*

Proof. Assume the contrary, that $k \geq n-1$ and $S = \{s_1, s_2, \ldots, s_n\}$ is a splitter set of G, a group of order $nk+1$. For each index j, $2 \leq j \leq n$, consider the two elements s_1 and s_j of S. Assume that each such j, condition (c.1) in Lemma 12.4 holds for s_1 and s_j; that is, there exist x_j and y_j, where $1 \leq x_j \leq n-2$, $1 \leq y_j \leq k$, such that $x_j s_1 + y_j s_j = 0$. Hence, there would be $n-1$ variables $x_1, x_2, \ldots, x_n$ with $n-2$ values from the interval $[1, n-2]$. This implies that two of these values of x_j are equal, say $x_u = x_v$, where $u \neq v$. Therefore, $y_u s_u = y_v s_v$, contradicting the assumption that s_u and s_v are in the splitter set.

Hence, an index j such that condition (c.2) in Lemma 12.4 holds must exist. This implies that G is a cyclic group, s_1 is a generator of G and $(1-n)s_1$ is in the splitter set.

Using the same arguments, with $(1-n)s_1$ playing the role of s_1 for the first of the two elements from the splitter set, implies that $(1-n)(1-n)s_1$ is also in the splitter set. Since s_1 is a generator of G and the order of G is larger than $(n-1)^2$ (since $k \geq n-1$), it follows that the elements s_1 and $(1-n)^2 s_1$ are distinct. Nonetheless, since

$$k(1-n)^2 \equiv k+2-n \pmod{kn+1},$$

it follows that

$$k(1-n)^2 s_1 = (k+2-n)s_1,$$

violating the fact that $(1-n)^2 s_1$ and s_1 are in the splitter set.

Thus, $k \leq n-2$. □

Recall that in the Hamming scheme, for a power of a prime q, a 1-perfect code of length n and its translates can be regarded as a tiling of the $(q-1,n)$-semi-cross. Since there exists such a code for $n=q+1$, it follows that there exists such a tiling for the $(q-1,q+1)$-semi-cross. If $p=q$ is a prime, then there exists such a code that can be described with a lattice tiling over $\mathbb{Z}_p^{p+1}$ that can be expanded to become a lattice over $\mathbb{Z}^{p+1}$. This implies that the bound of Theorem 12.5 cannot be improved in the general case. It can, however, be improved for specific values of n. When q is not a prime, the tiling, implied by a 1-perfect code, cannot be described by a lattice and hence it cannot be described with the group splitting technique. This implies that there are linear 1-perfect codes in the Hamming scheme that can be described by a lattice and that there are linear 1-perfect codes that cannot be described by a lattice. We would like also to know if there are related tilings that cannot be transferred into 1-perfect codes in the Hamming scheme.

12.3 Codes for Nonvolatile Memories and Quasi-Crosses

Flash memory was the fastest growing memory technology at the beginning of the 21st century. Flash memory cells use floating gate technology to store information using trapped charges. By measuring the charge level in a single flash memory cell and comparing it with a predetermined set of threshold levels, the charge level is quantized to one of m values, conveniently chosen to be $\mathbb{Z}_m$. While originally m was limited to 2, and each cell stored a single bit of information, new ***multilevel flash*** memory technology allows much larger values of m, thus storing $\log_2 m$ bits of information in each cell. It should be noted that other alternatives to the conventional multilevel modulation scheme have been suggested, such as, for example, rank modulation.

As is usually the case, the stored charge levels in flash cells suffer from noise that may affect the information retrieved from the cells. Many off-the-shelf coding solutions exist and have been applied for flash memory. The main problem with this approach, however, is the fact that these codes are not tailored to the specific errors occurring in flash memory and thus are wasteful. A more accurate model of the flash memory channel is, therefore, required to enable the design of better-suited codes.

The most notorious property of flash memory is its inherent asymmetry between cell programming, i.e., charge injection into cells and cell erasure, i.e., charge removal from cells. While the former is easy to perform on single cells, the latter works on large blocks of cells and physically damages the cells. Thus, when attempting to reach a target stored value in a cell, charge is slowly injected into the cell over several iterations. If the desired level has not been reached, then another round of charge injection is performed. If, however, the desired charge level has been passed, then there is no way to remove the excess charge from the cell without erasing an entire block of cells. In addition, the actions of cell programming and cell reading disturb adjacent cells by injecting extra unwanted charge into them. Because the careful iterative programming procedure employs small charge-injection steps, it follows that over-programming errors, as well as cell disturbs, are likely to have a bounded magnitude of error.

This technological constraint motivated the application of the asymmetric limited-magnitude error model to the case of flash memory. In this model, a transmitted vector $c \in \mathbb{Z}^n$ is received with an error ϵ as $y = c + \epsilon \in \mathbb{Z}^n$, where we say that e asymmetric limited-magnitude errors occurred with magnitude at most k if the error vector $\epsilon = (\epsilon_1, \ldots, \epsilon_n) \in \mathbb{Z}^n$ satisfies $0 \leq \epsilon_i \leq k$ for all i, and there are exactly e nonzero entries in ϵ.

The main drawback of the asymmetric limited-magnitude error model is the fact that not all error types were considered during the model formulation. Another type of common error in flash memories is due to retention, which is a slow process of charge leakage. As before, the magnitude of errors created by retention is limited; however, unlike over-programming and cell disturbs, which increase cell charge levels, retention errors reduce cell charge levels.

Therefore, a generalization to the error model, which is called ***the unbalanced limited-magnitude*** error model, was suggested. In this model, a transmitted vector $c \in \mathbb{Z}^n$ is now received with an error ϵ as the vector $y = c + \epsilon \in \mathbb{Z}^n$, where it is said that e unbalanced limited-magnitude errors occurred if the error vector $\epsilon = (\epsilon_1, \ldots, \epsilon_n) \in \mathbb{Z}^n$ satisfies $-k_- \leq \epsilon_i \leq k_+$ for all i, and there are exactly e nonzero entries in ϵ. Both k_+ and k_- are nonnegative integers, where we call k_+ the positive-error magnitude limit, and k_- the negative-error magnitude limit.

Henceforth, in this section only single error-correcting codes are considered. In general, assuming at most a single error occurs, the error sphere containing all possible received words $y = c + \epsilon$ forms a shape called a (k_+, k_-, n)***-quasi-cross***. This is a generalization of the (k, n)-semi-cross that is obtained when choosing $k_- = 0$, and the (k, n)-cross that is obtained when choosing $k_+ = k_- = k$. To avoid these two cases considered in Section 12.2, it is assumed in the current section and in the next one that $0 < k_- < k_+$. An example of a $(2, 1, 3)$-quasi-cross is depicted in Fig. 12.3.

In the unbalanced limited-magnitude-error channel model, the transmitted (or stored) word is a vector $v \in \mathbb{Z}^n$. A single error is a vector $\epsilon \in \mathbb{Z}^n$ whose entries are *zeroes* except for a single entry with a value belonging to the set

$$M = \{-k_-, \ldots, -2, -1, 1, 2, \ldots, k_+\}$$

where the integers $0 < k_- < k_+$ are the negative-error and positive-error magnitudes. For convenience we denote this set as $M = [-k_-, k_+]^-$. We define $\beta = k_-/k_+$ and call it the ***balance ratio***, where $0 < \beta < 1$.

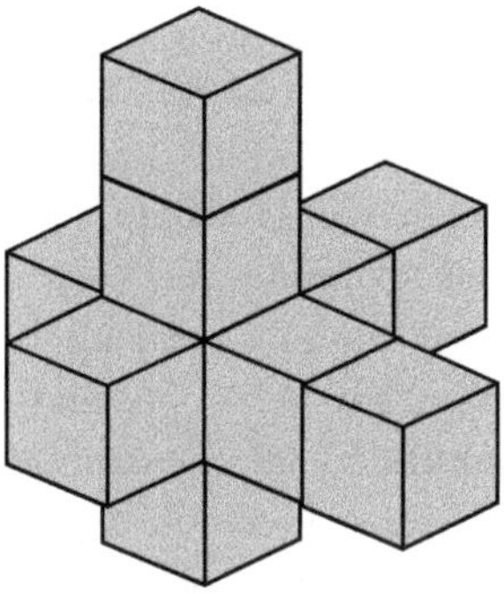

Fig. 12.3 A $(2,1,3)$-quasi-cross.

Given a transmitted vector $v \in \mathbb{Z}^n$, and provided that at most a single error occurred, the received word resides in the error sphere centered at v and is defined by

$$\mathcal{E}(v) \triangleq \{v\} \cup \{v + m \cdot \mathbf{e}_i \ : \ i \in [n], \ m \in M\} \ .$$

The shape $\mathcal{E}(0)$ is called a (k_+, k_-, n)***-quasi-cross***. By translation, $\mathcal{E}(v) = v + \mathcal{E}(0)$ for all $v \in \mathbb{Z}^n$. Let

$$Q \triangleq \{(x_1, \ldots, x_n) \ : \ 0 \leq x_i < 1, \ x_i \in \mathbb{R}\}$$

denote the ***unit cube*** centered at the origin. By abuse of terminology, the set of unit cubes $Q + \mathcal{E}(v)$, will also be called a (k_+, k_-, n)-quasi-cross centered at v for any $v \in \mathbb{Z}^n$. Note that the volume of $Q + \mathcal{E}(v)$ does not depend on the choice of v and it equal to $n(k_+ + k_-) + 1$.

12.4 Tiling with Quasi-Crosses

A v***-modular*** $B_h(M)$ ***sequence***, where $M \subseteq \mathbb{Z}^-$ is a subset S, $S \subseteq \mathbb{Z}_v^-$, of size n is represented by its characteristic vector of length v and weight n. The elements of $S = \{s_1, \ldots, s_n\}$, satisfy that all the sums $\sum_{i=1}^{h} m_i s_{j_i}$ are distinct, where $1 \leq j_1 < j_2 < \cdots < j_h \leq n$, and $m_i \in M$.

Thus, a v-modular $B_1(M)$ sequence is a group splitting of $\mathbb{Z}_v$ defined by M and S. These sequences are defined only by splitting a cyclic group.

When we have a v-modular $B_1(M)$ sequence S, i.e., a group splitting of $\mathbb{Z}_v$ by M and S, and, therefore, an $1 \times n$ parity-check matrix

$H = [s_1, s_2, \ldots, s_n]$, we can construct other packings, provided the elements of M are co-prime to v. This is done by constructing any $k \times n\frac{v^k-1}{v-1}$ parity-check matrix H' containing all distinct column vectors whose top nonzero element is from S. This is equivalent to a splitting of the noncyclic group $\mathbb{Z}_v^k$ by M and S being the columns of H'. Note that if H results in a tiling, then so does H'.

We shall now consider constructions of lattice tilings by (k_+, k_-, n)-quasi-crosses. We first examine the case of a constant balance ratio $0 < \beta < 1$ and show that for any rational β, there exist infinitely many triplets (k_+, k_-, n) such that $\beta = \frac{k_-}{k_+}$ and the (k_+, k_-, n)-quasi-crosses tile $\mathbb{Z}^n$. This is accomplished by constructions for all $k_+ + k_- = p - 1$, where p is a prime. We then focus on a particular case of $(2, 1, n)$-quasi-crosses and show an infinite family of tilings for them.

Construction 12.1. Let $0 < k_- < k_+$ be positive integers such that $k_= + k_- = p - 1$, where p is a prime. We set the multiplier set $M = [-k_-, k_+]^-$. Consider the cyclic group $G = \mathbb{Z}_{p^\ell}$, $\ell \in \mathbb{N}$. We split G using a splitter set S constructed recursively in the following manner:

$$S_1 = \{1\}$$

$$S_{i+1} = pS_i \cup \left\{s \in \mathbb{Z}_{p^{i+1}} \ : \ s \equiv 1 \pmod{p}\right\} .$$

The requested set is $S = S_\ell$.

Theorem 12.6. *The sets S and M from Construction 12.1 split $\mathbb{Z}_{p^\ell}$, forming a tiling by $(k_+, k_-, (p^\ell - 1)/(p-1))$-quasi-crosses and a p^ℓ-modular $B_1(M)$ sequence.*

Proof. The proof is by a simple induction. Obviously, M and $S_1 = \{1\}$ split $\mathbb{Z}_p$. Now assume M and S_i split $\mathbb{Z}_{p^i}$. Let us consider M, S_{i+1}, and $\mathbb{Z}_{p^{i+1}}$. We now show that if $ms = m's'$ in $\mathbb{Z}_{p^{i+1}}$, $m, m' \in M$, $s, s' \in S_{i+1}$, then $m = m'$ and $s = s'$.

For the first case, given any $s \in S_{i+1}$, p does not divide s, and given $m, m' \in M$, $m \neq m'$, since $M = [-k_-, k_+]^-$, it follows that $ms \neq m's$ since they leave different residues modulo p.

For the second case, let $s, s' \in S$, $s' \neq s'$, where p does not divide s, and let $m, m' \in M$, where m and m' are not necessarily distinct. If p divides s', then $ms \neq m's'$ since p does not divide ms but does divide $m's'$. We, therefore, assume that $s' \equiv 1 \pmod{p}$. If we write $s = qp + 1$ and $s' = q'p + 1$, $0 \leq q$, $q' \leq p^i - 1$, then $ms = m's'$ implies that $m = m'$

(by reduction modulo p). This implies that $mqp \equiv mq'p \pmod{p^{i+1}}$, but, $\gcd(m,p)=1$ and so $q \equiv q' \pmod{p^i}$, which (due to the range of q and q') implies that $q=q'$, i.e., $s=s'$.

For the last case, let $s,s' \in pS_i$. We note that the multiples of p in $\mathbb{Z}_{p^{i+1}}$ are isomorphic to $\mathbb{Z}_{p^i}$, and since M and S_i split $\mathbb{Z}_{p^i}$, for all $m,m' \in M$, it follows that if $ms=m's'$, then $m=m'$ and $s=s'$.

Finally, $|M| = p-1$, $|S_\ell| = (p^\ell-1)/(p-1)$, and hence $|M|\cdot|S_\ell|+1 = |\mathbb{Z}_{p^\ell}|$, implying that the splitting induces a tiling. □

Construction 12.2. Let $0 < k_- < k_+$ be positive integers such that $k_+ + k_- = p-1$, where p is a prime. We set the multiplier set $M = [-k_-, k_+]^-$. Consider the additive group of $\mathbb{F}_{p^\ell}$, $\ell \in \mathbb{N}$. Let α be a primitive element in $\mathbb{F}_{p^\ell}$ and define $S \triangleq \{P(\alpha) : P \in M_\ell^p[x]\}$, where $M_\ell^p[x]$ denotes the set of all monic polynomials of degree strictly less than $\ell-1$ over $\mathbb{F}_p$ in the indeterminate x.

Theorem 12.7. *The sets S and M from Construction 12.2 split the additive group of $\mathbb{F}_{p^\ell}$ and form a tiling by $(k_+, k_-, (p^\ell-1)/(p-1))$-quasi-crosses.*

Proof. Since α is primitive in $\mathbb{F}_{p^\ell}$, the elements $1, \alpha, \alpha^2, \ldots, \alpha^{\ell-1}$ form a basis of the additive group of $\mathbb{F}_{p^\ell}$ over $\mathbb{F}_p$. Since $M = \mathbb{F}_p^-$, it is easy to verify that $ms=m's'$, where $m,m' \in M$, $s,s' \in S$, implies that $m=m'$ and $s=s'$. Again, by counting the size of M and S, the splitting induces a tiling. □

Note that a special case of Construction 12.2 is the $[\frac{p^\ell-1}{p-1}, \frac{p^\ell-1}{p-1}-\ell, 3]$ Hamming code over $\mathbb{F}_p$.

From the next example we can observe that the lattice tilings of Construction 12.1 and Construction 12.2 are not equivalent.

Example 12.1. Consider two different 6-dimensional lattice tilings with $(k_+, k_-, n) = (3,1,6)$-quasi-crosses. Using Construction 12.1 we construct a lattice Λ_1 by splitting $\mathbb{Z}_{25}$ and getting a splitter set $S = \{1,5,6,11,16,21\}$, resulting in parity-check matrix

$$H_1 = [1\ 5\ 6\ 11\ 16\ 21]$$

over $\mathbb{Z}_{25}$. This produces a generator matrix for Λ_1

$$G_1 \triangleq \begin{bmatrix} 25 & 0 & 0 & 0 & 0 & 0 \\ 20 & 1 & 0 & 0 & 0 & 0 \\ 19 & 0 & 1 & 0 & 0 & 0 \\ 14 & 0 & 0 & 1 & 0 & 0 \\ 9 & 0 & 0 & 0 & 1 & 0 \\ 4 & 0 & 0 & 0 & 0 & 1 \end{bmatrix}.$$

We confirm that

$$\det G_1 = 25 = 6 \cdot (3+1) + 1 = n(k_+ + k_-) + 1,$$

making Λ_1 a tiling for $(3,1,6)$-quasi-crosses.

On the other hand, we can use Construction 12.2 to construct a lattice Λ_2. We split $\mathbb{F}_{25}$ to get a parity-check matrix

$$H_2 = \begin{bmatrix} 0 & 1 & 1 & 1 & 1 & 1 \\ 1 & 0 & 1 & 2 & 3 & 4 \end{bmatrix}$$

over $\mathbb{F}_5$. The corresponding generator matrix of the lattice Λ_2 is

$$G_2 = \begin{bmatrix} 5 & 0 & 0 & 0 & 0 & 0 \\ 0 & 5 & 0 & 0 & 0 & 0 \\ 4 & 4 & 1 & 0 & 0 & 0 \\ 3 & 4 & 0 & 1 & 0 & 0 \\ 2 & 4 & 0 & 0 & 1 & 0 \\ 1 & 4 & 0 & 0 & 0 & 1 \end{bmatrix}.$$

Again, we confirm that $\det G_2 = 25$. Note that the code $C_2 = \Lambda_2 \cap \mathbb{Z}_5^6$ is the $[6,4,3]$ Hamming code over $\mathbb{F}_5$.

Finally, to show that the lattices are not equivalent, it is readily verified that the minimal period of Λ_1 is $(25,5,25,25,25,25)$, while the minimal period of Λ_2 is $(5,5,5,5,5,5)$. Moreover, one can easily verify that Λ_1 cannot be reduced to a perfect code over $\mathbb{F}_5$.

The following theorem shows that there are infinitely many tilings by quasi-crosses of any given rational balance ratio.

Theorem 12.8. *For any given rational balance ratio* $\beta = \frac{k_-}{k_+}$*, where* $0 < \beta < 1$*, there exists an infinite sequence of quasi-crosses,* $\{(k_+^{(i)}, k_-^{(i)}, n^{(i)})\}_{i=1}^{\infty}$*, such that* $n^{(i)} < n^{(i+1)}$*,* $k_-^{(i)}/k_+^{(i)} = \beta$*, and there exists a tiling by* $(k_+^{(i)}, k_-^{(i)}, n^{(i)})$*-quasi-crosses, for all* $i \in \mathbb{N}$*.*

Proof. Given a rational $0 < \beta < 1$, let $k_+, k_- \in \mathbb{N}$ be such that $k_-/k_+ = \beta$. Denote $d = k_+ + k_-$ and consider the arithmetic progression $1, 1+d, 1+2d, \ldots, 1+id, \ldots$. Since $\gcd(1,d) = 1$, by the well-known Dirichlet's Theorem, the sequence contains infinitely many prime numbers. For any such prime, p, there exists $q \in \mathbb{N}$ such that $q \cdot k_+ + q \cdot k_- = p - 1$. We can apply Construction 12.1 and Construction 12.2 to form tilings by $(q \cdot k_+, q \cdot k_-, n)$-quasi-crosses with the required balance ratio and n unbounded. □

We now consider $(2,1,n)$-quasi-crosses tilings and their associated modular $B_1(M)$ sequences. The following construction is similar in flavor to Construction 12.1.

Construction 12.3. Let $k_+ = 2$, $k_- = 1$, and let the multiplier set be $M = \{-1, 1, 2\}$. Split the group $G = \mathbb{Z}_{4^\ell}$, $\ell \in \mathbb{N}$, using a splitter set S constructed recursively as follows:

$$S_1 = \{1\}$$

$$S_{i+1} = 4S_i \cup \{s \in \mathbb{Z}_{4^{i+1}} \ : \ s \equiv 1 \pmod 2,\ 2s < 4^{i+1}\}.$$

The requested set is $S = S_\ell$.

Theorem 12.9. *The sets S and M defined in Construction 12.3 split $\mathbb{Z}_{4^\ell}$, forming a tiling by $(2,1,(4^\ell - 1)/3)$-quasi-crosses and a 4^ℓ-modular $B_1(M)$ sequence.*

Proof. The proof is by induction, where the basis is formed from the sets M and S_1 which split $\mathbb{Z}_4$. Assume M and S_i split $\mathbb{Z}_{4^i}$ and consider M and S_{i+1}. For convenience, denote

$$S'_{i+1} = \{s \in \mathbb{Z}_{4^{i+1}} \ : \ s \equiv 1 \pmod 2,\ 2s < 4^{i+1}\}.$$

It is easy to verify that due to the restriction $2s < 4^{i+1}$, the elements of S'_{i+1} and the elements of $-S'_{i+1}$ are distinct, and together they contain all the 4^i odd integers in $\mathbb{Z}_{4^{i+1}}$. This implies that the elements of $2S'_{i+1}$ are also distinct and contain all the even integers in $\mathbb{Z}_{4^{i+1}}$ that have a reminder of 2 modulo 4.

We are left with all the multiples of 4 in $\mathbb{Z}_{4^{i+1}}$ that form a group isomorphic to $\mathbb{Z}_{4^i}$, and thus, by the induction hypothesis, are split by M and $4S_i$.

A simple counting argument shows that $|M| = 3$, $|S_\ell| = \frac{4^\ell - 1}{3}$ and, therefore, $|M| \cdot |S_\ell| + 1 = |\mathbb{Z}_{4^\ell}|$. It follows that M and S_ℓ split $\mathbb{Z}_{4^\ell}$ and form a tiling. □

One can further prove the following theorem.

Theorem 12.10. *A lattice tiling of $\mathbb{Z}^n$ with $(2,1,n)$-quasi-cross exists only with the parameters of Construction 12.3.*

We now elaborate on the idea behind Theorem 12.10. We observe that in this case, since the elements of M are not co-prime to 4, extending the 4^ℓ-modular $B_1(M)$ sequence to a parity-check matrix does not produce a valid tiling or even a packing. For example, if we take the trivial 4-modular $B_1(M)$ sequence, $\{1\}$ and attempt to create a parity-check matrix over $\mathbb{Z}_4$

$$H = \begin{bmatrix} 0\,1\,1\,1\,1 \\ 1\,0\,1\,2\,3 \end{bmatrix},$$

then we would find that M together with the columns of H is not a splitting of $\mathbb{Z}_4^2$ since $2 \cdot [1\ 0]^{\text{tr}} = 2 \cdot [1\ 2]^{\text{tr}}$ over $\mathbb{Z}_4$. Hence, the lattice formed by the parity-check matrix H is not a lattice packing of $(2,1,5)$-quasi-crosses.

Example 12.2. To find a tiling by $(2,1,5)$-quasi-crosses using Construction 12.3, we construct a lattice Λ by splitting $\mathbb{Z}_{16}$ with $S = \{1,3,4,5,7\}$. The parity-check matrix and generator matrix are

$$H = \begin{bmatrix} 1\,3\,4\,5\,7 \end{bmatrix},\ G = \begin{bmatrix} 16\ 0\,0\,0\,0 \\ 13\ 1\,0\,0\,0 \\ 12\ 0\,1\,0\,0 \\ 11\ 0\,0\,1\,0 \\ 9\ \ 0\,0\,0\,1 \end{bmatrix}.$$

Unlike Λ_2 from Example 12.1, which turned out to be inferred from a Hamming code, the lattice Λ is not related to 1-perfect code in the Hamming scheme. Its minimal period is $(16,16,4,16,16)$, and it contains a lattice point $(2,0,0,0,2)$ of Hamming weight two.

We now consider bounds on the parameters of lattice tilings by quasi-crosses, expressed in terms of the arm lengths of the quasi-crosses and the dimension of the tiling. Some of the theorems to follow may be viewed as extensions to the theorems on crosses and semi-crosses presented in Section 12.2. The first result is a generalization of Theorem 12.1.

Theorem 12.11. *For any $n \geq 2$, if*

$$\frac{2k_+(k_- + 1) - k_-^2}{k_+ + k_-} > n$$

then there is no tiling of (k_+, k_-, n)-quasi-crosses.

Proof. Given an integer $n \geq 2$, assume there exists a tiling of $\mathbb{Z}^n$ by (k_+, k_-, n)-quasi-crosses. Consider the plane $\{(x, y, 0, \dots, 0) \ : \ x, y \in \mathbb{Z}\}$. Clearly, translates of this plane tile $\mathbb{Z}^n$. Within this plane, we look at the subset

$$A = \{(x, y, 0, \dots, 0) \ : \ 0 \leq x, y \leq k_+ \text{ and } (x \leq k_- \text{ or } y \leq k_-)\}.$$

It is easy to verify that A cannot contain two points from two centers of (k_+, k_-, n)-quasi-crosses, or else the arms of two quasi-crosses would overlap. Thus, the density of the tiling (which is exactly $1/(n(k_+ + k_-) + 1)$) cannot exceed the reciprocal of the volume of A, i.e.,

$$\frac{1}{n(k_+ + k_-) + 1} \leq \frac{1}{(k_+ + 1)^2 - (k_+ - k_-)^2},$$

which implies the desired result. □

Corollary 12.2. *There is no tiling of $\mathbb{Z}^2$ by $(k_+, k_-, 2)$-quasi-crosses.*

Proof. It is easy to verify that for any $0 < k_- < k_+$,

$$\frac{2k_+(k_- + 1) - k_-^2}{k_+ + k_-} > 2.$$

□

The following theorem generalizes Theorem 12.4.

Theorem 12.12. *For any $n \geq 2$, if there exists a lattice tiling of $\mathbb{Z}^n$ by (k_+, k_-, n)-quasi-crosses, then $k_- \leq n - 1$.*

Proof. Let $0 < k_- < k_+$, and let $M = [-k_-, k_+]^-$. Assume there is a group splitting of an Abelian group G by M, with a splitter set $S = \{s_1, \dots, s_n\}$ that induces a lattice tiling with (k_+, k_-, n)-quasi-crosses, which implies that $|G| = n(k_+ + k_-) + 1$.

We first claim that for $2 \leq i \leq n$ there are integers x_i and y_i such that

$$k_+ + 1 \leq x_i \leq \left\lfloor \frac{n(k_+ + k_-) + 1}{k_- + 1} \right\rfloor,$$

$$|y_i| \leq k_-,$$

and

$$s_1 x_i + s_i y_i = 0.$$

To prove this claim we fix i and look at the integers

$$0 \leq a_1 \leq \left\lfloor \frac{n(k_+ + k_-) + 1}{k_- + 1} \right\rfloor, \quad 0 \leq a_2 \leq k_-$$

and the sums $s_1a_1 + s_ia_2$. Since

$$\begin{aligned}&\left(\left\lfloor \frac{n(k_+ + k_-)+1}{k_- + 1} \right\rfloor + 1\right)(k_- + 1)\\&\geq n(k_+ + k_-) + 1 - k_- + k_- + 1\\&= n(k_+ + k_-) + 2 > |G| \ ,\end{aligned}$$

it follows by the pigeonhole principle that there exist two distinct pairs (b_1, b_2), and (c_1, c_2), such that

$$s_1b_1 + s_ib_2 = 0 \qquad s_1c_1 + s_ic_2 = 0.$$

Assume w.l.o.g. that $b_1 \geq c_1$ and define

$$d_1 \triangleq b_1 - c_1 \qquad d_2 \triangleq b_2 - c_2 \ .$$

This implies that $s_1d_1 + s_id_2 = 0$, where $(d_1, d_2) \neq (0, 0)$. In addition,

$$0 \leq d_1 \leq \left\lfloor \frac{n(k_+ + k_-)+1}{k_- + 1} \right\rfloor, \qquad |d_2| \leq k_- \ .$$

If $0 \leq d_1 \leq k_+$ then $s_1d_1 = -s_id_2$, which contradicts the fact that S and M split G. Thus,

$$k_+ + 1 \leq d_1 \leq \left\lfloor \frac{n(k_+ + k_-)+1}{k_- + 1} \right\rfloor,$$

which proves the claim regarding the existence of x_i and y_i.

For the rest of the proof we distinguish between two cases depending on whether there exists a pair $x_i = x_j$ for $i \neq j$ or not.

Case 1. There exist $i \neq j$ such that $x_i = x_j$.

In this case,

$$0 = s_1x_i + s_iy_i = s_1x_j + s_jy_j,$$

which implies that $0 = s_iy_i = s_jy_j$; however, $-k_- \leq y_i,\ y_j \leq k_-$ and to avoid a contradiction to the group splitting, we must have that $y_i = y_j = 0$. This implies that $s_1x_i = 0$. Note that

$$-k_- \cdot s_1, \dots, -2s_1, -s_1, 0, s_1, 2s_1, \dots, k_+ \cdot s_1$$

are all distinct and hence the order of s_1 in G is at least $k_+ + k_- + 1$, but this order also has to divide x_i. Therefore,

$$k_+ + k_- + 1 \leq x_i \leq \left\lfloor \frac{n(k_+ + k_-)+1}{k_- + 1} \right\rfloor .$$

This implies that

$$k_- \leq n - 1 - \frac{k_-}{k_+ + k_-}$$

and since $0 < k_- < k_+$, it follows that $k_- \leq n - 2$.
Case 2. If $i \neq j$, then $x_i \neq x_j$.

Therefore, the number of distinct x_i's is at most the size of their integer interval and, therefore,

$$n - 1 \leq \left\lfloor \frac{n(k_+ + k_-) + 1}{k_- + 1} \right\rfloor - k_+ .$$

This implies that

$$k_- \leq n - 1 + \frac{1}{k_+ - 1} .$$

If $k_+ > 2$, then by the above, we have that $k_- \leq n-1$. If, however, $k_+ = 2$, then $k_- = 1$ and, obviously, $k_- \leq n - 1$. □

Corollary 12.3. *For any $n \geq 3$, if there exists a lattice tiling of $\mathbb{Z}^n$ by (k_+, k_-, n)-quasi-crosses and $k > \frac{n}{2} - 1$, then*

$$k_+ \leq \begin{cases} \frac{3n^2}{8} & n \text{ is even,} \\ \frac{3n^2-4n+1}{4} & n \text{ is odd.} \end{cases}$$

Proof. By Theorem 12.11, a necessary condition for such a lattice tiling to exist is that

$$\frac{2k_+(k_- + 1) - k_-^2}{k_+ + k_-} \leq n,$$

which implies that

$$k_+(2(k_- + 1) - n) \leq k_-^2 + nk_- .$$

If $k_- > \frac{n}{2} - 1$, the left-hand side of this equation is positive and hence

$$k_+ \leq \frac{k_-^2 + nk_-}{2(k_- + 1) - n} .$$

We need to maximize k_+ and by Theorem 12.12 we can restrict ourselves to $k_- \leq n - 1$. The maximum is attained at $k_- = \frac{n}{2}$ for n even, and at $k_- = \frac{n-1}{2}$ for n odd. Substituting the bound on k_+ gives the desired result. □

12.5 Tiling with Notched Cubes

After the discussion on tilings with crosses and semi-crosses in Section 12.2, and the generalization to quasi-crosses in Section 12.4, the concept of semi-cross will be generalized to another concept that can be called a ***chair*** or a ***notched cube***. We present tilings of $\mathbb{Z}^n$ with notched cubes by using generalized splitting first and by using lattice tiling later. The constructed tilings using the two methods is done without applying the translation between the two concepts implied by Lemmas 12.1 and 12.2, and Corollary 12.1.

An n-dimensional notched cube $\mathcal{S}_{L,K} \subset \mathbb{R}^n$, $L = (\ell_1, \ell_2, \ldots, \ell_n)$, $K = (k_1, k_2, \ldots, k_n) \in \mathbb{R}^n$, $0 < k_i < \ell_i$ for each i, $1 \leq i \leq n$, is an n-dimensional $\ell_1 \times \ell_2 \times \cdots \times \ell_n$ box from which an n-dimensional $k_1 \times k_2 \times \cdots \times k_n$ box was removed from one of its corners. Note that this definition implies that the n-dimensional notched cube $\mathcal{S}_{L,K}$ is not necessarily a discrete shape. Formally, it is defined by

$$\mathcal{S}_{L,K} = \{(x_1, x_2, \ldots, x_n) \in \mathbb{R}^n \ : \ 0 \leq x_i < \ell_i \ ,$$
$$\text{and there exists a } j,\ 1 \leq j \leq n, \text{ such that } x_j < \ell_j - k_j\}.$$

The following lemma on the volume of $\mathcal{S}_{L,K}$, $|\mathcal{S}_{L,K}|$, is an immediate consequence of the definition.

Lemma 12.5. *If* $L = (\ell_1, \ell_2, \ldots, \ell_n)$, $K = (k_1, k_2, \ldots, k_n)$ *are two vectors in* $\mathbb{R}^n$, *where* $0 < k_i < \ell_i$ *for each* i, $1 \leq i \leq n$, *then*

$$|\mathcal{S}_{L,K}| = \prod_{i=1}^{n} \ell_i - \prod_{i=1}^{n} k_i \ . \tag{12.2}$$

If $L = (\ell_1, \ell_2, \ldots, \ell_n)$, $K = (k_1, k_2, \ldots, k_n) \in \mathbb{Z}^n$, then the n-dimensional notched cube $\mathcal{S}_{L,K}$ is a discrete shape that is contained in $\mathbb{Z}^n$.

Remark 12.1. It is important to note that if $L = (\ell_1, \ell_2, \ldots, \ell_n)$, $K = (k_1, k_2, \ldots, k_n) \in \mathbb{R}^n$ are two integer vectors, then the two definitions, with the real numbers $\mathbb{R}$ and the integers $\mathbb{Z}$, coincide only if $\mathcal{S}_{L,K}$ is viewed as a collection of n-dimensional unit cubes that are connected by $(n-1)$-dimensional unit cubes, i.e., $\mathcal{S}_{L,K}$ is a discrete shape.

For $n = 2$, if $\ell_1 = \ell_2 = \ell$ and $k_1 = k_2 = \ell - 1$, then the notched cube coincides with a $(k-1, 2)$-semi-cross. Examples of a two-dimensional semi-cross and a three-dimensional notched cube are given in Fig. 12.4.

Now we present a construction of a tiling with n-dimensional notched cubes based on generalized splitting. The n-dimensional notched cubes that

Fig. 12.4 A $(3,2)$-semi-cross (a notched cube with $\ell = 4$) and a three-dimensional notched cube with $L = (5,4,3)$ and $K = (3,3,1)$.

are considered are discrete, i.e., $L, K \in \mathbb{Z}^n$. We start with a construction in which all the ℓ_i's are equal to ℓ, and all the k_i's are equal to $\ell - 1$. We generalize this construction to a case in which all the k_i's, with a possible exception of one, have multiplicative inverses in the related Abelian group.

Lemma 12.6. *Let $n \geq 2$, $\ell \geq 2$, be two integers and let G be the ring of integers modulo $\ell^n - (\ell-1)^n$, i.e., $\mathbb{Z}^{\ell^n-(\ell-1)^n}$. Then,*

(P1) *$\ell - 1$ and ℓ are elements of G^-.*
(P2) *$\alpha = \ell(\ell-1)^{-1}$ is an element of order n in G^-.*
(P3) *$1 + \alpha + \alpha^2 + \cdots + \alpha^{n-1}$ equals* zero *in G.*

Proof.

(P1) By definition, $\ell^n - (\ell-1)^n$ is *zero* in $G = \mathbb{Z}_{\ell^n-(\ell-1)^n}$. We also have that $\ell^n - (\ell-1)^n = \sum_{i=0}^{n-1} \binom{n}{i}(\ell-1)^i = 1 + (\ell-1)\sum_{i=1}^{n-1} \binom{n}{i}(\ell-1)^{i-1}$. It follows that $(\ell-1)(-\sum_{i=1}^{n-1} \binom{n}{i}(\ell-1)^{i-1}) = 1$ in G, and hence, $\ell - 1 \in G^-$. Since $\ell^n - (\ell-1)^n$ is *zero* in G, it follows that $\ell^n = (\ell-1)^n$, and hence $\ell \in G^-$ if and only if $\ell - 1 \in G^-$.

(P2) Clearly, $\alpha^n = \ell^n((\ell-1)^{-1})^n$ and since $\ell^n = (\ell-1)^n$, it follows that $\alpha^n = (\ell-1)^n(\ell-1)^{-n} = 1$. This also implies that α has a multiplicative inverse and hence $\alpha = \ell(\ell-1)^{-1} \in G^-$. Note that for each i, $1 \leq i \leq n-1$, we have $0 < \ell^i - (\ell-1)^i < \ell^n - (\ell-1)^n$. Therefore, $\ell^i \neq (\ell-1)^i$ in G and hence $\alpha^i = \ell^i((\ell-1)^{-1})^i \neq 1$. Thus, the order of α in G^- is n.

(P3) Clearly, $0 = \alpha^n - 1 = (\alpha - 1)(1 + \alpha + \alpha^2 + \cdots + \alpha^{n-1})$. By definition we have that $\alpha = \ell(\ell-1)^{-1}$ and hence $\alpha(\ell-1) = \ell$, $\alpha\ell - \alpha = \ell$, $\alpha - \alpha\ell^{-1} = 1$, $\alpha - 1 = \alpha\ell^{-1}$, $\alpha - 1 = (\ell-1)^{-1}$. Therefore, $0 = (\ell-1)^{-1}(1 + \alpha + \alpha^2 + \cdots + \alpha^{n-1})$, which implies that $1 + \alpha + \alpha^2 + \cdots + \alpha^{n-1} = 0$.

□

Theorem 12.13. *Let $n \geq 2$, $\ell \geq 2$, be two integers, $G = \mathbb{Z}_{\ell^n-(\ell-1)^n}$, and $\alpha = \ell(\ell-1)^{-1}$. Then $\mathcal{S}_{L,K}$, $L = (\ell, \ell, \dots, \ell)$, $K = (\ell-1, \ell-1, \dots, \ell-1)$, splits G with the splitting sequence $\beta = \beta_1, \beta_2, \dots, \beta_n$, defined by*

$$\beta_i = \alpha^{i-1}, \quad 1 \leq i \leq n .$$

Proof. We show by induction that every element in G can be expressed in the form $\mathcal{E} \cdot \beta$, for some $\mathcal{E} \in \mathcal{S}_{L,K}$.

The basis of induction is $0 = \mathbf{0} \cdot \beta$.

For the induction step we have to show that if $x \in G$ can be presented as $x = \mathcal{E} \cdot \beta$ for some $\mathcal{E} \in \mathcal{S}_{L,K}$ (i.e., $\mathcal{E} = (\varepsilon_1, \varepsilon_2, \dots, \varepsilon_n) \in \mathbb{Z}^n$, $0 \leq \varepsilon_i \leq \ell - 1$, $1 \leq i \leq n$, and for some j, $\varepsilon_j = 0$), then also $x+1$ can be presented in the same way. In other words, $x + 1 = \hat{\mathcal{E}} \cdot \beta$, where $\hat{\mathcal{E}} = (\hat{\varepsilon}_1, \hat{\varepsilon}_2, \dots, \hat{\varepsilon}_n) \in \mathcal{S}_{L,K}$.

If $\varepsilon_1 < \ell - 1$ and there exists $j \neq 1$ such that $\varepsilon_j = 0$, then

$$x + 1 = \hat{\mathcal{E}} \cdot \beta,$$

where $\hat{\mathcal{E}} = \mathcal{E} + \mathbf{e}_1$, $0 \leq \hat{\varepsilon}_i \leq \ell - 1$, $\hat{\varepsilon}_j = 0$ and the induction step is proved.

If $\varepsilon_1 = 0$ and there is no $j \neq 1$ such that $\varepsilon_j = 0$, then by (P3) of Lemma 12.6 we have that $\sum_{i=1}^{n} \beta_i = 0$ and hence

$$x + 1 = (\mathcal{E} + \mathbf{e}_1 - \mathbf{1}) \cdot \beta ,$$

i.e., $\hat{\mathcal{E}} = \mathcal{E} + \mathbf{e}_1 - \mathbf{1}$ is the required element of $\mathcal{S}_{L,K}$ and the induction step is proved.

Assume now that $\varepsilon_1 = \ell - 1$. Let j, $2 \leq j \leq n$ be the smallest index such that $\varepsilon_j = 0$.

$$x + 1 = \ell\beta_1 + \sum_{i=2}^{n} \varepsilon_i \beta_i.$$

Note that for each i, $1 \leq i \leq n-1$,

$$\ell\beta_i = \ell\ell^{i-1}((\ell-1)^{-1})^{i-1} = (\ell-1)\ell^i((\ell-1)^{-1})^i = (\ell-1)\beta_{i+1}.$$

Therefore,

$$x + 1 = (\ell - 1 + \varepsilon_2)\beta_2 + \sum_{i=3}^{n} \varepsilon_i \beta_i.$$

If $j = 2$, then $\hat{\mathcal{E}} = (0, \ell-1, \varepsilon_3, \dots, \varepsilon_n)$ and the induction step is proved. If $\varepsilon_2 > 0$, i.e., $j > 2$, then

$$x + 1 = (\varepsilon_2 - 1)\beta_2 + \ell\beta_2 + \sum_{i=3}^{n} \varepsilon_i \beta_i$$

$$= (\varepsilon_2 - 1)\beta_2 + (\ell - 1 + \varepsilon_3)\beta_3 + \sum_{i=4}^{n} \varepsilon_i\beta_i.$$

By iteratively continuing in the same manner we obtain

$$x + 1 = \sum_{i=2}^{j-1}(\varepsilon_i - 1)\beta_i + (\ell - 1 + \varepsilon_j)\beta_j + \sum_{i=j+1}^{n} \varepsilon_i\beta_i$$

and since $\varepsilon_j = 0$, we have that

$$\hat{\mathcal{E}} = (0, \varepsilon_2 - 1, \ldots, \varepsilon_{j-1} - 1, \ell - 1, \varepsilon_{j+1}, \ldots, \varepsilon_n)$$

and the induction step is proved.

Since $|G| = |\mathcal{S}_{L,K}|$, it follows that the set $\{\mathcal{E} \cdot \beta \ : \ \mathcal{E} \in \mathcal{S}_{L,K}\}$ has $|\mathcal{S}_{L,K}|$ elements. □

Corollary 12.4. *For each two integers $n \geq 2$ and $\ell \geq 2$ there exists a lattice tiling of $\mathbb{Z}^n$ with $\mathcal{S}_{L,K}$, $L = (\ell, \ell, \ldots, \ell)$, $K = (\ell - 1, \ell - 1, \ldots, \ell - 1)$.*

The next theorem and its proof are generalizations of Theorem 12.13 and its proof.

Theorem 12.14. *Let $L = (\ell_1, \ell_2, \ldots, \ell_n)$, $K = (k_1, k_2, \ldots, k_n)$ be two vectors in $\mathbb{Z}^n$ such that $0 < k_i < \ell_i$ for each i, $1 \leq i \leq n$. Let $\tau = \prod_{i=1}^{n} \ell_i$, $\kappa = \prod_{i=1}^{n} k_i$, $G = \mathbb{Z}_{\tau-\kappa}$ and assume that for each i, $2 \leq i \leq n$, $k_i \in G^-$. Then $\mathcal{S}_{L,K}$ splits G with the splitting sequence $\beta = \beta_1, \beta_2, \ldots, \beta_n$, defined by*

$$\beta_1 = 1$$
$$\beta_{i+1} = k_{i+1}^{-1}\ell_i\beta_i \quad 1 \leq i \leq n-1 \ .$$

Proof. First we show that $k_1\beta_1 = \ell_n\beta_n$. Since $\tau - \kappa$ equals zero in G, it follows that $\tau = \kappa$ in G and hence $k_1 = \ell_1\ell_2 \cdots \ell_n k_2^{-1}k_3^{-1} \cdots k_n^{-1}$. Therefore,

$$\ell_n\beta_n = \ell_n k_n^{-1}\ell_{n-1}\beta_{n-1} = \ \cdots$$

$$= \ell_n\ell_{n-1} \cdots \ell_1 k_n^{-1}k_{n-1}^{-1} \cdots k_2^{-1}\beta_1 = k_1\beta_1 \ .$$

As an immediate consequence of the definition we have that for each i, $1 \leq i \leq n-1$,

$$\ell_i\beta_i = k_{i+1}\beta_{i+1} \ .$$

We now show that

$$(L - K) \cdot \beta = 0. \tag{12.3}$$

$$(L-K)\cdot\beta=\sum_{i=1}^{n}(\ell_i-k_i)\beta_i=\sum_{i=1}^{n}(\ell_i\beta_i-k_i\beta_i)$$

$$=\ell_n\beta_n-k_n\beta_n+\sum_{i=1}^{n-1}(k_{i+1}\beta_{i+1}-k_i\beta_i)$$

$$=\ell_n\beta_n-k_n\beta_n+k_n\beta_n-k_1\beta_1=0.$$

Since $|\mathcal{S}_{L,K}|=|G|$, it follows that to prove Theorem 12.14, it is sufficient to show that each element in G can be expressed as $\mathcal{E}\cdot\beta$, for some $\mathcal{E}\in\mathcal{S}_{L,K}$. The proof will be done by induction.

The basis of induction is $0=\mathbf{0}\cdot\beta$.

In the induction step we show that if $x\in G$ can be presented as $\mathcal{E}\cdot\beta$ for some $\mathcal{E}\in\mathcal{S}_{L,K}$, then the same is true for $x+1$. In other words, $x+1=\hat{\mathcal{E}}\cdot\beta$, where $\hat{\mathcal{E}}=(\hat{\varepsilon}_1,\hat{\varepsilon}_2,\ldots,\hat{\varepsilon}_n)\in\mathcal{S}_{L,K}$.

Assume

$$x=\mathcal{E}\cdot\beta,$$

where $\mathcal{E}=(\varepsilon_1,\varepsilon_2,\ldots,\varepsilon_n)$, $0\le\varepsilon_i<\ell_i$ for each i, and there exists a j such that $\varepsilon_j<\ell_j-k_j$.

If $\varepsilon_1<\ell_1-k_1-1$ or if $\varepsilon_1<\ell_1-1$ and there exists $j\neq 1$ such that $\varepsilon_j<\ell_j-k_j$, then since $\beta_1=1$ it follows that

$$x+1=\hat{\mathcal{E}}\cdot\beta,$$

where $\hat{\mathcal{E}}=\mathcal{E}+\mathbf{e}_1$. Clearly, $0\le\hat{\varepsilon}_i\le\ell_i-1$; $\hat{\varepsilon}_1<\ell_1-k_1$ if $\varepsilon_1<\ell_1-k_1-1$ and otherwise $\hat{\varepsilon}_j<\ell_j-k_j$. Hence, the induction step is proved.

If $\varepsilon_1=\ell_1-k_1-1$ and there is no $j\neq 1$ such that $\varepsilon_j<\ell_j-k_j$, then by (12.3) we have that $(L-K)\cdot\beta=0$ and hence

$$x+1=(\mathcal{E}+\mathbf{e}_1-(L-K))\cdot\beta\ ,$$

i.e., $\hat{\mathcal{E}}=\mathcal{E}+\mathbf{e}_1-L+K$ is the required element of $\mathcal{S}_{L,K}$ and the induction step is proved.

Assume now that $\varepsilon_1=\ell_1-1$. Let $2\le j\le n$ be the smallest index such that $\varepsilon_j<\ell_j-k_j$.

$$x+1=\ell_1\beta_1+\sum_{i=2}^{n}\varepsilon_i\beta_i=(k_2+\varepsilon_2)\beta_2+\sum_{i=3}^{n}\varepsilon_i\beta_i.$$

If $j = 2$, then $\hat{\mathcal{E}} = (0, k_2 + \varepsilon_2, \varepsilon_3, \ldots, \varepsilon_n)$ and the induction step is proved. If $\varepsilon_2 \geq \ell_2 - k_2$, then

$$x + 1 = \ell_2\beta_2 + (\varepsilon_2 - (\ell_2 - k_2))\beta_2 + \sum_{i=3}^{n} \varepsilon_i\beta_i$$

$$= (\varepsilon_2 - (\ell_2 - k_2))\beta_2 + (k_3 + \varepsilon_3)\beta_3 + \sum_{i=4}^{n} \varepsilon_i\beta_i.$$

By iteratively continuing in the same manner we obtain

$$x + 1 = \sum_{i=2}^{j-1}(\varepsilon_i - (\ell_i - k_i))\beta_i + (k_j + \varepsilon_j)\beta_j + \sum_{i=j+1}^{n} \varepsilon_i\beta_i \ ,$$

and since $\varepsilon_j < \ell_j - k_j$, we have that

$$\hat{\mathcal{E}} = (0, \varepsilon_2 - \ell_2 + k_2, \ldots, \varepsilon_{j-1} - \ell_{j-1} + k_{j-1}, k_j + \varepsilon_j, \varepsilon_{j+1}, \ldots, \varepsilon_n)$$

is the element of $\mathcal{S}_{L,K}$, and the induction step is proved. □

Corollary 12.5. *Let $L = (\ell_1, \ell_2, \ldots, \ell_n)$, $K = (k_1, k_2, \ldots, k_n)$ be two vectors in $\mathbb{Z}^n$ such that $0 < k_i < \ell_i$ for each i, $1 \leq i \leq n$. Let $\tau = \prod_{i=1}^{n} \ell_i$ and assume that $\gcd(k_i, \tau) = 1$ for at least $n - 1$ of the k_i's. Then there exists a lattice tiling of $\mathbb{Z}^n$ with $\mathcal{S}_{L,K}$.*

Next, we consider a lattice tiling of $\mathbb{R}^n$ with $\mathcal{S}_{L,K} \subset \mathbb{R}^n$, where $L = (\ell_1, \ell_2, \ldots, \ell_n)$, $K = (k_1, k_2, \ldots, k_n) \in \mathbb{R}^n$. For the proof of the next theorem we need the following lemma.

Lemma 12.7. *If $X = (x_1, x_2, \ldots, x_n) \in \mathbb{R}^n$, then, $\mathcal{S}_{L,K} \cap (X + \mathcal{S}_{L,K}) \neq \varnothing$ if and only if $|x_i| < \ell_i$, for $1 \leq i \leq n$, and there exist integers j and r, $1 \leq j, r \leq n$, such that $x_j < \ell_j - k_j$ and $-(\ell_r - k_r) < x_r$.*

Proof. Assume first that $\mathcal{S}_{L,K} \cap (X + \mathcal{S}_{L,K}) \neq \varnothing$, i.e., there exists $(a_1, a_2, \ldots, a_n) \in \mathcal{S}_{L,K} \cap (X + \mathcal{S}_{L,K})$. By the definition of $\mathcal{S}_{L,K}$ it follows that

$$0 \leq a_i < \ell_i \ , \quad \text{for each } i, \ 1 \leq i \leq n \ , \tag{12.4}$$

and there exists a j such that

$$a_j < \ell_j - k_j \ . \tag{12.5}$$

Similarly, for $X + \mathcal{S}_{L,K}$ we have

$$x_i \leq a_i < x_i + \ell_i \ , \quad \text{for each } i, \ 1 \leq i \leq n \ , \tag{12.6}$$

and there exists an r such that

$$a_r < x_r + \ell_r - k_r \ . \tag{12.7}$$

It follows from (12.4) and (12.6) that $x_i \le a_i < \ell_i$ and $-\ell_i \le a_i - \ell_i < x_i$ for each i, $1 \le i \le n$. Hence, $|x_i| < \ell_i$ for each i, $1 \le i \le n$. From (12.5) and (12.6) it follows that that $x_j \le a_j < \ell_j - k_j$. Finally, (12.4) and (12.7) imply that $x_r > a_r - (\ell_r - k_r) \ge -(\ell_r - k_r)$.

Now let $X = (x_1, x_2, \dots, x_n) \in \mathbb{R}^n$ such that $|x_i| < \ell_i$ for each i, $1 \le i \le n$, and there exist j, r such that $x_j < \ell_j - k_j$ and $x_r > -(\ell_r - k_r)$. Consider the point $A = (a_1, a_2, \dots, a_n) \in \mathbb{R}^n$, where $a_i = \max\{x_i, 0\}$.

By definition, for each i, $1 \le i \le n$,

$$0 \le a_i < \ell_i$$

and $a_j < \ell_j - k_j$. Hence, $A \in \mathcal{S}_{L,K}$.

Clearly, if $x_i < 0$, then $a_i = 0$ and if $x_i \ge 0$, then $a_i = x_i$. In both cases, since $0 < x_i + \ell_i$, it follows that we have

$$x_i \le a_i < x_i + \ell_i \ .$$

We also have $0 < x_r + \ell_r - k_r$, and, therefore, $x_r \le a_r < x_r + \ell_r - k_r$. Hence, $A \in X + \mathcal{S}_{L,K}$.

Thus, $A \in \mathcal{S}_{L,K} \cap (X + \mathcal{S}_{L,K})$, i.e., $\mathcal{S}_{L,K} \cap (X + \mathcal{S}_{L,K}) \neq \varnothing$. □

The next theorem is a generalization of Corollary 12.5.

Theorem 12.15. *Let $L = (\ell_1, \ell_2, \dots, \ell_n)$, $K = (k_1, k_2, \dots, k_n) \in \mathbb{R}^n$, $0 < k_i < \ell_i$, for all $1 \le i \le n$. If Λ is the lattice generated by the matrix*

$$\mathbf{G} \triangleq \begin{bmatrix} \ell_1 & -k_2 & 0 & 0 & \dots & 0 \\ 0 & \ell_2 & -k_3 & 0 & \dots & 0 \\ \vdots & \vdots & \ddots & \ddots & \ddots & \vdots \\ 0 & \dots & 0 & \ell_{n-2} & -k_{n-1} & 0 \\ 0 & 0 & \dots & 0 & \ell_{n-1} & -k_n \\ -k_1 & 0 & \dots & 0 & 0 & \ell_n \end{bmatrix},$$

then Λ is a lattice tiling of $\mathbb{R}^n$ with $\mathcal{S}_{L,K}$.

Proof. It is easy to verify that $|\det \mathbf{G}| = \prod_{i=1}^n \ell_i - \prod_{i=1}^n k_i$ and hence $V(\Lambda) = |\mathcal{S}_{L,K}|$. We will use Lemma 11.2 to show that Λ is a tiling of $\mathbb{R}^n$ with $\mathcal{S}_{L,K}$. For this, it is sufficient to show that Λ is a packing of $\mathbb{R}^n$ with $\mathcal{S}_{L,K}$.

Let $X \in \Lambda$, $X \neq \mathbf{0}$, and assume to the contrary that $\mathcal{S}_{L,K} \cap (X + \mathcal{S}_{L,K}) \neq \varnothing$. Since $X \in \Lambda$, it follows that there exist integers $\lambda_0, \lambda_1, \lambda_2, \ldots, \lambda_n = \lambda_0$, not all zeros, such that $x_i = \lambda_i \ell_i - \lambda_{i-1} k_i$, for every i, $1 \leq i \leq n$. By Lemma 12.7, we have that for each i, $1 \leq i \leq n$,

$$-\ell_i < \lambda_i \ell_i - \lambda_{i-1} k_i < \ell_i \ ,$$

i.e.,

$$\frac{\lambda_{i-1} k_i}{\ell_i} - 1 < \lambda_i < \frac{\lambda_{i-1} k_i}{\ell_i} + 1 \ .$$

Since λ_i is an integer, it follows that $\lambda_i = \left\lfloor \frac{\lambda_{i-1} k_i}{\ell_i} \right\rfloor$ or $\lambda_i = \left\lceil \frac{\lambda_{i-1} k_i}{\ell_i} \right\rceil$. For each i, $0 \leq i \leq n-1$, if $\lambda_i = \rho \geq 0$, then since $k_{i+1} < \ell_{i+1}$ we have that

$$0 \leq \left\lfloor \frac{\rho k_{i+1}}{\ell_{i+1}} \right\rfloor \leq \lambda_{i+1} \leq \left\lceil \frac{\rho k_{i+1}}{\ell_{i+1}} \right\rceil \leq \rho \ .$$

Hence,

$$0 \leq \lambda_{i+1} \leq \lambda_i \ . \tag{12.8}$$

Similarly, if $\lambda_i \leq 0$, we have that

$$\lambda_i \leq \lambda_{i+1} \leq 0 \ .$$

If $\lambda_0 \geq 0$, then by (12.8) we have

$$\lambda_0 = \lambda_n \leq \lambda_{n-1} \leq \cdots \leq \lambda_1 \leq \lambda_0 \ ,$$

and hence $\lambda_i = \rho$ for each i, $1 \leq i \leq n$. Similarly, we have $\lambda_i = \rho$ for each i, $1 \leq i \leq n$ if $\lambda_0 \leq 0$. If $\rho > 0$, then since ρ is an integer we have that $x_i = \rho(\ell_i - k_i) \geq \ell_i - k_i$, for each i, $1 \leq i \leq n$. Hence, there is no j such that $x_j < \ell_j - k_j$, which contradicts Lemma 12.7. Similarly, if $\rho < 0$, then for each i, $1 \leq i \leq n$, $x_i = \rho(\ell_i - k_i) \leq -(\ell_i - k_i)$, and hence there is no r such that $x_r > -(\ell_r - k_r)$, which contradicts Lemma 12.7. Therefore, $\rho = 0$, i.e., for each i, $0 \leq i \leq n$, $\lambda_i = 0$, a contradiction. Hence, Λ is a lattice packing of $\mathbb{R}^n$ with $\mathcal{S}_{L,K}$

Thus, by Lemma 11.2, Λ is a lattice tiling of $\mathbb{R}^n$ with $\mathcal{S}_{L,K}$. □

Remark 12.2. Note that the construction, implied by Theorem 12.15 and based on lattices, covers all the parameters of integers that are not covered by generalized splitting.

12.6 Notes

Tiling is a geometric concept and a comprehensive excellent book on this topic is [Stein and Szabó (1994)]. Two-dimensional tiling of various shapes are analyzed in [Grünbaum and Shephard (1987)]. The two-dimensional shapes which consists of squares of the same size (unit) connected in their a complete unit edge are called polyominoes. A fascinating book on their related mathematical problems, puzzles, and packings, was written by [Golomb (1996)]. This book was the motivated by the combinatorics of Lee spheres. This book has motivated the popular computer game known as Tetris. Solomon W. Golomb received the American National Medal of Science for his advances in mathematics and communications at an awards ceremony held at the White House. In his speech before the ceremony President Barack Obama mentioned that the combinatorics of Solomon W. Golomb inspired the game of Tetris.

Section 12.1. The idea of group splitting was defined in [Hajós (1942)] and discussed in [Stein (1967a,b); Hickerson (1983); Stein (1984, 1986)]. Generalized splitting was defined in [Buzaglo and Etzion (2013a)]. The methods were used for cross and semi-cross [Stein (1967a, 1984)], and quasi-cross [Schwartz (2012)]. Similar sequences known as $B_h[\ell]$ sequences (and related to v-modular $B_h(M)$ sequences) were defined in [Kløve, Bose, and Elarief (2011)] and further discussed in this paper and in [Kløve, Luo, Naydenova, and Yari (2011)] for construction of codes that correct asymmetric errors with limited magnitude. These $B_h[\ell]$ sequences are modification of the well-known Sidon sequences and their generalisations [O'Bryant (2004)]. Finally, generalized splitting is also a generalization of a method discussed in [Varshamov (1964, 1965)].

Section 12.2. Packings and tilings with semi-crosses were considered in [Stein (1967a, 1984); Hickerson and Stein (1986)] and further analyzed for error-correction of asymmetric errors in flash memories in [Kløve, Bose, and Elarief (2011)]. In [Stein (1967b, 1984)], the packings and tilings of crosses were considered. Theorem 12.1 was given by [Stein (1967b)]. Theorem 12.4 was formulated by [Szabó (1984)] and the proof provided in this section of the chapter was taken from [Stein and Szabó (1994)]. Lemma 12.4 and Theorem 12.5 are the work of [Stein (1985)] with some of the proof done by Hickerson. More recent work of tilings with crosses was done by [Horak and AlBdaiwi (2012a); Horak and Hromada (2014)].

Section 12.3. The analysis for quasi-crosses was presented by [Schwartz

(2012)], where the connection to coding for flash memories was considered. The material in this section was taken from [Schwartz (2012)].

Section 12.4. The term quasi-cross was given by [Schwartz (2012)] who also considered the lattice tilings based on this shape and the results that are presented in this chapter are taken from this paper. Nonexistence results for tilings with quasi-crosses were presented in [Schwartz (2014)]. The quasi-cross was further generalized by [Wei and Schwartz (2020)] to accommodate other shapes that represent errors in the asymmetric limited-magnitude channel. A comprehensive work on the related tilings for these shapes were presented in this paper. Some nonexistence lattice tilings for some of these shapes were proved in [Buzaglo and Etzion (2013a)]. A proof for Dirichlet's Theorem used in Theorem 12.8 can be found in [Selberg (1949)]. A very simple and short proof to the instance of the theorem which was used in the chapter can be found in [Niven and Zuckerman (1980), p. 226]. Finally, Theorem 12.10 was also proved in [Schwartz (2012)].

Section 12.5. The problem of lattice tiling for $\mathbb{Z}^n$ by notched cubes was considered first in [Stein (1990)] and later in [Schmerl (1994)] and [Kolountzakis (1998)]. The tilings presented in this section of the chapter, which are based on generalized splitting and lattice tiling, are due to [Buzaglo and Etzion (2013a)].

There are other interesting shapes that can be considered in this context. The last interesting shape that we want to mention is the $(0.5, n)$-cross, also called a *half-cross*. This shape is like a cross, but instead of having arms of integer length, it has arms of length 0.5 in its n dimensions, one in the negative direction and one in the positive direction. In order to make it a discrete shape, it is scaled by two in each dimension to obtain a shape denoted by Υ_n in which the center of the $(0.5, n)$-cross is transferred into an n-dimensional cube with sides of size two. The size of this cube is 2^n, and each arm of length 0.5 is transferred into 2^{n-1} unit cubes. Therefore, the size of Υ_n is $2^n(n+1)$. Examples of a $(0.5, 3)$-cross and Υ_3 are given in Fig. 12.5.

Tiling $\mathbb{Z}^n$ by half-crosses is highly related to perfect codes. The following results were obtained in [Buzaglo and Etzion (2013b)].

Theorem 12.16. *If $\mathcal{T}$ is an integer lattice tiling with translates of Υ_n, then either $n = 2^t - 1$ or $n = 3^t - 1$ for some $t > 0$.*

Theorem 12.17. *There exists an integer tiling of Υ_n, where $n = 2^t - 1$,* *constructed from a binary 1-perfect code of length* $n = 2^t - 1$. *There exists an integer tiling of* Υ_n, *where* $n = 3^t - 1$, *constructed from a ternary 1-perfect code of length* $\frac{n}{2} = \frac{3^t-1}{2}$. *Lattice tilings, in these cases, are constructed from the related linear codes.*

Corollary 12.6. *The* $(0.5, n)$*-cross has a tiling if and only if either* $n = 2^t - 1$ *or* $n = 3^t - 1$ *for some* $t > 0$.

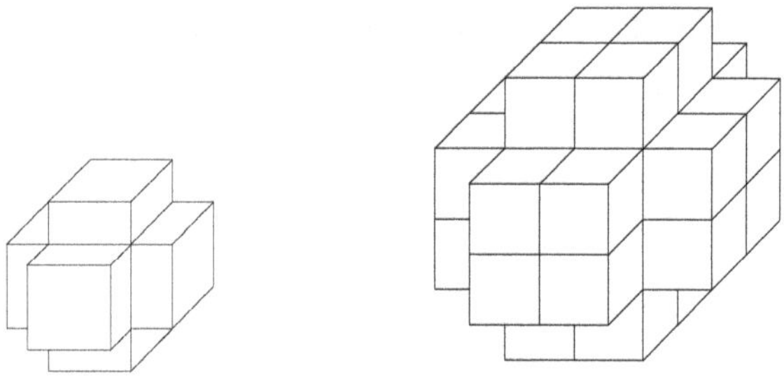

Fig. 12.5 A $(0.5, 3)$-cross on the left side and Υ_3 on the right side.

The half-cross can be viewed as a generalization of a Lee sphere. A sequence of such generalizations for which one is the half-cross were presented in [Etzion (2002)]. The tiling of $\mathbb{Z}^n$ by translates of Υ_n is a perfect dominating set in the n-cube Q_n (which is equivalent to $\mathbb{F}_2^n$). The definition of a perfect dominating set can be found in [Weichsel (1994)] and was further generalized and investigated in [Araujo, Dejter, and Horak (2014)], where the connection to Υ_n was explored.

Chapter 13

Codes in Other Metrics

This chapter is devoted to perfect codes in various metrics that are completely unrelated. Moreover, each metric is completely different from the metrics discussed so far. Each of these metrics has some special properties that motivate its presentation in this book. Section 13.1 is devoted to the deletion channel. If both deletions and insertions are permitted in the channel, then a metric is defined based on the minimum number of deletions and insertions. If only deletions (insertions, respectively) can occur in the channel, there is no metric associated with this scenario, although the set of codes defined by deletions (insertions, respectively) is a subset of the codes defined by both deletion and insertions. Moreover, a code $\mathcal{C}$ corrects e deletions if and only if $\mathcal{C}$ corrects e insertions. Nevertheless, surprisingly, 1-perfect codes with different sizes exist in the deletion channel (but not the insertion channel). The Hamming scheme can be generalized to many different metrics called poset metrics and these metrics are discussed in Section 13.2. All 1-perfect codes for this very large family of metrics will be fully characterized. In computer systems errors can come in bursts. For such errors, burst-correcting codes were designed. Section 13.3 considers perfect codes that can correct one such burst of length two. Other types of bursts are also considered in this section. In Section 13.4 we briefly discuss the Kendall τ-metric defined on the set of permutations S_n. This metric is right distance invariant but not left distance invariant.

13.1 Perfect Deletion-Correcting Codes

The deletion channel is one of the most important channels in coding theory. Assume that a word $x = (x_1, \ldots, x_n)$ of length n, over $\mathbb{Z}_q$, was transmitted and during transmission the symbol at position i was not received, i.e., the received word of length $n-1$ was $(x_1,\ldots,x_{i-1},x_{i+1},\ldots,x_n)$. This is the behavior of the deletion channel, i.e., some symbols can be dropped during transmission. Alternatively, there is a possibility that after $(x_1,\ldots,x_n)$ was transmitted, a new symbol z was inserted between position i and position $i+1$, i.e., the received word of length $n+1$ was $(x_1,\ldots,x_i,z,x_{i+1},\ldots,x_n)$, demonstrating to us the behavior of the insertion channel. The first scenario is for one deletion and the second scenario is for one insertion. Clearly, this can be generalized when in the deletion channel we have more than one deletion and in the insertion channel we have more than one insertion. Both the deletion channel and the insertion channel do not define a metric. This is immediately observed since for a given two words x and y, the word y is not necessarily can be obtained from the word x by a sequence of deletions (or insertions). For example x can be the all-zero word of length n and y can be the set of all-one word of length m. Clearly, y cannot be obtained from x by a sequence of deletions (or insertions). It can be obtained by a sequence of n deletions and m insertions. It is possible, however, to define a metric based on the minimum number of deletions and insertions that are required to change a word x of length n_1 into a word y of length n_2, $0 \leq n_1, n_2 \leq n$ for some integer n (it can be also defined without a restriction on the length of the words). In such a channel, containing both deletions and insertions, we can define a distance on the space $\bigcup_{i=0}^{n} \mathbb{Z}_q^i$, where the distance between $x, y \in \bigcup_{i=0}^{n} \mathbb{Z}_q^i$ is the minimum number of deletions and insertions required to change x into y. This definition of distance is a metric that is interesting on its own and has many applications. It is not difficult to show that there are no nontrivial perfect codes in this metric. Nevertheless, in this section we are interested in and consider only deletion errors, where perfect codes can be defined although no metric is defined. Finally, it is worth to mention that a code $\mathcal{C} \subset \mathbb{F}_q^n$ can correct e deletions if and only if $\mathcal{C}$ can correct e insertions.

In the deletion channel, where only deletion can occur, if a codeword x was sent and a word y was received after a sequence of deletions, our goal is to recover the codeword x from the received word y. There is no deletion metric, but there are deletion balls and deletion spheres. The deletion sphere with radius e centered at $x \in \mathbb{Z}_q^n$, $\mathcal{D}_e(x)$, contains all the words of length $n-e$ that can be obtained from x using exactly e deletions. The deletion ball $\mathcal{B}_e(x)$ is a union of deletion spheres $\mathcal{B}_e(x) \triangleq \bigcup_{i=0}^{e} \mathcal{D}_i(x)$. As a consequence, using simple counting arguments on the covered words of each length, one can easily verify that there is no nontrivial perfect code if at most e deletions occurred. When exactly e deletions occurred, by abuse of notation the deletion ball with radius e around x will be equal to the deletion sphere $\mathcal{D}_e(x)$. Therefore, we define the ***deletion sphere (ball)*** with radius e around a word x of length n to be $\mathcal{D}_e(x)$. A set of words of length n forms a code $\mathcal{C}$ that can correct e deletions if the deletion balls (spheres) with radius e are pairwise disjoint. A code $\mathcal{C}$ of length n is an e-perfect code in the deletion channel if the balls (spheres) with radius e around the codewords of $\mathcal{C}$ form a partition of $\mathbb{Z}_q^{n-e}$. Before presenting a construction for 1-perfect codes, it should be noted that the size of a deletion sphere with radius one around a word x depends on the word x.

In the rest of this section, we restrict ourselves to binary codes. The following code, $\text{VT}_t(n)$, is called the Varshamov–Tenengolts code.

Definition 13.1. For $0 \leq t \leq n$, the code $\text{VT}_t(n)$ consists of binary vectors $(x_1, x_2, \ldots, x_n)$ defined by

$$\text{VT}_t(n) \triangleq \left\{ (x_1, x_2, \ldots, x_n) \in \mathbb{Z}_2^n \ : \ \sum_{i=1}^{n} i \cdot x_i \equiv t \ (\text{mod } n+1) \right\} . \quad (13.1)$$

Theorem 13.1. *For each t, $0 \leq t \leq n$, and each word $y \in \mathbb{Z}_2^{n-1}$, there exists exactly one codeword $x \in VT_t(n)$ such that $y \in \mathcal{D}_1(x)$.*

Proof. The proof is given by presenting a decoding of each received word y of length $n-1$ into the unique codeword x of length n from which one symbol was deleted to receive y.

Suppose a codeword $x = (x_1, x_2, \ldots, x_n) \in \text{VT}_t(n)$ is transmitted, the symbol x_j from the j-th coordinate was deleted, and $y = (y_1, y_2, \ldots, y_{n-1})$ was received. Let ℓ_0 be the number of *zeroes* to the left of the j-th coordinate in x, ℓ_1 be the number of *ones* to the left of the j-th coordinate in x, r_0 be the number of *zeroes* to the right of the j-th coordinate in x, and r_1 be the number of *ones* to the right of the j-th coordinate in x. Clearly, $j = \ell_0 + \ell_1 + 1$, $n = \ell_0 + \ell_1 + 1 + r_0 + r_1$, and $\text{wt}(x) = \ell_1 + r_1 + x_j$.

Let $w = \ell_1 + r_1$ be the weight of y and let $\Upsilon \triangleq \sum_{i=1}^{n-1} i \cdot y_i$. If $x_j = 0$, then this definition implies that

$$\Upsilon = \sum_{i=1}^{n-1} i \cdot y_i = \sum_{i=1}^{j-1} i \cdot y_i + \sum_{i=j}^{n-1} i \cdot y_i = \sum_{i=1}^{j-1} i \cdot x_i + j \cdot x_j + \sum_{i=j+1}^{n} (i-1) \cdot x_i$$

$$= \sum_{i=1}^{n} i \cdot x_i - \sum_{i=j+1}^{n} x_i = \sum_{i=1}^{n} i \cdot x_i - r_1 .$$

Similarly, if $x_j = 1$, then

$$\Upsilon = \sum_{i=1}^{n-1} i \cdot y_i = \sum_{i=1}^{j-1} i \cdot y_i + \sum_{i=j}^{n-1} i \cdot y_i = \sum_{i=1}^{j-1} i \cdot x_i + \sum_{i=j+1}^{n} (i-1) \cdot x_i$$

$$= \sum_{i=1}^{n} i \cdot x_i - j - \sum_{i=j+1}^{n} x_i = \sum_{i=1}^{n} i \cdot x_i - (j + r_1) .$$

Clearly, the sum $j + r_1 = 1 + \ell_0 + \ell_1 + r_1 = 1 + w + \ell_0$ is larger than w, but not larger than $n + 1$.

The difference between $\sum_{i=1}^{n} i \cdot x_i$ (from which (13.1) is computed) and Υ is at most w if $x_j = 0$. This difference is larger than w, but at most $n+1$, if $x_j = 1$. Moreover, difference between different values of Υ is less than $n+1$. This implies that each deletion yields another value of Υ modulo $n + 1$.

Therefore, given any received word $y \in \mathbb{Z}_2^{n-1}$, there exists a unique codeword $x \in \mathbb{Z}_2^n$ from which y was received. □

Corollary 13.1. *For each t, $0 \leq t \leq n$, the code $VT_t(n)$ is a 1-perfect code of length n.*

The proof of Theorem 13.1 can be used to know which codeword x was submitted given the received word y, i.e., to define a decoding algorithm. Since $w = \ell_1 + r_1$, it follows that if the difference between $\sum_{i=1}^{n} i \cdot x_i$ and Υ modulo $n + 1$ is less than or equal to w, then a *zero* was deleted. If the difference is greater than w modulo $n+1$, then a *one* was deleted. Moreover, since this difference is r_1 if $x_j = 0$ and $1 + w + \ell_0$ if $x_j = 1$, it follows that ℓ_0, ℓ_1, r_0, r_1 can be computed and the position of the deleted symbol can be located. Additionally, the definition of the code $\text{VT}_t(n)$, $0 \leq t \leq n$, and Corollary 13.1 imply that the set of 1-perfect codes, $\{\text{VT}_t(n) \ : \ 0 \leq t \leq n\}$ forms a partition of $\mathbb{Z}_2^n$. This is very similar to codes in other metric, where a perfect code and its translates form a partition of the space.

Example 13.1. When $n = 2$, then we have that $\text{VT}_0(2) = \{00, 11\}$, $\text{VT}_1(2) = \{10\}$, and $\text{VT}_2(2) = \{01\}$. Note, the $\text{VT}_0(2)$ is also a 1-perfect insertion code, but there is no other such code for $n \geq 2$.

Problem 13.1. Do there exists other binary 1-perfect deletion-correcting codes?

Problem 13.2. Do there exists q-ary 1-perfect deletion-correcting codes for some $q > 2$?

13.2 Perfect Poset-Correcting Codes

The poset (partially ordered set) metric is a generalization of the Hamming metric. Let $(P, \leq)$ be an arbitrary finite poset, where the number of points in P is n, and the partial order relation is denoted by $\leq$. Let $[n]$ be the the set of points in P. If $A \subseteq [n]$, then $\langle A \rangle$ denotes the smallest ***ideal*** in P that contains A, i.e.,

$$\langle A \rangle \triangleq \{\alpha \, : \, (\exists \beta)(\beta \in A, \; \alpha \leq \beta)\}.$$

Note that the notation is the same as for linear span, but the notation for the smallest ideal is used only in this paragraph. We define the *P-weight* of x, $\text{wt}_P(x)$, $x \in \mathbb{F}_q^n$, to be the cardinality of $\langle \text{supp}(x) \rangle$, i.e., $\text{wt}_P(x) = |\langle \text{supp}(x) \rangle|$. For two words $x, y \in \mathbb{F}_q^n$, the P-distance, $d_P(x, y)$, is defined by $d_P(x, y) = \text{wt}_P(x - y)$. An (n, M, d) P-code $\mathcal{C}$ over $\mathbb{F}_q$ is a subset of size M in $\mathbb{F}_q^n$, such that for each two codewords $x, y \in \mathcal{C}$, we have that $d_P(x, y) \geq d$, i.e., the minimum P-distance of the code is d. If P is an antichain (isolated points), then these definitions coincide with those of the Hamming metric. For each n, there are finitely many posets, but in total there are infinitely many posets to handle and it is impossible to partition them into a finite number of types, where all the posets in one type can be handled together for the analysis of the related perfect codes. We restrict ourselves to 1-perfect codes for all metrics defined by posets.

Each poset defines a metric and also some of them, but not all of them, define association schemes. It is not difficult to verify that some poset metrics do not define a scheme. For example, consider the poset defined on $\mathbb{F}_2^n$ by $P \triangleq [n]$, where $n \geq 4$, $1 \leq 2$, $x \leq x$ for each $x \in P$, and the other pairs of elements are unrelated. Consider the three words $x = (0, 0, 0, \ldots, 0, 0, 0)$, $y = (0, 1, 0, \ldots, 0, 0, 0)$, and $z = (0, 0, 0, \ldots, 0, 1, 1)$. Clearly, $d_P(x, y) = d_P(x, z) = 2$. There is no $u \in \mathbb{F}_2^n$ such that $d_P(u, x) = 1$ and $d_P(u, z) = 2$. If, however, $u = (1, 0, 0, \ldots, 0, 0, 0)$, then $d_P(u, x) = 1$ and $d_P(u, y) = 2$, implying the nonexistence of unique intersecting numbers and hence P is not an association scheme.

As we will see, a 1-perfect code for each poset metric P defined by a poset P, called a ***P-metric***, is easily characterized by the number of words of weight one in P. Let m be the number of such words of weight one in P. Such words are called ***minimal elements*** in the partially ordered set. Each of these m words is associated with exactly one element of $[n]$. For each such element $\alpha \in [n]$ there is no $\beta \in [n]$, $\beta \neq \alpha$, such that $\alpha \geq \beta$.

Lemma 13.1. *The size of a 1-ball, in a P-metric over $\mathbb{F}_q$ with m words of* *weight one, does not depend on the center of the ball, and this size is equal to* $1 + (q-1)m$.

Proof. The fact that the size of the ball does not depend on its center is easily verified from the simple fact that $d_P(x, y) = \text{wt}_P(x - y)$ which implies that the sphere of radius e around a word $x_1 \in \mathbb{F}_q^n$ has the same size as the size of the sphere of radius e around a word $x_2 \in \mathbb{F}_q^n$. Hence, w.l.o.g. we can compute the size of a ball centered at the all-zero word. There is exactly one word with P-weight zero. Each nonzero alphabet symbol can be used in the m positions. There are m positions associated with the words of weight one and $q-1$ nonzero alphabet symbols for a total of $(q-1)m$ words of weight one. Thus, the size of a ball with radius one is $1 + (q-1)m$. □

If a 1-perfect P-code $\mathcal{C}$ exists, then the size of the ball with radius one should divide the size of the space $\mathbb{F}_q^n$, where $|\mathbb{F}_q^n| = q^n$. It follows that $1 + (q-1)m = p^t$, where $q = p^\ell$ and p is a prime. This implies that $(p^\ell - 1)m = p^t - 1$, i.e., $m = \frac{p^t-1}{p^\ell-1}$ and hence ℓ must divides t. Therefore, $p^t = p^{\ell r} = q^r$, i.e., the size of a ball with radius one is $1 + (q-1)m = q^r$ and the size of $\mathcal{C}$ is q^{n-r}. Thus, we have

Theorem 13.2. *If* $\mathcal{C}$ *is a 1-perfect* P*-code over* $\mathbb{F}_q$ *of length* n *with* m *words of weight one in the metric* P*, then* $m = \frac{q^r-1}{q-1}$ *and* $|\mathcal{C}| = q^{n-r}$.

Finally, we characterize 1-perfect P-codes. The first lemma is a simple observation.

Lemma 13.2. *If* $\mathcal{C}$ *is a 1-perfect* P*-code with minimum* P*-distance three, then the balls with radius one centered at the codewords of* $\mathcal{C}$ *are disjoint.*

Theorem 13.3. *Let* $\mathcal{C}$ *be a 1-perfect* P*-code of length* n *with* m *words of weight one in the metric* P*, where* $m = \frac{q^r-1}{q-1}$ *and* $\mathcal{C}$ *has* q^{n-r} *codewords. If the minimum* P*-distance of* $\mathcal{C}$ *is three, then* $\mathcal{C}$ *is a 1-perfect* P*-code.*

Proof. By Lemma 13.2, in a P-code $\mathcal{C}$ with minimum P-distance three, the 1-balls centered at the codewords of $\mathcal{C}$ are disjoint. By Theorem 13.2, we have that a 1-perfect P-code $\mathcal{C}$ of length n has q^{n-r} codewords. Since by Lemma 13.1, the size of a ball with radius one does not depend on its center, it follows that a P-code $\mathcal{C}$ of length n with m words of weight one in the metric P, where $m = \frac{q^r-1}{q-1}$ and $\mathcal{C}$ has q^{n-r} codewords and minimum P-distance three is a 1-perfect code. □

Constructions of a 1-perfect code for a poset P with m words of weight one in the P-metric, is rather simple. We construct a linear 1-perfect Hamming code $\mathcal{C}$ of length m over $\mathbb{F}_q$ on the coordinates that correspond to these m elements of $[n]$ associated with the words of weight one. Let $\mathcal{T}$ be the set of all q^{n-m} words whose support is contained in the $n-m$ points which are not associated with words of weight one. Clearly, $\mathcal{T}$ is a linear code and $\langle \mathcal{C} \rangle \cap \langle \mathcal{T} \rangle = \{\mathbf{0}\}$, where $\langle \mathcal{C} \rangle$ is the linear span of $\mathcal{C}$ and $\langle \mathcal{T} \rangle$ is the linear span of $\mathcal{T}$. The set $\mathcal{C} + \mathcal{T} \triangleq \{c + t \ : \ c \in \mathcal{C}, \ t \in \mathcal{T}\}$ is a 1-perfect code in the metric P. The parity-check matrix of the code consists of m columns of the 1-perfect code over $\mathbb{F}_q$ and $n-m$ columns of *zeroes*.

Example 13.2. Let $q = 2$, $m = 15$, and $n = 25$, and consider any poset P that consists of three points associated with words of weight one with *ones* in positions 1, 4, 6, 7, 8, 11, 14, 15, 16, 18, 20, 21, 22, 23, and 25. The 4×25 parity-check matrix of any such 1-perfect P-code is

$$\begin{bmatrix} 0&0&0&0&0&0&0&0&0&0&0&0&0&0&1&1&0&1&0&1&1&1&1&0&1 \\ 0&0&0&0&0&0&1&1&0&0&1&0&0&1&0&0&0&0&0&0&1&1&1&0&1 \\ 0&0&0&1&0&1&0&0&0&0&1&0&0&1&0&0&0&1&0&1&0&0&1&0&1 \\ 1&0&0&0&0&1&0&1&0&0&0&0&0&1&0&1&0&0&0&1&0&1&0&0&1 \end{bmatrix}.$$

We have chosen to analyze 1-perfect codes on poset metric. The analysis of e-perfect codes, where $e > 1$, in poset metric is more complicated and will not be considered in this section.

Problem 13.3. Can the set of perfect codes in the poset metric be classified in a finite number of types? Any nontrivial classification will be of some interest.

13.3 Perfect Burst-Correcting Codes

In many memory systems an error is not restricted to one position and an error event can cause a burst of errors in adjacent positions. In this section we consider linear codes that can correct one burst whose length is at most b, called a ***b-burst***, i.e., the first symbol and the last symbol in the sequence of errors occur in a window whose size is at most b, where some symbols in this window can be in error and some are not. A code that corrects a b-burst is a code that is capable of correcting a set of errors that occur within b consecutive positions.

A code is called a b-***burst-correcting code*** if it can correct any single burst of length b (or shorter). When the burst can occur between symbols at the end of the codeword and symbols at the start of the codeword, it is called a ***cyclic burst***. A code is called a ***cyclic*** b-***burst-correcting code*** if it can correct any single burst (including a cyclic burst) of length b (or shorter). Combining the definitions of a linear code and a b-burst correcting code, we have the following simple observation.

Lemma 13.3. *A linear code can correct any burst of length at most b if and only if no codeword is the sum of two bursts of length at most b.*

Proof. Assume first that a code $\mathcal{C}$ can correct any burst of length at most b. Assume the contrary, that a codeword $c \in \mathcal{C}$ is the sum of two such bursts β_1 and β_2, i.e. $c = \beta_1 + \beta_2$. If such a burst is the received word, then both $\mathbf{0}$ and c could have been the submitted codewords, a contradiction. Thus, no codeword is a sum of two bursts.

Assume now that in a code $\mathcal{C}$ no codeword is a sum of two bursts of length at most b. Now assume the contrary, that the code cannot correct some burst of length at most b. This implies that there exists a codeword $c \in \mathcal{C}$ and a burst β for which $c + \beta$ cannot be decoded. Hence, there exists another burst β' such that $c + \beta + \beta'$ is another codeword $c' \in \mathcal{C}$. Therefore, $c' - c = \beta + \beta'$ and since $c' - c \in \mathcal{C}$, the sum of the two bursts is a codeword, a contradiction. Thus, the code $\mathcal{C}$ can correct any burst of length at most b. □

Corollary 13.2. *An $[n, k]$ b-burst-correcting code satisfies $n - k \geq 2b$.*

Proof. The column vectors of the $r \times n$ parity-check matrix H contain distinct nonzero column vectors that span a subspace whose dimension is $r = n - k$. Any b columns of H that correspond to a b-burst span a b-subspace. If a nonzero column vector v is in the span of two disjoint b-bursts x and y, then we will not be able to distinguish in which one of them the burst of errors occurred. Hence, two disjoint b-bursts should form two disjoint b-subspaces. Two b-subspaces in a $(2b - 1)$-space cannot be disjoint and, therefore, $n - k = r \geq 2b$. □

For the rest of this section we consider only binary codes. First, we consider the number of possible distinct errors that can be caused by a b-burst. Note that for each b consecutive positions, there are $2^b - 1$ possible errors, which implies that there are $2^b - 1$ different syndromes that are associated with any b-burst in the parity-check matrix of the code. For a binary $[n, n-r]$ cyclic b-burst-correcting code, to count the number of possible errors that can be caused by b-bursts, consider the first position of each possible b-burst. Each of the n positions can be the first position in the b-burst. With this first position in the b-burst, each of the next $b-1$ positions can be in error. Therefore, there are 2^{b-1} possible errors associated with each of these n positions that start a b-burst. Hence, there are $n2^{b-1}$ distinct errors associated with all the cyclic b-bursts, each of which corresponds to a distinct nonzero syndrome of the code's parity-check matrix. Since the number of nonzero vectors of length r is $2^r - 1$, it follows that $2^r - 1 \geq n2^{b-1}$. This inequality, along with the fact that n is an integer, implies that

$$n \leq 2^{r-b+1} - 1. \tag{13.2}$$

A cyclic b-burst-correcting code that attains (13.2) with equality is said to be an ***optimum code***. If $b \geq 2$, then $n2^{b-1} \neq 2^r - 1$, i.e., $n < 2^{r-b+1} - 1$, and hence such a code cannot be perfect. A simple parity-matrix for such a $[2^r - 1, 2^r - r - 2]$ code can be obtained for $b = 2$. For any primitive element α in $\mathbb{F}_{2^r}$ we form the following $(r+1) \times (2^r - 1)$ parity-check matrix.

$$H = \begin{bmatrix} \alpha^0 & \alpha^1 & \alpha^2 & \cdots & \alpha^{2^r-2} \\ 1 & 1 & 1 & \cdots & 1 \end{bmatrix},$$

Note that all column vectors of length $r+1$ ending with a *one* are distinct and the sum of any two distinct adjacent columns ends with a *zero*. These sums are distinct since $\alpha^i + \alpha^{i+1} \neq \alpha^j + \alpha^{j+1}$ for $1 \leq i < j \leq 2^r - 2$. The only column vector of length $r+1$, ending with a *one*, which is not a syndrome associated with a 2-burst is the vector that starts with r *zeroes* and ends with a *one*.

This idea can be easily generalized for an even redundancy $r+2$ and $b = 3$. Let α be a primitive element in $\mathbb{F}_{2^r}$ such that $1 + \alpha \neq \alpha^{3i+2}$ for each i. It is known that such a primitive element always exists. Consider now the following $(r+2) \times (2^r - 1)$ parity-check matrix.

$$H = \begin{bmatrix} \alpha^0 & \alpha^1 & \alpha^2 & \alpha^3 & \alpha^4 & \alpha^5 & \cdots & \alpha^{2^r-4} & \alpha^{2^r-3} & \alpha^{2^r-2} \\ 0 & 1 & 1 & 0 & 1 & 1 & \cdots & 0 & 1 & 1 \\ 1 & 0 & 1 & 1 & 0 & 1 & \cdots & 1 & 0 & 1 \end{bmatrix},$$

To prove that the code $\mathcal{C}$ for which H is its parity-check matrix form a $[2^r - 1, 2^r - r - 3]$ 3-burst-correcting code, we have to show that all the

$2^{r+2} - 4$ cyclic syndromes associated with bursts of length 3 of H are distinct. First, note that $\alpha^i + \alpha^{i+1} \neq \alpha^j + \alpha^{j+1}$, $\alpha^i + \alpha^{i+2} \neq \alpha^j + \alpha^{j+2}$, and $\alpha^i + \alpha^{i+1} + \alpha^{i+2} \neq \alpha^j + \alpha^{j+1} + \alpha^{j+2}$ for $1 \leq i < j \leq 2^r - 2$. All the syndromes ending with two *zeroes* are constructed by summing all combinations of three consecutive columns from H. Next, it can be verified from the cyclic appearances of columns ending with 01, or 10, or 11, that to avoid a repeat in a syndrome we must have that $1 + \alpha = \alpha^\ell$, where $\ell \not\equiv 2 \pmod 3$ and $1 + \alpha^2 = \alpha^\ell$, where $\ell \not\equiv 1 \pmod 3$. These two requirements are equivalent in $\mathbb{F}_{2^r}$ and are satisfied by the choice of α. Thus, $\mathcal{C}$ is an optimum cyclic 3-burst-correcting code. This idea can be generalized further to construct optimum cyclic b-burst-correcting code for each $b > 3$.

For an $[n, n-r]$ b-burst-correcting code (which is not cyclic), there are $(n-b+1)2^{b-1}$ possible distinct errors associated with b-bursts that start in any of the first $n-b+1$ positions. There are another possible $2^{b-1} - 1$ errors associated with the $(b-1)$-burst of the last $b-1$ positions. Therefore, the total number of distinct syndromes should be exactly

$$(n-b+1)2^{b-1} + 2^{b-1} - 1 = (n-b+2)2^{b-1} - 1 \ .$$

Again, since the number of nonzero vectors of length r is $2^r - 1$, it follows that $2^r - 1 \geq (n-b+2)2^{b-1} - 1$, i.e.,

$$n \leq 2^{r-b+1} + b - 2 \ . \tag{13.3}$$

A b-burst-correcting code that attains (13.3) with equality is a ***perfect b-burst-correcting code***. Now, we show that for each $r \geq 5$, there exists a perfect $[2^{r-1}, 2^{r-1} - r]$ 2-burst-correcting code. For a parity-check matrix $H = [h_1, h_2, \cdots, h_n]$ of an $[n, k]$ code, let $\mathcal{S}(H)$ denote the set of syndromes obtained from a single column and two adjacent columns of H, i.e.,

$$\mathcal{S}(H) = \{h_i \ : \ 1 \leq i \leq n\} \cup \{h_i + h_{i+1} \ : \ 1 \leq i \leq n-1\}.$$

By Corollary 13.2 there is no 2-burst-correcting code for redundancy less than 4. It can easily be verified by simple backtracking that an [8,4] 2-burst-correcting code does not exist. Perfect 2-burst-correcting codes for redundancies 5 and 6, i.e., a perfect [16,11] 2-burst-correcting code and a perfect [32,26] 2-burst-correcting code, respectively, were obtained by computer search. Their parity-check matrices are given below.

$$H_5 \triangleq \begin{bmatrix} 1000101010000101 \\ 0100001001101011 \\ 0010101000110010 \\ 0001000110101101 \\ 0000010101010110 \end{bmatrix},$$

$$H_6 \triangleq \begin{bmatrix} 10001001110101110011110010011100 \\ 01000101000110110000100101101111 \\ 00101001000000001011011010101010 \\ 00010100100100000101101011011111 \\ 00000010010010101010111011111110 \\ 00000000001001010101010101010101 \end{bmatrix}.$$

Given an $r \times 2^{r-1}$ parity-check matrix H for a perfect 2-burst-correcting code, we construct the following four matrices.

$$\tilde{H}_1 = \begin{bmatrix} H \\ T_1 \end{bmatrix} \quad \text{where } T_1 = \begin{bmatrix} 1\,1\,1\,1 \cdots 1\,1 \\ 0\,1\,0\,1 \cdots 0\,1 \end{bmatrix}$$

$$\tilde{H}_2 = \begin{bmatrix} H \\ T_2 \end{bmatrix} \quad \text{where } T_2 = \begin{bmatrix} 0\,0\,0\,0 \cdots 0\,0 \\ 0\,0\,0\,0 \cdots 0\,0 \end{bmatrix}$$

$$\tilde{H}_3 = \begin{bmatrix} H \\ T_3 \end{bmatrix} \quad \text{where } T_3 = \begin{bmatrix} 1\,0\,1\,0 \cdots 1\,0 \\ 1\,1\,1\,1 \cdots 1\,1 \end{bmatrix}$$

$$\tilde{H}_4 = \begin{bmatrix} H \\ T_4 \end{bmatrix} \quad \text{where } T_4 = \begin{bmatrix} 0\,1\,0\,1 \cdots 0\,1 \\ 1\,0\,1\,0 \cdots 1\,0 \end{bmatrix}$$

Lemma 13.4. *Each column vector of length $r+2$ that does not start with r zeroes appears exactly once either as a column vector of one of the $\tilde{H}_i$'s or as a sum of two adjacent column vectors from one of the $\tilde{H}_i$'s.*

Proof. The proof follows immediately from the following observations.

- In each position, for the related column vector, the four T_i's have exactly all the four possible 2-tuples.

- In each pair of adjacent positions, the adjacent column vectors in the four T_i's sum exactly to all the four possible 2-tuples.
- Each column vector of length r appears exactly once either as a column or as a sum of two adjacent columns from H.

□

For the $r \times n$ matrix

$$F = \left[f_1 \ f_2 \ f_3 \cdots \ f_{n-2} \ f_{n-1} \ f_n \right],$$

let

$$\mu(F) = \left[f_1 + f_2 \ f_2 \ f_3 \cdots \ f_{n-2} \ f_{n-1} \ f_n \right],$$

$$\gamma(F) = \left[f_1 \ f_2 \ f_3 \cdots \ f_{n-2} \ f_{n-1} \ f_{n-1} + f_n \right],$$

$$\lambda(F) = \left[f_1 + f_2 \ f_2 \ f_3 \cdots \ f_{n-2} \ f_{n-1} \ f_{n-1} + f_n \right].$$

For the $r \times n$ parity-check matrix $H = [h_1 \ h_2 \ \cdots \ h_n]$, let H^R denote the ***reverse*** of the matrix, i.e.,

$$H^R = \left[h_n \ \cdots \ h_2 \ h_1 \right].$$

The following lemma is a simple observation

Lemma 13.5. *For any $r \times n$ parity-check matrix H, where $n \geq 4$ we have that*

$$\mathcal{S}(H) = \mathcal{S}(\mu(H)) = \mathcal{S}(\gamma(H)) = \mathcal{S}(\lambda(H)) = \mathcal{S}(H^R).$$

We are now in a position to state the main result of this section.

Theorem 13.4. *If H is an $r \times 2^{r-1}$ parity-check matrix of a perfect 2-burst-correcting code, then the matrix*

$$\tilde{H} = \left[\tilde{H}_1 \ \gamma(\tilde{H}_2^R) \ \lambda(\tilde{H}_3) \ \mu(\tilde{H}_4^R) \right] \tag{13.4}$$

is an $(r+2) \times 2^{r+1}$ parity-check matrix of a perfect 2-burst-correcting code.

Proof. Clearly, it is sufficient to prove that every nonzero column vector of length $r+2$ is either a column of $\tilde{H}$ or the sum of two adjacent columns of $\tilde{H}$.

Let $\tilde{x} = (x_1, \dots, x_r, x_{r+1}, x_{r+2})$ be a nonzero vector of length $r+2$ and let $x = (x_1, \dots, x_r)$ be the related vector of length r. We distinguish between two cases depending on whether x is the all-zero vector or not.

Case 1. If x is a nonzero vector.

This implies by Lemmas 13.4 and 13.5 that $\tilde{x}^{\text{tr}}$ is either a column of $\tilde{H}$ or the sum of two adjacent columns of $\tilde{H}$.

Case 2. x is the all-zero vector.

- If $(x_{r+1}\ x_{r+2}) = (1\ 1)$, then $\tilde{x}^{\text{tr}}$ is the sum of the last column of $\tilde{H}_1$ and the first column of $\gamma(\tilde{H}_2^R)$.
- If $(x_{r+1}\ x_{r+2}) = (1\ 0)$, then $\tilde{x}^{\text{tr}}$ is the sum of the last column of $\gamma(\tilde{H}_2^R)$ and the first column of $\lambda(\tilde{H}_3)$.
- If $(x_{r+1}\ x_{r+2}) = (0\ 1)$, then $\tilde{x}^{\text{tr}}$ is sum the last column of $\lambda(\tilde{H}_3)$ and the first column of $\mu(\tilde{H}_4^R)$.

□

Combining Theorem 13.4, the perfect [16, 11] 2-burst-correcting code, and the perfect [32, 26] 2-burst-correcting code, respectively, whose parity-check matrices are H_5 and H_6, respectively, we have that

Corollary 13.3. *For each $r \geq 5$, there exists a perfect $[2^{r-1}, 2^{r-1} - r]$ 2-burst-correcting code.*

Having presented perfect 2-burst-correcting codes, it is natural to ask whether there exist perfect b-burst-correcting codes for $b > 2$.

Problem 13.4. Do there exist perfect b-burst-correcting codes for $b > 2$?

As noted before, there are no perfect [8, 4] 2-burst-correcting codes. This leads to another interesting question.

Problem 13.5. Do there exist perfect $[n, n-r]$ b-burst-correcting codes for some $r = 2b$?

Finally, we return to nonbinary codes and consider the following natural question.

Problem 13.6. Generalize the results and the problems in this section for a nonbinary alphabet or prove that such perfect b-burst-correcting codes, where $b > 1$, do not exist.

There are other problems concerning perfect codes which are arising from burst errors. Assume that the channel accepts binary words of a given length n and the most common error is a sequence of positions which are all in errors. In other words, if k errors occur, then they occur in k consecutive locations. Assume further that the errors can be occurred cyclically, i.e., such a sequence of k errors can start at the end of a codeword and can end at the beginning of the codeword. Such a burst of errors will be called a *full-burst.* A code capable of correcting such a burst of up to b errors will be called a *b-full-burst-correcting code* and if it can correct also such cyclic errors, then it will be called a *cyclic b-full-burst-correcting code.* The sphere packing bound for such a linear code is given in the following theorem.

Theorem 13.5. *If C is a cyclic $[n, k]$ code correcting one full-burst of length b then $n \leq \frac{2^{n-k}-1}{b}$.*

Proof. Clearly, the full-burst can start in n possible distinct positions of the codeword and it length can be any integer between 1 and b. Therefore, the number of distinct syndromes must be $n \cdot b$, while the number of possible nonzero syndromes is $2^{n-k} - 1$. Thus, $n \cdot b \leq 2^{n-k} - 1$ and the claim of the theorem follows. □

A cyclic b-full-burst correcting code is a perfect code if it attains the bound of Theorem 13.5 with equality. Are there such perfect codes? Note, that b must be odd and also the length of such a code must be odd. The first set of parameters which form a candidate for such a perfect code is when $n - k$ is even and $b = 3$.

Problem 13.7. Does there exist a perfect cyclic $[n, k]$ b-full-burst-correcting code? In particular does there exist such a code when $b = 3$ and $n - k$ is even?

A second set of possible parameters for a perfect cyclic $[n, k]$ b-full-burst-correcting code is when $n = 2^r + 1$, $b = 2^r - 1$, and $n - k = 2r$. Unfortunately, for such set of parameters there is no perfect code.

Theorem 13.6. *There is no perfect cyclic $[n, k]$ b-full-burst-correcting code when $n = 2^r + 1$ and $b = 2^r - 1$.*

Proof. Assume the contrary that such a perfect code exists and let $H = [h_1 h_2 \cdots h_n]$ be its parity-check matrix. Let $t = \sum_{i=1}^{n} h_i$ and distinguish now between two cases depending whether $t = 0$ or $t \neq 0$.

Case 1. If $t = 0$, then $\sum_{i=1}^{n-2} h_i = h_{n-1} + h_n$ and hence we have a repeated syndrome, a contradiction.

Case 2. If $t \neq 0$, then we can pair the set of syndromes in a set $\{\{s_j, s_\ell\} : s_j + s_\ell = t\}$. Consider $s_1 = h_1$ and the pair $\{s_1, s_i\}$ such that $s_1 + s_i = t$. If the columns of the full-burst associated with s_i contain the column h_1, then the sum of all the columns not in s_i form a full-burst whose syndrome equals to the all-zero vector, a contradiction. Hence, s_i does not contain h_1 which implies that the sum of the ℓ columns which do not contain h_1 and also do not contain any column which are associated with s_i, is the all-zero vector. Now, these columns contains either consecutive columns in H or two sequences of consecutive columns in H. If these columns are consecutive, then they are associated with a full-burst whose syndrome equals to the all-zero vector, a contradiction. If these columns are not consecutive, then they are associated with two full-bursts with the same syndrome, a contradiction.

Thus, the contradictions in the two cases implies that there is no perfect cyclic $[n, k]$ b-full-burst-correcting code when $n = 2^r + 1$ and $b = 2^r - 1$. □

13.4 Notes

The three distance measures which are discussed in this chapter are just a drop in the sea of distance measures. An encyclopedia of distances was written by [Deza and Deza (2009)]. Many of these distance measures are metrics and on many of them perfect codes were discussed throughout the years. As there is no preference to some of them on the other we will not mention any of them. We just say that the field of perfect codes is considerably richer than what was presented in this book. Our target was to introduce the most important metrics with respect to perfect codes. There are many metrics on spaces which were introduced and got lot of attention due to modern application. One such example is the set S_n of all $n!$ permutations on $[n]$. There are several metrics defined on this space. This space has several applications in coding theory, but it got lot of attention in the introduction of flash memories and the rank-modulation scheme introduced in [Jiang, Mateescu, Schwartz, and Bruck (2009)]. Coding for rank modulation was considered in several papers, e.g., [Jiang, Schwartz, and Bruck (2010); Barg and Mazumdar (2010); Tamo and Schwartz (2010); En Gad, Langberg, Schwartz, and Bruck (2011); Mazumdar, Barg, and Zémor (2013); Buzaglo and Etzion (2015); Buzaglo, Yaakobi, Etzion, and Bruck

(2016)]. One of the metric defined on S_n is the Kendall τ-metric.

We denote a permutation $\sigma \in S_n$ by $\sigma = [\sigma(1), \sigma(2), \ldots, \sigma(n)]$. For two permutations $\sigma, \pi \in S_n$, their multiplication $\pi \circ \sigma$ is defined as the composition of σ on π, namely, $\pi \circ \sigma(i) = \sigma(\pi(i))$, for all $1 \leq i \leq n$. Under this operation, the set S_n is a noncommutative group, known as the symmetric group of order $n!$. We denote by $\varepsilon \triangleq [n]$ the identity permutation of S_n. Given a permutation $\sigma \in S_n$, an ***adjacent transposition***, $(i, i+1)$, for some $1 \leq i \leq n-1$, is an exchange of the two adjacent elements $\sigma(i)$ and $\sigma(i+1)$ in σ. The result is the permutation $\pi = [\sigma(1), \ldots, \sigma(i-1), \sigma(i+1), \sigma(i), \sigma(i+2), \ldots, \sigma(n)]$. Observe that the notation $(i, i+1)$ is also used for the cycle decomposition of the permutation $[1, 2, \ldots, i-1, i+1, i, i+2, \ldots, n]$ and the permutation π can also be written as $\pi = (i, i+1) \circ \sigma$. In other words, left multiplication by $(i, i+1)$ exchanges the elements in positions $i, i+1$. Right multiplication by $(i, i+1)$ exchanges the elements $i, i+1$. Two adjacent transpositions $(i, i+1)$ and $(j, j+1)$ are called ***disjoint*** if either $i+1 < j$ or $j+1 < i$. For two permutations $\sigma, \pi \in S_n$, the Kendall τ-distance between σ and π, $d_K(\sigma, \pi)$, is defined as the minimum number of adjacent transpositions needed to transform σ into π. For $\sigma \in S_n$, the Kendall τ-weight of σ, $w_K(\sigma)$, is defined as the Kendall τ-distance between σ and the identity permutation ε. The following expression for $d_K(\sigma, \pi)$ is well known

$$d_K(\sigma, \pi) = |\{(i, j) \ : \ \sigma^{-1}(i) < \sigma^{-1}(j) \wedge \pi^{-1}(i) > \pi^{-1}(j)\}|. \tag{13.5}$$

For a permutation $\sigma = [\sigma(1), \sigma(2), \ldots, \sigma(n)] \in S_n$, the ***reverse*** of σ is the permutation $\sigma^R \triangleq [\sigma(n), \sigma(n-1), \ldots, \sigma(2), \sigma(1)]$. It follows from (13.5) that for every $\sigma, \pi \in S_n$, $d_K(\sigma, \pi) \leq \binom{n}{2}$ and $d_K(\sigma, \pi) = \binom{n}{2}$ if and only if $\pi = \sigma^R$. The following lemma is an immediate consequence from the expression to compute the Kendall τ-distance given in (13.5).

Lemma 13.6. *For every $\sigma, \pi \in S_n$,*

$$d_K(\sigma, \pi) + d_K(\sigma^r, \pi) = d_K(\sigma, \sigma^r) = \binom{n}{2}.$$

The Kendall τ-metric is not left distance invariant, but it is right distance invariant and hence it satisfies the local inequality lemma which implies that it also satisfies the code-anticode bound, which implies that we can consider diameter perfect codes in S_n with the Kendall τ-metric. Several interesting results on perfect codes in S_n with the Kendall τ-metric were proved in [Buzaglo and Etzion (2015)].

Theorem 13.7. *There is no 1-perfect code in S_n with the Kendall τ-metric, where $n > 4$ is a prime or $n \in \{4, 6, 8, 9, 10\}$.*

Problem 13.8. Prove the nonexistence of 1-perfect code in S_n with the Kendall τ-metric for parameters which are not covered by Theorem 13.7.

Problem 13.9. Prove the nonexistence of e-perfect code in S_n with the Kendall τ-metric, where $e > 1$?

Contrary to Theorem 13.7, if we also add the transposition $(n, 1)$ (cyclic adjacent transposition) then there exists a 1-perfect code in S_5 which contains exactly 20 codewords.

$$\begin{array}{llll}
(0,1,2,3,4), & (0,2,4,1,3), & (0,3,1,4,2), & (0,4,3,2,1), \\
(1,2,3,4,0), & (2,4,1,3,0), & (3,1,4,2,0), & (4,3,2,1,0), \\
(2,3,4,0,1), & (4,1,3,0,2), & (1,4,2,0,3), & (3,2,1,0,4), \\
(3,4,0,1,2), & (1,3,0,2,4), & (4,2,0,3,1), & (2,1,0,4,3), \\
(4,0,1,2,3), & (3,0,2,4,1), & (2,0,3,1,4), & (1,0,4,3,2).
\end{array}$$

Theorem 13.8.

- *For each $\sigma \in S_n$, the set $\{\sigma, \sigma^R\}$ is a D-diameter perfect code in S_n with the Kendall τ-metric, where $\sigma \in S_n$ and $D = \binom{n}{2} - 1$.*
- *If $2e+1 = \binom{n}{2}$, then $\{\sigma, \sigma^R\}$ is an e-perfect code in S_n with the Kendall τ-metric, where $\sigma \in S_n$.*

We end this part of our discussion with the following list of research problems.

Problem 13.10. Prove the nonexistence of perfect codes in S_n, using the Kendall τ-metric, for more values of n and/or other distances.

Problem 13.11. Do there exist more D-diameter perfect codes in S_n with the Kendall τ-metric, for $2 \leq D < \binom{n}{2} - 1$? We conjecture that the answer is no.

Problem 13.12. What is the size of a maximum size anticode in S_n with diameter D using the Kendall τ-metric?

Section 13.1. The deletion channel has been considered extensively since it has many applications in storage devices and other areas of information theory. The Varshamov–Tenengolts codes were introduced in [Varshamov and Tenengolts (1965)] for correction of asymmetric errors. It was observed in [Levenshtein (1965a,b)] that these codes can be used for correction of deletions or insertions and they are perfect deletion codes. The deletion channel is one of the most difficult ones to analyze. We do not even know whether the Varshamov–Tenengolts codes are optimal. A comprehensive work on single-deletion-correcting codes was done by [Sloane (2002)]. There are many interesting results in this paper and especially connections to shift-register sequences and some important open problems. It was proved in the paper that for each $2 \leq i \leq n$,

$$\mathrm{VT}_0(n) \geq \mathrm{VT}_i(n) \geq \mathrm{VT}_1(n) \ .$$

Problem 13.13. Prove that the code $\mathrm{VT}_0(n)$ is the largest single-deletion-correcting code of length n, or disprove this conjecture.

Constructions of perfect deletion codes derived from combinatorial designs and especially $(n-2)$-perfect deletion codes of length n, where n is very small, were constructed in [Bours (1995); Mahmoodi (1998); Yin (2001); Shalaby, Wang and Yin (2002); Klein (2004); Wang and Yin (2006); Wang (2008); Chee, Ge and Ling (2010); Wei and Ge (2015)]. Nevertheless, e-perfect codes with small $e > 1$ and large length words are not known.

Finally, the metric based on deletions and insertions, where $d_\ell(x, y)$, $x, y \in \mathbb{Z}_q^n$ is half the minimum number of deletions and insertions required to change x into y is of a great interest. This distance is called the ***Levenstein distance*** (note that in some papers the Levenstein distance is defined differently and it is different from this metric). One interesting property of this metric is that there exist a few balls of radius e that are anticodes with diameter $2e$, but they are not maximal anticodes and, as a consequence, not maximum size anticodes. We remind the reader that the code-anticode bound is not relevant for the deletion channel and hence we will not elaborate more on this phenomenon.

Section 13.2. Posets codes were introduced in [Bruladi, Graves, and Lawrence (1995)] based on previous work [Niederreiter (1987, 1991, 1992)]. A generalization of the posets to directed graphs was considered in [Etzion, Firer and Machado (2018)]. The work of [Bruladi, Graves, and Lawrence (1995)] also started the research on perfect poset codes. They considered the case where the poset is one chain and the case where the poset consists of two chains of equal size. The distance measure defined by a poset is a metric, but it is not always an association scheme [Barg, Felix, Firer, and Spreafico (2014)].

Generally, perfect poset codes of various types were extensively studied. For example, such codes were considered in [Ahn, Kim, Kim, and Kim (2003); Hyun and Kim (2004); Lee (2004); Hyun and Kim (2008); Jang, Kim, Oh, and Rho (2008); Dass, Sharma and Verma (2017); Hyun, Kim, and Park (2019); Dass, Sharma, and Verma (2020)]. The posets in these papers are of different types an the perfect codes have different radii.

Section 13.3. Corollary 13.2, known as the Reiger bound, was proved in [Reiger (1960)]. This bound is a Singleton bound for b-burst correcting codes. It was generalized for two-dimensional codes by [Bossert and Sidorenko (1996)]. Codes that attain this two-dimensional bound for correcting rectangles were constructed in [Boyarinov (2006)]. The first construction of optimum codes was for b-burst-correcting codes, where $b = 2$ and any redundancy $r \geq 3$, was presented in [Abramson (1959)]. This result was extended for $b = 3$ and $b = 4$ by [Elspas and Short (1962)]. In [Abdel-Ghaffar, McEliece, Odlyzko, and van Tliborg (1986)], it was proved that infinitely many optimum codes exist for each $b \geq 2$. As observed, it is clear that if a code attains (13.2), then $2^r - 1 > n2^{b-1}$, i.e., not all nonzero column vectors of length r are syndromes obtained from the bursts of length b. For example, when $b = 2$, not all nonzero column vectors of length r are syndromes obtained either from a single column or from two adjacent columns, i.e., one column vector of length r is not such syndrome. Therefore, these optimum codes are not perfect codes. In [Abdel-Ghaffar, McEliece, Odlyzko, and van Tliborg (1986)] it is also proved that for each even r, there exists a primitive element α in the field $\mathbb{F}_{2^r}$ such that $1 + \alpha \neq \alpha^{3i+2}$ for all i. Generalization for nonbinary alphabet was done in [Abdel-Ghaffar (1988)]. Generalization of optimum codes, which correct cyclic b-bursts, to multidimensional arrays were considered in [Breitbach, Bossert, Zyablov, and Sidorenko (1998); Etzion and Yaakobi (2009)]. The results on perfect b-burst-correcting codes were presented in [Etzion (2001a)]. The b-burst-correcting codes can be generalized in such a way that the bursts are inside bytes. Such perfect codes were considered also in [Etzion (2001a)]. Other optimal codes for correcting single errors and detecting adjacent errors were presented and analyzed in [Etzion (1992); Biberstein and Etzion (2000)]. For more information on byte-oriented error-correcting codes, burst-correcting codes, and their applications, the reader is referred to [Chen (1983, 1986); Rao and Fujiwara (1989)].

Finally, the [16, 11] perfect 2-burst-correcting code and the [32, 26] perfect 2-burst-correcting code presented in [Etzion (2001a)] were found by Marina Biberstein.

Bibliography

A, N. Q., Gyorfi, L., and Massey, J. L. (1992). On the existence of optimum cyclic burst-correcting codes over GF(q), *IEEE Trans. on Infor. Theory* **38**, pp. 940–949.

Abdel-Ghaffar, K. A. S. (1988). On the existence of optimum cyclic burst-correcting codes over GF(q), *IEEE Trans. on Infor. Theory* **34**, pp. 329–332.

Abdel-Ghaffar, K. A. S., McEliece, R. J., Odlyzko, A. M., and van Tliborg, H. C. A. (1986). On the existence of optimum cyclic burst correcting codes, *IEEE Trans. on Infor. Theory* **32**, pp. 768–775.

Abramson, N. M. (1959). A class of systematic codes for nonindependent errors, *IRE Trans. Infor. Theory.* **5**, pp. 150–157.

Ahn. J., Kim, H. K., Kim, J. S., and Kim, M. (2003). Classification of perfect linear codes with crown poset structure *Discrete Math.* **268**, pp. 21–30.

Ahlswede, R., Aydinian, H. K. and Khachatrian L. H. (2001). On perfect codes and related concepts, *Designs, Codes and Crypto.* **22**, pp. 221–237.

Ahlswede, R. and Blinovsky, V. (2008). *Lecture on Advances in Combinatorics* (Springer, Berlin, Heidelberg).

Ahlswede, R. and Khachatrian, L. H. (1997). The complete intersection theorem for systems of finite sets, *European J. Combin.* **18**, pp. 125–136.

Ahlswede, R. and Khachatrian, L. H. (1998). The diametric theorem in Hamming spaces – optimal anticodes, *Advances in Applied Math.* **20**, pp. 429–449.

Albert, A. A. and Sandler, R. (1968). *An Introduction to Finite Projective Planes* (Holt, Reinhart and Winston, London).

Alltop, W. O. (1976). Binary codes with improved minimum weights, *IEEE Trans. on Inform. Theory* **22**, pp. 241–243.

Alon, N., Ben-Eliezer, O., Shangguan, C., and Tamo, I. (2020). The hat guessing number of graphs, *J. of Combin. Theory, Ser. B*, **144**, pp. 119–149.

Amrani, O., Be'ery, Y., Vardy, A., Sun, F.-W., and van Tilborg, H. C. A. (1994). The Leech lattice and the Golay: bounded-distance decoding and multilevel constructions, *IEEE Trans. on Infor. Theory* **40**, pp. 1030–1043.

Anbar, N., Bartoli, D., Giulietti, and Platoni, I. (2014). Small complete caps from singular cubics, *J. of Combin. Designs* **22**, pp. 409–424.

Assmus Jr., E. and Key, J. D. (1990). Affine and projective planes, *Discrete Math.* **83**, pp. 161–187.

Araujo, C., Dejter, I., and Horak, P. (2014). A generalization of Lee codes, *Designs, Codes and Crypto.* **70**, pp. 77–90.

Aravamuthan, S. and Lodha, S. (2006). Covering codes for hats-on-a-line, *The Electronic J. of Combin.* **13**, #R21.

Astola, J. (1982a). An Elias-type bound for Lee codes over large alphabets and its application to perfect codes, *IEEE Trans. Inf. Theory* **28**, pp. 111–113.

Astola, J. (1982b). A note on perfect Lee codes over small alphabets, *Discrete Applied Math.* **4**, pp. 227–228.

Avgustinovich, S. V., Heden, O., and Solovieva F. I. (2006). On intersection problem for perfect binary codes, *Designs, Codes and Crypto.* **39**, pp. 317–322.

Avgustinovich, S. V. and Solovieva, F. I. (1996a). Existence of nonsystematic perfect binary codes, *Proc. 5th Internat, workshop on Algebraic and combinatorial Coding Theory,* pp. 15–19, Cozopol, Bulgaria.

Avgustinovich, S. V. and Solovieva, F. I. (1996b). On nonsystematics perfect binary codes, *Problemy Peredachi Infor.* **32**, pp. 47–50.

Baker, R. D. van Lint, J. H. and Wilson, R. M. (1983). On the Preparata and Goethals codes, *IEEE Trans. on Infor. Theory* **29**, pp. 342–345.

Ball, S. (2012). On sets of vectors of a finite vector space in which every subset of basis size is a basis, *J. European Math. Soc.* **14**, pp. 733–748.

Ball, S. and de Beule, J. (2012). On sets of vectors of a finite vector space in which every subset of basis size is a basis II, *Designs, Codes and Crypto.* **65**, pp. 5–14.

Bannai, E. (1977). Codes in bi-partite distance-regular graphs, *J. London Math. Soc.* **2**, pp. 197–202.

Bannai, E. and Noda, R. (2016). Some bounds for the number of blocks III, *Discrete Math.* **339**, pp. 2313–2328.

Baranyai, Zs. (1975). On the factorization of the complete uniform hypergraph, in *Infinite and Finite sets,* A. Hajnal, T. Rado, and V. T. Sós, Eds., pp. 91–108, 1975 (North-Holland, Amsterdam).

Barg, A. Felix, L. V. Firer, M. and Spreafico, M. V.P. (2010). Linear codes on posets with extension property, *Discrete Math.* **317**, pp. 1–13.

Barg, A. and Mazumdar, A. (2010). Codes in permutations and error correction for rank modulation, *IEEE Trans. on Infor. Theory* **56**, pp. 3158–3165.

Bar-Yahalom, E. and Etzion, T. (1997). Intersection of isomorphic linear codes, *J. Combin. Theory, Ser. A* **80**, pp. 247–256.

Berlkamp, E. R. (1968). *Algebraic Coding Theory* (McGraw-Hill, New York).

Baumert, L. D. (1971). *Cyclic Difference Sets* (Springer, Berlin, Heidelberg, New York).

Belov, B. I., Logachev, V. N., and Sandimirov V. P. (1974). Construction of a class of linear binary codes achieving the Varshamov-Griesmer bound, *Problems of Infor. Transmission* **10**, pp. 211–217.

Bespalov, E. (2020) On the non-existence of extended perfect codes and some perfect colorings, *arxiv.org/abs/2008.13260.*

Best, M. R. (1982). *A contribution for the nonexistence of perfect codes,* Ph.D. thesis, University of Amsterdam, Amsterdam, The Netherlands.

Best, M. R. (1983). Perfect codes hardly exist, *IEEE Trans. on Infor. Theory* **20**, pp. 349–351.

Best, M. R. and Brouwer, A. E. (1977). The triply shortened binary Hamming code is optimal, *Discrete Mathematics* **17**, pp. 235–245.

Beth, T., Jungkickel, D., and Lenz, H. (1999). *Design Theory* (Cambridge Univ. Press, United Kingdom).

Beutelspacher, A. and Ueberberg, J. (1991). A characteristic property of geometric t-spreads in finite projective spaces, *European J. Combin.* **12**, pp. 277–281.

Bhattacharya, S. (2020). Periodicity and decidability of tilings of $\mathbb{Z}^2$, *American J. of Math.* **142**, pp. 255–266.

Biberstein, M. and Etzion, T. (2000). Optimal codes for single-error correction, double-adjacent-error detection, *IEEE Trans. on Infor. Theory* **46**, pp. 2188–2193.

Bierbrauer, J., Marcugini, S., and Pambianco F. (1973). The smallest size of a complete cap in PG(3, 7), *Discrete Math.* **306**, pp. 1257–1263.

Biggs, N. L. (1973). Perfect codes in graphs, *J. Combin. Theory, Ser. B.* **15**, pp. 289–296.

Biggs, N. L. (1974). Perfect codes and distance-transitive graphs, in *Combinatorics*, T. P. McDonough and V. C. Mavron, Eds., London Math. Soc. Lecture Note 13, pp. 1–8 (Cambridge Univ. Press, United Kingdom).

Biggs, N. (1993). *Algebraic Graph Theory* (Cambridge Univ. Press, United Kingdom).

Bitan, S. and Etzion, T. (1995). Constructions for optimal constant weight cyclically permutable codes and difference families, *IEEE Trans. on Infor. Theory* **41**, pp. 77–87.

Blackburn, S. R., Etzion, T., Martin K. M., and Paterson, M. B. (2010). Two-dimensional patterns with distinct differences – constructions, bounds, and maximal anticodes, *IEEE Trans. on Infor. Theory* **56**, pp. 1216–1229.

Blackmore, T. (1999). Every binary $(2^m - 2, 2^{2^m-2-m}, 3)$ code can be lengthened to form a perfect code of length $2^m - 1$, *IEEE Trans. on Inf. Theory* **45**, pp. 698–700.

Blahut, R. E. (1983). *Theory and Practice of Error Control Codes* (Addison-Wesley, Reading, MA).

Blake, I. F. and Mullin R. C. (1975). *The Mathematical Theory of Coding* (Academic Press).

Blanchard, J. L. (1995). A construction for orthogonal arrays with strength $t \geq 3$, *Discrete Math.* **137**, pp. 35–44.

Blaum, M., Bruck, S., and Vardy, A. (1996). MDS arrays code with independent parity symbols, *IEEE Trans. Inf.Theory* **42**, pp. 529–542.

Blaum, J. Bruck, J. and Vardy, A. (1998). Interleaving schemes for multidimensional cluster errors *IEEE Trans. on Infor. Theory* **44**, pp. 730–743.

Blinco, A. D., El-Zanati, S. I., Seelinger, G. F., Sissokho, P. A., Spence, L. E., and Vanden Eynden, C. (2008). On Vector space partitions and uniformly resolvable designs, *Designs, Codes and Crypto.* **48**, pp. 69–77.

Bose, R. C. and Mesner, D. M. (1959). On linear associative algebras corresponding to association schemes of partially balanced designs, *Annals of Mathematical Statistics* **30**, pp. 21–38.

Bose, R. C. and Nair, K. R. (1939). Partially balanced designs, *Sankhya* **4**, pp. 337–372.

Bose, R. C. and Shimamoto, T. (1952). Classification and analysis of partially balanced incomplete block designs with two associate classes, *J. of the American Statistical Association* **47**, pp. 151–184.

Bose, R. C., Shrikhande, S. S., and Parker, E. T. (1960). Further results on the construction of mutually orthogonal Latin squares and the falsity of Euler's conjecture, *Canad. J. Math.* **1** pp. 189–203.

Bossert, M. and Sidorenko, V. (1996). Singleton-type bound for blot-correcting codes, *IEEE Trans. on Infor. Theory* **42**, pp. 1021–1023.

Bours, P. A. H. (1995). On the construction of perfect deletion-correcting codes using design theory, *Designs, Codes and Crypto.* **6**, pp. 5–20.

Boyarinov, I. M. (2006). Two-dimensional array codes correcting rectangular burst errors, *Problems Infor. Transm.* **42**, pp. 26–43.

Braun, M., Etzion, T., Östergård, P. R. J., Vardy, A., and Wassermann, A. (2016). Existence of q-analogs of Steiner systems, *Forum of Mathematics, Pi* **4**, pp. 1–14.

Breitbach, M. Bossert, M. Zyablov, V. and Sidorenko, V. (1998). Array codes correcting a two-dimensional cluster of errors, *IEEE Trans. on Infor. Theory* **44**, pp. 2025–2031.

Brouwer, A. E., Cohen, A. M., and Neumaier, A. (1989). *Distance-Regular Graphs* (Springer-Verlag, Berlin).

Brouwer, A. E., Shearer, J. B., Sloane, N. J. A., and Smith, W. D. (1990). A new table of constant weight codes, *IEEE Trans. on Infor. Theory* **36**, pp. 1334–1380.

Brouwer, A. E. and Tolhuizen, L. M. G. M. (1993). A sharpening of the Johnson bound for binary linear codes and the nonexistence of linear codes with Preparata parameters, *Designs, Codes and Crypto.* **3**, pp. 95–98.

Browning, K. A., Dillon, J. F., McQuistan, M. T., and Wolfe, A. J. (2009). An APN permutation in dimension six, in: G. McGuire, G. L. Mullen, D. Panario, and I. E. Shparlinski, editors, *Finite Fields: Theory and Applications.* Proc. 9th International Conf. on Finite Fields and Applications (Dublin, July 13-17, 2009), Contemporary Mathematics vol. 518, American Mathematical Society, Providence, RI, 2010, pp. 33–42.

Bruck, R. H. and Ryser, H. J. (1949). The non-existence of certain finite projective planes, *Canad. J. Math.* **1**, pp. 88–93.

Bruladi, R. A., Graves, J. S., and Lawrence, K. M. (1991). Codes with poset metric, *Discrete Math.* **147**, pp. 57–72.

Bruladi, R. A., Pless, V. S., and Wilson, R. M. (1989). Short codes with a given covering radius, *IEEE Trans. on Infor. Theory* **35**, pp. 99–109.

Buratti, M. (1993). Improving two theorems of Bose on difference families, *J. of Combin. Designs* **3**, pp. 15–24.

Bush, K. A. (1952). Orthogonl arrays of index unity, *Ann. Math. Stat.* **23**, pp. 426–434.

Butler, S., Hajiaghayi, M. T., Kleinberg, R. D., and Leighton, T. (2009). Hat guessing games, *SIAM Reviews* **51**, pp. 399–413.

Buzaglo, S. and Etzion, T. (2013a). Tilings with n-dimensional chairs and their applications to asymmetric codes, *IEEE Trans. on Infor. Theory* **59**, pp. 1573–1582.

Buzaglo, S. and Etzion, T. (2013b). Tilings by $(0.5, n)$-crosses and perfect codes, *SIAM J. on Discrete Math.* **27**, pp. 1067–1081.

Buzaglo, S. and Etzion, T. (2015). Bounds on the size of permutation codes with the Kendall's τ-metric, *IEEE Trans. on Infor. Theory* **61**, pp. 3241–3250.

Buzaglo, S., Yaakobi, E., Etzion, T., and Bruck, J. (2016). Systematic error-correcting codes for permutations and multi-permutations, *IEEE Trans. on Infor. Theory* **62**, pp. 3113–3124.

Cai, H., Chrisnata, J., Etzion, T., Schwartz, M., and Wachter-Zeh, A. (2020). Network-coding solutions for minimal combination networks and their sub-networks, *IEEE Trans. on Infor. Theory* **66**, pp. 6786–6798.

Cameron, P. (1974a). Generalisation of Fisher's inequality to fields with more than one element, in *Combinatorics*, T. P. McDonough and V. C. Mavron, Eds., London Math. Soc. Lecture Note 13, pp. 9–13 (Cambridge Univ. Press, United Kingdom).

Cameron, P. (1974b). Locally symmetric designs, *Geometriae Dedicata* **3**, pp. 65–76.

Cameron, P. J., Thas, J. A., and Payne, S. E. (1976). Polarities of generalized hexagons and Perfect Codes, *Geometriae Dedicata* **59**, pp. 525–528.

Cao, H., Ji, L. and Zhu, L. (2007). Constructions for generalized Steiner systems GS(3,4,v,2), *Designs, Codes and Crypto.* **45**, pp. 185–197.

Carlet, C., Charpin, P., and Zinoviev, V. (1998). Codes, bent functions and permutations suitable for DES-like cryptosystems, *Designs, Codes and Crypto.* **15**, pp. 125–156.

Chee, Y. M., Dau, S. H., Ling, A. C. H. and Ling, S.(2008). The sizes of optimal q-ary codes of weight three and distance Four: A complete solution, *IEEE Trans. on Infor. Theory* **54**, pp. 1291–1295.

Chee, Y. M., Dau, S. H., Ling, A. C. H., and Ling, S. (2010). Linear size optimal q-ary constant weight codes and constant composition codes, *IEEE Trans. on Infor. Theory* **56**, pp. 140–151.

Chee, Y. M., Etzion, T., Kiah, H. M., and Vardy, A. (2018). Cooling codes: thermal-management coding for high-performance interconnects, *IEEE Trans. on Infor. Theory* **64**, pp. 3062–3085.

Chee, Y. M., Ge, G., and Ling, A. C. H. (2010). Spectrum of sizes for perfect deletion-correcting codes, *SIAM J. on Discrete Math.* **24**, pp. 33–55.

Chee, Y. M., Ge, G., Zhang, H., and Zhang, X. (2015). Hanani triple packings and optimal q-ary codes of constant weight three, *Designs, Codes and Crypto.* **75**, pp. 387–403.

Chee, Y. M. and Ling, S. (2007). Constructions for q-ary constant-weight codes, *IEEE Trans. on Infor. Theory* **53**, pp. 135–146.

Chen, C. L. (1983). Error-correcting codes with byte error detection capability, *IEEE Trans. on Computers* **32**, pp. 615–621.

Chen, C. L. (1986). Byte oriented error-correcting code for semiconductor memory systems, *IEEE Trans. on Computers* **35**, pp. 646–648.

Chen, K., Ge, G., and Zhu, L. (1999). Generalized steiner triple systems with group size five, *J. of Combin. Designs* **7**, pp. 441–452.

Chen, K., Ge, G., and Zhu, L. (2000). Starters and related codes, *J. of Statistical Planning and Inference* **86**, pp. 379–395.

Chihara, L. (1987). On the zeros of the Askey-Wilson polynomials, with applications to coding theory, *SIAM J. Math. Anal.* **6**, pp. 191–207.

Chouinard II, L. G. (1983). Partitions of the 4-subsets of a 13-set into disjoint projective planes, *Discrete Math.* **45**, pp. 297–300.

Chung, H. and Kumar, P. V. (1990). Optical orthogonal codes – New bounds and an optimal construction, *IEEE Trans. on Infor. Theory* **36**, pp. 866–873.

Chung, F. R. K., Salehi, J. A., and Wei, V. K. (1989). Optical orthogonal codes: Design, analysis, and applications, *IEEE Trans. on Infor. Theory* **35**, pp. 595–604.

Cohen, G., Honkala, I., Litsyn, L., and Lobstein A. (1997). *Covering Codes* (North-Holland, Amsterdam).

Cohen, G. D., Karpovsky, M. G., Mattson Jr., H. F., and Schatz, J. R. (1991). Covering radius – survey and recent results, *IEEE Trans. on Infor. Theory* **31**, pp. 328–343.

Cohen, G., Litsyn, S., Vardy, A., and Zémor, G. (1996). Tilings of binary spaces, *SIAM J. on Discrete Math.* **9**, pp. 393–412.

Colbourn, C. J. and Dinitz, J. H. (2007). *Handbook of Combinatorial Designs* (Chapman and Hall/CRC Press, Boca Raton, FL).

Conway, J. and Sloane, N. J. A. (1986). Soft decoding techniques for codes and lattices, including the Golay code and the Leech lattice, *IEEE Trans. on Infor. Theory* **32**, pp. 41–50.

Conway, J. H. and Sloane, N. J. A. (1988). *Sphere Packings, Lattices, and Groups* (Springer-Verlag, New York).

Czerwinski, T. and Oakden, D. (1992). Association schemes and t-designs in regular semilattices, *J. of Combin. Theory, Ser. A* **59**, pp. 193–217.

Danev, D. and Dodunekov, S. (2008). A family of ternary quasi-perfect BCH codes, *Designs, Codes and Crypto.* **49**, pp. 265–271.

Danev, D., Dodunekov, S., and Radkova, D. (2011). A family of constacyclic ternary quasi-perfect codes with covering radius 3, *Designs, Codes and Crypto.* **59**, pp. 111–118.

Dass, B. K., Sharma, N., and Verma, R.(2017). Perfect codes in poset spaces and poset block spaces, *Finite Fields and Their Applications* **46**, pp. 90–106.

Dass, B. K., Sharma, N., and Verma, R. (2020). MDS and I-perfect poset block codes, *Science Direct* **62**.

Davydov, A. A., Faina, G., Marcugini, S., and Pambianco, F. (2009). On sizes of complete caps in projective spaces $\mathrm{PG}(n,q)$ and arcs in $\mathrm{PG}(2,q)$, *J. of Geometry* **94**, pp. 31–58.

Davydov, A. A., Giulietti, M., Marcugini, S., and Pambianco, F. (2010). New inductive constructions of complete caps in PG(N, q), q even, *J. of Combin. Designs* **18**, pp. 177–201.

Davydov, A. A. and Tombak, L. M. (1989a). Quasi-perfect linear binary codes with distance 4 and complete caps in projective geometry, *Probl. Infor. Transm.* **25**, pp. 265–275.

Davydov, A. A. and Tombak, L. M. (1989b). Quasiperfect linear binary codes with distance 4 and complete caps in projective geometry, *Probl. Peredachi Infor.* **25**, pp. 11–23.

Davydov, A. A. and Tombak, L. M. (1991). An alternative to the Hamming code in the class of SEC-DED codes in semiconductor memory, *IEEE Trans. on Infor. Theory* **37**, pp. 897–902.

Delsarte, P. (1973). An algebraic approach to association schemes of coding theory, *Philips J. Res.* **10**, pp. 1–97.

Delsarte, P. (1974). Association schemes and t-designs in regular semilattices, *J. of Combin. Theory, Ser. A* **20**, pp. 230–243.

Delsarte, P. (1978). Bilinear forms over a finite field, with applications to coding theory, *J. of Combin. Theory, Ser. A.* **25**, pp. 226–241.

Delsarte, P. and Goethals, J. M. (1975). Unrestricted codes with the Golay parameters are unique, *Discrete Math.* **12**, pp. 211–224.

Delsarte, P. and Levenshtein V. I. (1998). Association schemes and coding theory, *IEEE Trans. on Infor. Theory* **44**, pp. 2477–2504.

Dembowski, P. (1968). *Finite Geometries* (Springer, Berlin).

Dempwolff, U. (1994). Translation planes of order 27, *Designs, Codes and Crypto.* **27**, pp. 105–121.

Deza, M. M. and Deza, E. (2009). *Encyclopedia of Distances,* (Springer-Verlag, Berlin).

Dinh, H. Q. and López-Permouth, S. R. (2004). Cyclic and negacyclic codes over finite chain rings, *IEEE Trans. on Infor. Theory* **50**, pp. 1728–1744.

Ding, C. (2014). *Codes from Difference Sets* (World Scientific, Singapore).

Drápal, A. (2002). Yet another approach to the extended ternary Golay, *Discrete Math.* **256**, pp. 459–464.

Dodunekov, S. M. and Encheva, S. B. (1993). Uniqueness of some linear sub-codes of the extended binary Golay code, *Problemy Peredachi Informatsii* **29**, pp. 45-51. Translated in: *Problems in Inform. transm.* **29**, pp. 38–43.

Elspas, B. and Short, R. A. (1962). A note on optimum burst-error-correcting codes, *IRE Trans. on Infor. Theory* **8**, pp. 39–42.

El-Zanati, S. I., Heden, O., Seelinger, G. F., Sissokho, P. A., Spence, L. E., and Vanden Eynden, C. (2010). Partitions of the 8-dimensional vector space over GF(2), *J. of Combin. Designs* **18**, pp. 462–474.

El-Zanati, S., Jordon, H., Seelinger, G., Sissokho, P. and Spence, L. (2008). The maximum size of partial 3-spread in a finite vector space over GF(2), *Designs, Codes and Crypto.* **54**, pp. 101–107.

El-Zanati, S. I., Seelinger, G. F., Sissokho, P. A., Spence, L. E., and Vanden Eynden, C. (2007). Partitions of finite vector spaces Into subspaces, *J. of Combin. Designs* **16**, pp. 329–341.

El-Zanati, S. I., Seelinger, G. I., Sissokho, P. I., Spence, L. E. and Vanden Eynden,

C. (2009). On partitions of finite vector spaces of low dimension over GF(2), *Discrete Math.* **309**, pp. 4727–4735.

En Gad, E., Langberg, M., Schwartz, M. and Bruck, J. (2011). Constant-weight Gray codes for local rank modulation, *IEEE Trans. on Infor. Theory* **57**, pp. 7431–7442.

Ericson, T. and Levenshtein, I. L. (1994). Superimposed codes in the Hamming space, *IEEE Trans. on Infor. Theory* **40**, pp. 1882–1893.

Etienne, G. (1987). Perfect codes and regular partitions in graphs and groups, *European J. Combin.* **8**, pp. 139–144.

Etzion, T. (1992). Optimal codes for correcting single errors and detecting adjacent errors, *IEEE Trans. on Infor. Theory* **38**, pp. 1357–1360.

Etzion, T. (1996a). On the nonexistence of perfect codes in the Johnson scheme, *SIAM J. Discrete Math.* **9**, pp. 201–209.

Etzion, T. (1996b). On threshold schemes from large sets, *J. of Combin. Designs* **4**, pp. 323–338.

Etzion, T. (1996c). Nonequivalent q-ary perfect codes, *SIAM J. on Discrete Math.* **9**, pp. 413–423.

Etzion, T. (1997). Optimal constant weight codes over Z_k and generalized designs, *Discrete Math.* **169**, pp. 55–82.

Etzion, T. (1998). Perfect byte-correcting codes, *IEEE Trans. on Infor. Theory.* **44**, pp. 3140–3146.

Etzion, T. (2001a). Constructions for perfect 2-burst-correcting codes, *IEEE Trans. on Infor. Theory* **47**, pp. 2553–2555.

Etzion, T. (2001b). On perfect codes in the Johnson scheme, *Discrete Math. and Theoretical Computer Science* on Codes and Association Schemes **56**, pp. 125–130.

Etzion, T. (2002). Tilings with generalized Lee spheres, in *Mathematical Properties of Sequences and other Combinatorial Structures*, J.-S No, H.-Y Song, T. Helleseth, and P. V. Kumar, Eds., New York: Springer-Verlag, 2002, pp. 181–198.

Etzion, T. (2007). Configuration distribution and designs of codes in the Johnson scheme, *J. of Combin. Designs* **15**, pp. 15–34.

Etzion, T. (2011). Product constructions for perfect Lee codes, *IEEE Trans. on Infor. Theory* **57**, pp. 7473–7481.

Etzion, T. (2021). Non-binary diameter perfect constant-weight codes, *IEEE Trans. on Infor. Theory*, to appear.

Etzion, T., Firer, M., and Machado, A. (2018). Metrics based on finite directed graphs and coding invariants, *IEEE Trans. on Infor. Theory* **64**, pp. 2398–2409.

Etzion, T., Gorla, E., Ravagnani, A., and Wachter-Zeh, A. (2016). Optimal Ferrers diagram rank-metric codes, *IEEE Trans. on Infor. Theory* **62**, pp. 1616–1630.

Etzion, T. and Greenberg, G. (1993). Constructions for perfect mixed Codes and other covering codes, *IEEE Trans. on Infor. Theory.* **39**, pp. 209–214.

Etzion, T. and Hartman, A. (1991). Towards a large set of Steiner quadruple systems, *SIAM J. on Discrete Math.* **4**, pp. 182–195.

Etzion, T. and Hooker, N. (2018). Residual q-Fano planes and related structures, *The Electronics J. of Combin.* #P2.3.

Etzion, T. and Mounits, B. (2005). Quasi-perfect codes with small distance, *IEEE Trans. on Infor. Theory* **51**, pp. 3938–3946.

Etzion, T. and Schwartz, M. (2004) Perfect constant-weight codes, *IEEE Trans. on Infor. Theory* **50**, pp. 2156–2165.

Etzion, T. and Silberstein, N. (2009). Error-correcting codes in projective spaces via rank-metric Codes and Ferrers Diagrams, *IEEE Trans. on Infor. Theory* **55**, pp. 2909–2919.

Etzion, T. and Silberstein, N. (2013). Codes and designs related to lifted MRD codes, *IEEE Trans. on Infor. Theory* **59**, pp. 1004–1017.

Etzion, T. and Storme, L. (2016). Galois geometries and coding theory, *Designs, Codes and Crypto.* **78**, pp. 311–350.

Etzion, T. and van Lint, J. (2001). On perfect constant weight codes, *Discrete Math. and Theoretical Computer Science* on Codes and Association Schemes **56**, pp. 131–134.

Etzion, T. and Vardy, A. (1994). Perfect binary codes: constructions, properties, and enumeration, *IEEE Trans. on Infor. Theory* **40**, pp. 754–763.

Etzion, T. and Vardy, A. (1998). On perfect codes and tilings: problems and solutions, *SIAM J. on Discrete Math.* **11**, pp. 203–223.

Etzion, T. and Vardy, A. (2002). Two-dimensional interleaving schemes with repetitions: constructions and bounds, *IEEE Trans. on Infor. Theory* **48**, pp. 428–457.

Etzion, T. and Vardy, A. (2011). Error-correcting codes in projective space, *IEEE Trans. on Infor. Theory* **57**, pp. 1165–1173.

Etzion, T., Vardy, A., and Yaakobi, E. (2013). Coding for the Lee and Manhattan metrics with weighing matrices, *IEEE Trans. Inform. Theory* **59**, pp. 6712–6723.

Etzion, T. and Wachter-Zeh, A. (2018). Vector network coding based on subspace codes outperforms scalar linear network coding, *IEEE Trans. on Infor. Theory* **64**, pp. 2460–2473.

Etzion, T. and Yaakobi, E. (2009). Error-correction of multidimensional bursts, *IEEE Trans. on Infor. Theory* **55**, pp.961–976.

Etzion, T. and Zhang, H. (2019). Grassmannian codes with new distance measures for network coding, *IEEE Trans. on Infor. Theory* **65**, pp. 4131–4142.

Etzion, T. and Zhou, J. (2021) Large sets with multiplicity, *Designs, Codes and Crypto.* **89**, pp. 1661–1690.

Euler, L. (1750-51). Consideratio quarumdam serierum quae singularibus proprietatibus sunt praeditae, *Novi Commentarii Academiae Scientiarum Petropolitanae* **3**, pp. 10–12, 86–108; Opera Omnia, Ser. I, vol. 14, B. G. Teubner, Leipzig, 1925, pp. 516–541.

Euler, L. (1782). Recherches sur une nouvelle espèce de quarrés magiques, *Verh. Zeeuwsch. Genootsch. Wetensch. Vlissengen* **9**, pp. 85–239.

Feige, U. (2010). On Optimal Strategies for a Hat Game on Graphs, *SIAM J. Discrete Math.* **24**, pp. 782–791.

Fine, N. J. (1947). Binomial coefficients modulo a prime, *Amer. Math. Monthly*

54, pp. 589–592.

Fon-Der-Flaass, D. G. (2007). A bound on correlation immunity, *Siberian Electron. Math. Rep.* **4**, pp. 133–135.

Frankl, P. and Wilson, R. M. (1986). The Erdös-Ko-Rado theorem for vector spaces, *J. of Combin. Theory, Ser. A* **43**, pp. 228–236.

Gabidulin, E. M. (1985). Theory of codes with maximum rank distance, *Probl. Infor. Transmiss.* **21**, pp. 1–12.

Gabidulin, E. M., Davydov, A. A., and Tombak, L. M. (1991). Linear codes with covering radius 2 and other new covering codes, *IEEE Trans. on Infor. Theory* **37**, pp. 219–224.

Gadouleau, M. (2018). Finite dynamical systems, hat games, and coding theory, *SIAM J. Discrete Math.* **32**, pp. 1922–1945.

Gadouleau, M. and Georgiou, N. (2015). New constructions and bound for Winkler's hat game, *SIAM J. Discrete Math.* **29**, pp. 823–834.

Ge, G. (2000). Generalized Steiner triple systems with group size $g \equiv 1, 5 \pmod 6$, *Australasian J. of Combin.* **21**, pp. 37–47

Ge, G. (2002). Further results on the existence of generalized Steiner triple systems with group size $g \equiv 1, 5 \pmod 6$, *Australasian J. of Combin.* **25**, pp. 19–27.

Geramita, A. V. and Seberry, J. (1979). *Orthogonal Designs: Quadratic Forms and Hadamard Matrices* (Mercel Deckker, New York).

Gijswijt, D., Schrijver, A. and Tanaka, H. (2006). New upper bounds for nonbinary codes based on the Terwilliger algebra and semidefinite programming, *J. Combin. Theory, Ser. A* **113**, pp. 1719–1731.

Giulietti, M. (2000). Small complete caps in PG$(2, q)$ for q an odd square, *J. of Geometry* **69**, pp. 110–116.

Giulietti, M. (2007a). Small complete caps in PG(N, q), q even, *J. of Combin. Designs* **15**, pp. 420–436.

Giulietti, M. (2007b). Small complete caps in Galois affine spaces, *J. Algebr. Combin.* **25**, pp. 149–168.

Giulietti, M. and Pasticci, F. (2007). Quasi-perfect linear codes with minimum distance 4, *IEEE Trans. on Infor. Theory* **53**, pp. 1928–1935.

Glock, S., Kühn, D., Lo, A., and Osthus, D. (2016). The existence of designs via iterative absorption, *arxiv.org/abs/1611.06827*.

Godsil, C. and Royle, G. F. (2001). *Algebraic Graph Theory* (Springer, New York).

Goethals, J. M. (1971). On the Golay perfect binary code, *J. of Combin. Theory* **112**, pp. 178–186.

Goethals, J. M. and Snover, S. L. (1972). Nearly perfect binary codes *Discrete Math.* **3**, pp.65–88.

Golay, M. J. E. (1949). Notes on digital coding, *Proc. IEEE* **27**, p. 657.

Goldberg, D. Y. (1986). Reconstructing the ternary Golay code, *J. of Combin. Theory, Ser. A* **42**, pp. 296–299.

Golomb, S. W. (2017). *Shift Register Sequences,* 3rd revised edn. (World Scientific, Singapore).

Golomb, S. W. (1996). *Polyominoes,* revised and expanded second edn. (Princeton Science Library, Princeton).

Golomb, S. W. and Posner, E. C. (1964). Rook domains, Latin square, affine planes and error-distributing codes, *IEEE Trans. on Infor. Theory* **10**, pp. 196–208.

Golomb, S. W. and Welch L. R. (1970). Perfect codes in the Lee metric and the packing of polyominos, *SIAM J. Applied Math.* **18**, pp. 302–317.

Gordon, D. M. (2006). Perfect single error-correcting codes in the Johnson scheme, *IEEE Transactions on Information Theory.* **52**, pp. 4670–4672.

Graham, R. L., Knuth, D. E., and Patashnik, O. (1994). *Concrete Mathematics: A Foundation for Computer Science* (Addison-Wesley, Reading, MA).

Graham, R. L. and Sloane, N. J. A. (1980). Lower bounds for constant weight codes, *IEEE Trans. on Infor. Theory* **26**, pp. 37–43.

Graham, R. L. and Sloane, N. J. A. (1985). On the covering radius of codes, *IEEE Trans. on Infor. Theory* **31**, pp. 385–401.

Gravier, S., Mollard, M., and Payan, C. (1998). On the nonexistence of 3-dimensional tiling in the Lee metric, *European J. Combin.* **19**, pp. 567–572.

Gravier, S., Mollard, M., and Payan, C. (2001). On the nonexistence of 3-dimensional tiling in the Lee metric II, *Discrete Math.* **235**, pp. 157–157.

Griesmer, J. H. (1960). A bound for error-correcting codes, *IBM J. Res. Develop.* **4**, pp. 532–542.

Grünbaum, B. and Shephard, G. C. (1987). *Tilings and Patterns* (W. H. Freeman & Company, New York).

Guo, W., Kasala, S., Rao, M. B., and Tucker B. (2007). The hat problem and some variations, in *Advances in Distribution Theory, Order Statistics, and Inference. Statistics for Industry and Technology*, Balakrishnan, N., Sarabia, J. M., and Castillo, E., Eds., pp. 459–479 (Birkhäuser, Boston).

Hadamard, J. (1893). Résolution d'une question relative aux déterminants, *Bull. Sci. Math.* **17**, pp. 240–248.

Hajós, G. (1942). Uber einfache und mehrfache Bedeckung des n-dmensionalen Raumes mit einem Wrfelgitter, *Math. Zeit.* **47**, pp. 427–467.

Hamada, N., Helleseth, T., and Ytrehus, Ø. (1993). A new class of nonbinary codes meeting the Griesmer bound, *Discrete Applied Mathematics* **47**, pp. 219–226.

Hamada, N. and Tamari, F. (1982). Construction of optimal linear codes using flats and spreads in a finite projective geometry, *European J. Combin.* **3**, pp. 129–141.

Hamming, R. W. (1950). Error detecting and error correcting codes, *Bell System Technical J.* **29**, pp. 147–160.

Hammond, P. (1976). Nearly perfect codes in distance-regular graphs, *Discrete Math.* **14**, pp. 41–56.

Hammond, P. (1982). On the non-existence of perfect and nearly perfect codes, *Discrete Math.* **39**, pp. 105–109.

Hammond, P. and Smith, D. H. (1975). Perfect codes in the graphs O_k, *J. Combin. Theory, Ser. B* **19**, pp. 239–255.

Hammons, A. R., Kumar, P. V., Calderbank, A. R., Sloane, N. J. A., and Solé, P. (1994). The $\mathbb{Z}_4$-linearity of Kerdock, Preparata, Goethhals, and related codes, *IEEE Trans. on Infor. Theory* **40**, pp. 301–319.

Hanani, H. (1960). On quadruple systems, *Canad. J. Math.* **12**, pp. 145–147.
Hanani, H. (1975). Balanced incomplete block designs and related designs, *Discrete Math.* **11**, pp. 255–369.
Hedayat, A. S., Sloane, N. J. A., and Stufken, J. (1999). *Orthogonal Arrays – Theory and Applications* (Springer).
Heden, O. (1974). Perfect codes in antipodal distance-transitive graphs, *Math. Scand.* **35**, pp. 29–37.
Heden, O. (1975). A generalized Lloyd theorem and mixed perfect codes, *Math. Scand.* **37**, pp. 13–26.
Heden, O. (1977). A new construction of group and nongroup perfect codes, *Infor. and Control* **34**, pp. 314–323.
Heden, O. (1977). A survey of perfect codes, *Advances in Math. of Commun.* **2**, pp. 223–247.
Heden, O. (2009a). Necessary and sufficient conditions for the existence of a class of partitions of a finite vector space, *Designs, codes and Crypto.* **53**, pp. 69–73.
Heden, O. (2009b). On the length of the tail of a vector space partition, *Discrete Math.* **309**, pp. 6169–6180.
Heden, O. and Roos, C. (2011). The non-existence of some perfect codes over non-prime power alphabets, *Discrete Math.* **311**, pp. 1344–1348.
Helleseth, T. (1981). A characterization of codes meeting the Griesmer bound, *Infor. and Control* **50**, pp. 128–159.
Helleseth, T. (1983). New constructions of codes meeting the Griesmer bound, *IEEE Trans. on Infor. Theory* **29**, pp. 434–439.
Helleseth, T. and van Tilborg, H. C. A. (1981). A new class of codes meeting the Griesmer bound, *IEEE Trans. on Infor. Theory* **27**, pp. 548–555.
Herzog, M. and Schönheim, J. (1971). Linear and nonlinear single-error correcting perfect mixed codes, *Infor. and Control* **18**, pp. 364–368.
Herzog, M. and Schönheim, J. (1972). Group partition, factorization and the vector covering problem, *Canad. Math. Bull* **15**, pp. 207–214.
Hickerson, D. (1983). Splitting of finite groups, *Pacific J. Math.* **107**, pp. 147–171.
Hickerson, D. and Stein, S. (1986). Abelian groups and packing by semi-crosses, *Pacific J. Math.* **122**, pp. 95–109.
Hill, R. (1978). Caps and codes, *Discrete Math.* **22**, pp. 111–137.
Hiramine, Y. (2005). A conjecture on affine planes of prime order, *J. of Combin. Theory, Ser. A* **52**, pp. 44–50.
Hirschfeld, J. W. P. (1998). *Projective Geometries over Finite Fields* (Oxford Science Publications, United Kingdom).
Hirschfeld, J. W. P. and Storme, L. (1998). The packing problem in statistics, coding theory and finite projective spaces, *J. Statist. Plann. Infer.* **72**, pp. 355–380.
Hong, S. J. and Patel, A. M. (1972). A general class of maximal codes for computer applications, *IEEE Trans. on Computers* **21**, pp. 1322–1331.
Honkala, I. (1991). On (k, t)-subnormal covering codes, *IEEE Trans. on Infor. Theory* **37**, pp. 1203–1206.
Horak, P. (2009a). On perfect Lee codes, *Discrete Math.* **309**, pp. 5551–5561.

Horak, P. (2009b). Tilings in Lee metric, *European J. of Combin.* **30**, pp. 480–489.

Horak, P. and AlBdaiwi, B. F. (2012a). Non-periodic tilings of R^n by crosses, *Discrete Comput. Geometry* **47**, pp. 1–16.

Horak, P. and AlBdaiwi, B. F. (2012b). Diameter perfect Lee codes, *IEEE Trans. on Infor. Theory* **58**, pp. 5490–5499.

Horak, P. and Grošek, O. (2014). A new approach towards the Golomb—Welch conjecture, *European J. Combin.* **38**, pp. 12–22.

Horak, P. and Hromada, V. (2014). Tiling R^5 by crosses, *Discrete Comput. Geometry* **51**, pp. 269–284.

Horak, P. and Kim, D. (2018). 50 years of Golomb–Welch conjecture, *IEEE Trans. on Infor. Theory* **64**, pp. 3048–3061.

Hughes, D. R. and Piper, F. C. (1973). *Projective Planes* (Springer-Verlag, New York).

Hyun, J. Y. and Kim, H. K. (2004). The poset structures admitting the extended binary Hamming code to be a perfect code, *Discrete Math.* **288**, pp. 37–47.

Hyun, J. Y. and Kim, H. K. (2008). Maximum distance separable poset codes, *Designs, Codes and Crypto.* **48**, pp. 247–261.

Hyun, J. Y., Kim, H. K. and Park, J. R. (2019). Weighted Posets and Digraphs Admitting the Extended Hamming Code to be a Perfect Code, *IEEE Trans. on Infor. Theory* **65**, pp. 4464–4672.

Jang, C., Kim, H. K., Oh, D. Y. and Rho, Y. (2008). The poset structures admitting the extended binary Golay code to be a perfect code, *Discrete Math.* **308**, pp. 4057–4068.

Ji, L. (2005) A new existence proof for large sets of disjoint Steiner triple systems, *J. of Combin. Theory, Ser. A* **112**, pp. 308–327.

Ji, L., Wu, D., and Zhu, L.(2005). Existence of generalized Steiner systems $\mathrm{GS}(2,4,v,2)$, *Designs, Codes and Crypto.* **36**, pp. 83–99.

Jiang, A., Mateescu, R., Schwartz, M., and Bruck, J. (1998). Rank modulation for flash memories, *IEEE Trans. on Infor. Theory* **55**, pp. 2659–2673.

Jiang, A., Schwartz, M. and Bruck, J. (2010). Correcting charge-constrained errors in the rank-modulation scheme, *IEEE Trans. on Infor. Theory* **56**, pp. 2112–2120.

Johnson, S. M. (1962). A new upper bound for error-correcting codes, *IEEE Trans. on Infor. Theory* **8**, pp. 203–207.

Johnson, S. M. (1972). Upper bounds for constant weight error correcting codes, *Discrete Math.* **3**, pp. 109–124.

Jungnickel D. (1992). Difference sets, in *Contemporary Design Theory: a collection of surveys*, Dinitz, J. H. and Stinson, D. R., Eds., pp. 241–324 (Wiley, New York).

Kabatiansky, G. A. and Panchenko V. I. (1988). Packings and coverings of Hamming space with unit spheres, *Problems Peredach. Infor.* **24**, pp. 3–16.

Kantor, W. M. (1983). On the inequivalence of generalized Preparata codes, *IEEE Trans. on Infor. Theory* **29**, pp. 345–348.

Keedwell, A. D. and Dénes, J. (2015). *Latin Squares and their Applications* (North-Holland, Amsterdam).

Keevash, P. (2014) The existence of designs, *arxiv.org/abs/1401.3665.*

Keevash, P. (2018) The existence of designs II, *arxiv.org/abs/1802.05900.*

Kharaghani, H. and Tayfeh-Rezaie, B. (2004). A Hadamard matrix of order 428, *J. of Combin. Designs* **13**, pp. 435–448.

Khare, A. (2009). Vector spaces as unions of proper subspaces, *Linear Algebra and its Applications* **431**, pp. 1681–1686.

Kiermaier, M. and Laue, R. (2015). Derived and residual subspace designs, *Advance of Math. in Commun.* **9**, pp. 105–115.

Kim, D. (2017). Nonexistence of perfect 2-error-correcting Lee codes in certain dimensions, *European J. Combin.* **63**, pp. 1–5.

Klein, A. (2004). On Perfect Deletion-Correcting Codes, *J. of Combin. Designs* **12**, pp. 72–77.

Kløve, T., Bose, B., and Elarief, N. (2011). Systematic, single limited magnitude error correcting codes for flash memories, *IEEE Trans. Infor. Theory* **57**, pp. 4477–4487.

Kløve, K., Luo, J., Naydenova, I., and Yari, S. (2011). Some codes correcting asymmetric errors of limited magnitude, *IEEE Trans. Infor. Theory* **57**, pp. 7459–7472.

Knuth, D. E. (1986). Efficient balance codes, *IEEE Trans. on Infor. Theory* **40**, pp. 51–53.

Koelink, E. and van Assche, W. (2009). Leonhard Euler and a q-analogue of the logarithm, *Proc. Amer. Math. Soc.* **137**, pp. 1663–1676.

Koetter, R. and Kschischang, F. R. (2008). Coding for errors and erasures in random network coding, *IEEE Trans. on Infor. Theory* **54**, pp. 3579–3591.

Kolountzakis, M. (1998). Lattice tilings by cubes: Whole, notched and extended, *The Electronic J. of Combin.* **5**, pp. 1–11.

Kratochvil, J. (1985). 1-Perfect codes over self-complementary graphs, *Commentationes Mathematicae Universitatis Carolinae* **26**, pp. 589–595.

Kratochvil, J. (1986). Perfect codes over graphs, *J. Combonatorial Theory, Ser. B* **40**, pp. 224–228.

Kratochvil, J. (1988). Perfect codes in general graphs, *Colloquia Mathematica Societatie Janos Bolyai.* **52**, pp. 357–364.

Krotov, D. S. (2001). $\mathbb{Z}_4$-linear Hadamard and extended perfect codes, *Electronic Notes Discrete Math.* **6**, pp. 107–112.

Krotov, D. S. (2001). Inductive constructions of perfect ternary constant-weight codes with distance 3, *Probl. Inform. Transm.* **37**, pp. 1–9.

Krotov, D. S. (2008). On diameter perefect constant-weight ternary codes, *Discrete Math.* **308**, pp. 3104–3114.

Krotov, D. S., Östergård, P. R. J., and Pottonen, O. (2011). On optimal binary one-error-correcting codes of length $2^m - 4$ and $2^m - 3$, *IEEE Trans. on Infor. Theory* **57**, pp. 6771–6779.

Krotov, D. S., Östergård, P. R. J., and Pottonen, O. (2016). Non-existence of a ternary constant weight $(16, 5, 15; 2048)$ diameter perfect code, *Advances of Math. in Commun.* **10**, pp. 393–399.

Lagarias, J. C. and Wang, Y. (1996). Tiling the line with translates of one tile, *Invent. Math.* **124**, pp. 341–365.

Lam, C. W. H., Kolesova, G., and Thiel, L. (1991). A computer search for finite projective planes of order 9, *Discrete Math.* **92**, pp. 187–195.

Lam, C. W. H., Thiel, L., and Seiercz, S. (1989). The nonexistence of finite projective planes of order 10, *Canad. J. Math.*. **41**, pp. 1117–1123.

Landsberg, G. (1893). Über eine Anzahlbestimmung und eine damit zusammenhängende Reihe, *J. Reine Angew. Math.* **111**, pp. 87–88.

Lee, C. Y. (1958). Some properties of nonbinary error-correcting codes, *IRE Trans. on Infor. Theory* **4**, pp. 72–82.

Lee, Y. (2004). Projective systems and Perfect codes with a poset metric, *Finite fields and their Applications* **10**, pp. 105–112.

Le Lionnais, F. (1983). *Les nombres remarquables* (Hermann, Paris).

Lenstra Jr., H. W. (1972). Two theorems on perfect codes, *Discrete Math.* **3**, pp. 125–132.

Lenstra Jr., H. W. and Seroussi, G. (2002). On hats and other covers, *IEEE International Symposium on Information Theory (ISIT)*.

Lepistö, T. (1981). A modification of the Elias-bound and nonexistence theorems for perfect codes in the Lee-metric, *Infor. and Control* **49**, pp. 109–124.

Leung, K. H. and Zhou, Y. (2020) No lattice tiling of $\mathbb{Z}^n$ by Lee spheres of radius 2, *J. of Combin. Theory, Ser. A* **171**, no. 105157.

Levenshtein, V. I. (1961). The application of Hadamard matrices to a problem in coding, *Problemy Kibernetiki* **5**, pp. 123–136. English translation in *Problems of Cybernetics* 5 (1964), pp. 166-184.

Levenshtein, V. I. (1965a). Binary codes capable of correcting deletions, insertions and reversals (in Russian), *Doklady Akademii Nauk SSSR* **163**, pp. 845–848. English translation in *Soviet Physics Dokl.* **10**, pp. 707–710, 1966.

Levenshtein, V. I. (1965b). Binary codes capable of correcting spurious insertions and deletions of ones (in Russian), *Problemy Peredachi Informatsii* **1**, pp. 12–25. English translation in *Problems of Infor. Trans.* **1**, pp. 8–17.

Li, C. and Helleseth, T. (2016). Quasi-perfect linear codes from planar and APN functions, *Crypto. Commun.* **8**, pp. 215–227.

Lidl, R. and Niederreiter, H. (1997). *Introduction to Finite Fields and Their Applications* (Cambridge Univ. Press, United Kingdom).

Lin, S. and Costello Jr., D. J. (2004). *Error Correcting Coding: Fundamentals and Applications* (Prentice-Hall, Upper Saddle River, NJ).

Lindström, K. (1975a). The nonexistence of unknown nearly perfect binary codes, *Ann. Univ. Turku, Ser. A* **169**, pp. 3–28.

Lindström, B. (1975b). Group partitions and mixed perfect codes, *Canad. Math. Bull.* **18**, pp. 57–60.

Lindström, K. (1977). All nearly perfect codes are known, *Infor. and Control* **35**, pp. 40–47.

Loidreau, P. (2014). Asymptotic behaviour of codes in rank metric over finite fields, *Designs, Codes and Crypto.* **71**, pp. 105–118.

Lu, J. X. (1983). On large sets of disjoint Steiner triple systems, I – III, *J. of Combin. Theory, Ser. A* **34** pp. 140–146, 157–155, 156–182.

Lu, J. X. (1984). On large sets of disjoint Steiner triple systems, I – III, *J. of Combin. Theory, Ser. A* **37**, pp. 136–163, 164–188, 189–192.

Lunardon, G. (1999). Normal spreads, *Geometriae Dedicata,* **75**, pp. 245–261.

MacWilliams, F. J. and Sloane, N. J. A. (1977). *The Theory of Error-Correcting Codes* (North-Holland, Amsterdam).

Mahmoodi, A. (1998). Existence of perfect 3-deletion-correcting codes, *Designs, Codes, and Crypto.* **14**, pp. 81–87.

Malyugin, S. A. (2010). On nonsystematic perfect codes over finite fields, *J. Appl. Industr. Math.* **4**, pp. 218–230.

Martin, W. J. (1992). *Completely regular subsets*, Ph.D. thesis, University of Waterloo, ON, Canada.

Martin, W. J. and Zhu, X. J. (1995). Anticodes for the Grassmann and bilinear forms graphs, *Designs, Codes, Crypto.* **6**, pp. 73–79.

Mazumdar, A., Barg, A., and Zémor, G. (2013). Constructions of rank modulation codes, *IEEE Trans. on Infor. Theory* **59**, pp. 1018–1029.

Minkowski, H. (1904). Dichteste gitterformige lagerung kongruenter korper, *Nachrichten Ges. Wiss. Gottingen*, pp. 311–355.

Mollard, M. (1986). A generalized parity function and its use in the construction of perfect codes, *SIAM J. Alg. Disc. Methods* **7**, pp. 113–115.

Mollard, M. (2011). On perfect codes in cartesian products of graphs, *European J. of Combin.* **32**, pp. 398–403.

Moreno O. (1983). Further results on quasi-perfect codes related to the Goppa codes, *Congressus Numerantium* **40**, pp. 249–256.

Moreno, O., Zhang, Z., Kumar, P. V., and Zinoviev, V. A. (1993). New constructions of optimal cyclically permutable constant weight codes, *IEEE Trans. on Infor. Theory* **41**, pp. 448–455.

Mounits, B., Etzion, T., and Litsyn, S.(2002). Improved upper bounds on the sizes of codes, *IEEE Trans. on Infor. Theory* **48**, pp. 880–886.

Mounits, B., Etzion, T., and Litsyn, S. (2007). New upper bounds on codes via association schemes and linear programming, *Advances of Math. in Commun.* **1**, pp. 173–195.

Niederreiter, H. (1987). Point sets and sequences with small discrepancy, *Monatsh. Math.* **104**, pp. 273–337.

Niederreiter, H. (1991). A combinatorial problem for vector spaces over finite field, *Discrete Math.* **96**, pp. 221–228.

Niederreiter, H. (1992). Orthogonal arrays and other combinatorial aspects in the theory of uniform point distributions in unit cubes, *Discrete Mathematics* **106/107**, pp. 361–367.

Niven, I. and Zuckerman, H. S. (1980). *An Introduction to the Theory of Numbers*, fourth edition, (John Wiley & Sons, New York).

Nordstrom, A. W. and Robinson, J. P. (1967). An optimum nonlinear code, *Infor. and Control* **11**, pp. 613–616.

O'Bryant, K. (2004). A complete annotated bibliography of work related to Sidon sequences, *The Electronic J. of Combin.* **DS11**, pp. 1–39.

Östergård, P. R. J. and Potteonen, O. (2009). The perfect binary one-error-correcting codes of length 15: Part I—classification. *IEEE Trans. on Infor. Theory* **55**, pp.4657–4660.

Östergård, P. R. J. and Potteonen, O. (2011). Two optimal one-error-correcting codes of length 13 that are not doubly shortened perfect codes, *Designs,*

Codes and Crypto. **59**, pp. 281–285.

Östergård, P. R. J., Potteonen, O., and Phelps, T. (2010). The perfect binary one-error-correcting codes of length 15: Part II—properties, *IEEE Trans. on Infor. Theory* **56**, pp. 2571–2582.

Östergård, P. R. J. and Svanström, M. (2002). Ternary constant weight weight codes, *Electron. J. Combin.* **9**, R41.

Östergård, P. R. J. and Vardy, A. (2004). Resolving the existence of full-rank tilings of binary Hamming spaces, *SIAM J. Discrete Math.* **18**, pp. 382–387.

Pasquier, G. (1980). The binary Golay code obtained from an extended cyclic code over $\mathbb{F}_8$, *European J. Combinatorics* **1**, pp. 369–370.

Paterson, M. B. and Stinson D. R. (2010). Yet another hat games, *The Electronic J. of Combin.* **17**, #R86.

Peng, X. H. and Farrell, P. G. (2006). On Construction of the (24,12,8) Golay codes, *IEEE Trans. of Infor. Theory* **52**, pp. 3669–3675.

Perkins, S., Sakhnovich, A. L., and Smith, D. H. (2006). On an upperbound for mixed error-correcting codes, *IEEE Trans. of Infor. Theory* **52**, pp. 708–712.

Peterson, W. W. (1961). *Error-Correcting Codes* (MIT Press, Cambridge, MA).

Phelps, K. T. (1983). A combinatorial construction of perfect codes, *SIAM J. Algebraic Discrete Methods* **4**, pp. 398–403.

Phelps, K. T. (1984a). A general product construction for error-correcting codes, *SIAM J. Algebraic Discrete Methods* **5**, pp. 224–228.

Phelps, K. T. (1984b). A product construction for perfect codes over arbitrary alphabets, *IEEE Trans. Infor. Theory* **30**, pp. 769–771.

Phelps, K. T. and Levan, M. (1995). Kernels of nonlinear Hamming codes, *Designs, Codes and Crypto.* **6**, pp. 247–257.

Phelps, K. T. and Levan, M. (1999). Nonsystematic perfect codes, *SIAM J. Discrete Math.* **12**, pp. 27–34.

Phelps, K. T., Rifà, J., and Villanueva, M. (2005). Kernels and p-kernels of p^r-ary 1-perfect codes, *Designs, Codes and Crypto.* **37**, pp. 243–261.

Phelps, K. T., and Villanueva, M. (2002a). Rank of q-ary 1-perfect codes, *Designs, Codes and Crypto.* **27**, pp. 139–144.

Phelps, K. T., and Villanueva, M. (2002b). On perfect codes: rank and kernel, *Designs, Codes and Crypto.* **27**, pp. 183–194.

Phelps, K., and Yin, C. (1997). Generalized Steiner systems with block size three and group size $g \equiv 3 \pmod 6$, *J. of Combin. Designs* **5**, pp. 417–432.

Pless, V. (1968). On the uniqueness of the Golay codes, *J. of Combin. Theory, Ser. A* **5**, pp. 215–228.

Pless, V. (1994). Decoding the Golay code, *IEEE Trans. on Infor. Theory* **32**, pp. 561–567.

Pless, V. (1989). *Introduction to the Theory of Error-Correcting Codes* (John Wiley and Sons Inc., New York).

Pless, V. (1992). More on the uniqueness of the Golay codes, *Discrete Math.* **106-107**, pp. 391–398.

Pless, V. S., and Huffman, W. C., Eds. (1989). *Handbook on Coding Theory*

(North-Holland, Amsterdam, The Netherlands).

Plotkin, M. (1960). Binary codes with specified minimum distances, *IEEE Trans. on Infor. Theory* **6**, pp. 445–450.

Post, K. A. (1975). Nonexistence theorems on perfect Lee codes over large alphabets, *Infor. and Control* **29**, pp. 369–380.

Preparata, F. P. (1968). A class of optimum non-linear double-error-correcting codes, *Infor. and Control* **13**, pp. 378–400.

Raghavarao, D. (1971). *Constructions and Combinatorial Problems in the Design of Experiments* (John Wiley, New York).

Rao, T. R. N. and Fujiwara, E. (1989). *Error-Control Coding for Computer Systems* (Prentice-Hall, London).

Raviv, N., Silberstein, N., and Etzion, T. (2017). Constructions of high-rate minimum storage regenerating codes over small fields, *IEEE Trans. on Infor. Theory* **63**, pp. 2015–2037.

Reiger, S. H. (1960). Codes for correction of 'clustered' errors, *IRE Trans. on Infor. Theory* **6**, pp. 16–21.

Reuvers, H. F. H. (1977). *Some non-existence theorems for perfect codes over arbitrary alphabets,* Ph.D. thesis, Eindhoven University of Technology, Eindhoven, The Netherlands.

Robinson, R. M. (1954). Mersenne and Fermat numbers, *Proc. Amer. Math. Soc.* **5** pp. 842–846.

Romanov, A. M. (2019). On non-full-rank perfect codes over finite fields, *Designs, codes and Crypto.* **87**, pp. 995–1003.

Roos, C. (1983). A note on the existence of perfect constant weight codes, *Discrete Math.* **47** pp. 121–123.

Roth, R. M. (1991). Maximum-rank array codes and their application to crisscross error correction, *IEEE Trans. on Infor. Theory.***37**, pp. 328–336.

Roth, R. M. (2005). *Introduction to Coding Theory* (Cambridge Univ. Press, United Kingdom).

Roth, R. M. and Siegel P. H. (1994). Lee-metric BCH codes and their application to constrained and partial-response channels, *IEEE Trans. on Infor. Theory* **40**, pp. 1083–1096.

Selberg, A. (1949). An elementary proof of Dirichlet's theorem about primes in an arithmetic progression, *Ann. of Math.* **50**, pp. 297–304.

Schmerl, J. H. (1994). Tiling space with notched cubes, *Discrete Math.* **133**, pp. 225–235.

Schönheim, J. (1984). Mixed codes, *Proc.Calgary Internat. Conf. on Combinatorial Structures and their Applications* , pp. 385 (Gordon and Breach, New York).

Schwartz, M. (2004). *Tilings, anticodes, and two-dimensional coding*, Ph.D. thesis, Technion, Haifa, Israel.

Schwartz, M. (2012). Quasi-cross lattice tilings with applications to flash memory, *IEEE Trans. on Infor. Theory* **58**, pp. 2397–2405.

Schwartz, M. (2014). On the non-existence of lattice tilings by quasi-crosses, *European J. of Combin.* **36**, pp. 130–142.

Schwartz, M. and Etzion, T. (2002). Codes and anticodes in the Grassman graph,

J. of Combin. Theory, Ser. A **97**, pp. 27–42.

Schwartz, M. and Tamo, I. (2011). Optimal permutation anticodes with the infinity norm via permanents of (0,1)-matrices, *J. of Combin. Theory, Ser. A* **118**, pp. 1761–1774.

Seelinger, G. I., Sissokho, P. I., Spence, L. E. and Vanden Eynden, C. (2012a). Partitions of finite vector spaces over GF(2)into subspaces of dimensions 2 and s, *Finite Fields and Their Applications* **18**, pp. 1114–1132.

Seelinger, G. I., Sissokho, P. I., Spence, L. E. and Vanden Eynden, C. (2012b). Partitions of $V(n,q)$ into 2- and s-dimensional subspaces, *J. of Combin. Designs* **20**, pp. 467–482.

Semakov, N. V. and Zinoviev, V. A. (1969). Complete and quasi-complete balanced codes, *Problems of Infor. Trans.* **5**, pp. 11–13.

Segre, B. (1955). Curve razionali normali e k-archi negli spazi finiti, *Ann. Mat. Pura Appl.* **39**, pp. 357–379.

Shalaby, N., Wang, J. and Yin, J. (2002). Existence of perfect 4-deletion-correcting codes with length six, *Designs, Codes and Cryptogr.* **27**, pp. 145–156.

Shannon, C. (1948). A mathematical theory of communication, *Bell System Technical J.* **27**, pp. 379–423, 623–656.

Shapiro, H. S. and Slotnick, D. L. (1959). On the mathematical theory of error correcting codes, *IBM J. Res Develop.* **3**, pp. 25–37.

Shimabukuro, O. (2005). On the nonexistence of perfect codes in J$(2w+p^2,w)$, *Ars Combinatoria* **75** pp. 129–134.

Silberstein, N. (2007). *Properties of codes in the Johnson scheme*, M.Sc. thesis, Technion, Haifa, Israel.

Silberstein, N. and Etzion, T. (2010). On perfect codes in the Johnson graph, *12th International Workshop on Algebraic and Combinatorial Coding Theory (ACCT2010)* Novosibirsk, Russia (September 2010).

Silberstein, N. and Etzion, T. (2011). Enumerative coding for Grassmannian space, *IEEE Trans. on Infor. Theory* **57**, pp. 365–374.

Silberstein, N., Etzion, T., and Schwartz, M. (2019). Locality and availability of array codes constructed from subspaces, *IEEE Trans. on Infor. Theory* **65**, pp. 2648–2660.

Silva, D. and Kschischang, F. R. (2008). On metrics for error correction in network coding, *IEEE Trans. Infor. Theory* **55**, pp. 5479–5490.

Silva, D., Kschischang, F. R., and Koetter, R. (2008). A rank-metric approach to error control in random network coding, *IEEE Trans. Infor. Theory* **54**, pp. 3951–3967.

Singleton, R. C. (1964). Maximum distance q-nary codes, *IEEE Trans. on Infor. Theory* **10**, pp. 116–118.

Sloane, N. J. A. (2002). On single-deletion-correcting codes, in: K. T. Arasu, Á. Seress, Eds., Codes and Designs – Ray-Chaudhuri Festschrift, pp. 273–292.

Sloane, N. J. A., Reddy, S. M., and Chen, C. L. (1972). New binary codes, *IEEE Trans. on Infor. Theory* **18**, pp. 503–510.

Smith, D. H. (1980). Perfect Codes in the graphs O_k and $L(O_k)$, *Glasgow Math. J.* **21**, pp. 169–172.

Snyders, J. and Be'ery, Y. (1989). Maximum likelihood soft decoding of binary block codes and decoders for the Golay codes, *IEEE Trans. on Infor. Theory* **37**, pp. 963–975.

Solomon, G. and Stiffler, J. J. (1965). Algebraically punctured cyclic codes, *Infor. and Control* **8**, pp. 170–179.

Solovieva, F. I. (1981). On binary nongroup codes, *Methodi Diskr, Analiza* **37**, pp. 65–76.

Solovieva, F. I. (1989). A class of binary perfect codes generated by q-ary codes, *Methodi. Diskr. Analiza.* **48**, pp. 70–72.

Solovieva, F. I. (1994). A combinatorial construction of perfect binary codes, *Proc. 4th Internat Workshop on Algebric and combinatorial Coding Theory,* Novgorod, pp. 171–174.

Špacapan, S. (2007). Nonexistence of face-to-face four-dimensional tilings in the Lee metric, *Eur. J. Combin.* **28**, 1, pp. 127–133.

Stein, S. (1967a). Algebraic tiling, *Amer. Math. Mon.* **81**, pp. 445–462.

Stein, S. (1967b). Factoring by subsets, *Pacific J. Math.* **22**, pp. 523–541.

Stein, S. (1984). Packings of R^n by certain error spheres, *IEEE Trans. Inf. Theory* **30**, pp. 356–363.

Stein, S. (1985). Lattice-tiling by certain star bodies, *Studia Sci. Math. Hung.* **20**, pp. 71–76.

Stein, S. (1986). Tiling, packing, and covering by clusters, *Rocky Mountain J. Math.* **16**, pp. 277–321.

Stein, S. (1990). The notched cube tiles $\mathbb{R}^n$, *Discrete Math.* **80**, pp. 335–337.

Stein, S. K., and Szabó, S. (1994). *Algebra and Tiling* (The Mathematical Association of America).

Stinson, D. R. (1984). A short proof of the nonexistence of a pair of orthogonal Latin squares of order six, *J. of Combin. Theory, Ser. A* **36**, pp. 373–376.

Struik, R. (1994). *Covering codes*, Ph.D. thesis, Eindhoven University of Technology, Eindhoven, The Netherlands.

Svanström, M. (1999). A class of perfect ternary constant-weight codes, *Designs, Codes and Crypto.* **18**, pp. 223–229.

Svanström, M. (1999). *Ternary codes with weight constraints*, Ph.D. thesis, Linköping University, Linköping, Sweden.

Sylvester, J. (1884). Mathematical questions, with their solutions, *Educational Times* **4**, p. 21.

Szabó, S. (1984). A bound on k for tiling by (k, n)-crosses and semicrosses, *Acta Math. Acad. Sci. Hung.* **44**, pp. 97–99.

Szegedy, M. (1998). Algorithms to tile the infinite grid with finite clusters, *Proc. 39th Annu. Symp. Found. Comput. Sci.*, pp. 137–145.

Tamo, I. and Schwartz, M. (2010). Correcting limited-magnitude errors in the rank-modulation scheme, *IEEE Trans. Infor. Theory* **56**, pp. 2551–5560.

Tarry, G. (1900). Le problème des 36 officers, *C. R. Assoc. Fr. Au. Sci.* **1**, pp. 122–123.

Tarry, G. (1901). Le problème des 36 officers, *C. R. Assoc. Fr. Au. Sci.* **2**, pp. 170–203.

Teirlinck, L. (1991). A completion of Lu's determination of the spectrum for large sets of disjoint Steiner triple systems, *J. of Combin. Theory, Ser. A* **57**, pp. 302–305.

Teirlinck, L. (1994). Some new 2-resolvable Steiner quadruple systems, *Designs, Codes and Crypto.* **4**, pp. 5–10.

Thas, J. A. (1977). Two infinite classes of perfect codes in metrically regular graphs, *J. Combin. Theory, Ser. B* **23**, pp. 236–238.

Thomas, S. (1996). Designs and partial geometries over finite fields, *Geometriae Dedicata* **63**, pp. 247–253.

Tietäväinen , A. (1970). On the nonexistence of perfect 4-Hamming-error-correcting codes, *Ann. Acad. Sci. Fennicae, Ser. A. I.* **485**, pp. 1–6.

Tietäväinen , A. (1974). A short proof for the non-existence of unknown perfect codes over GF(q), $q > 2$, *Ann Acad. Sci. Fennicae, Ser. A. I.* **580**, pp. 1–6.

Tietäväinen, A. and Perko, A. (1971). There are no unknown perfect binary codes, *Ann Univ. Turko. Ser A. I.* **148**, pp. 3–10.

Ulrich, W. (1957). Non-binary error-correction codes, *Bell Syst. Tech. J.* **36**, pp. 1341–1388.

van Lint, J. H. (1970a). On the nonexistence of perfect 2- and 3-Hamming error correcting codes over GF(q), *Infor. and Control* **16**, pp. 396–401.

van Lint, J. H. (1970b). On the nonexistence of perfect 5-, 6- and 7-Hamming error correcting codes over GF(q), Report 70-WSK-06, Eindhoven University of Technology, The Netherlands.

van Lint, J. H. (1971a). Nonexistence theorems for perfect error- correcting codes, *Computers in Algebra and Number Theory,* **IV**, SIAM-AMS Proceedings.

van Lint, J. H. (1971b). *Coding Theory,* (Springer-Verlag, New York).

van Lint, J. H. (1974). Recent results on perfect codes and related topics, *Combinatorics* **1**, Hall and Van Lint, Eds., pp. 158–178, Mathematical Centre, Amsterdam.

van Lint, J. H. (1975). A survey of perfect codes, *Rocky Mountain J. of Math.* **5**, pp. 199–224.

van Lint, J. H. and Tolhuizen, L. (1999). On perfect ternary constant weight codes, *Designs, Codes and Crypto.* **18**, pp. 231–234.

van Wee, G. J. M. (1991). On the non-existence of certain perfect mixed codes, *Discrete Math.* **87**, pp. 323–326.

Varshamov, R. R. (1964). On the theory of asymmetric codes, *Rep. Acad. Sci. USSR* **157**, pp. 546–548.

Varshamov, R. R. (1965). On some specifics error-correcting linear codes, *Rep. Acad. Sci. USSR* **164**, pp. 757–760.

Varshamov, R. R. and Tenengolts, G. M. (1965). Codes which correct single asymmetric errors (in Russian), *Avtomatika i Telemekhanika* **26**, pp. 288–292. English translation in *Automation and Remote Control* **26**, pp. 286–290.

Vasil'ev, J. L. (1962). On nongroup close-packed codes, *Probl. Kibemet.* **8**, pp. 375–378. See also in In: Blake, I. F. (Ed.) Algebraic Coding Theory: History and Development, Dowden, Hutchinson and Ross, 1973, pp. 351–357.

Vardy, A. and Be'ery, Y. (1991). More efficient soft decoding of the Golay codes, *IEEE Trans. on Infor. Theory* **37**, pp. 667–672.

Wagner, T. J. (1966). A search technique for quasi-perfect codes, *Infor. and Control* **9**, pp. 94–99.

Wallis, W. D. (1997). *One-factorizations* (Kluwer Academic Publisher, Dordrecht, Boston, London).

Wang, J. (2015). Some combinatorial constructions for optimal perfect deletion-correcting codes, *Designs, Codes and Crypto.* **48**, pp. 331–347.

Wang, J. and Yin, J. (2006). Constructions for perfect 5-deletion-correcting codes of length 7, *IEEE Trans. on Infor. Theory* **52**, pp. 3676–3685.

Wei, H. and Ge, G. (2015). Spectrum of sizes for perfect 2-deletion-correcting codes of length 4, *Designs, Codes and Crypto.* **74**, pp. 127–151.

Wei, H. and Schwartz, M. (2020). On tilings of asymmetric limited-magnitude balls, *arxiv/2006.00198.*

Weichsel, P. M. (1994). Dominating sets in n-cubes, *J. of Graph Theory* **18**, pp. 479–488.

Wilson, R. M. (1972a). An existence theory for pairwise balanced designs I. Composition theorems and morphisms, *J. of Combin. Theory, Ser. A* **13**, pp. 220–245.

Wilson, R. M. (1972b). An existence theory for pairwise balanced designs II. The structure of PBD-closed sets and the existence conjectures, *J. of Combin. Theory, Ser. A* **13**, pp. 246–273.

Wilson, R. M. (1972c). Cyclotomy and difference families in elementary abelian groups, *J. Number Theory* **4**, pp. 17–47.

Wilson, R. M. (1975). An existence theory for pairwise balanced designs III: Proof of the existence conjectures, *J. of Combin. Theory, Ser. A* **18**, pp. 71–79.

Wilson, S. B., and Phelps, K. T. (1999). Constant weight codes and group divisible designs, *Designs, Codes and Crypto.* **16**, pp. 11–27.

Wu, D. and Fan, P. (2009). Construction of optimal quaternary constant weight codes via group divisible designs, *Discrete Math.* **309**, pp. 6009–6013.

Wu, D., Ge, G. and Zhu, L. (2001). Generalized steiner systems $\mathrm{GS}_4(2,4,v,g)$ for $g = 2,3,6$ *J. of Combin. Designs* **9**, pp. 401–423.

Wu, D. and Zhu, L. (2001). Generalized Steiner systems $\mathrm{GS}(2,4,v,2)$ with v a prime power $\equiv 7 (\mathrm{mod}\ 12)$, *Designs, Codes and Crypto.* **24**, pp. 69–80.

Yaabobi, E. and Bruck, J. (2019). On the uncertainty of information retrieval in associative memories, *IEEE Trans. on Infor. Theory* **65**, pp. 2155–2165.

Yin, J. (2001). A combinatorial construction for perfect deletion-correcting codes *Designs, Codes and Crypto.* **23**, pp. 99–110.

Zhang, Y., Etzion, T., and Yaakobi, E. (2020). Bounds on the length of functional PIR and batch codes, *IEEE Trans. on Infor. Theory* **66**, pp. 4917–4934.

Zaremba, S. K. (1950). A covering theorem for abelian group, *J. London Math. Soc.* **26**, pp. 242–246.

Zaremba, S. K. (1952). Covering problems concerning abelian groups, *J. London Math. Soc.* **27**, pp. 242–246.

Zhang, H. and Ge, G. (2013). Optimal quaternary constant-weight codes with weight four and distance five, *IEEE Trans. on Infor. Theory* **59**, pp. 1617–1629.

Zhang, H. and Ge, G. (2017). Perfect and quasi-perfect codes under the ℓ_p metric,

IEEE Trans. on Infor. Theory **63**, pp. 4325–4331.

Zhang, H., Zhang, X. and Ge, G. (2012). Optimal ternary constant-weight codes with weight 4 and distance 5, *IEEE Trans. on Infor. Theory* **58**, pp. 2706–2718.

Zinoviev, V. A. and Leontiev, V. K. (1973). The nonexistence of perfect codes over Galois fields, *Probl. Control Infor. Theory* **2**, pp. 123–132 (pp. 16–24 in English Translation).

Lightning Source UK Ltd.
Milton Keynes UK
UKHW020003051022
409929UK00003B/355

9 789811 255878